D1297078

Fundamentals of Fire Phenomena

James G. Quintiere
University of Maryland, USA

JOHN WILEY & SONS, LTD

Telephone (+44) 1243 779777

Email (for orders and customer service enquiries): cs-books@wiley.co.uk
Visit our Home Page on www.wiley.com

Other Wiley Editorial Offices

John Wiley & Sons Inc., 111 River Street, Hoboken, NJ 07030, USA

Jossey-Bass, 989 Market Street, San Francisco, CA 94103-1741, USA

Wiley-VCH Verlag GmbH, Boschstr. 12, D-69469 Weinheim, Germany

John Wiley & Sons Australia Ltd, 42 McDougall Street, Milton, Queensland 4064, Australia

John Wiley & Sons (Asia) Pte Ltd, 2 Clementi Loop #02-01, Jin Xing Distripark, Singapore 129809

John Wiley & Sons Canada Ltd, 22 Worcester Road, Etobicoke, Ontario, Canada M9W 1L1

Wiley also publishes its books in a variety of electronic formats. Some content that appears in print may not be available in electronic books.

Library of Congress Cataloging-in-Publication Data

British Library Cataloguing in Publication Data

A catalogue record for this book is available from the British Library

ISBN-13 978-0-470-09113-5 (HB)
ISBN-10 0-470-09113-4 (HB)

Typeset in 10/12pt Times by Thomson Press (India) Limited, New Delhi
Printed and bound in Great Britain by Antony Rowe Ltd, Chippenham, Wiltshire
This book is printed on acid-free paper responsibly manufactured from sustainable forestry in which at least two trees are planted for each one used for paper production.

To those that have brought the logic of science to fire –
I have tried to make them part of this book.

Contents

Preface

Many significant developments in fire science have emanated from New Jersey. Howard Emmons, the father of US fire research, was born in Morristown and went to school at Stevens Institute of Technology in Jersey City; Dick Magee (flame spread measurements) also taught at Stevens; John deRis (opposed flow flame spread) lived in Englewood; Gerry Faeth (turbulent flames) was raised in Teaneck; Frank Steward (porous media flame spread) and Forman Williams (modeling) hailed from New Brunswick; Bob Altenkirch (microgravity flame spread) is President of New Jersey Institute of Technology (formerly NCE); the FAA has their key fire laboratory for aircraft safety in Atlantic City (Gus Sarkos, Dick Hill and Rich Lyon); Yogesh Jaluria (natural convection in fires) teaches at Rutgers; Glenn Corbett lives in Waldwick and teaches fire science at John Jay College; Jim Milke (fire protection) was raised in Cherry Hill; Irv Glassman (liquid flames spread) taught at Princeton University, inspiring many to enter fire research: Fred Dryer (kinetics), Thor Eklund (aircraft), Carlos Fernandez-Pello (flame spread), Takashi Kashiwagi (material flammability), Tom Ohlemiller (smoldering), Kozo Saito (liquid flame spread), Bill Sirignano (numerical modeling) and more; even Chiang L. Tien (radiation in fire), the former Chancellor of the University of California (Berkeley), graduated from Princeton and almost proceeded to join NYU Mechanical Engineering before being attracted to Berkeley.

I was born in Passaic, New Jersey, studied at NCE and graduated with a PhD in Mechanical Engineering from NYU. I guess something in the water made me want to write this book.

The motivation and style of this text has been shaped by my education. My training at NYU was focused on the theoretical thermal sciences with no emphasis on combustion. My mentor, W. K. Mueller, fostered careful analysis and a rigor in derivations that is needed to reveal a true understanding of the subject. That understanding is necessary to permit a proper application of the subject. Without that understanding, a subject is only utilized by using formulas or, today, computer programs. I have tried to bring that careful and complete approach to this text. It will emphasize the basics of the subject of fire, and the presentation is not intended to be a comprehensive handbook on the subject. Theory at the level of partial differential equations is avoided in favor of a treatment based on control volumes having uniform properties. Therefore, at most, only ordinary differential equations are formulated. However, a theoretical approach to fire is not sufficient for the solution of problems. Theory cannot allow solutions to phenomena in this complex field, but it does establish a basis for experimental correlations. Indeed, the term *fire modeling* applies more to the formulas that have developed from experimental correlations than from theoretical principles alone. It will be seen, throughout this text, and particularly in Chapter 12 on scaling, that theory strengthens and generalizes experimental correlations. Until computers are powerful enough to solve the basic known equations that apply to fire with appropriate data, we will have to rely on this empirical approach. However, as in other complex fields, correlations refined and built upon with theoretical understanding are invaluable engineering problem-solving tools.

I honed my knowledge of fire by having the opportunity to be part of the National Bureau of Standards, NBS (now the National Institute of Standards and Technology, NIST), in an era of discovery for fire that began in the early 1970s. In 1974 the Center for Fire Research (CFR) was established under John Lyons at NBS, joining three programs and establishing a model for a nation's investment in fire research. That combined program, influenced by the attention to standards from Alex Robertson and Irwin Benjamin, innovations such as oxygen consumption calorimetry (Bill Parker, C. Huggett and V. Babauskas) and others, brought a change in the way fire was looked at. Many had thought that science did not apply; fire was too complex, or rules were essential enough. At that same time, the NSF RANN program for fire research at $2 million was led by Ralph Long who had joined the CFR. Professors Hoyt Hottel and Howard Emmons had long championed the need for fire research in the US and now in the 1970s it was coming to fruition. In addition, the Factory Mutual Research Corporation (FMRC), influenced by Emmons, formed a basic research group, first under John Rockett, nearly by Philip Thomas and mostly carried out by Ray Friedman. The FM fire program became a model of engineering science for fire, bringing such stalwarts as G. Heskestad, J. deRis, A. Tewarson and more to the field.

The CFR staff numbered close to 120 at its peak and administered then over $4 million in grants to external programs. The President's National Commission on Fire Prevention and Control published its report of findings, *America Burning* (1973), which gave a mandate for fire research and improved technology for the fire service. This time of plenty in the 1970s gave us the foundation and legacy to allow me to write this book. My association at CFR with H. Baum, T. Kashiwagi, M. Harkleroad, B. McCaffrey, H. Nelson, W. Parker, J. Raines, W. Rinkinen and K. Steckler all importantly added to my learning experience. The interaction with distinguished academics in the country from some of the leading institutions, and the pleasure to engage visiting scientists – X. Bodart, M. Curtat, Y. Hasemi, M. Kokkala, T. Tanaka, H. Takeda and K. Saito – all brought something to me. This was the climate at CFR and it was not unique to me; it was the way it was for those who were there.

In the beginning, we at CFR reached out to learn and set an appropriate course. Understanding fire was 'in', and consumer safety was 'politically correct'. Interactions with those that knew more were encouraged. John Lyons arranged contact with the Japanese through the auspices of the US–Japan Panel on Natural Resources (UJNR). Visits with the distinguished program at the Fire Research Station (FRS) in the UK were made. The FRS was established after World War II and at its peak in the 1960s under Dennis Lawson had Philip H. Thomas (modeling) and David Rasbash (suppression) leading its two divisions. Their collective works were an example and an essential starting point to us all. Japan had long been involved in the science of fire. One only had to experience the devastation to its people and economy from the 1923 Tokyo earthquake to have a sensitivity to fire safety. That program gave us S. Yokoi (plumes), K. Kawagoe (vent flows), T. Hirano (flame spread) and many more. The UJNR program of exchange in fire that continued for about the next 25 years became a stimulus for researchers in both countries. The decade of 1975 to about 1985 produced a renaissance in fire science.

An intersection of people, government support and international exchange brought fire research to a level of productive understanding and practical use. This book was inspired and influenced by that experience and the collective accomplishments of those that took part. For that experience I am grateful and indebted. I hope my style of presentation will properly represent those that made this development possible.

This book is intended as a senior level or graduate text following introductory courses in thermodynamics, fluid mechanics, and heat and mass transfer. Students need general calculus with a working knowledge of elementary ordinary differential equations. I believe the presentation in this text is unique in that it is the first fire text to emphasize combustion aspects and to demonstrate the continuity of the subject matter for fire as a discipline. It builds from chemical thermodynamics and the control volume approach to establish the connection between premixed and diffusion flames in the fire growth sequence. It culminates in the system dynamics of compartment fires without embracing the full details of zone models. Those details can be found in the text *Enclosure Fire Dynamics* by B. Karlsson and J. G. Quintiere. The current text was influenced by the pioneering work of Dougal Drysdale, *An Introduction to Fire Dynamics*, as that framed the structure of my course in that subject over the last 20 years. It is intended as a pedagogical exposition designed to give the student the ability to look beneath the engineering formulas. In arriving at the key engineering results, sometimes an extensive equation development is used. Hopefully some students may find this development useful, while others might find it distracting. Also the text is not a comprehensive representation of the literature, but will cite key contributions as illustrative or essential for the course development.

I am indebted to many for inspiration and support in the preparation of this text, but the tangible support came from a few. Professor Kristian Hertz was kind enough to offer me sabbatical support for three months in 1999 when I began to formally write the chapters of this text at the Denmark Technical University (DTU). That time also marked the initiation of the MS degree program in fire engineering at the DTU to provide the needed technical infrastructure to support Denmark's legislation on performance codes for fire safety. Hopefully, this book will help to support educational programs such as that at the DTU, although other engineering disciplines might also benefit by adding a fire safety element to their curriculum. The typing of this text was made tedious by the addition of equation series designed to give the student the process of theoretical developments. For that effort, I am grateful to Stephanie Smith for sacrificing her personal time to type this manuscript. Also, Kate Stewart stepped in at the last minute to add help to the typing, especially with the problems and many illustrations. Finally, the Foundation of the Society of Fire Protection Engineering provided some needed financial support to complete the process. Hopefully this text will benefit their members. I would be remiss if I did not acknowledge the students who suffered through early incomplete versions, and tedious renditions on the black-board. I thank those students offering corrections and encouragement; and a special appreciation to my graduate students who augmented my thinking and gave me illustrative results for the text. Lastly, I am especially indebted to John L. Bryan for believing in me enough to give me an opportunity to teach and to join a special Department in 1990.

James G. Quintiere
The John L. Bryan Chair
in Fire Protection Engineering
University of Maryland
College Park, Maryland, USA

Nomenclature

A	Area, pre-exponential factor
b	S–Z variable (Chapter 9), plume width
B	Spalding B number, dimensionless plume width
Bi	Biot number
c	Specific heat
C	Coefficient
CS	Control surface
CV	Control volume
d	Distance
D	Diffusion coefficient, diameter
Da	Damkohler number
E	Energy, activation energy
f	Mixture fraction
F	Radiation view factor, force
Fr	Froude number
g	Gravitational acceleration
Gr	Grashof number
h	Heat or mass transfer coefficient, enthalpy per unit mass
H	Enthalpy
h_{fg}	Heat of vaporization
j	Number of flow surfaces
k	Conductivity
l	Length
L	Heat of gasification
Le	Lewis number = Pr/Sc
m	Mass
M	Molecular weight
n	Number of droplets
$\mathbf{n}$	Unit normal vector
N	Number of species in mixture
Nu	Nusselt number
p	Pressure
Pr	Prandtl number

q	Heat
Q	Fire energy (or heat)
r	Stoichiometric mass oxygen to fuel ratio
R	Universal gas constant
Re	Reynolds number
s	Stoichiometric air to fuel mass ratio, entropy per unit mass
S	Surface area
S_{u}	Laminar burning speed
Sc	Schmidt number
Sh	Sherwood number
t	Time
T	Temperature
v	Velocity
V	Volume
w	Velocity in the z direction
W	Width, work
x	Coordinate
X	Mole fraction, radiation fraction (with designated subscript)
y	Coordinate
y_i	Mass yield of species i per mass of fuel lost to vaporization
Y	Mass fraction
z	Vertical coordinate
α	Thermal diffusivity
β	Coefficient of volumetric expansion
γ	Ratio of specific heats, variable defined in Chapters 7 and 9
δ	Distance, Damkohler number (Chapter 5)
Δ	Difference, distance
Δh_{c}	Heat of combustion
Δh_{v}	Heat of vaporization
ε	Emissivity
θ	Dimensionless temperature, angle
$\theta_{\mathrm{FO,f}}$	$r_{\mathrm{o}}(B+1)/[B\ (r_{\mathrm{o}}+1)]$ (Chapter 9)
κ	Radiation absorption coefficient
λ	Wavelength
μ	Viscosity
ν	Kinematic viscosity $=\mu/\rho$
ν_i	Stoichiometric coefficient
ξ	Dimensionless distance
π	3.1416...
Π	Dimensionless group
ρ	Density
σ	Stefan–Boltzman constant$= 5.67 \times 10^{-11}$ kW/m^2 K^4
τ	Dimensionless time
τ_{o}	$c_p(T_{\mathrm{v}}-T_\infty)/L$
ϕ	Dimensionless temperature, equivalence ratio

Subscripts

a, air	Air
ad	Adiabatic
b	Burning
B	Balloon
BL	Boundary layer
c	Combustion, convection, critical
d	Diffusion
dp	Dew point
e	External
f	Flame
F	Fuel
g	Gas
i	Species
l	Liquid
L	Lower limit
m	Mass transfer
o	Initial, vent opening
O_2, ox	Oxygen
p	Products, pressure, pyrolysis front
py	Pyrolysis
Q	Quenching
r	Radiation
R	Reaction
s	Surface
st	Stoichiometric
t	Total
T	Thermal
u	Unburned mixture
U	Upper limit
v	Vaporization
w	Wall, water
x	Coordinate direction
∞	Ambient

Other

$\dot{a}$	Rate of a
$\tilde{a}$	Molar a
a'	a per unit length
a''	a per unit area
a'''	a per unit volume
$\boldsymbol{a}$	Vector a

1

Introduction to Fire

1.1 Fire in History

Fire was recognized from the moment of human consciousness. It was present at the creation of the Universe. It has been a part of us from the beginning. Reliance on fire for warmth, light, cooking and the engine of industry has faded from our daily lives of today, and therefore we have become insensitive to the behavior of fire. Mankind has invested much in the technology to maintain fire, but relatively little to prevent it. Of course, dramatic disastrous fires have been chronicled over recorded history, and they have taught more fear than complete lessons. Probably the low frequency of individually having a bad fire experience has caused us to forget more than we have learned. Fire is in the background of life in the developed world of today compared to being in the forefront of primitive cultures. Fire rarely invades our lives now, so why should we care about it?

As society advances, its values depend on what is produced and those sources of production. However, as the means to acquire products becomes easier, values turn inward to the general societal welfare and our environment. Uncontrolled fire can devastate our assets and production sources, and this relates to the societal costs of fire prevention and loss restoration. The effects of fire on people and the environment become social issues that depend on the political ideology and economics that prevail in the state. Thus, attention to fire prevention and control depend on its perceived damage potential and our social values in the state. While these issues have faced all cultures, perhaps the twentieth century ultimately provided the basis for addressing fire with proper science in the midst of significant social and technological advances, especially among the developed countries.

In a modern society, the investment in fire safety depends on the informed risk. Reliable risk must be based on complete statistics. An important motivator of the US government's interest to address the large losses due to fire in the early 1970s was articulated in the report of the National Commission on Fire Prevention and Control (*America Burning* [1]). It stated that the US annually sustained over $11 billion in lost

Fundamentals of Fire Phenomena James G. Quintiere

resources and 12 000 lives were lost due to fire. Of these deaths, 3500 were attributed to fires in automobiles; however, a decimal error in the early statistics had these wrong by a factor of 10 (only 350 were due to auto fires). Hence, once funding emerged to address these issues, the true annual death rate of about 8000 was established (a drop in nearly 4000 annual deaths could be attributed to the new research!). The current rate is about 4000 (4126 in 1998 [2]). A big impact on this reduction from about 8000 in 1971 to the current figure is most likely attributed to the increasing use of the smoke detector from nearly none in 1971 to over 66 % after 1981 [3]. The fire death rate reduction appears to correlate with the rate of detector usage. In general, no clear correlation has been established to explain the fire death rates in countries, yet the level of technological and political change seem to be factors.

While the estimated cost of fire (property loss, fire department operations, burn injury treatment, insurance cost and productivity loss) was $11.4 billion in the US for 1971 [1], it is currently estimated at 0.875 % of the Gross Domestic Product (GDP) ($\$11.7 \times 10^{12}$) or about $102 billion [2]. The current US defense spending is at 3.59 % of GDP. The fire cost per GDP is about the same for most developed countries, ranging from 0.2 to 1.2 % with a mean for 23 countries at 0.81 % [2]. The US is among the highest at per capita fires with 6.5/1000 people and about 1.7 annual deaths per 10^5 people. Russia tops the latter category among nations at 9.87. This gives a perceived risk of dying by fire in an average lifetime (75 years) at about 1 in 135 for Russia and 1 in 784 for the US. Hence, in the lifetime of a person in the US, about 1 in 800 people will die by fire, about 10 in 800 by auto accidents and the remainder (789) will most likely die due to cancer, heart disease or stroke. These factors affect the way society and governments decide to invest in the wellbeing of its people.

Fire, like commercial aircraft disasters, can take a large quantity of life or property cost in one event. When it is realized that 15 to 25 % of fires can be attributed to arson [2], and today terrorism looms high as a threat, the incentive to invest in improved fire safety might increase. The 9/11 events at the World Trade Center (WTC) with a loss of life at nearly 3000 and a direct cost of over $10 billion cannot be overlooked as an arson fire. In the past, such catastrophic events have only triggered 'quick fixes' and short-term anxiety. Perhaps, 9/11 – if perceived as a fire safety issue – might produce improved investment in fire safety. Even discounting the WTC as a fire event, significant disasters in the twentieth century have and should impact on our sensitivity to fire. Yet the impact of disasters is short-lived. Events listed in Table 1.1 indicate some significant fire events of the twentieth century. Which do you know about? Related illustrations are shown in Figures 1.1 to 1.4. It is interesting to note that the relatively milder (7.1 versus 7.7) earthquake centered in Loma Prieta in 1989 caused an identical fire in San Francisco's Marina district, as shown in Figure 1.2(b).

1.2 Fire and Science

Over the last 500 years, science has progressed at an accelerating pace from the beginnings of mathematical generality to a full set of conservation principles needed to address most problems. Yet fire, one of the earliest tools of mankind, needed the last 50 years to give it mathematical expression. Fire is indeed complex and that surely helped to retard its scientific development. But first, what is fire? How shall we define it?

Table 1.1 Selected fire disasters of the twentieth century

Year	Event	Source
1903	Iroquois Theater fire, Chicago, 620 dead	—
1904	Baltimore conflagration, \$50 million (\$821 million, 1993)	[5]
1906	San Francisco earthquake (those that were there, call it 'the fire'), 28 000 buildings lost, 450 dead, \$300 million due to fire (\$5 billion, 1993), \$15 million due to earthquake damage	[4]
1923	Tokyo earthquake, 150 000 dead and 700 000 homes destroyed mostly by fire, 38 000 killed by a fire tornado (whirl) who sought refuge in park	[4]
1934	Morro Castle ship fire, 134 dead, Asbury Park, NJ	[4]
1937	Hindenburg airship fire, Lakehurst, NJ	[4]
1947	*SS Grandcamp* and Monsanto Chemical Co. (spontaneous ignition caused fire and explosion), Texas City, TX, \$67 million (\$430 million, 1993)	[5]
1967	Apollo One spacecraft, 3 astronauts died, fire in 100 % oxygen	[4]
1972	Managua, Nicaragua earthquake, 12 000 dead, fire and quake destroyed the entire city	[4]
1980	MGM Grand Hotel, Las Vegas, NV, 84 dead	[4]
1983	Australian bushfires, 70 dead, 8500 homeless, 150 000 acres destroyed, ~\$1 billion	[4]
1991	One Meridian Plaza 38-story high-rise, Philadelphia, 3 fire fighters killed, \$500 million loss as structural damage caused full demolition	—
1991	Oakland firestorm, \$1.5 billion loss	[5]
1995	Milliken Carpet Plant, Le Grange, GA, 660 00 ft^2 destroyed, full-sprinklers, \$500 million loss	—
2001	World Trade Center twin towers, ~3000 died, ~\$10 billion loss, insufficient fire protection to steel structure	—

A flame is a chemical reaction producing a temperature of the order of at least 1500 K and generally about 2500 K at most in air. Fire is generally a turbulent ensemble of flames (or flamelets). A flamelet or laminar flame can have a thickness of the order of 10^{-3} cm and an exothermic production rate of energy per unit volume of about 10^8 W/cm^3. However, at the onset of ignition, the reaction might only possess

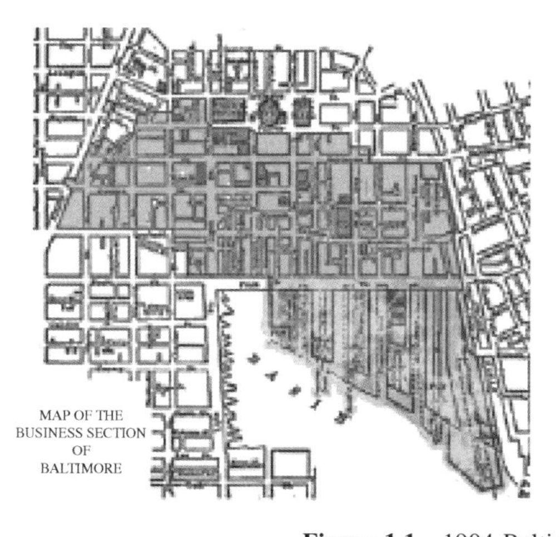

Figure 1.1 1904 Baltimore conflagration

(a)

(b)

Figure 1.2 (a) 1906 San Francisco earthquake. (b) area of fire damage for San Francisco, 1906

about 10^{-2} W/cm^3. This is hardly perceptible, and its abrupt transition to a full flame represents a jump in thermal conditions, giving rise to the name thermal explosion.

A flame could begin with the reactants mixed (premixed) or reactants that might diffuse together (diffusion flame). Generally, a flame is thought of with the reactants in the gas phase. Variations in this viewpoint for a flame or fire process might occur and are defined in special terminology. Indeed, while flame applies to a gas phase reaction, fire,

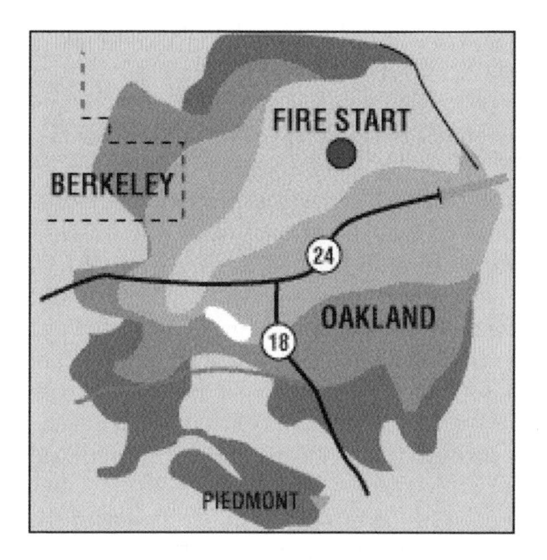

Figure 1.3 Oakland Hills fire storm, 1991

Figure 1.4 Collapse of the south tower at the World Trade Center

and its synonym combustion, refers to a broader class of reactions that constitute a significant energy density rate. For example, smoldering is a combustion reaction (that could occur under temperatures as low as 600 K) between oxygen in air and the surface of a solid fuel. The combustion wave propagation through dynamite might be termed by some as an explosion, yet it is governed by premixed flame theory. Indeed, fire or combustion might more broadly represent an exothermic chemical reaction that results from a runaway rate caused by temperature or catalytic effects. Note that we have avoided the often-used definition of fire as 'a chemical reaction in air giving off heat and light'. However, a flame may not always be seen; e.g. an H_2 flame would be transparent to the eye and not easily seen. A flame could be made adiabatic, and therefore heat is not given off. This could occur within the uniform temperature soot-laden regions of a large fire. Moreover, oxygen in air might not be the only oxidizer in a reaction termed combustion or fire. In general, we might agree that a flame applies to gas phase combustion while fire applies to all aspects of uncontrolled combustion.

The science of fire required the development of the mathematical description of the processes that comprise combustion. Let us examine the historical time-line of those necessary developments that began in the 1600s with Isaac Newton. These are listed in Table 1.2, and pertain to macroscopic continuum science (for examples see Figures 1.5 and 1.6). As with problems in convective heat and mass transfer, fire problems did not require profound new scientific discoveries after the general conservation principles and constitutive relations were established. However, fire is among the most complex of transport processes, and did require strategic mathematical formulations to render solutions. It required a thorough knowledge of the underlying processes to isolate its dominant elements in order to describe and effectively interpret experiments and create general mathematical solutions.

Table 1.2 Developments leading to fire science

Time	Event	Key initiator
~1650	Second law of motion (conservation of momentum)	Isaac Newton
1737	Relationship between pressure and velocity in a fluid	Daniel Bernoulli
~1750	First law of thermodynamics (conservation of energy)	Rudolph Clausius
1807	Heat conduction equation (Fourier's law)	Joseph Fourier
1827	Viscous equations of motion of a fluid	Navier
1845		Stokes
~1850	Chemical History of the Candle Lectures at the Royal Society	Michael Faraday
1855	Mass diffusion equation (Fick's law)	A. Fick
1884	Chemical reaction rate dependence on temperature	S. Arrhenius
~1900	Thermal radiation heat transfer	Max Planck
1928	Solution of diffusion flame in a duct	Burke and Schumann
~1940	Combustion equations with kinetics	Frank-Kamenetskii
~1930		Semenov
~1950		Zel'dovich
~1950	Convective burning solutions	H. Emmons, D. B. Spalding
~1960	Fire phenomena solutions	P. H. Thomas
~1970	Leadership in US fire research programs	R. Long, J. Lyons

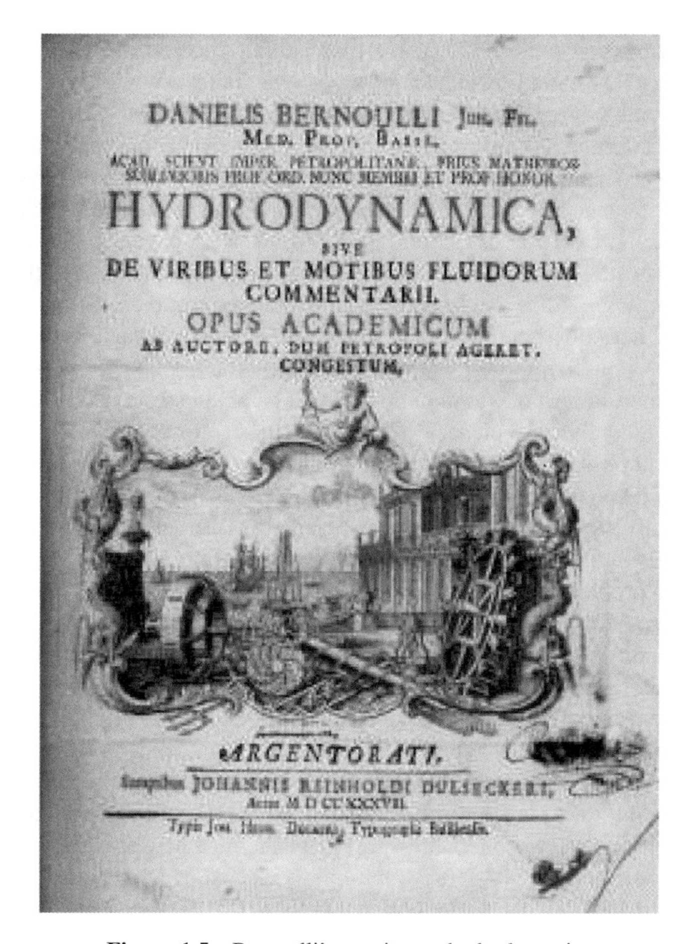

DANIELIS BERNOULLI Joh. Fil.
Med. Prof. Basil.
ACAD. SCIENT. IMPER. PETROPOLITANAE, PRIUS MATHESEOS SUBLIMIORIS PROF. ORD. NUNC MEMBRI ET PROF. HONOR.
HYDRODYNAMICA,
SIVE
DE VIRIBUS ET MOTIBUS FLUIDORUM
COMMENTARII.
OPUS ACADEMICUM
AB AUCTORE, DUM PETROPOLI AGERET, CONGESTUM.

ARGENTORATI,
Sumptibus JOHANNIS REINHOLDI DULSECKERI,
Anno M D CC XXXVIII.
Typis Joh. Henr. Deckeri, Typographi Basiliensis.

Figure 1.5 Bernoulli's treatise on hydrodynamics

Figure 1.6 Significant scientists in (a) heat transfer (Jean Baptiste Joseph Fourier), (b) chemistry (Svante August Arrhenius, Nobel Prize in Chemistry 1903) and (c) combustion (Nikolay Semenov, Nobel Prize in Chemistry 1956)

While Newton forged the basis of the momentum equation, he also introduced the concept of mass, essentially defined by $m = F/a$. Joule established the equivalence between work and energy, and the first law of thermodynamics (energy conservation) introduces the concept of energy as the change due to heat loss and work done by a system. Heat is then defined as that energy transferred from the system due to temperature difference. Of course, the second law of thermodynamics (Clausius) does not enter into solution methods directly, but says a system cannot return to its original state without a cost. That cost involves transfers of mass and energy, and these transport fluxes become (approximately) related to the gradient of their associated intensive properties; e.g. heat flux is proportional to the temperature gradient as in Fourier's law and species mass flux is proportional to its concentration gradient as in Fick's law. The nonlinear transport of heat due to radiation provides a big obstacle to obtaining analytical solutions for fire; and turbulence, despite the exact formulation of this unsteady phenomena through the Navier–Stokes equations over 150 years ago, precludes general solutions to fluid flow problems. Hence, skill, ingenuity and broad knowledge are needed to formulate, solve and understand fire phenomena. The Russian scientists Semenov [6], Frank-Kamenetskii [7] and Zel'dovich *et al.* [8] provide much for the early formulation of combustion problem solutions. These problems addressed the role of kinetics and produced the Zel'dovich number, $(E/RT)/(1 + c_pT/\Delta h_c)$, E/RT being the Arrhenius parameter. Spalding [9] and Emmons [10] laid a foundation for solutions to diffusive burning of the condensed phase. This foundation served as a guidepost to other far more complex problems involving soot, radiation and flame spread. The Spalding B number is a key dimensionless group that emerges in these problems, and represents the ratio of energy released in combustion to that needed for fuel vaporization. The energy production rate by fire (more specifically its rate of change of enthalpy due to chemical changes at 25 °C and 1 atmosphere pressure) will be termed here as firepower. Many have used the term heat release rate (HRR), but this is viewed as a misnomer by this author as the firepower has the same chemical production rate that would be attributed to a combustion-based power plant, and is not identical to the heat transferred in the reaction. Such misnomers have occurred throughout thermodynamics as represented in the terms: heat capacity, heat of vaporization and heat of combustion – all related to enthalpies. These misnomers might be attributed to remnants of the caloric theory of heat in which heat was viewed as being part of matter.

1.3 Fire Safety and Research in the Twentieth Century

At the start of the twentieth century there were significant activities in US fire safety. The National Fire Protection Association (NFPA) was founded in 1897 principally to address sprinkler use and its standardization. The Underwriter's Laboratory (UL) was founded in 1894 by William H. Merrill to address electrical standards and testing. In 1904, Congress created the National Bureau of Standards (NBS), later becoming the National Institute of Standards and Technology (NIST) in 1987. After the Baltimore conflagration (1904), Congress directed NBS to address the structural fire safety of buildings. That program began in 1914 under Simon Ingberg, and had a profound influence on standards and testing dealing with structural fire protection over the next 60 years. It is noteworthy to consider the words of the NBS Director to Congress in 1914 when US fire losses were reported at more than ten time those of Europe: 'The greatest fire loses are in the cities having laws and regulations' [11].

Sixty years later *America Burning* cited 'One basic need is to strengthen the grounding of knowledge about fire in a body of scientific and engineering theory, so that real world problems can be dealt with through predictive analyses' [1]. The result of that President's Commission report led to the consolidation of the three existing fire programs at NBS at that time: building fire safety (I. Benjamin and D. Gross), fire research and fire fighting (J. Rockett and A. F. Robertson) and flammable fabrics (J. Clark and C. Huggett). The consolidation forged the Center for Fire Research (CFR) under John Lyons in 1974. It included the development of a library for fire science (M. Rappaport and N. Jason) – a necessity for any new research enterprise. The CFR program focused on basic research and important problems of that time. They involved clothing fabric and furnishing fire standards, the development of smoke detectors standards, floor covering flammability and special standards for housing and health care facilities. Its budget was about $5 million, with more than half allocated as internal NBS funds. It co-opted the NSF RANN (Research for Applied National Needs) fire grants program along with Ralph Long, its originator. This program was endowed solely for basic fire research at $2 million. Shortly after 1975, the Products Research Committee (PRC), a consortium of plastic producers, contributed about $1 million annually to basic research on plastic flammability for the next five years. This industry support was not voluntary, but mandated by an agreement with the Federal Trade Commission that stemmed from a complaint that industry was presenting faulty fire safety claims with respect to test results. This total budget for fire research at NBS of about $8 million in 1975 would be about $30 million in current 2004 dollars. Indeed, the basic grants funding program that originating from the NSF is only about $1 million today at NBS, where inflation would have its originally allocated 1974 amount at about $7.5 million today. Thus, one can see that the extraordinary effort launched for fire research in the 1970s was significant compared to what is done today in the US. While US fire research efforts have waned, international interests still remain high.

Fire research in the USA during the 1970s was characterized by a sense of discovery and a desire to interact with centers of knowledge abroad. For years, especially following World War II, the British and the Japanese pursued research on fire with vigor. The Fire Research Station at Borehamwood (the former movie capital of the UK) launched a strong research program in the 1950s with P. H. Thomas, a stalwart of fire modeling [12], D. Rasbash (suppression), M. Law (fire resistance) and D. L. Simms (ignition). The Japanese built a strong infrastructure of fire research within its schools of architecture, academia in general and government laboratories. Kunio Kawagoe was the ambassador of this research for years (about 1960 to 1990). He is known for contributions to compartment fires. Others include S. Yokoi (plumes), T. Hirano (flame spread), T. Wakamatsu (smoke movement), T. Tanaka (zone modeling), Y. Hasemi (fire plumes), T. Jin (smoke visibility), K. Akita (liquids) and H. Takeda (compartment fire behavior). In Sweden, beginning in the early 1970s, Professor O. Pettersson engaged in fire effects on structures, fostering the work of S. E. Magnusson and founding the fire engineering degree program at Lund University. Vihelm Sjolin, coming from the Swedish civil defense program, provided leadership to fire researchers like Kai Odeen (compartment fires) and Bengt Hagglund (radiation, fire dynamics), and later directed a $1 million Swedish basic research program in fire during the 1980s. Today Sweden has strong fire research efforts at Lund University and at the SP Laboratory in Boras (U. Wickstrom and

B. Sundstrom). Programs remain active around the world: PR China (W. C. Fan), Finland (M. Kokkala and O. Keski-Rinkinen), France (P. Joulain and P. Vantelon) and the United Kingdom (G. Cox, D. Drysdale, J. Torero, G. Makhviladze, V. Molkov, J. Shields and G. Silcock) to cite only a few.

Enabled by ample funding, enthusiasm for problems rich in interdisciplinary elements and challenged by new complex problems, many new researchers engaged in fire research in the 1970s. They were led by Professor Howard Emmons (Harvard), who had been encouraged to enter the field of fire research by Professor Hoyt Hottel (MIT). They both brought stature and credibility to the US fire research interest. Emmons also advised the Factory Mutual Research Corporation (FMRC) on its new basic program in fire research. The FM program inspired by Jim Smith was to serve insurance interests – a significant step for industry. That spawned a remarkably prolific research team: R. Friedman, J. deRis, L. Orloff, G. Markstein, M. Kanury, A. Modak, C. Yao, G. Heskestad, R. Alpert, P. Croce, F. Taminini, M. Delichatsios, A. Tewarson and others. They created a beacon for others to follow.

Under NBS and NSF funding, significant academic research efforts emerged in the US at this time. This brought needed intellect and skills from many. It was a pleasure for me to interact with such dedicated and eager scientists. Significant programs emerged at UC Berkeley (C. L. Tien, P. Pagni, R. B. Williamson, A. C. Fernandez-Pello and B. Bressler), Princeton (I. Glassman, F. Dryer, W. Sirignano and F. A. Williams), MIT (T. Y. Toong and G. Williams), Penn State (G. M Faeth), Brown (M. Sibulkin), Case-Western Reserve (J. Tien and J. Prahl) and, of course, Harvard (H. Emmons and G. Carrier). Many students schooled in fire were produced (A. Atreya, I. Wichman, K. Saito, D. Evans and V. Babrauskas – to name only a few). All have made their mark on contributions to fire research. Other programs have contributed greatly: the Naval Research Laboratory (H. Carhart and F. Williams), FAA Technical Center (C. Sarkos, R. Hill, R. Lyon and T. Eklund), SWRI (G. Hartzell and M. Janssens), NRC Canada (T. Harmathy, T. T. Lie and K. Richardson), IITRI (T. Waterman) and SRI (S. Martin and N. Alvares). This collective effort reached its peak during the decade spanning 1975 to 1985. Thereafter, cuts in government spending took their toll, particularly during the Reagan administration, and many programs ended or were forced to abandon their basic research efforts. While programs still remain, they are forced into problem-solving modes to maintain funds for survival. Indeed, the legacy of this fruitful research has been the ability to solve fire problems today.

1.4 Outlook for the Future

In 2002 NIST was directed by Congress under $16 million in special funding to investigate the cause of the 9/11 World Trade Center (WTC) building failures. The science of fire, based on the accomplishment of the fertile decade (1975 to 1985), has increasingly been used effectively in fire investigations. The Bureau of Alcohol, Tobacco and Firearms (BATF), having the responsibility of federal arson crimes, has built the largest fire laboratory in the world (2003). The dramatic events of 9/11 might lead to a revitalized public awareness that fire science is an important (missing) element in fire safety regulation tenets. The NIST investigation could show that standards used in regulation for fire safety might be insufficient to the public good. Students of this text will acquire the ability to work problems and find answers, related to aspects of the 1993 bombing and fire at the WTC, and the 9/11 WTC attacks. Some of these problems address issues sometimes misrepresented or missed by media reporting. For example, it can clearly be shown that the jet fuel in the aircraft attacks on the WTC only provided an ignition source of several minutes, yet there are still reports that the jet fuel, endowed with special intensity, was responsible for the collapse of the steel structures. It is generally recognized that the fires at the WTC were primarily due to the normal furnishing content of the floors, and were key in the building failures. It is profound that NIST is now investigating such a significant building failure due to fire as the mandate given by Congress to the NBS founding fire program in 1914 – 90 years ago – that mandate was to improve the structural integrity of buildings in fire and to insure against collapse.

While much interest and action have been taken toward the acceptance of performance codes for fire safety in, for example, Japan, Denmark, Australia, New Zealand, the US and the rest of the world, we have not fully created the infrastructure for the proper emergence of performance codes for fire. Of course, significant infrastructure exists for the process of creating codes and standards, but there is no overt agency to insure that fire science is considered in the process. The variations in fire test standards from agency to agency and country to country are a testament to this lack of scientific underpinning in the establishment of regulations. Those that have expertise in fire standards readily know that the standards have little, if any, technical bases. Yet these standards have been generally established by committees under public consensus, albeit with special interests, and their shortcomings are not understood by the general public at large.

Even without direct government support or international harmonization strategies, we are evolving toward a better state of fire standards with science. Educational institutions of higher learning probably number over 25 in the world currently. That is a big step since the founding of the BS degree program in fire protection engineering at the University of Maryland in 1956 (the first accredited in the US). This was followed by the first MS program, founded by Dave Rasbash in 1975, at the University of Edinburgh. It is very likely that it will be the force of these programs that eventually lead to the harmonization of international fire standards using science as their underlying principles. This will likely be an uneven process of change, and missteps are likely; to paraphrase Kristian Hertz (DTU, Denmark): 'It is better to use a weak scientific approach that is transparent enough to build upon for a standard practice, than to adopt a procedure that has no technical foundation, only tradition.' Hopefully, this text will help the student contribute to these engineering challenges in fire safety standards. Fire problems will continually be present in society, both natural and manmade, and ever changing in a technological world.

1.5 Introduction to This Book

This book is intended to be pedagogical, and not inclusive as a total review of the subject. It attempts to demonstrate that the subject of fire is a special engineering discipline built on fundamental principles, classical analyses and unique phenomena. The flow of the subject matter unfolds in an order designed to build upon former material to advance properly to the new material. Except for the first and last chapters, instruction should follow the order of the text. In my teaching experience, a one-semester 15-week, 45-hour course, covering most of Chapters 2 through 11, are possible with omission of specialized or advanced material. However, Chapter 11 on compartment fires is difficult to cover in any depth because the subject is too large, and perhaps warrants a special course that embraces associate computer models. The text by Karlsson and Quintiere, *Enclosure Fire Dynamics*, is appropriate for such a course (see Section 1.5.4).

The current text attempts to address the combustion features of fire engineering in depth. That statement may appear as an oxymoron since fire and combustion are the same, and one might expect them to be fully covered. However, fire protection engineering education has formerly emphasized more the protection aspect than the fire component. Figure 1.7 shows a flow chart that represents how the prerequisite

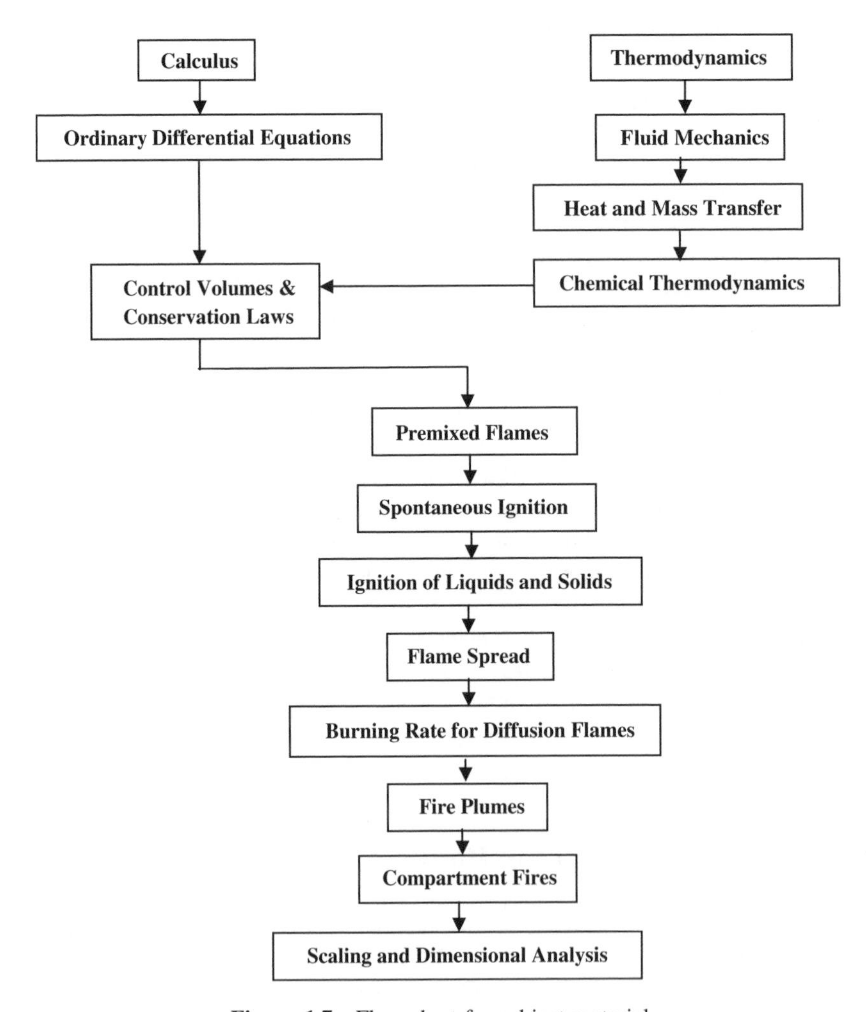

Figure 1.7 Flow chart for subject material

material relates to the material in the text. Standard texts can be consulted for the prerequisite subjects. However, it is useful to present a brief review of these subjects to impart the minimum of their facets and to stimulate the student's recollections.

1.5.1 Thermodynamics

Thermodynamics is the study of matter as a macroscopic continuum. Mass is the tangible content of matter, identified through Newton's second law of motion, $m = F/a$. A system of fixed mass which could consist of a collection of different chemical states of matter (species) must remain at a constant total energy if it is isolated from its surroundings. For fire systems only internal energy (U) will be significant (excluding kinetic, electrical, etc.). When a system interacts with its surroundings, the first law of thermodynamics is expressed as

$$U_2 - U_1 = Q + W \quad \text{or as a rate expression} \quad \frac{\mathrm{d}U}{\mathrm{d}t} = \dot{Q} + \dot{W} \tag{1.1}$$

The internal energy changes from 1 to 2 and heat (Q) and work (W) are considered positive when added to the system. Work is defined as the product of the resultant of forces and their associated distances moved. Heat is energy transferred solely due to temperature differences. (Sometimes, the energy associated with species diffusion is included in the definition of heat.) The first law expressed in rate form requires the explicit introduction of the transport process due to gradients between the system and the surroundings. Despite these gradients implying nonequilibrium, thermodynamic relationships are assumed to hold at each instant of time, and at each point in the system.

The second law of thermodynamics allows us to represent the entropy change of a system as

$$\mathrm{d}S = \frac{1}{T}\mathrm{d}U + \frac{p}{T}\mathrm{d}V \tag{1.2}$$

This relationship is expressed in extensive properties that depend on the extent of the system, as opposed to intensive properties that describe conditions at a point in the system. For example, extensive properties are made intensive by expressing them on a per unit mass basis, e.g. $s = S/m$ density, $\rho = 1/v$, $v = V/m$. For a pure system (one species), Equation (1.2) in intensive form allows a definition of thermodynamic temperature and pressure in terms of the intensive properties as

$$\begin{aligned} \text{Temperature}, T &= \left(\frac{\partial u}{\partial s}\right)_v \\ \text{Pressure}, p &= \left(\frac{\partial s}{\partial v}\right)_u \end{aligned} \tag{1.3}$$

The thermodynamic pressure is equal to the mechanical pressure due to force at equilibrium. While most problems of interest possess gradients in the intensive properties (i.e. T, p, Y_i), local thermodynamic equilibrium is implicitly assumed. Thus, the equation of state, such as for an ideal gas

$$p = \frac{RT\rho}{M} \tag{1.4}$$

holds for thermodynamic properties under equilibrium, although it may be applied over variations in time and space. Here M is the molecular weight, ρ is the density and R is the universal gas constant. For ideal gases and most solids and liquids, the internal energy and enthalpy $(h = u + p/\rho)$ can be expressed as

$$\mathrm{d}u = c_v\,\mathrm{d}T \quad \text{and} \quad \mathrm{d}h = c_p\,\mathrm{d}T \quad \text{respectively} \tag{1.5}$$

where c_v and c_p are the specific heats of constant volume and constant pressure. It can be shown that for an ideal gas, $R = c_p - c_v$.

1.5.2 *Fluid mechanics*

Euler expressed Newton's second law of motion for a frictionless (inviscid) fluid along a streamline as

$$\int_1^2 \rho\frac{\partial v}{\partial t}\mathrm{d}s + \left(\rho\frac{v^2}{2} + p + \rho g z\right)_2 - \left(\rho\frac{v^2}{2} + p + \rho g z\right)_1 = 0 \tag{1.6}$$

The velocity (v) is tangent along the streamline and z is the vertical height opposite in sign to the gravity force field, $g = 9.81\,\mathrm{N/kg}$ $(1\,\mathrm{N} = 1\,\mathrm{kg\,m/s^2})$. For a viscous fluid, the right-hand side (RHS) is negative and in magnitude represents the power dissipation per unit volume due to viscous effects. The RHS can also be represented as the rate of shaft work added to the streamline or system in general.

Pressure is the normal stress for an inviscid fluid. The shear stress for a Newtonian fluid is given as

$$\tau = \mu\frac{\partial v}{\partial n} \tag{1.7}$$

where μ is the viscosity and n is the normal direction. Shear stress is shown for a streamtube in Figure 1.8. Viscous flow is significant for a small Reynolds number

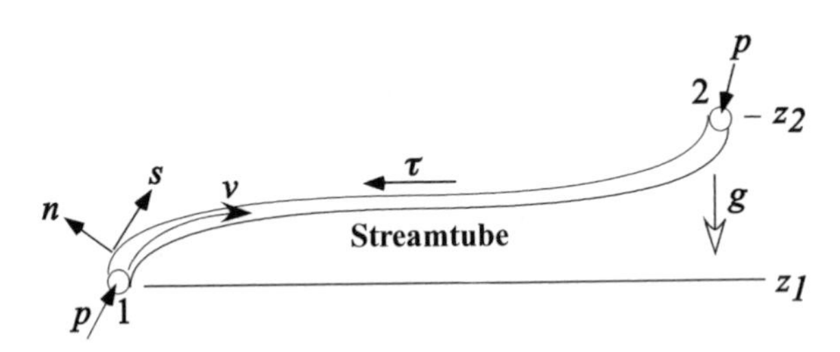

Figure 1.8 Flow along a streamline

($Re = \rho v l / \mu$), where l is a characteristic length. Near surfaces, l might be thought of as the boundary-layer thickness where all of the viscous effects are felt and Re is locally small. However, if l is regarded as a characteristic geometric length, as the chord of an airfoil, Re might be found to be large. This means that the bulk flow away from the surface is nearly inviscid. Flows become unstable at large Re, evolving into a chaotic unsteady flow known as turbulence. Turbulence affects the entire flow domain except very close to the surface, where it is suppressed and laminar flow prevails.

1.5.3 Heat and mass transfer

Heat can flow due to the temperature difference in two distinct mechanisms: (1) radiation and (2) conduction. Radiation is energy transfer by electromagnetic waves and possesses frequency (or wavelength). Radiant energy can be generated in many different ways, which usually identifies its application, such as radio waves, microwaves, gamma rays, X-rays, etc. These different forms are usually identified by their wavelength ranges. Planck's law gives us the emitted radiation flux due to temperature alone as

$$\dot{q}'' = \varepsilon \sigma T^4 \tag{1.8}$$

where ε is the emissivity of the radiating body, varying between 0 and 1. The emissivity indicates the departure of an ideal (blackbody) emitter. For solids or liquids under fire heating conditions, although ε depends on wavelength, it can generally be regarded as a constant close to 1. For gases, ε will be finite only over discrete wavelengths, but for soot distributed in combustion product gas, ε is continuous over all wavelengths. It is this soot in flames and combustion products that produces the bulk of the radiation. In general, for smoke and flames the emissivity can be represented for each wavelength as

$$\varepsilon = 1 - \mathrm{e}^{-\kappa l} \tag{1.9}$$

The absorption coefficient κ depends in general on wavelength (λ). For large κl, ε approaches 1 – a perfect emitter. This limit generally occurs at turbulent flames of about 1 m in thickness.

The other mode of heat transfer is conduction. The conductive heat flux is, by Fourier's law,

$$\dot{q}'' = -k\frac{\partial T}{\partial n} \approx k\frac{\Delta T}{l} \tag{1.10}$$

For laminar conditions of slow flow, as in candle flames, the heat transfer between a fluid and a surface is predominately conductive. In general, conduction always prevails, but in the unsteadiness of turbulent flow, the time-averaged conductive heat flux between a fluid and a stationary surface is called convection. Convection depends on the flow field that is responsible for the fluid temperature gradient near the surface. This dependence is contained in the convection heat transfer coefficient h_c defined by

$$\dot{q}'' = -k\left(\frac{\partial T}{\partial n}\right)_s \equiv h_c(T - T_s) \tag{1.11}$$

Convection is primarily given in terms of correlations of the form

$$Nu \equiv \frac{h_c l}{k} = C\,Re^n Pr^m \quad \text{or} \quad Nu \equiv \frac{h_c l}{k} = C\,Gr^n Pr \tag{1.12}$$

where the Grashof number, Gr, replaces the Re in the case of natural convection, which is common to most fire conditions. The Grashof number is defined as

$$Gr \equiv \frac{\beta g(T - T_s)l^3}{\nu^2} \tag{1.13}$$

where β is the coefficient of volume expansion, $1/T$ for a gas, and Pr is the Prandtl number, $c_p\mu/k$. These correlations depend on the nature of the flow and the surface geometry.

Mass transfer by laminar (molecular) diffusion is directly analogous to conduction with the analog of Fourier's law as Fick's law describing the mass flux (mass flow rate per unit area) of species i due to diffusion:

$$\rho_i V_i = \dot{m}_i'' = -\rho D \frac{\partial Y_i}{\partial n} \tag{1.14}$$

where V_i is the diffusion velocity of the ith species, ρ is the mixture density and D_i is the diffusion coefficient and Y_i is the mass fraction concentration of the ith species. The corresponding analog for mass flux in convection is given in terms of the mass transfer coefficient, h_m, as

$$\dot{m}_i'' = h_m(Y_i - Y_{i,s}) \tag{1.15}$$

The convective mass transfer coefficient h_m can be obtained from correlations similar to those of heat transfer, i.e. Equation (1.12). The Nusselt number has the counterpart Sherwood number, $Sh \equiv h_m l/D_i$, and the counterpart of the Prandtl number is the Schmidt number, $Sc = \mu/\rho D$. Since $Pr \approx Sc \approx 0.7$ for combustion gases, the Lewis number, $Le = Pr/Sc = k/\rho D c_p$ is approximately 1, and it can be shown that $h_m = h_c/c_p$. This is a convenient way to compute the mass transfer coefficient from heat transfer results. It comes from the Reynolds analogy, which shows the equivalence of heat transfer with its corresponding mass transfer configuration for $Le = 1$. Fire involves both simultaneous heat and mass transfer, and therefore these relationships are important to have a complete understanding of the subject.

1.5.4 Supportive references

Standard texts on the prerequisites abound in the literature and any might serve as refreshers in these subjects. Other texts might serve to support the body of material in this text on fire. They include the following combustion and fire books:

Cox, G. (ed.), *Combustion Fundamentals of Fire*, Academic Press, London, 1995.
Drysdale, D., *Introduction to Fire Dynamics*, 1st edn., John Wiley & Sons, Ltd, Chichester, 1985.
Glassman, I., *Combustion*, Academic Press, New York, 1977.

Kanury, A. M., *Introduction to Combustion Phenomena*, Gordon and Breach, New York, 1975.
Karlsson B. and Quintiere J. G., *Enclosure Fire Dynamics*, CRC Press, Boca Raton, Florida, 2000.
Turns, S. R., *An Introduction to Combustion*, 2nd edn, McGraw-Hill, Boston, 2000.
Williams, F. A., *Combustion Theory*, 2nd edn, Addison-Wesley Publishing Co., New York, 1985.

References

1. Bland, R. E. (Chair), *America Burning*, The National Commission on Fire Prevention and Control, US Government Printing Office, Washington DC, 1973.
2. Brushlinski, N., Sokolov, S. and Wagner, P., *World Fire Statistics at the End of the 20th Century*, Report 6, Center of Fire Statistics of CTIF, International Technical Committee for the Prevention and Extinction of Fire, Moscow and Berlin, 2000.
3. Gomberg, A., Buchbinder, B. and Offensend, F. J., *Evaluating Alternative Strategies for Reducing Residential Fire Loss – The Fire Loss Model*, National Bureau of Standards, NBSIR 82-2551, August 1982.
4. *The World's Worst Disasters of the Twentieth Century*, Octopus Books, London, 1999.
5. Sullivan, M. J., Property loss rises in large-loss fires, *NFPA J.*, November/December 1994, 84–101.
6. Semenov, N., *Chemical Kinetics and Chain Reactions*, Moscow, 1934.
7. Frank-Kamenetskii, D. A., *Diffusion and Heat Transfer in Chemical Kinetics*, Plenum Press, New York, 1969.
8. Zel'dovich, Y. B., Barenblatt, G. I., Librovich, V. B. and Makhviladze, G. M., *The Mathematical Theory of Combustion and Explosion*, Nauka, Moscow, 1980.
9. Spalding, D. B., The combustion of liquid fuels, *Proc. Comb. Inst.*, 1953, **4**, pp. 847–64.
10. Emmons, H. W., The film combustion of liquid fuel, *Z. Angew Math. Mech.*, 1956, **36**, 60–71.
11. Cochrane, R. C., *Measures for Progress – A History of the National Bureau of Standards*, US Department of Commerce, 1966, p. 131.
12. Rasbash, D. J. and Alvarez, N. J. (eds), Advances in fire physics – a symposium to honor Philip H. Thomas, *Fire Safety J.*, 1981, **3**, 91–300.

Problems

1.1 You may have seen the History Channel documentary or other information pertaining to the making of the World Trade Towers. Write a short essay, no more than two pages, on how you would have protected the Towers from collapse. You are to use practical fire safety technology, supported by scientific reasoning and information. You need to specify clearly what you would have done to prevent the Towers from collapse, based on fire protection principles, and you must justify your position. You can use other sources, but do not prepare a literature report; I am looking for your thinking.

1.2 Why do fire disasters seem to repeat despite regulations developed to prevent them from occurring again? For example, contrast the recent Rhode Island nightclub fire with the infamous Coconut Grove and Beverly Hill Supper Club fires. Can you identify other fires in places of entertainment that were equally deadly?

1.3 Look up the significant people that contributed to the development of fire science. Pick one, and write a one-page description of the contribution and its significance.

1.4 Examine the statistics of fire. Is there any logical reason why the US death rate is among the highest and why Russia's increased so markedly after the fall of the Soviet Union?

1.5 The Oakland Hills fire in 1991 was a significant loss. Is the repetition of such fires due to earthquakes or the interface between the urban and wildlife domains? What is being done to mitigate or prevent these fires?

1.6 How does science enter into any of the codes and standards that you are familiar with?

2

Thermochemistry

2.1 Introduction

Thermochemistry is the study of chemical reactions within the context of thermodynamics. Thermodynamics is the science of equilibrium states of matter which follow the conservation of mass and energy. Applications of the second law of thermodynamics can establish criteria for equilibrium and determine the possible changes of state. In general, changes of state take place over time, and states can vary over matter as spatial gradients. These gradients of states do not constitute equilibrium in the thermodynamic sense, yet it is justifiable to regard relationships between the thermodynamic properties to still apply at each point in a gradient field. This concept is regarded as local equilibrium. Thus the properties of temperature and pressure apply as well as their state relationships governed by thermodynamics. However, with regard to chemical properties, in application to fire and combustion, it is not usually practical to impose chemical equilibrium principles. Chemical equilibrium would impose a fixed distribution of mass among the species involved in the reaction. Because of the gradients in a real chemically reacting system, this equilibrium state is not reached. Hence, in practical applications, we will hypothesize the mass distribution of the reacting species or use measured data to prescribe it. The resultant states may not be consistent with chemical equilibrium, but all other aspects of thermodynamics will still apply. Therefore, we shall not include any further discussion of chemical equilibrium here, and the interested student is referred to any of the many excellent texts in thermodynamics that address this subject.

Where applications to industrial combustion systems involve a relatively limited set of fuels, fire seeks anything that can burn. With the exception of industrial incineration, the fuels for fire are nearly boundless. Let us first consider fire as combustion in the gas phase, excluding surface oxidation in the following. For liquids, we must first require evaporation to the gas phase and for solids we must have a similar phase transition. In the former, pure evaporation is the change of phase of the substance without changing its composition. Evaporation follows local thermodynamics equilibrium between the gas

Fundamentals of Fire Phenomena James G. Quintiere

and liquid phases. Heat is required to change the phase. Likewise, heat is required in the latter case for a solid, but now thermal decomposition is required. Except for a rare instance, of a subliming fuel, the fuel is not likely to retain much of its original composition. This thermal decomposition process is called pyrolysis, and the decomposed species can vary widely in kind and in number.

Fuels encountered in fire can be natural or manmade, the latter producing the most complications. For example, plastics constitute a multitude of synthetic polymers. A polymer is a large molecule composed of a bonded chain of a repeated chemical structure (monomer). For example, the polymers polyethylene and polymethyl-methylmethracrylate (PMMA) are composed of the monomers C_2H_4 and $C_5H_8O_2$ respectively. The C_2H_4, standing alone, is ethylene, a gas at normal room temperatures. The properties of these two substances are markedly different, as contrasted by their phase change temperatures:

	Ethylene	Polyethylene
Melt temperature(°C)	-169.1	~ 130
Vaporization temperature(°C)	-103.7	~ 400

For ethylene, the phase transition temperatures are precise, while for polyethylene these are very approximate. Indeed, we may have pyrolysis occurring for the polymer, and the phase changes will not then be so well defined. Moreover, commercial polymers will contain additives for various purposes which will affect the phase change process. For some polymers, a melting transition may not occur, and instead the transition is more complex. For example, wood and paper pyrolyze to a char, tars and gases. Char is generally composed of a porous matrix of carbon, but hydrogen and other elements can be attached. Tars are high molecular weight compounds that retain a liquid structure under normal temperatures. The remaining products of the pyrolysis of wood are gases consisting of a mixture of hydrocarbons, some of which might condense to form an aerosol indicative of the 'white smoke' seen when wood is heated. The principal ingredient of wood is cellulose, a natural polymer whose monomer is $C_6H_{10}O_5$. Wood and other charring materials present a very complex path to a gaseous fuel. Also wood and other materials can absorb water from the atmosphere. This moisture can then change both the physical and chemical properties of the solid fuel. Thus, the pyrolysis process is not unique for materials, and their description for application to fire must be generally empirical or very specific.

2.2 Chemical Reactions

Combustion reactions in fire involve oxygen for the most part represented as

$$\text{Fuel(F)} + \text{oxygen}(O_2) \rightarrow \text{products(P)}$$

The oxygen will mainly be derived from air. The fuel will usually consist of mostly carbon (C), hydrogen (H) and oxygen (O) atoms in a general molecular structure, $F: C_xH_yO_z$. Fuels could also contain nitrogen (N), e.g. polyurethane, or chlorine (Cl), e.g. polyvinylchloride $(C_2H_3Cl)_n$.

We use F as a representative molecular structure of the fuel in terms of its atoms and P, a similar description for the product. Of course, we can have more than one product, but symbolically we only need to represent one here. The chemical reaction can then be described by the chemical equation as

$$\nu_F F + \nu_0 O_2 + \nu_D D \rightarrow \nu_P P + \nu_D D \tag{2.1}$$

where the ν's are the stoichiometric coefficients. The stoichiometric coefficients are determined to conserve atoms between the left side of the equation, composing the reactants, and the right side, the products. In general, $\nu_P P$ represents a sum over $\nu_{P,i}\, P_i$ for each product species, i. Fuel or oxygen could be left over and included in the products, as well as an inert diluent (D) such as nitrogen in air, which is carried without change from the reactants to the products.

A complete chemical reaction is one in which the products are in their most stable state. Such a reaction is rare, but, in general, departure from it is small for combustion systems. For example, for the fuel $C_xH_yO_z$, the complete products of combustion are CO_2 and H_2O. Departures from completeness can lead to additional incomplete products such as carbon monoxide (CO), hydrogen (H_2) and soot (mostly C). For fuels containing N or Cl, incomplete products of combustion are more likely in fire, yielding hydrogen cyanide (HCN) and hydrogen chloride (HCl) gases instead of the corresponding stable products: N_2 and Cl_2. In fire, we must rely on measurements to determine the incompleteness of the reaction. Where we might ignore these incomplete effects for thermal considerations, we cannot ignore their effects on toxicity and corrosivity of the combustion products.

A complete chemical reaction in which no fuel and no oxygen is left is called a stoichiometric reaction. This is used as a reference, and its corresponding stoichiometric oxygen to fuel mass ratio, r, can be determined from the chemical equation. A useful parameter to describe the state of the reactant mixture is the equivalence ratio, ϕ, defined as

$$\phi = \left(\frac{\text{mass of available fuel}}{\text{mass of available oxygen}}\right) r \tag{2.2}$$

If $\phi < 1$, we have burned all of the available fuel and have leftover oxygen. This state is commonly called 'fuel-lean'. On the other hand, if $\phi > 1$, there is unburned gaseous fuel and all of the oxygen is consumed. This state is commonly called 'fuel-rich'. When we have a fire beginning within a room, ϕ is less than 1, starting out as zero. As the fire grows, the fuel release can exceed the available oxygen supply. Room fires are termed ventilation-limited when $\phi \geq 1$.

The chemical equation described in Equation (2.1) can be thought of in terms of molecules for each species or in terms of mole (n). One mole is defined as the mass (m) of the species equal to its molecular weight (M). Strictly, a mole (or mol) is a gram-mole (gmole) in which m is given in grams. We could similarly define a pound-mole (lb-mole), which would contain a different mass than a gmole. The molecular weight is defined as the sum of the atomic weights of each element in the compound. The atomic weights have been decided as a form of relative masses of the atoms with carbon (C) assigned the value of 12, corresponding to six protons and six neutrons.* Thus, a mole is a relative

*The average atomic weight for carbon is 12.011 because about 1 % of carbon found in nature is the isotope, carbon 13, having an extra neutron in the nucleus.

mass of the molecules of that species, and a mole and molecular mass are proportional. The constant of proportionality is Avogadro's number, N_0, which is 6.022×10^{23} molecules/mole. For example, the mass of a methane molecule is determined from CH_4 using the atomic weights C = 12.011 and H = 1.007 94:

$$[(1)(12.011) + 4(1.007\,94)]\frac{g}{\text{mole}} \times \frac{1\ \text{mole}}{6.022 \times 10^{23}\ \text{molecule}} = 2.66 \times 10^{-23}\frac{g}{\text{molecule}}$$

The chemical equation then represents a conservation of atoms, which ensures conservation of mass and an alternative view of the species as molecules or moles. The stoichiometric coefficients correspond to the number of molecules or moles of each species.

Example 2.1 Ten grams of methane (CH_4) are reacted with 120 g of O_2. Assume that the reaction is complete. Determine the mass of each species in the product mixture.

Solution The chemical equation is for the stoichiometric reaction:

$$CH_4 + \nu_0 O_2 \rightarrow \nu_{P_1} CO_2 + \nu_{P_2} H_2O$$

where ν_F was taken as 1. Thus,

$$\begin{array}{lll} \text{Conserving C atoms}: & \nu_{P_1} = 1 & \\ \text{Conserving H atoms}: & 4 = 2\nu_{P_2} & \text{or} \quad \nu_{P_2} = 2 \\ \text{Conserving O atoms}: & 2\nu_0 = 2 + 2 & \text{or} \quad \nu_0 = 2 \end{array}$$

Hence,

$$CH_4 + 2\,O_2 \rightarrow CO_2 + 2\,H_2O$$

as the stoichiometric reaction.

The reactants contain 10 g of CH_4 and 120 g of O_2. The stoichiometric ratio is

$$r = \frac{m_{O_2}}{m_{CH_4}} \quad \text{for the stoichiometric reaction}$$

Since the stoichiometric reaction has 1 mole (or molecule) of CH_4 and 2 moles of O_2, and the corresponding molecular weights are

$$M_{CH_4} = 12 + 4(1) = 16\,\text{g/mole}$$

as C has an approximate weight of 12 and H is 1, and

$$M_{O_2} = 2(16) = 32\,\text{g/mole}$$

we compute

$$r = \frac{(2\,\text{moles}\,O_2)(32\,\text{g/mole}\,O_2)}{(1\,\text{mole}\,CH_4)(16\,\text{g/mole}\,CH_4)} = 4\,\text{g}\,O_2/\text{g}\,CH_4$$

The equivalence ratio is then

$$\phi = \left(\frac{10\,\mathrm{g\,CH_4}}{120\,\mathrm{g\,O_2}}\right)(4) = \frac{1}{3} < 1$$

Therefore, we will have O_2 left in the products. In the products, we have

$$\begin{aligned} m_{O_2} &= 120\,\mathrm{g} - m_{O_2,\mathrm{reacted}} \\ &= 120\,\mathrm{g} - (10\,\mathrm{g\,CH_4})\left(\frac{4\mathrm{g\,O_2}}{1\,\mathrm{g\,CH_4}}\right) = 80\,\mathrm{g\,O_2} \\ m_{CO_2} &= (10\,\mathrm{g\,CH_4})\left(\frac{1\,\mathrm{mole\,CO_2}}{1\,\mathrm{mole\,CH_4}}\right)\left(\frac{44\mathrm{g/mole\,CO_2}}{16\,\mathrm{g/mole\,CH_4}}\right) = 27.5\,\mathrm{g\ CO_2} \\ m_{H_2O} &= (10\,\mathrm{g\,CH_4})\left(\frac{2\,\mathrm{moles\,H_2O}}{1\,\mathrm{mole\,CH_4}}\right)\left(\frac{18\,\mathrm{g/mole\,H_2\,O}}{16\,\mathrm{g/mole\,CH_4}}\right) = 22.5\,\mathrm{g\,H_2O} \end{aligned}$$

Note that mass is conserved as 130 g was in the reactant as well as the product mixture.

2.3 Gas Mixtures

Since we are mainly interested in combustion in the gas phase, we must be able to describe reacting gas mixtures. For a thermodynamic mixture, each species fills the complete volume (V) of the mixture. For a mixture of N species, the mixture density (ρ) is related to the individual species densities (ρ_i) by

$$\rho = \sum_{i=1}^{N} \rho_i \tag{2.3}$$

Alternative ways to describe the distribution of species in the mixture are by mass fraction,

$$Y_i = \frac{\rho_i}{\rho} = \frac{m_i/V}{m/V} = \frac{m_i}{m} \tag{2.4}$$

and by mole fraction,

$$X_i = \frac{n_i}{n} \tag{2.5}$$

where n is the sum of the moles for each species in the mixture. By the definition of molecular weight, the molecular weight of a mixture is

$$M = \sum_{i=1}^{N} X_i M_i \tag{2.6}$$

It follows from Equations (2.4) and (2.6) that the mass and mole fractions are related as

$$Y_i = \frac{n_i M_i}{nM} = \frac{X_i M_i}{M} \tag{2.7}$$

Example 2.2 Air has the approximate composition by mole fraction of 0.21 for oxygen and 0.79 for nitrogen. What is the molecular weight of air and what is its corresponding composition in mass fractions?

Solution By Equation (2.6),

$$M_a = \left(0.21 \frac{\text{moles O}_2}{\text{mole air}}\right)(32\,\text{g/mole O}_2) + \left(0.79 \frac{\text{moles } N_2}{\text{mole air}}\right)(28\,\text{g/mole N}_2)$$

$$M_a = 28.84\,\text{g/mole air}$$

By Equation (2.7),

$$Y_{O_2} = \frac{(0.21)(32)}{(28.84)} = 0.233$$

and

$$Y_{N_2} = 1 - Y_{O_2} = 0.767$$

The mixture we have just described, even with a chemical reaction, must obey thermodynamic relationships (except perhaps requirements of chemical equilibrium). Thermodynamic properties such as temperature (T), pressure (p) and density apply at each point in the system, even with gradients. Also, even at a point in the mixture we do not lose the macroscopic identity of a continuum so that the point retains the character of the mixture. However, at a point or infinitesimal mixture volume, each species has the same temperature according to thermal equilibrium.

It is convenient to represent a mixture of gases as that of perfect gases. A perfect gas is defined as following the relationship

$$p = \frac{nRT}{V} \tag{2.8}$$

where R is the universal gas constant,

$$8.314\,\text{J/mole K}$$

It can be shown that specific heats for a perfect gas only depend on temperature.

A consequence of mechanical equilibrium in a perfect gas mixture is that the pressures developed by each species sum to give the mixture pressure. This is known as Dalton's law, with the species pressure called the partial pressure, p_i:

$$p = \sum_{i=1}^{N} p_i \tag{2.9}$$

Since each species is a perfect gas for the same volume (V), it follows that

$$\frac{p_i}{p} = \frac{n_i RT/V}{nRT/V} = \frac{n_i}{n} = X_i \tag{2.10}$$

This is an important relationship between the mole fraction and the ratio of partial and mixture pressures.

Sometimes mole fractions are called volume fractions. The volume that species i would occupy if allowed to equilibrate to the mixture pressure is

$$V_i = \frac{n_i RT}{p}$$

From Equation (2.8) we obtain

$$\frac{V_i}{V} = \frac{n_i}{n} = X_i$$

Hence, the name volume fraction applies.

2.4 Conservation Laws for Systems

A system is considered to be a collection of matter fixed in mass. It can exchange heat and work with its surroundings, but not mass. The conservation laws are expressed for systems and must be adjusted when the flow of matter is permitted. We will return to the flow case in Chapter 3, but for now we only consider systems.

The conservation of mass is trivial to express for a system since the mass (m) is always fixed. For a change in the system from state 1 to state 2, this is expressed as

$$m_1 = m_2 \tag{2.11a}$$

or considering variations over time,

$$\frac{\mathrm{d}m}{\mathrm{d}t} = 0 \tag{2.11b}$$

Alternatively, the rate or time-integrated form applies to the system.

The conservation of momentum or Newton's second law of motion is expressed for a system as

$$\boldsymbol{F} = \frac{\mathrm{d}(m\boldsymbol{v})}{\mathrm{d}t} \tag{2.12}$$

where $\boldsymbol{F}$ is the vector sum of forces acting on the system and $\boldsymbol{v}$ is the velocity of the system. If $\boldsymbol{v}$ varies over the domain of the system, we must integrate $\rho\boldsymbol{v}$ over the system volume before we take the derivative with time. Implicit in the second law is that $\boldsymbol{v}$ must be measured with respect to a 'fixed' reference frame, such as the Earth. This is called an inertial frame of reference. Although we will not use this law often in this book, it will be important in the study of fire plumes in Chapter 10.

Since our system contains reacting species we need a conservative law for each species. Consider the mass of species $i(m_i)$ as it reacts from its reactant state (1) to its product state (2). Then since the mass of i can change due to the chemical reaction with $m_{i,r}$ being produced by the reaction

$$m_{i,2} = m_{i,1} + m_{i,r} \tag{2.13}$$

or as a rate

$$\frac{\mathrm{d}m_i}{\mathrm{d}t} = \dot{m}_{i,r} \tag{2.14}$$

The right-hand side of Equation (2.14) is the mass rate of production of species i due to the chemical reaction. This term represents a generation of mass. However, for the fuel or oxygen it can be a sink, having a negative sign. From Example 2.1, $m_{O_2,2}$ is 80 g, $m_{O_2,1}$ is 120 g and $m_{O_2,r}$ is -40 g.

The conservation of energy or the first law of thermodynamics is expressed as the change in total energy of the system, which is equal to the net heat added to the system from the surroundings (Q) minus the net work done by the system on the surroundings (W). Energy (E) is composed of kinetic energy due to macroscopic motion and internal energy (U) due to microscopic effects. These quantities, given as capital letters, are extensive properties and depend on the mass of the system. Since each species in a mixture contributes to U we have for a system of volume (V),

$$U = V\sum_{i=1}^{N} \rho_i u_i \tag{2.15}$$

where u_i is the intensive internal energy given as U_i/m_i. Since the mass of the system is fixed from Equations (2.4) and (2.5), it can be shown that

$$u = \sum_{i=1}^{N} Y_i u_i \tag{2.16}$$

or

$$\tilde{u} = \sum_{i=1}^{N} X_i \tilde{u}_i \tag{2.17}$$

where the designation ($\tilde{}$) means per mole. Furthermore, since u and $\tilde{u}$ are intensive properties valid at a point, the relationships given in Equations (2.16) and (2.17) always hold at each point in a continuum. In all our applications, kinetic energy effects will be negligible, so an adequate expression for the first law is

$$U_2 - U_1 = Q - W \tag{2.18}$$

2.4.1 Constant pressure reaction

For systems with a chemical reaction, an important consideration is reactions that occur at constant pressure. This could represent a reaction in the atmosphere, such as a fire, in which the system is allowed to expand or contract according to the pressure of the surrounding atmosphere. Figure 2.1 illustrates this process for a system contained in a cylinder and also bounded by a frictionless and massless piston allowed to move so that the pressure is always constant on each side.

We consider the work term to be composed of that pertaining to forces associated with turning 'paddlewheel' shafts and shear stress (W_s), and work associated with normal pressure forces (W_p). For most applications in fire, W_s will not apply and therefore we will ignore it here. For the piston of face area, A, the work due to pressure (p) on the surroundings, moving at a distance $(x_2 - x_1)$, is

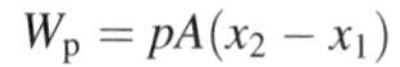

$$W_p = pA(x_2 - x_1)$$

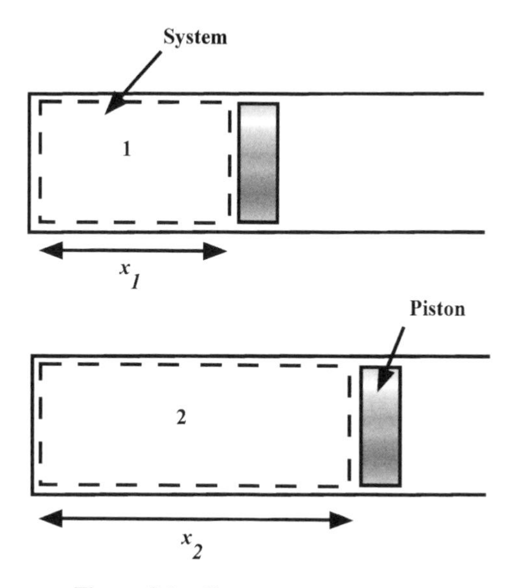

Figure 2.1 Constant pressure process

Since the volume of the system can be expressed by Ax,

$$W_p = p(V_2 - V_1)$$

or

$$W_p = p_2V_2 - p_1V_1$$

since $p_1 = p_2 = p$. Then the first law (Equation (2.18)) becomes

$$U_2 - U_1 = Q - (p_2V_2 - p_1V_1)$$

The property enthalpy (H), defined as

$$H = U + pV \tag{2.19}$$

is now introduced. Thus, for a constant pressure reacting system, without shaft or shear work, we have

$$\boxed{H_2 - H_1 = Q} \tag{2.20}$$

2.4.2 *Heat of combustion*

In a constant pressure combustion system in which state 1 represents the reactants and state 2 the products, we expect heat to be given to the surroundings. Therefore, Q is negative and so is the change in enthalpy according to Equation (2.20). A useful property of the reaction is the heat of combustion, which is related to this enthalpy change. We define the heat of combustion as the positive value of this enthalpy change per unit mass or mole of fuel reacted at 1 atm and in which the temperature of the system before and after the reaction is 25 °C. It is given as

$$\Delta h_c = -\frac{(H_2 - H_1)}{m_{F,r}} \tag{2.21a}$$

and

$$\Delta \tilde{h}_c = -\frac{(H_2 - H_1)}{n_{F,r}} \tag{2.21b}$$

with H_2 and H_1 evaluated at 1 atm, 25 °C. Often the heat of combustion is more restrictive in its definition, applying only to the stoichiometric (complete) reaction of the fuel. Sometimes the heat of combustion is given as a negative quantity which applies to Equation (2.21) without the minus sign before the parentheses. Since combustion gives rise to heat transfer, it is more rational to define it as positive. Also it should become apparent that the heat of combustion represents the contribution of energy from the sole process of rearranging the atoms into new molecules according to the chemical equation.

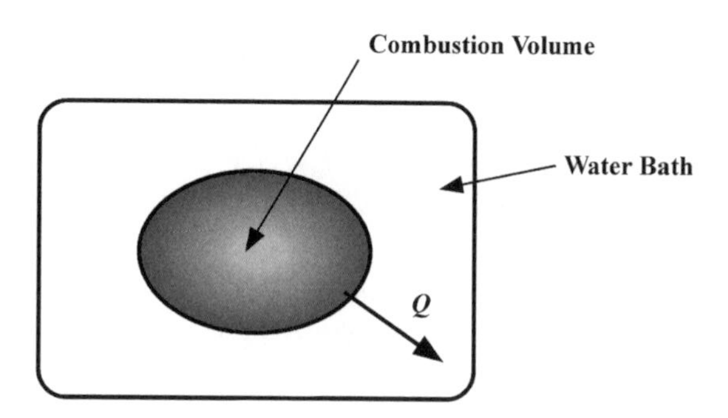

Figure 2.2 Idealized oxygen bomb calorimeter

In other words, there is no energy contribution due to a temperature change since it remains constant at 25 °C.

The heat of combustion of solids or liquids is usually measured in a device known as an oxygen bomb calorimeter. Such a device operates at a constant volume between states 1 and 2, and its heat loss is measured by means of the temperature rise to a surrounding water-bath. This is schematically shown in Figure 2.2. The combustion volume is charged with oxygen and a special fuel is added to ensure complete combustion of the fuel to be measured. Since the process is at constant volume (V), we have

$$U_2 - U_1 = Q$$

from Equation (2.18). By the definition of enthalpy, we can write

$$H_2 - H_1 = Q + (p_2 - p_1)V$$

By analysis and by measurement, we can determine $H_2 - H_1$ and express it in terms of values at 25 °C and 1 atm. Since Q is large for combustion reactions, these pressure corrections usually have a small effect on the accuracy of the heat of combustion determined in this way.

2.4.3 *Adiabatic flame temperature*

The adiabatic flame temperature is defined as the maximum possible temperature achieved by the reaction in a constant pressure process. It is usually based on the reactants initially at the standard state of 25 °C and 1 atm. From Equation (2.20), the adiabatic temperature (T_{ad}) is determined from the state of the system at state 2 such that

$$H_2(T_{ad}) = H_1(25\,^\circ\text{C} \quad \text{and} \quad 1\,\text{atm}) \tag{2.22}$$

We shall now see how we can express H_2 in terms of the product mixture. It might be already intuitive, but we should realize that T_{ad} depends on the state of the reactants. In addition to the fuel, it depends on whether oxygen is left over, and if a diluent, such as nitrogen, must also be heated to state 2. Reactants that remain in the product state reduce

T_{ad} since the energy associated with the combustion has been used to heat these species along with the new ones produced.

2.5 Heat of Formation

The molar heat of formation is the heat required to produce one mole of a substance from its elemental components at a fixed temperature and pressure. The fixed temperature and pressure is called the standard state and is taken at 25 °C and 1 atm. The heat of formation is designated by the symbol $\Delta h^{\circ}_{f,i}$ for the ith species and is given in energy units per unit mass or mole of species i. Usually, substances composed of a single element have $\Delta h^{\circ}_{f,i}$ values at zero, but this is not always true since it depends on their state. For example, carbon as graphite is 0, but carbon as a diamond is 1.88 kJ/mole. Diamond has a different crystalline structure than the graphite, and thus it takes 1.88 kJ/mole to produce 1 mole of C (diamond) from 1 mole of C (graphite). This is a physical, not chemical, change of the molecular structure. Again, we are complying with the first law of thermodynamics in this process, but we do not know if or how the process can proceed.

An alternative, but equivalent, way to define the heat of formation of a species is to define it as the reference enthalpy at the standard state (25 °C, 1 atm). Thus,

$$h_i(25\,^{\circ}\mathrm{C}, 1\ \mathrm{atm}) \equiv \Delta h^{\circ}_{f,i} \tag{2.23}$$

Strictly, this definition only applies to substances in equilibrium at the standard state.

For a perfect gas, and approximately for solids and liquids at small changes from 1 atm pressure, the enthalpy is only a function of temperature. It can be written in terms of specific heat at constant pressure, c_p, as

$$\boxed{h_i(T) = \Delta h^{\circ}_{f,i} + \int_{25\,^{\circ}\mathrm{C}}^{T} c_{p,i}\, \mathrm{d}T} \tag{2.24}$$

Strictly, this definition only applies to substances in equilibrium at the standard state. For the high temperatures achieved in combustion, c_p should always be regarded as a function of temperature. Some values are shown in Table 2.1. Values for the molar heats of formation for several substances are listed in Table 2.2. More extensive listings have been compiled and are available in reference books. The student should recognize that the values have been determined from measurements and from the application of Equation (2.20). Some examples should show the utility and interpretation of the heat of formation.

Table 2.1 Specific heat at constant pressure, $\tilde{c}_{p,i}$(J/K mole) (abstracted from Reference [1])

Temperature (K)	H_2(g)	C(s)	CH_4(g)	C_2H_4(g)	O_2(g)	N_2(g)	H_2O(g)	CO_2(g)	CO(g)
298	28.8	8.5	35.8	42.9	29.4	29.1	33.6	37.1	29.1
600	29.3	16.8	52.2	70.7	32.1	30.1	36.3	47.3	30.4
1000	30.2	21.6	71.8	110.0	34.9	32.7	41.3	54.3	33.2
1600	32.7	24.2	88.5	112.1	36.8	35.1	48.1	58.9	35.5
2500	35.8	25.9	98.8	123.1	38.9	36.6	53.9	61.5	36.8

(g) gaseous state, (s) graphite, solid state.

Table 2.2 Heat of formation $\Delta \tilde{h}_f^\circ$ in kJ/mole (at 25 °C and 1 atm)[a] (abstracted from Reference [2])

Substance	Formula	State	$\Delta \tilde{h}_f^\circ$(kJ/mole)
Oxygen	O_2	g	0
Nitrogen	N_2	g	0
Graphite	C	s	0
Diamond	C	s	1.88
Carbon dioxide	CO_2	g	−393.5
Carbon monoxide	CO	g	−110.5
Hydrogen	H_2	g	0
Water	H_2O	g	−241.8
Water	H_2O	l	−285.9
Chlorine	Cl_2	g	0
Hydrogen chloride	HCl	g	−92.3
Hydrogen cyanide	HCN	g	+135.1
Methane	CH_4	g	−74.9
Propane	C_3H_8	g	−103.8
n-Butane	C_4H_{10}	g	−124.7
n-Heptane	C_7H_{16}	g	−187.8
Benzene	C_6H_6	g	+82.9
Formaldehyde	CH_2O	g	−115.9
Methanol	CH_4O	g	−201.2
Methanol	CH_4O	l	−238.6
Ethanol	C_2H_6O	l	−277.7
Ethylene	C_2H_4	g	52.5

[a] Values for gaseous substances not in equilibrum at the standard state have been determined from the liquid and the heat of vaporization.

2.6 Application of Mass and Energy Conservation in Chemical Reactions

Example 2.3 Consider the formation of benzene and carbon dioxide from their elemental substances. Find the heat required per unit mole at the standard state: 25 °C, 1 atm.

Solution For benzene, formed from graphite and hydrogen gas,

$$6\,\mathrm{C} + 3\,\mathrm{H_2} \rightarrow \mathrm{C_6H_6}$$

is the chemical equation. By the application of Equation (2.20), we express the total enthalpies in molar form as

$$\widetilde{H}_1 = (6\,\text{moles C})(\widetilde{h}_\mathrm{C}(25\,^\circ\mathrm{C})) + (3\,\text{moles}\,H_2)(\widetilde{h}_{\mathrm{H_2}}(25\,^\circ\mathrm{C}))$$

and

$$\widetilde{H}_2 = (1\,\text{mole}\,\mathrm{C_6H_6})(\widetilde{h}_{\mathrm{C_6H_6}}(25\,^\circ\mathrm{C}))$$

Since the enthalpies are the heats of formation by Equation (2.23), we find from Table 2.2 that

$$Q = \Delta\widetilde{h}_f^\circ \text{ for benzene}$$

or

$$Q = +82.9 \text{ kJ/mole of benzene}$$

and heat is required.

In a similar fashion, obtaining carbon dioxide from graphite (carbon) and oxygen

$$C + O_2 \rightarrow CO_2$$

gives −393.5 kJ/mole for CO_2 for Q, or heat is lost. This reaction can also be viewed as the combustion of carbon. Then by Equation (2.21b), this value taken as positive is exactly the heat of combustion of carbon; i.e. $\Delta\widetilde{h}_c = 393.5\,\text{kJ/mole}$ carbon.

Example 2.4 Suppose we have the stoichiometric reaction for carbon leading to CO_2, but in the final state after the reaction the temperature increased from 25 to 500 °C. Find the heat lost.

Solution Since state 1 is at 25 °C for C and O_2, we obtain as before

$$\widetilde{H}_1 = 0$$

However, at state 2, we write, from Equation (2.24),

$$\widetilde{H}_2 = n_2\widetilde{h}_2 = (1 \text{ mole } CO_2)\left(\Delta\widetilde{h}_f^\circ + \int_{25\,^\circ C}^{500\,^\circ C} c_p \, dT\right)$$

From Tables 2.1 and 2.2, we take a mean value of $\widetilde{c}_p = 47\,\text{J/mole K}$ and obtain

$$\widetilde{H}_2 = (1\,\text{mole})\left[-393.5 \text{ kJ/mole} + 47 \times 10^{-3} \text{ kJ/mole K}(475\,K)\right]$$

or

$$\widetilde{H}_2 = -371.2\,\text{kJ/mole}$$

Therefore, the heat lost is 371.2 kJ/mole.

Example 2.5 One mole of CO burns in air to form CO_2 in a stoichiometric reaction. Determine the heat lost if the initial and final states are at 25 °C and 1 atm.

Solution The stoichiometric reaction is

$$CO + \text{air} \rightarrow CO_2 + N_2$$

where all of the CO and all of the O_2 in the air are consumed. It is convenient to represent air as composed of 1 mole $O_2 + 3.76$ moles of N_2, which gives 4.76 moles of the mixture, air. These proportions are the same as $X_{O_2} = 0.21$ and $X_{N_2} = 0.79$. The chemical equation is then

$$CO + \frac{1}{2}(O_2 + 3.76\ N_2) \rightarrow CO_2 + \frac{3.76}{2} N_2$$

We designate state 1 as reactants and state 2 as the product mixture. By Equation (2.20),

$$Q = H_2 - H_1$$

Using molar enthalpies, we develop the total enthalpy of each mixture as

$$H = \sum_{i=1}^{N} n_i \widetilde{h}_i$$

where $\widetilde{h}_i$ is found from Equation (2.24) and using Tables 2.1 and 2.2,

$$H_1 = (1\,\text{mole CO})(-110.5\ \text{kJ/mole}) + \frac{1}{2}(0) = -110.5\,\text{kJ}$$

$$H_2 = (1\,\text{mole CO}_2)(-393.5\ \text{kJ/mole}) + \left(\frac{3.76}{2}\right)(0) = -393.5\,\text{kJ}$$

$$Q = -393.5 - (-110.5) = -283.0\,\text{kJ}$$

From our definition of the heat of combustion, the result in Example 2.5 is unequivocally the heat of combustion of CO, $\Delta\widehat{h}_c = 283\,\text{kJ/mole CO}$, or $\Delta\widetilde{h}_c = 283/(12+16) = 10.1\,\text{kJ/g CO}$. From Equations (2.21) and (2.24), we can generalize this result as

$$\Delta\widetilde{h}_c = \left(\sum_{i=1}^{N} \nu_i\ \Delta\widetilde{h}^{\circ}_{f,i}\right)_{\text{Reactants}} - \left(\sum_{i=1}^{N} \nu_i\ \Delta\widetilde{h}^{\circ}_{f,i}\right)_{\text{Products}} \tag{2.25}$$

where the ν_i's are the molar stoichiometric coefficients with ν_F is taken as one. Usually, tabulated values of Δh_c for fuels are given for a stoichiometric (complete) reaction. A distinction is sometimes given between H_2O being a liquid in the product state ('gross' value) or a gas ('net' value). The gross heat of combustion is higher than the net because some of the chemical energy has been used to vaporize the water. Indeed, the difference between the heats of formation of water in Table 2.2 for a liquid (−285.9 kJ/mole) and a gas (−241/8 kJ/mole) is the heat of vaporization at 1 atm and 25 °C:

$$\Delta h_v = \frac{(285.9 - 241.8)\,\text{kJ/mole}}{18\,\text{g/mole H}_2\text{O}} = 2.45\,\text{kJ/g H}_2\text{O}$$

Example 2.6 Consider the same problem as in Example 2.5, but now with the products at a temperature of 625 °C.

Solution To avoid accounting for c_p variations with temperature, we assume constant average values. We select approximate average values from Table 2.1 as

$$c_{p,N_2} = 30\,\mathrm{J/mole\,N_2\ K}$$

$$c_{p,CO_2} = 47\,\mathrm{J/mole\,CO_2\ K}$$

Then for H_2 we write

$$H_2 = \sum_{i=1}^{2} \nu_i \left[\Delta h_{f,i} + c_{p_i}(T_2 - 25\,^\circ\mathrm{C})\right]$$

where we sum over the products CO_2 and N_2:

$$\begin{aligned}
H_2 = {} & (1\,\text{mole}\,\mathrm{CO_2})\left[-393.5\,\mathrm{kJ/mole\,CO_2} + 0.047\,\mathrm{kJ/mole\,K}(600\,\mathrm{K})\right] \\
& + \left(\frac{3.76}{2}\,\text{moles}\,\mathrm{N_2}\right)\left[0 + 0.030\ \mathrm{kJ/mole\,K}(600\,\mathrm{K})\right] \\
= {} & (-393.5) + \left[(1)(0.047) + \left(\frac{3.76}{2}\right)(0.030)\right](600) \\
= {} & (-393.5) + (0.10325)(600) \\
= {} & -331.55\,\mathrm{kJ}
\end{aligned}$$

$$Q = H_2 - H_1 = (-331.55) - (-110.5) = -221.05\,\mathrm{kJ}$$

For Example 2.6 we see that the heat lost due to the chemical reaction is composed of two parts: one comes from the heats of formation and the other from energy stored in the products mixture to raise its temperature. It can be inferred from these examples that we can write

$$\boxed{\begin{aligned}(-Q) = \nu_F\,\Delta\widetilde{h}_c + & \sum_{i=1}^{N,\text{Reactants}} n_i \widetilde{c}_{p_i}(T_1 - 25\,^\circ\mathrm{C}) \\ & - \sum_{i=1}^{N,\text{Products}} n_i \widetilde{c}_{p_i}(T_2 - 25\,^\circ\mathrm{C})\end{aligned}} \tag{2.26}$$

where ν_F is the moles (or mass) of fuel reacted and n_i are the moles (or mass) of each species in the mixture before (1, reactants) and after (2, products) the reaction. If we further consider specific heat of the mixture, since from Equations (2.17), (2.19), (2.24) and (2.25)

$$\widetilde{c}_p = \left(\frac{\partial \widetilde{h}}{\partial T}\right)_p = \frac{\partial}{\partial T}\left(\sum X_i h_i\right) = \sum X_i \widetilde{c}_{p_i}$$
$$= \left(\frac{1}{n}\right) \sum n_i \widetilde{c}_{p_i} \tag{2.27}$$

then

$$(-Q) = \nu_{\mathrm{F}}\, \Delta \widetilde{h}_{\mathrm{c}} + n_1 \widetilde{c}_{p_1}(T_1 - 25\,^{\circ}\mathrm{C}) - n_2 \widetilde{c}_{p_2}(T_2 - 25\,^{\circ}\mathrm{C}) \tag{2.28}$$

In other words, the heat lost in the chemical reaction is given by a chemical energy release plus a 'sensible' energy associated with the heat capacity and temperature change of the mixture states.

2.7 Combustion Products in Fire

Up to now we have only considered prescribed reactions. Given a reaction, the tools of thermodynamics can give us the heat of combustion and other information. In a combustion reaction we could impose conditions of chemical equilibrium, or ideally complete combustion. While these approximations can be useful, for actual fire processes we must rely on experimental data for the reaction. The interaction of turbulence and temperature variations can lead to incomplete products of combustion. For fuels involving $C_xH_yO_z$ we might expect that

$$C_xH_yO_z + O_2 \rightarrow CO_2,\ CO,\ H_2O,\ H_2,\ \text{soot},\ CH$$

where soot is mostly carbon and by CH we mean the residual of hydrocarbons. Unless one can show a chemical similarity among the various burning conditions, the combustion products will be dependent on the experimental conditions.

Tewarson [3] has studied the burning characteristics of a wide range of materials. He has carried out experiments mainly for horizontal samples of 10 cm × 10 cm under controlled radiant heating and air supply conditions. Chemical measurements to determine the species in the exhaust gases have identified the principal products of combustion. Measurement of the mass loss of the sample during combustion will yield the mass of gas evolved from the burning sample. This gas may not entirely be fuel since it could contain evaporated water or inert fire retardants. It is convenient to express the products of combustion, including the chemical energy liberated, in terms of this mass loss. This is defined as a yield, i.e.

$$\text{yield of combustion product } i,\ y_i \equiv \frac{\text{mass of species } i}{\text{mass loss of sample}}$$

Values of yields for various fuels are listed in Table 2.3. We see that even burning a pure gaseous fuel as butane in air, the combustion is not complete with some carbon monoxide, soot and other hydrocarbons found in the products of combustion. Due to the incompleteness of combustion the 'actual' heat of combustion (42.6 kJ/g) is less than the 'ideal' value (45.4 kJ/g) for complete combustion to carbon dioxide and water. Note that although the heats of combustion can range from about 10 to 50 kJ/g, the values expressed in terms of oxygen consumed in the reaction (Δh_{O_2}) are fairly constant at 13.0 ± 0.3 kJ/g O_2. For charring materials such as wood, the difference between the actual and ideal heats of combustion are due to distinctions in the combustion of the volatiles and subsequent oxidation of the char, as well as due to incomplete combustion. For example,

$$\text{Wood} + \text{heat} \longrightarrow \text{volatiles} + \text{char}$$

$$\text{Volatiles} + \text{air} \longrightarrow \text{incomplete products}, \Delta h_{c,vol}$$

$$\text{Char} + \text{air} \longrightarrow \text{incomplete products}, \Delta h_{c,char}$$

$$\Delta h_{c,wood} = \Delta h_{c,vol} + \Delta h_{c,char}$$

where here the heats of combustion should all be considered expressed in terms of the mass of the original wood. For fuels that completely gasify to the same chemical state as in a phase change for a pure substance, the yield is equivalent to the stoichiometric coefficient. However, for a fuel such as wood, which only partially gasifies and the gaseous products are no longer the original wood chemical, we cannot equate yield to a stoichiometric coefficient. The use of 'yield' is both practical and necessary in characterizing such complex fuels.

Let us examine the measured benzene reaction given in Table 2.3. From the ideal complete stoichiometric reaction we have

$$C_6H_6 + \frac{15}{2} O_2 \rightarrow 6\,CO_2 + 3\,H_2O$$

The ideal heat of combustion is computed from Equation (2.25) using Table 2.2 as the net value (H_2O as a gas)

$$\begin{aligned}\Delta \tilde{h}_c &= [(1)(+82.9)] - [(6)(-393.5) + (3)(-241.8)] \\ &= 3169.3\,\text{kJ/mole}\,C_6H_6\end{aligned}$$

or

$$\Delta h_c = (3169.3)/(78\text{g/mole}) = 40.6\,\text{kJ/g}$$

This is higher than the value given in Table 2.3 (i.e., 40.2 kJ/g) since we used benzene from Table 2.2 as a gas and here it was burned as a liquid. The difference in energy went into vaporizing the liquid benzene and was approximately 0.4 kJ/g. (See Table 6.1)

Table 2.3 Products of combustion in fire with sufficient air (abstracted from Tewarson [3])

Substance	Formula	Heat of combustion Δh_c		Heat of combustion per mass of O_2 Δh_{O_2}	Actual yields (g products/g mass lost)			
		Ideal (kJ/g fuel burned)	Actual (kJ/g mass loss)	(kJ/g O_2 consumed)	CO_2	CO	Soot	Other hydrocarbon
Methane (g)	CH_4	50.1	49.6	12.5	2.72	—	—	—
Propane (g)	C_3H_8	46.0	43.7	12.9	2.85	0.005	0.024	0.003
Butane (g)	C_4H_{10}	45.4	42.6	12.6	2.85	0.007	0.029	0.003
Methanol (l)	CH_4O	20.0	19.1	13.4	1.31	0.001	—	—
Ethanol (l)	C_2H_6O	27.7	25.6	13.2	1.77	0.001	0.008	0.001
n-Heptane (l)	C_7H_{16}	44.6	41.2	12.7	2.85	0.010	0.037	0.004
Benzene (g)	C_6H_6	40.2	27.6	13.0	2.33	0.067	0.181	0.018
Wood (red oak)	$CH_{1.7}O_{0.72}$	17.1	12.4	13.2	1.27	0.004	0.015	0.001
Nylon	$(C_6H_{11}NO)_n$	30.8	27.1	11.9	2.06	0.038	0.075	0.016
PMMA	$(C_5H_8O_2)_n$	25.2	24.2	13.1	2.12	0.010	0.022	0.001
Polyethyene, PE	$(C_2H_4)_n$	43.6	30.8	12.8	2.76	0.024	0.060	0.007
Polypropylene, PS	$(C_3H_6)_n$	43.4	38.6	12.7	2.79	0.024	0.059	0.006
Polystyrene, PS	$(C_8H_8)_n$	39.2	27.0	12.7	2.33	0.060	0.164	0.014

Let us continue further with this type of analysis, but now consider the actual reaction measured. We treat soot as pure carbon and represent the residual hydrocarbons as C_xH_y. Then the equation, without the proper coefficients, is

$$C_6H_6 + \nu_{O_2} O_2 \rightarrow CO_2,\ CO, H_2O,\ C,\ C_xH_y$$

Since benzene is a pure substance and we continue to treat it as a gas, we can regard the yields in this case as stoichiometric coefficients. From the measured yields in Table 2.3 and the molecular weights

	C_6H_6	O_2	CO_2	CO	C	C_xH_y
M_i (g/mole)	78	32	44	28	12	$12x + y$

We calculate the stoichiometric efficients, ν_i:

$$\begin{aligned} CO_2: &\quad 2.33\text{g}CO_2/\text{g}C_6H_6\left(\frac{78}{44}\right) &= 4.13 \\ CO: &\quad (0.067)\left(\frac{78}{28}\right) &= 0.187 \\ C: &\quad (0.181)\left(\frac{78}{12}\right) &= \underline{1.177} \\ &\qquad\qquad \text{Total} & 5.494 \approx 5.49 \end{aligned}$$

Since we must have six carbon atoms, we estimate $x = 0.51$ and balance the H and O atoms:

$$\begin{aligned} H: &\quad 6 = 2\nu_{H_2O} + y \\ O: &\quad 2\nu_{O_2} = (2)(4.13) + (0.187) + \nu_{H_2O}(1) \end{aligned}$$

Selecting arbitrarily $y = 1$, we obtain

$$\begin{aligned} C_6H_6 + 5.47\ O_2 \rightarrow\ & 4.13\ CO_2 + 0.187\ CO \\ & + 1.18\ C + 2.5\ H_2O + C_{0.51}\,H \end{aligned}$$

Since we do not know the composition of the hydrocarbons, we cannot deal correctly with the equivalent compound $C_{0.51}H$. This is likely to be a mixture of many hydrocarbons including H_2. However, for purposes of including its effect on the actual heat of combustion, we will regard it as $\frac{1}{4}$ C_2H_4, ethylene, as an approximation. Then, as before, we compute the heat of combustion:

$$\begin{aligned} \Delta\tilde{h}_c &= [(1)(+82.9)] - [(4.13)(-393.5) + (0.187)(-110.5) \\ &\quad + (1.18)(0) + (2.5)(-241.8) + (0.25)(+52.5)] \\ \Delta\tilde{h}_c &= 2320.1\ \text{kJ/mole}\ C_6H_6 \end{aligned}$$

or

$$\Delta \tilde{h}_c = 29.7\ \text{kJ/g}\ C_6H_6$$

This is very close to the value given in Table 2.3 determined by measurements (i.e. 27.6 kJ/g).

One method of obtaining the chemical energy release in a fire is to measure the amount of oxygen used. Alternatively, some have considered basing this on the CO_2 and CO produced. Both methods are based on the interesting fact that calculation of the heat of combustion based on the oxygen consumed, or alternatively the CO_2 and CO produced, is nearly invariant over a wide range of materials. Huggett [4] demonstrated this for typical fire reactions, and Table 2.3 illustrates its typical variation. For the stoichiometric benzene reaction, it is listed in Table 2.3 as 13.0 kJ/g O_2. For the actual benzene reaction from our estimated result, we would obtain

$$\Delta h_{O_2} = (29.7\ \text{kJ/g}\ C_6H_6)\left(\frac{78\ \text{g}\ C_6H_6}{(5.47)(32)\ \text{g}\ O_2}\right) = 13.2\ \text{kJ/g}\ O_2$$

This result illustrates that even for realistic reactions, the Δh_{O_2} appears to remain nearly constant.

Table 2.3 data appear to remain invariant as long as there is sufficient air. By this we mean that with respect to the stoichiometric reaction there is more air than required. In terms of the equivalence ratio, we should have $\phi < 1$. When ϕ is zero we have all oxygen and no fuel and when $\phi = 1$ we have the stoichiometric case with no oxygen or fuel in the products. However, when $\phi > 1$ we see large departures to the data in Table 2.3. In general, CO soot and hydrocarbons increase; Δh_c accordingly decrease. Tewarson [3] vividly shows in Figures 2.3 and 2.4 how the Δh_c and the yield of CO changes for six

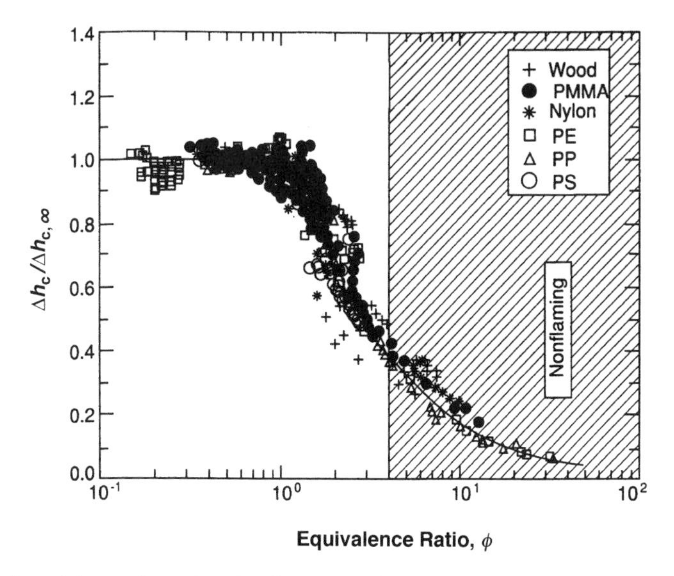

Figure 2.3 Actual heat of combustion in terms of the equivalence ratio where $\Delta h_{c,\infty}$ is the reference value given in Table 2.3 [3]

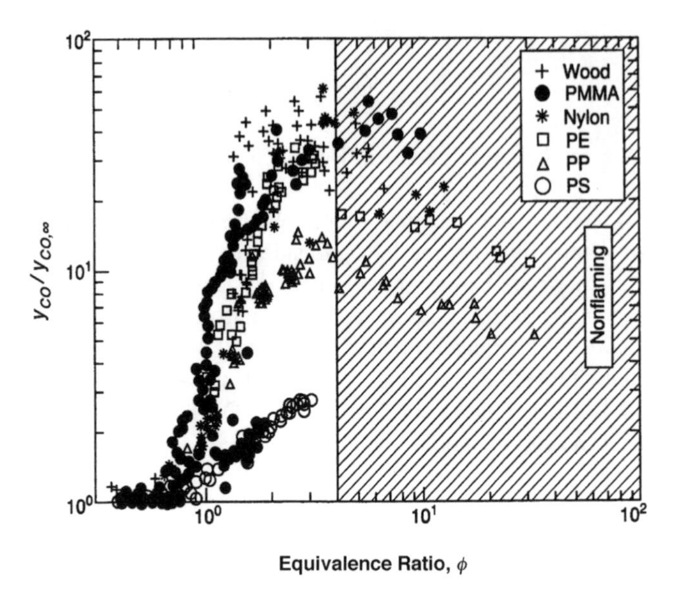

Figure 2.4 Yield of CO in terms of the equivalence ratio where the normalize yield, $y_{\text{CO},\infty}$, is given in Table 2.3 [3]

realistic fuels listed in Table 2.3. These results have importance in modeling ventilation-limited fires and determining their toxic hazard. These data are from the FMRC flammability apparatus [3] using the 10 cm × 10 cm horizontal sample, and for enclosure fires [5].

The flaming results extend to $\phi = 4$ in Figures 2.3 and 2.4, at which point gas phase combustion appears to cease. However, combustion must continue since the heat of combustion remains nonzero. This is due to oxidation of the remaining solid fuel. If we consider wood, it would be the oxidation of the surface char composed primarily of carbon. From Example 2.3, we obtain the heat of combustion for carbon (going to CO_2) as 32.8 kJ/g carbon. From Figure 2.4, we see a significant production of carbon monoxide at $\phi = 4$, and therefore it is understandable that Figure 2.3 yields a lower value: $\Delta h_c/\Delta h_{c,\infty} = 0.4$ with $\Delta h_{c,\infty}$ from Table 2.3 of 12.4 kJ/g, or Δh_c at $\phi = 4$ is 4.96 kJ/g. Here the actual heat of combustion of the char (under nonflaming conditions) at $\phi = 4$ is 4.96 kJ/g of mass lost while its ideal value is 32.8 kJ/g of carbon reacted. The 'actual' value is not based on stoichemistry of the fuel burned, but on the mass evolved to the gaseous state. The mass evolved may not be equal to the mass of fuel burned.

Therefore, we have shown that thermodynamic principles and properties can be used to describe combustion reactions provided their stoichiometry is known. Since measurements in fire are based on the fuel mass lost, yields are used as empirical properties to describe the reaction and its products. When fire conditions become ventilation-limited ($\phi \geq 1$), the yield properties of a given fuel depend on ϕ. Although the generality of the results typified by Figures 2.3 and 2.4 have not been established, their general trends are accepted.

References

1. *JANNAF Thermochemical Tables*, 3rd edn., American Institute of Physics for the National Bureau of Standards, Washington, DC, 1986.
2. Rossini, F.D. *et al.* (eds), *Selected Values of Chemical Thermodynamic Properties*, National Bureau of Standards Circular 500, 1952.
3. Tewarson, A., Generation of heat and chemical compounds in fires, in *The SFPE Handbook of Fire Protection Engineering*, 2nd edn., (eds P.J. DiNenno *et al.*), Section 3, Chapter 4, National Fire Protection Association, Quincy, Massachusetts, 1995, pp. **3**-53 to **3**-124.
4. Huggett, C., Estimation of rate of heat release by means of oxygen consumption, *J. Fire and Flammability*, 1980, **12**, 61–5.
5. Tewarson, A., Jiang, F. H. and Morikawa, T., Ventilation-controlled combustion of polymers, *Combustion and Flame*, 1993, **95**, 151–69.

Problems

2.1 Calculate the vapor densities (kg/m^3) of pure carbon dioxide (CO_2), propane (C_3H_8) and butane (C_4H_{10}) at 25 °C and pressure of 1 atm. Assume ideal gas behavior.

2.2 Assuming ideal gas behavior, what will be the final volume if 1 m^3 of air is heated from 20 to 700 °C at constant pressure?

2.3 Identify the following thermodynamic properties for liquid mixtures:

(a) Thermal equilibrium between two systems is given by the equality of ___________.

(b) Mechanical equilibrium between two systems is given by the equality of ___________.

(c) Phase equilibrium between two systems of the same substance is given by the equality of _________________.

2.4 Show that the pressure rise (p–p_o) in a closed rigid vessel of volume (V) for the net constant heat addition rate ($\dot{Q}$) is

$$\frac{p - p_o}{p_o} = \frac{\dot{Q}t}{\rho_o V c_v T_o}$$

where ρ_o is the initial density, p_o is the initial pressure, T_o is the initial temperature, t is time and c_v is the specific heat at constant volume. The fluid in the vessel is a perfect gas with constant specific heats.

2.5 Formaldehyde (CH_2O) burns to completeness in air. Compute:

(a) stoichiometric air to fuel mass ratio;

(b) mole fraction of fuel in the reactant mixture for an equivalence ratio (ϕ) of 2;

(c) mole fraction of fuel in the product mixture for $\phi = 2$.

2.6 Hydrogen reacts with oxygen. Write the balanced stoichiometric chemical equation for the complete reaction.

2.7 The mole fraction of argon in a gas mixture with air is 0.1. The mixture is at a pressure of 1000 Pa and 25 °C. What is the partial pressure of the argon?

2.8 Compute the enthalpy of formation of propane at 25 °C from its chemical reaction with oxygen and its ideal heat of combustion given in Table 2.3.

2.9 Consider the complete stoichiometric reaction for the oxidation of butane:

$$C_4H_{10} + \_O_2 \rightarrow \_CO_2 + \_H_2O$$

Calculate the heat of formation using data in Tables 2.2 and 2.3.

2.10 Compute the heat transferred in the oxidation of 1 mole of butane to carbon monoxide and water, with reactants and products at 25 °C. Use Table 2.2.

$$n\text{-}C_5H_{12} + 1\frac{1}{2}\,O_2 \rightarrow 5CO + 6\,H_2O$$

2.11 The products for the partial combustion of butane in oxygen were found to contain CO_2 and CO in the ratio of 4:1. What is the actual heat released per mole of butane burned if the only other product is H_2O? The reactants are at 25 °C and the products achieve 1000 °C. Use average estimates for specific heats. For butane use 320 J/K mole.

2.12 Consider heptane burned in air with reactants and products at 25 °C. Compute the heat release per gram of oxygen consumed for (a) CO_2 formed and (b) only CO formed. The other product of combustion is water vapor.

2.13 (a) Write the balanced chemical equation for stoichiometric combustion of benzene (C_6H_6). Assume complete combustion. Calculate the mass of air required to burn a unit mass of combustible.

(b) For a benzene–air equivalence ratio of 0.75, write the balanced chemical equation and calculate the adiabatic flame temperature in air. The initial temperature is 298 K and pressure is 1 atm. Assume complete combustion.

(c) Calculate the mole fraction for each product of combustion in part (b).

(d) Benzene often burns incompletely. If 20 % of the carbon in the benzene is converted to solid carbon and 5 % is being converted to CO during combustion, the remainder being converted to CO_2, calculate the heat released per gram of oxygen consumed. How does this compare with the value in Table 2.3?

2.14 (a) For constant pressure processes show that the net heat released for reactions I and II in succession is the same as for reaction III. The initial and final temperatures are 25 °C. Use data in the text.

I. $C(s) + \frac{1}{2}\,O_2(g) \rightarrow CO(g)$

II. $CO(g) + \frac{1}{2}\,O_2(g) \rightarrow CO_2(g)$

III. $C(s) + O_2(g) \rightarrow CO_2(g)$

(b) In reaction II, for oxygen in the reactants:

(i) What is the mole fraction of O_2?

(ii) What is the mass fraction of O_2?

(iii) What is the partial pressure of O_2 if the system pressure is 2 atm.

2.15 Carbon monoxide burns in air completely to CO_2.

(a) Write the balanced chemical equation for this reaction.

(b) Calculate the stoichiometric fuel to air ratio by mass.

(c) For this reaction, at a constant pressure of 1 atm, with the initial and final temperatures of 25 °C, calculate the change in enthalpy per unit mass of CO for a stoichiometric fuel–air mixture.

(d) What is the quantity in (c) called?

(e) If the final temperature of this reaction is 500 °C instead, determine the heat lost from the system per unit mass of CO.

(f) Calculate the equivalence ratio for a mixture of five moles of CO and seven moles of air. What species would we expect to find in the products other than CO_2?

2.16 Determine the heat of combustion of toluene and express it in proper thermodynamic form. The heat of formation is 11.95 kcal/mole.

2.17 Hydrogen stoichiometrically burns in oxygen to form water vapor. Let $c_{p,i} = 1.5$ J/g K for all species.

(a) For a reaction at 25 °C, the enthalpy change of the system per unit mass of H_2 is called ____________________.

(b) For a reaction at 25 °C, the enthalpy change of the system per unit mass of H_2O is called ____________________.

(c) Compute the enthalpy change of the system per unit mass of water vapor at 1000 °C.

(d) The initial temperature is 25 °C and the temperature after the reaction is 1000 °C. Compute the heat transfer for the system per unit mass of H_2 burned. Is this lost or added?

(e) Repeat (d) for an initial temperature of 300 °C.

2.18 Compute the heat of combustion per gram mole of acetonitrile.

Acetonitrile ($C_2H_3N_{(g)}$) burns to form hydrogen cyanide ($HCN_{(g)}$), carbon dioxide and water vapor.

Heat of formation in kcal/gmol	
Hydrogen cyanide:	32.3
Acetonitrile:	21.0
Water vapor:	−57.8
Carbon dioxide:	−94.1
Oxygen:	0.0

2.19 An experimentalist assumes that, when wood burns, it can be approximated by producing gaseous fuel in the form of formaldehyde and char in the form of carbon.

Heat of formation in kcal/gmol	
Formaldehyde (CH_2O):	−27.7
Carbon (C):	0
Oxygen (O_2):	0
Nitrogen (N_2):	0
Water vapor (H_2O):	−57.8
Carbon dioxide (CO_2):	−94.1

Assume complete combustion in the following:

(a) Compute the heat of combustion of CH_2O in kJ/kg.

(b) Compute the heat of combustion of C in kJ/kg.

(c) What heat of combustion would an engineer use to estimate the energy release of wood during flaming combustion?

(d) The char yield of wood is 0.1 g C/g wood. Compute the heat of combustion of the solid wood after all of the char completely oxidizes, in kJ/g of wood.

2.20 Determine by calculation the enthalpy of formation in kJ/mole of CH_4 given that its heat of combustion is 50.0 kJ/g at 25 °C. The heat of formation for carbon dioxide is –394 kJ/mole and water vapor is –242 kJ/mole.

2.21 Nylon ($C_6H_{11}NO$) burns as a gas in air. There is 8 times the amount of stoichiometric air available by moles. The reaction takes place at constant pressure. The initial fuel temperature is 300 °C and the air is at 25 °C. The reaction produces CO_2, H_2O (gas) and HCN in the molar ratio, $HCN/CO_2 = 1/5$.

Other properties: Heat of formation of nylon = −135 kJ/mole

Species	Molar specific heats at constant pressure (J/K mole)
Nylon (gas)	136
H_2O	50
CO_2	60
HCN	90
N_2	35
O_2	35

(a) Compute the heat of combustion for this reaction using data in Table 2.2.

(b) Compute the temperature of the final state of the products for an adiabatic process.

2.22 Nylon burns to form carbon dioxide, water vapor and hydrogen cyanide, as shown in the stoichiometric equation below:

$$C_6H_{11}NO_2 + 6.5\,O_2 \rightarrow 5\,CO_2 + 5\,H_2O + HCN$$

The heat of combustion of nylon is 30.8 kJ/g. Find the heat of formation of the nylon in kJ/mole.

2.23 Neoprene (C_4H_5Cl) is burned in air at constant pressure with the reactants and products at 25 °C. The mass yields of some species are found by measurement in terms of g/g_F. These product yields are CO at 0.1, soot (taken as pure carbon, C) at 0.1 and gaseous unburned hydrocarbons (represented as benzene, C_6H_6) at 0.03. The remaining products are water as a gas, carbon dioxide and gaseous hydrogen chloride (HCl).

(a) Write the chemical equation with molar stoichiometric coefficients, taking air as 1 mole of $O_2 + 3.77$ moles of N_2 (i.e. 4.77 moles of air).

(b) Calculate the heat of formation of the neoprene (in kJ/mole) if it is known that the heat of combustion for reaction (a) is 11 kJ/g neoprene.

(c) If the reaction were complete to only stable gaseous products, namely water vapor, carbon dioxide and chlorine (Cl_2), compute the heat of combustion in kJ/g.

(d) For both reactions (a) and (c), compute the energy release per unit mass of oxygen consumed, i.e. [H(reactants)$-H$(products)]/mass of oxygen.

Atomic weights: C,12; H,1; O,16; N,14; Cl,35.5, and use Table 2.2.

2.24 Formaldehyde is stoichiometrically burned at constant pressure with oxygen (gas) to completion. CO_2 and H_2O, condensed as a liquid, are the sole products. The initial temperature before the reaction is 50 °C and the final temperature after its completion is 600 °C. Find, per mole of fuel, the heat transferred in the process, and state whether it is added or lost. Assume a constant specific heat of 35 J/mole K for all the species. Use Table 2.2 for all of your data.

2.25 The heat of combustion for the reaction times the mass of fuel reacted is equal to:

(a) the chemical energy released due to realignment of the molecular structures;

(b) the heat lost when the reactants and the products are at 25 °C;

(c) the chemical energy when the fuel concentration in the reactants is at the upper flammable limit.

2.26 Cane sugar, $C_{12}H_{22}O_{11}$, reacts with oxygen to form char, gaseous water and carbon dioxide. Two moles of char are formed per mole of sugar cane. Assume the char is pure carbon (graphite). Compute the heat of combustion on a mass basis (kJ/g) for the sugar cane in this reaction. The heat of formation of the sugar cane is −2220 kJ/mole. Cite any data taken from Tables 2.1 and 2.2.

2.27 Hydrogen gas (H_2) reacts completely with air in a combustor to form water vapor with no excess air. The combustor is at steady state. If the flow rate of the hydrogen is 1 kg/s, what is the heat transfer rate to or from the combustor? The heat of combustion of hydrogen gas is 121 kJ/g, the initial temperature of the air and hydrogen is 600 K and the exiting temperature of the gases leaving the combustor is 3000 K. Assume constant and equal specific heats of constant pressure, 1.25 kJ/kg K. Show all work, diagrams and reasoning, and indicate whether the heat is from or to the combustor wall.

2.28 Polystryrene can be represented as C_8H_8. Its heat of combustion for a *complete* reaction to carbon dioxide and water vapor is 4077 kJ/mole.

(a) Compute the heat of formation of C_8H_8 in kJ/mole.

(b) Compute the heat of combustion of C_8H_8 in kJ/g for the *incomplete* reaction in air with water vapor in the products as shown below:

$$C_8H_8 + \left(\frac{17}{2}\right)[O_2 + 3.77\,N_2] \rightarrow 6\,CO_2 + CO + C + 4\,H_2O + \left(\frac{17}{2}\right)(3.77)\,N_2$$

2.29 Polystyrene $(C_8H_8)_n$ reacts with the oxygen in air; 4.77 moles of air can be represented as $O_2 + 3.77\,N_2$.

(1) 1 C_8H_8 + ___ (O_2 + 3.77 N_2) → ___ CO_2 + ___ $H_2O(g)$ + ___ N_2

(2) 1 C_8H_8 + ___ (O_2 + 3.77 N_2) → ___ CO_2 + ___ $H_2O(g)$ + ___ N_2 + ___ CO + ___ C

For the incomplete reaction, (2), the molar ratios of the incomplete products are given as

$$CO/\ CO_2 = 0.07 \text{ and } C/\ CO_2 = 0.10$$

(a) Balance each chemical equation.

(b) Compute the heat of combustion for each, on a mass basis.

(c) Compute the heat of combustion per unit mass of oxygen used for each reaction.

(d) Compute the adiabatic flame temperature for each.

(e) Compute the adiabatic flame temperature for each reaction in pure oxygen.

Use the information in the chapter 2 to work the problem. You can assume constant specific heats of the species at an appropriate mean temperature (use 1600 K). Also, the heat of formation of polystyrene is – 38.4 kJ/mole.

2.30 (a) Determine the heat of combustion of ethylene, C_2H_4, for a complete reaction to carbon dioxide and water vapor.

(b) Determine its adiabatic flame temperature for a complete reaction with pure oxygen. Assume again water vapor in the products. Use information in Tables 2.1 and 2.2.

2.31 Calculate the adiabatic flame temperatures for the following mixtures initially at 25 °C:

(a) stoichiometric butane–oxygen mixture;

(b) stoichiometric butane–air mixture;

(c) 1.8 % butane in air.

Use specific heat values of 250 J/K mole for butane and 36 J/K mole for oxygen and nitrogen.

2.32 Calculate the adiabatic flame temperature (at constant pressure) for ethane C_2H_6 in air:

(a) at the lower flammability limit ($X_L = 3.0\%$) and

(b) at the stoichiometric mixture condition.

Assume constant specific heats for all species (1.1 kJ/kg K) and use the data from Tables 2.3 and 4.5. The initial state is 1 atm, 25 °C and the initial mixture density is 1.2 kg/m^3.

2.33 A gaseous mixture of 2 % (by volume) acetone and 4 % ethanol in air is at 25 °C and a pressure of 1 atm.

Data

Acetone (C_3H_6O) has Δh_c = 1786 kJ/g mol

Ethanol (C_2H_5OH) has Δh_c = 1232 kJ/g mol

Atomic weights: H = 1 mole, C = 12 moles, O = 16 moles and N = 14 moles

Specific heats, $c_{p,i}$ = 1 kJ/kg K, constant for each species

(a) For a constant pressure reaction, calculate the partial pressure of the oxygen in the product mixture.

(b) Determine the adiabatic flame temperature of this mixture.

(c) If this mixture was initially at 400 °C, what will the resultant adiabatic flame temperature be?

2.34 Polyacrylonitrile (C_3H_3N) burns to form vapor, carbon dioxide and nitrogen. The heat of formation of the polyacrylonitrile is +15.85 kcal/g mol (1 cal= 4.186 kJ). Use data from Tables 2.1 and 2.2; use specific heat values at 1000 K.

(a) Write the balanced chemical equation for the stoichiometric combustion in oxygen.

(b) Determine the heat of combustion of the polyacrylonitrile.

(c) Write the balanced chemical equation for the stoichiometric combustion in air.

(d) Determine the adiabatic flame temperature if the fuel burns stoichiometrically in air.

2.35 Toluene (C_7H_8) as a gas burns stoichiometrically in air to completion (i.e. forming carbon dioxide and water vapor).

Data

Heat of formation of toluene is +11.95 kcal/g mol

Species	Specific heat (cal/g mol K) at 1500 K
Oxygen	8.74
Water (gas)	11.1
Carbon dioxide	14.0

(a) For products and reactants at 25 °C, compute the heat loss per unit gram mole of toluene consumed.

(b) What is the quantity in (a) called?

(c) For reactants at 25 °C, compute the adiabatic flame temperature.

2.36 Check all correct answers below:

(a) For reacting systems, the conservation of mass implies:

the preservation of atoms,

the conservation of moles,

the conservation of molecules,

the conservation of species,

the equality of mass for the reactants and products in a closed system.

(b) Stoichiometric coefficients represent:

the number of molecules for each species in the reaction,

the number of moles for each species in the reaction,

the number of atoms for each species in the reaction,

the number of grams for each species in the reaction,

the number before the chemical species formulas in a balanced chemical equation.

3

Conservation Laws for Control Volumes

3.1 Introduction

A control volume is a volume specified in transacting the solution to a problem typically involving the transfer of matter across the volume's surface. In the study of thermodynamics it is often referred to as an open system, and is essential to the solution of problems in fluid mechanics. Since the conservation laws of physics are defined for (fixed mass) systems, we need a way to transform these expressions to the domain of the control volume. A system has a fixed mass whereas the mass within a control volume can change with time.

Control volumes can address transient problems, allowing us to derive ordinary differential equations that govern uniform property domains. This approach leads to a set of equations that govern the behavior of room fires. This is commonly called a zone model. A control volume can also consist of an infinitesimal or differential volume from which the general partial differential conservation equations can be derived. The solution to such equations falls within the subject of computational fluid mechanics or computational fluid dynamics (CFD) modeling. We will not address the latter. Instead we will use control volume analysis to fit the physics of the problem, and derive the appropriate equations needed. This should give to the student more insight into what is happening and what is relevant.

We will take a general and mathematical approach in deriving the conservation laws for control volumes. Some texts adopt a different strategy and the student might benefit from seeing alternative approaches. However, once derived, the student should use the control volume conservation laws as a tool in problem solving. To do so requires a clear understanding of the terms in the equations. This chapter is intended as a reference for the application of the control volume equations, and serves as an extension of thermochemistry to open systems.

Fundamentals of Fire Phenomena James G. Quintiere

3.2 The Reynolds Transport Theorem

The Reynolds transport theorem is a general expression that provides the mathematical transformation from a system to a control volume. It is a mathematical expression that generally holds for continuous and integrable functions. We seek to examine how a function $f(x, y, z, t)$, defined in space over x, y, z and in time t, and integrated over a volume, V, can vary over time. Specifically, we wish to examine

$$\frac{\mathrm{d}}{\mathrm{d}t}\iiint_{V(t)} f(x, y, z, t)\,\mathrm{d}V \tag{3.1}$$

where the volume, V, can vary with time. Since we integrate over V before taking the derivative in time, the integral and therefore the result is only a function of time.

A word of caution should be raised. The formidable process of conducting such spatial integration should be viewed as principally symbolic. Although the mathematical operations will be valid, it is rare that in our application they will be so complicated. Indeed, in most cases, these operations will be very simple. For example, if $f(x, y, z, t)$ is uniform in space, then we simply have, for Equation (3.1),

$$\frac{\mathrm{d}(fV)}{\mathrm{d}t} = V\frac{\mathrm{d}f}{\mathrm{d}t} + f\frac{\mathrm{d}V}{\mathrm{d}t}$$

Let us examine this more generally for $f(x, y, z, t)$ and a specified moving volume, $V(t)$.

The moving volume can be described in terms of surface, S, to which each point on it is given a velocity, $\boldsymbol{v}$. That velocity is also defined over the spatial coordinates and time $\boldsymbol{v}(x, y, z, t)$. To complete the description of V in time, we need to not just describe its surface velocity but also the orientation of the surface at each point. This is done by defining a unit vector $\boldsymbol{n}$ which is always oriented outward from the volume and normal to the surface. This is illustrated in Figure 3.1, where ΔS represents a small surface element. As we shrink to the point (x, y, z), we represent ΔS as a differential element, dS.

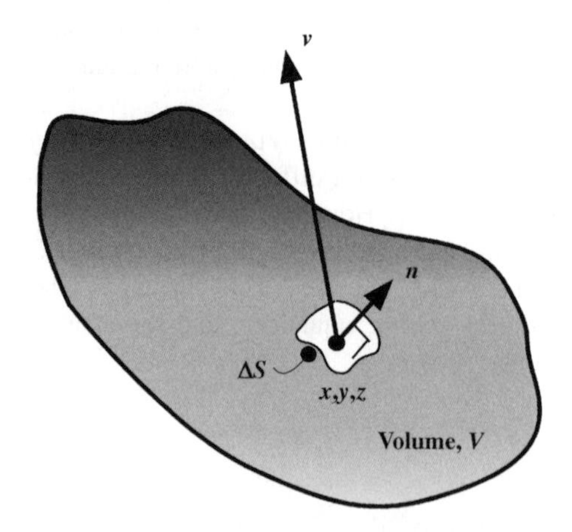

Figure 3.1 A description for a moving volume, V

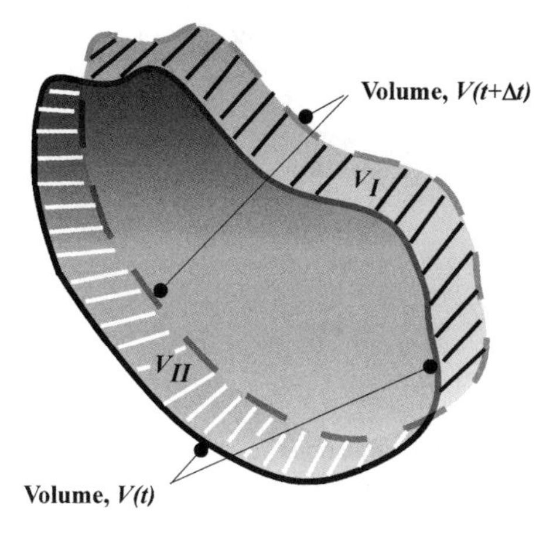

Figure 3.2 Change in volume V over Δt

Let us consider what happens when V moves and we perform the operations in Equation (3.1). This is depicted in Figure 3.2. Here we have displayed the drawing in two dimensions for clarity, but our results will apply to the three-dimensional volume. In Figure 3.2 the volume is shown for time t, and a small time increment later, Δt, as a region enclosed by the dashed lines. The shaded regions V_{I} and V_{II} are the volumes gained and lost respectively. For these volumes, we can represent Equation (3.1) by its formal definition of the calculus as

$$\frac{\mathrm{d}}{\mathrm{d}t}\iiint_{V(t)} f(x,y,z,t)\mathrm{d}V$$
$$= \lim_{\Delta t\to 0}\frac{\iiint_{V(t+\Delta t)} f(x,y,z,t+\Delta t)\mathrm{d}V - \iiint_{V(t)} f(x,y,z,t)\mathrm{d}V}{\Delta t} \tag{3.2}$$

From Figure 3.2, the first integral on the right-hand side of Equation (3.2) can be expressed as

$$\iiint_{V(t+\Delta t)} f(x,y,z,t+\Delta t)\mathrm{d}V = \iiint_{V(t)} f(x,y,z,t+\Delta t)\mathrm{d}V + \iiint_{V_{\mathrm{I}}(t+\Delta t)} f(x,y,z,t+\Delta t)\mathrm{d}V - \iiint_{V_{\mathrm{II}}(t+\Delta t)} f(x,y,z,t+\Delta t)\mathrm{d}V \tag{3.3}$$

For volumes V_{I} and V_{II} we consider the integrals as sums over small volumes, ΔV_j, defined by the motion of their corresponding surface elements, ΔS_j. Figure 3.3 provides a (two-dimensional) geometric representation of this process in which ΔS_j moves over ΔS

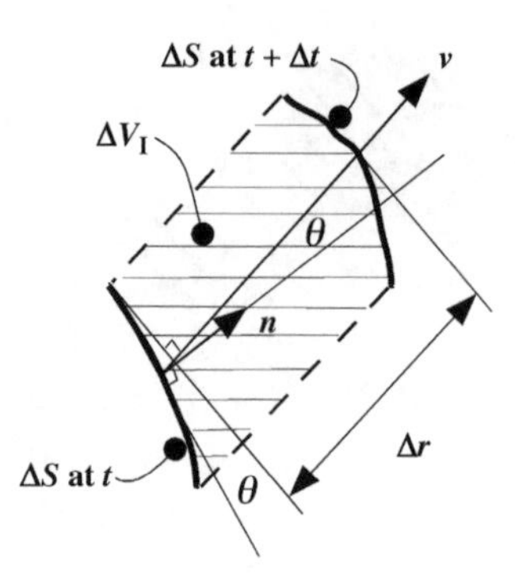

Figure 3.3 Relationship between V_{I} and ΔS

to define an intercepted volume ΔV_{I_j}. Temporarily, we drop the j-subscript. The motion of ΔS is Δr in the direction of $\boldsymbol{v}$ where

$$\Delta r = v\,\Delta t \tag{3.4}$$

with v the magnitude of the velocity vector. By geometry, the volume of a rectilinear region is the product of its cross-sectional area perpendicular to its length times its length. Since two angles are equal whose legs are mutually perpendicular, with this angle being θ in Figure 3.3,

$$\Delta V_{\mathrm{I}} = (\Delta S)(\cos\theta)\Delta r \tag{3.5}$$

From the definition of the vector dot product, Equations (3.4) and (3.5) can be combined to give

$$\Delta V_{\mathrm{I}} = (\Delta S)\left(\left(\frac{\boldsymbol{v}}{v}\right)\cdot\boldsymbol{n}\right)(v\,\Delta t) = (\boldsymbol{v}\cdot\boldsymbol{n})(\Delta S)(\Delta t) \tag{3.6}$$

Since $\boldsymbol{v}$ is always more than 90° out of phase with $\boldsymbol{n}$ for V_{II}, the minus sign in Equation (3.3) for the integral is directly accounted for by the application of Equation (3.6) to ΔV_{II} as well. It then follows that

$$\iiint_{V_{\mathrm{I}}} f(x, y, z, t+\Delta t)\mathrm{d}V - \iiint_{V_{\mathrm{II}}} f(x, y, z, t+\Delta t)\mathrm{d}V$$
$$= \sum_{j=1}^{N} f(x, y, z, t+\Delta t)\boldsymbol{v}\cdot n\Delta S_j\Delta t \tag{3.7}$$

Combining Equation (3.7) with Equation (3.3), Equation (3.2) can be written as

$$\frac{\mathrm{d}}{\mathrm{d}t}\iiint_{V(t)} f(x, y, z, t)\mathrm{d}V = \lim_{\Delta t\to 0}\frac{\iiint_{V(t)}[f(x, y, z, t+\Delta t) - f(x, y, z, t)]\mathrm{d}V}{\Delta t}$$
$$+ \lim_{\Delta t\to 0}\frac{\sum_{j=1}^{N} f(x, y, z, t+\Delta t)\boldsymbol{v}\cdot\boldsymbol{n}\Delta S_j\Delta t}{\Delta t} \tag{3.8}$$

From the calculus, the limit process formally gives

$$\frac{\mathrm{d}}{\mathrm{d}t}\iiint_{V(t)} f(x,y,z,t)\mathrm{d}V = \iiint_{V(t)} \left(\frac{\partial f}{\partial t}\right)_{x,y,z} \mathrm{d}V + \iiint_{S(t)} f(x,y,z,t)\boldsymbol{v}\cdot\boldsymbol{n}\mathrm{d}S \quad (3.9)$$

where $S(t)$ is the surface enclosing the volume $V(t)$.

Equation (3.9) is the Reynolds transport theorem. It displays how the operation of a time derivative over an integral whose limits of integration depend on time can be distributed over the integral and the limits of integration, i.e. the surface, S. The result may appear to be an abstract mathematical operation, but we shall use it to obtain our control volume relations.

3.3 Relationship between a Control Volume and System Volume

Suppose we now assign a physical meaning to the velocity $\boldsymbol{v}$, representing it as the velocity of matter in the volume, V. Then if V always contains the same mass, it is a system volume. The properties defined for each point of the system represent those of a continuum in which the macroscopic character of the system is retained as we shrink to a point. Properties at a molecular or atomic level do not exist in this continuum context. Furthermore, since the system volume is fixed in mass, we can regard volume V to always enclose the same particles of matter as it moves in space. Each particle retains its continuum character and thermodynamic properties apply.

Let us see how to represent changes in properties for a system volume to property changes for a control volume. Select a control volume (CV) to be identical to volume $V(t)$ at time t, but to have a different velocity on its surface. Call this velocity, $\boldsymbol{w}$. Hence, the volume will move to a different location from the system volume at a later time. For example, for fluid flow in a pipe, the control volume can be selected as stationary ($\boldsymbol{w} = 0$) between locations 1 and 2 (shown in Figure 3.4, but the system moves to a new location later in time. Let us apply the Reynolds transport theorem, Equation (3.9), twice: once to a system volume, $V(t)$, and second to a control volume, CV, where CV and V are identical at time t. Since Equation (3.9) holds for any well-defined volume and surface velocity distribution, we can write for the system

$$\frac{\mathrm{d}}{\mathrm{d}t}\iiint_{V(t)} f\,\mathrm{d}V = \iiint_{V(t)} \left(\frac{\partial f}{\partial t}\right) \mathrm{d}V + \iint_{S(t)} f\boldsymbol{v}\cdot\boldsymbol{n}\,\mathrm{d}S \quad (3.10)$$

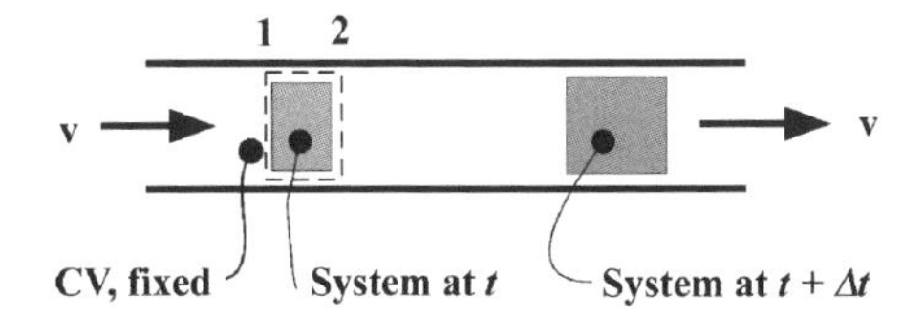

Figure 3.4 Illustration of a fixed control volume for pipe flow and a system

and for the control volume

$$\frac{\mathrm{d}}{\mathrm{d}t}\iiint_{CV} f\,\mathrm{d}V = \iint_{CV}\left(\frac{\partial f}{\partial t}\right)\mathrm{d}V + \iint_{CS} f\boldsymbol{w}\cdot\boldsymbol{n}\,\mathrm{d}S \tag{3.11}$$

The first terms on the right-hand side of Equations (3.10) and (3.11) are identical since $\mathrm{CV} = V(t)$ at time t. Furthermore, since $S(t) = CS$(control surface) at time t, on substituting Equation (3.11) from Equation (3.10) and rearranging, we obtain

$$\boxed{\frac{\mathrm{d}}{\mathrm{d}t}\iiint_{V(t)} f\,\mathrm{d}V = \frac{\mathrm{d}}{\mathrm{d}t}\iiint_{\mathrm{CV}} f\,\mathrm{d}V + \iint_{\mathrm{CS}} f(\boldsymbol{v}-\boldsymbol{w})\cdot\boldsymbol{n}\,\mathrm{d}S} \tag{3.12}$$

This is the relationship we seek. It says that the rate of change of a property f defined over a system volume is equal to the rate of change for the control volume plus a correction for matter that carries f in or out. This follows since $\boldsymbol{v}-\boldsymbol{w}$ is the relative velocity of matter on the boundary of the control volume. If $\boldsymbol{v}-\boldsymbol{w}=0$, no matter crosses the boundary. As we proceed in applying Equation (3.12) to the conservation laws and in identifying a specific property for f, we will bring more meaning to the process. We will consider the system to be composed of a fluid, but we need not be so restrictive, since our analysis will apply to all forms of matter.

3.4 Conservation of Mass

In general, we consider a fluid system that consists of a reacting mixture. Let ρ, the density of the mixture, be identified for f. Then from Equation, (2.11b) and (3.12), the conservation of mass for a control volume is

$$\boxed{\frac{\mathrm{d}}{\mathrm{d}t}\iiint_{\mathrm{CV}} \rho\,\mathrm{d}V + \iint_{\mathrm{CS}} \rho(\boldsymbol{v}-\boldsymbol{w})\cdot\boldsymbol{n}\,\mathrm{d}S = 0} \tag{3.13}$$

The surface integral represents the net flow rate of mass out of the control volume. This is easily seen from Figure 3.5, where $\boldsymbol{v}-\boldsymbol{w}$ is the relative velocity at the fluid giving a flow

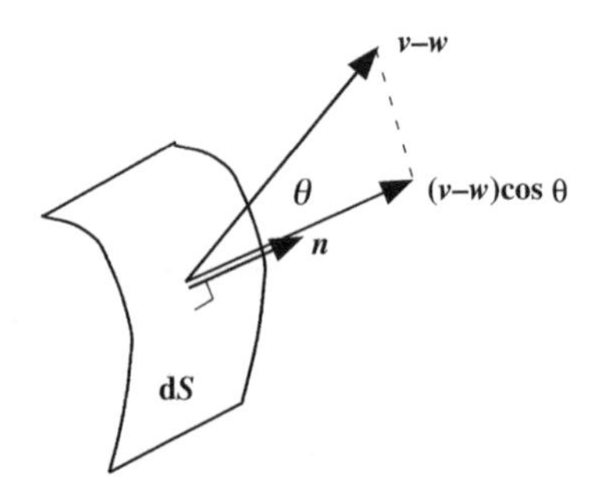

Figure 3.5 Flow rate from a control volume

of fluid particles across ΔS. Only the component of the particle motion along $\boldsymbol{n}$ can cross the boundary. We have already seen that $\boldsymbol{v} - \boldsymbol{w} \cdot \boldsymbol{n}(\Delta S)(\Delta t)$ is the volume of fluid crossing ΔS in time Δt by Equation (3.6). It follows that

$$\Delta \dot{m} = \rho(\boldsymbol{v} - \boldsymbol{w}) \cdot \boldsymbol{n}\, \Delta S \tag{3.14}$$

is the mass flow rate of the fluid through ΔS. Equation (3.13) can be expressed in words as

$$\begin{bmatrix} \text{Rate of change for} \\ \text{mass within CV} \end{bmatrix} + \begin{bmatrix} \text{sum of all mass} \\ \text{flow rates out} \end{bmatrix} - \begin{bmatrix} \text{sum of all mass} \\ \text{flow rates in} \end{bmatrix} = 0$$

An alternative form to Equation (3.13) for discrete regions of mass exchange, j, is

$$\boxed{\left(\frac{\mathrm{d}m}{\mathrm{d}t}\right)_{\mathrm{CV}} + \left(\sum_{j=1} \dot{m}_j\right)_{\mathrm{out}} - \left(\sum_{j=1} \dot{m}_j\right)_{\mathrm{in}} = 0} \tag{3.15}$$

where the sum is implied to be over all flow paths from and to the control volume.

Example 3.1 A balloon filled with air at density ρ allows air to escape through an opening of cross-sectional area, S_0, with a uniform velocity, v_0, given by a measurement device on the balloon. Derive an expression for the volume V_{B} of the balloon over time.

Solution Select the control volume to always coincide with the surface of the balloon and to be coincident with the plane of the exit velocity as shown in Figure 3.6. From Equation (3.13),

$$\frac{\mathrm{d}}{\mathrm{d}t} \iiint \rho\, \mathrm{d}V = \rho \frac{\mathrm{d}}{\mathrm{d}t} V_{\mathrm{B}}$$

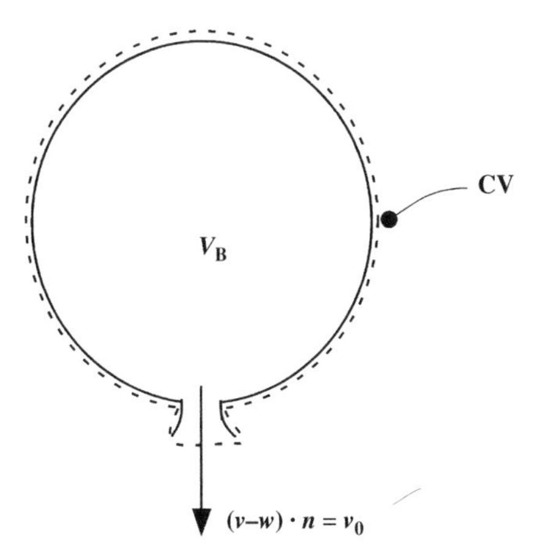

Figure 3.6 Shrinking balloon problem

regarding ρ as constant for air, and

$$\iint \rho(\boldsymbol{v} - \boldsymbol{w}) \cdot \boldsymbol{n}\, \mathrm{d}S = \rho v_0 S_0$$

since only the exit surface contributes with $\boldsymbol{v} - \boldsymbol{w}$ on the remainder of the control surface. The density and the exit term are constant, so by integration with $V_{\mathrm{B,o}}$, the initial volume,

$$V_{\mathrm{B}} = V_{\mathrm{B,0}} - v_0 S_0 t$$

3.5 Conservation of Mass for a Reacting Species

To consider the control volume form of the conservation of mass for a species in a reacting mixture volume, we apply Equation (2.14) for the system and make the conversion from Equation (3.12). Here we select $f = \rho_i$, the species density. In applying Equation (3.13), $\boldsymbol{v}$ must be the velocity of the species. However, in a mixture, species can move by the process of diffusion even though the bulk of the mixture might be at rest. This requires a more careful distinction between the velocity of the bulk mixture and its individual components. Indeed, the velocity $\boldsymbol{v}$ given in Equation (3.13) is for the bulk mixture. Diffusion velocities, $\boldsymbol{V}_i$, are defined as relative to this bulk mixture velocity $\boldsymbol{v}$. Then, the absolute velocity of species i is given as

$$\boldsymbol{v}_i = \boldsymbol{v} + \boldsymbol{V}_i \tag{3.16a}$$

with the bulk velocity defined as

$$\rho \boldsymbol{v} \equiv \sum_{i=1}^{N} \rho_i\, \boldsymbol{v}_i \tag{3.16b}$$

where N is the number of species in the mixture. The mass flow rate per unit area, or species mass flux, due to diffusion is given by Fick's law:

$$\rho_i \boldsymbol{V}_i = -\rho D_i \nabla Y_i \tag{3.17}$$

where D_i is a diffusion coefficient for species i in the mixture. This expression of mass transfer of a species is analogous to Fourier's law for the heat flux due to conduction.

In order always to include species i in the system before converting to a control volume, $\boldsymbol{v}$ must be selected as $\boldsymbol{v}_i$. Then we obtain from Equations (2.14) and (3.12),

$$\boxed{\begin{aligned} &\frac{\mathrm{d}}{\mathrm{d}t}\iiint_{\mathrm{CV}} \rho_i\, \mathrm{d}V + \iint_{\mathrm{CS}} \rho_i(\boldsymbol{v} - \boldsymbol{w}) \cdot \boldsymbol{n}\, \mathrm{d}S \\ &\quad + \iint_{\mathrm{CS}} \rho_i \boldsymbol{V}_i \cdot \boldsymbol{n} \mathrm{d}S = \dot{m}_{i,\mathrm{r}} \end{aligned}} \tag{3.18}$$

In words, this reads as

$$\begin{bmatrix}\text{Rate of}\\ \text{change}\\ \text{of mass}\\ \text{for}\\ \text{species } i \text{ in}\\ \text{the CV}\end{bmatrix} + \begin{bmatrix}\text{net bulk}\\ \text{mass flow}\\ \text{rate of}\\ \text{species } i\\ \text{from the}\\ \text{CV}\end{bmatrix} + \begin{bmatrix}\text{net mass}\\ \text{flow rate}\\ \text{of species } i\\ \text{due to def-}\\ \text{fusion from}\\ \text{the CV}\end{bmatrix} = \begin{bmatrix}\text{rate of pro-}\\ \text{duction of}\\ \text{mass for}\\ \text{species } i \text{ due}\\ \text{to chemical}\\ \text{reaction within}\\ \text{the CV}\end{bmatrix}$$

Since the chemical reaction may not necessarily be uniformly dispersed over the control volume, it is sometimes useful to represent it in terms of a local reaction rate, $\dot{m}'''_{i,\mathrm{r}}$, or the mass production rate of species i per unit volume. Then

$$\dot{m}_{i,\mathrm{r}} = \iiint_{\mathrm{CV}} \dot{m}'''_{i,\mathrm{r}} \mathrm{d}V \tag{3.19}$$

Usually for applications to combustors or room fires, diffusion effects can be ignored at surfaces where transport of fluid occurs, but within diffusion flames these effects are at the heart of its transport mechanism.

A more useful form of Equation (3.18) is

$$\left(\frac{\mathrm{d}m_j}{\mathrm{d}t}\right)_{\mathrm{CV}} + \sum_{j,\,\text{net out}} \dot{m}_{i,j} + \sum_{j,\,\text{net out}} \dot{m}_{i,\mathrm{d},j} = \dot{m}_{i,\mathrm{r}} \tag{3.20}$$

where the sums are over all j flow paths; $\dot{m}_{i,j}$ represents the bulk flow rate of species i and $\dot{m}_{i,\mathrm{d},j}$ represents the diffusion flow rate at each flow path. The diffusion mass flux is given by Equation (3.17). Where gradients of the species concentration (Y_i) are zero across a flow path, then $\dot{m}_{i,\mathrm{d}}$ is zero.

When concentrations are uniform throughout the control and uniform at entering flow paths, Equation (3.20) can be expressed in terms of Y_i. The mass of species in the control volume can be given as

$$m_i = m\,Y_i \tag{3.21}$$

where m is the mass in the CV and Y_i is its mass fraction. At entering j flow paths the mass fraction for each species i is represented as $Y_{i,j}$. Then Equation (3.20) becomes

$$\frac{\mathrm{d}(Y_i m)}{\mathrm{d}t} + \left(\sum_j \dot{m}_j Y_i\right)_{\text{out}} - \left(\sum_j \dot{m}_j Y_{i,j}\right)_{\text{in}} + \sum_{j,\,\text{net out}} \dot{m}_{i,\mathrm{d},j} = \dot{m}_{i,\mathrm{r}} \tag{3.22}$$

By multiplying Equation (3.15) by Y_i and then subtracting it from Equation (3.22), we obtain for this uniform concentration case

$$\boxed{m\frac{\mathrm{d}Y_i}{\mathrm{d}t} + \sum_{j,\,\text{in only}} \dot{m}_j (Y_i - Y_{i,j}) + \sum_{j,\,\text{net out}} \dot{m}_{i,\mathrm{d},j} = \dot{m}_{i,\mathrm{r}}} \tag{3.23}$$

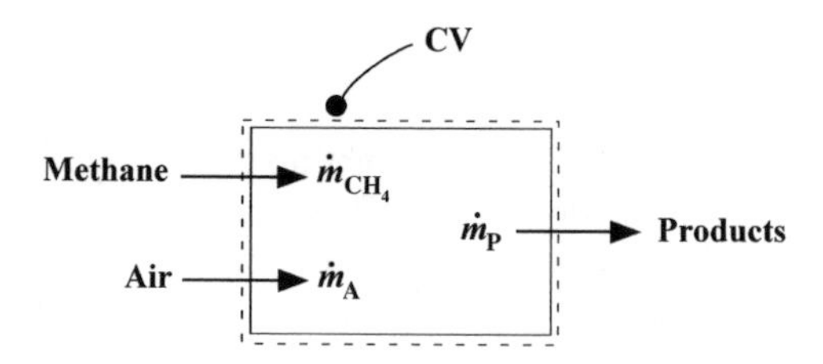

Figure 3.7 Combustor problem

Note that the bulk out-flow contribution has cancelled.

Example 3.2 Consider a combustor using air and methane under steady conditions. A stoichiometric reaction is assumed and diffusion effects are to be neglected. The reactants and products are uniform in properties across their flow streams. Find the mass fraction of carbon dioxide in the product stream leaving the combustor.

Solution The combustor is shown in Figure 3.7 with the control volume. The stoictiometric reaction is

$$CH_4 + 2(O_2 + 3.76\,N_2) \rightarrow CO_2 + 2\,H_2O + 7.52\,N_2$$

Since the conservation equations are in terms of mass, it is useful to convert the stoichiometric coefficients to mass instead of moles. The condition of steady state means that all properties within the control volume are independent of time. Even if the spatial distribution of properties varies within the control volume, by the steady state condition the time derivative terms of Equations (3.13) and (3.18) are zero since the control volume is also not changing in size. Only when the resultant integral is independent of time can we ignore these time derivative terms. As a consequence, the conservation of mass becomes

$$\dot{m}_P - \dot{m}_{CH_4} - \dot{m}_A = 0$$

and the conservation of CO_2 becomes

$$\dot{m}_P Y_{CO_2} = \dot{m}_{CO_2,r}$$

Since we have a stoichiometric reaction, all of the entering methane is consumed:

$$\dot{m}_{CH_4} = (-\dot{m}_{CH_4,r})$$

Recall that in our convention $\dot{m}_{i,r}$ is positive in sign when it is produced.

From Example 2.2, the molecular weight of air is 28.84 g/mole air. By the chemical equation

$$\dot{m}_A = \frac{(2)(4.76 \text{ moles air})(28.84 \text{ g/mole air})}{(1\text{mole } CH_4)(16 \text{ g/mole } CH_4)} \dot{m}_{CH_4,r} = 17.16\, \dot{m}_{CH_4}$$

Also

$$\dot{m}_{CO_2,r} = \frac{(1\text{mole } CO_2)(44 \text{ g/mole } CO_2)}{(1\text{mole } CH_4)(16 \text{ g/mole } CH_4)}(-\dot{m}_{CH_4,r})$$
$$= (2.75)(-\dot{m}_{CH_4,r})$$

Combining

$$Y_{CO_2} = \frac{\dot{m}_{CO_2,r}}{\dot{m}_P}$$
$$= \frac{(2.75)(-\dot{m}_{CH_4,r})}{\dot{m}_A + \dot{m}_{CH_4}}$$
$$= \frac{(2.75)\dot{m}_{CH_4}}{(18.16)\dot{m}_{CH_4}}$$
$$Y_{CO_2} = 0.151$$

3.6 Conservation of Momentum

The conservation of momentum or Newton's second law applies to a particle or fixed set of particles, namely a system. The velocity used must always be defined relative to a fixed or inertial reference plane. The Earth is a sufficient inertial reference. Therefore, any control volume associated with accelerating aircraft or rockets must account for any differences associated with how the velocities are measured or described. We will not dwell on these differences, since we will not consider such noninertial applications.

The law in terms of a control volume easily follows from Equations (2.12) and (3.12) for the bulk system with f selected as ρv_x, taking v_x as the component of $\boldsymbol{v}$ in the x direction:

$$\boxed{\sum F_x = \frac{\mathrm{d}}{\mathrm{d}t}\iiint\limits_{\mathrm{CV}} \rho v_x \,\mathrm{d}V + \iint\limits_{\mathrm{CS}} \rho v_x(\boldsymbol{v} - \boldsymbol{w}) \cdot \boldsymbol{n}\,\mathrm{d}S} \tag{3.24}$$

In words,

$$\begin{bmatrix}\text{Net sum of all}\\ \text{forces acting on}\\ \text{the fluid in the}\\ \text{CV in the } x\\ \text{direction}\end{bmatrix} = \begin{bmatrix}\text{rate of change}\\ \text{of momentum in}\\ \text{the } x \text{ direc-}\\ \text{tion for the}\\ \text{fluid in the CV}\end{bmatrix} + \begin{bmatrix}\text{net rate of}\\ \text{momentum in}\\ \text{the } x \text{ direc-}\\ \text{tion flowing}\\ \text{out of the CV}\end{bmatrix}$$

Example 3.3 Determine the force needed to hold a balloon filled with hot air in place. Consider the balloon to have the same exhaust conditions as in Example 3.1 and the hot air density, ρ, is constant. The cold air density surrounding the balloon is designated as ρ_∞. Assume the membrane of the balloon to have zero mass.

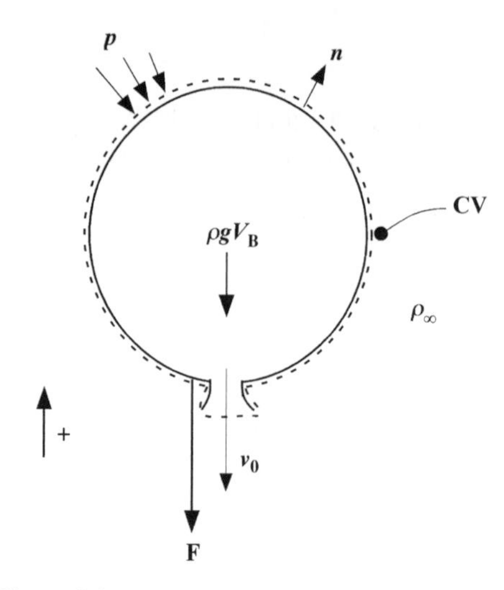

Figure 3.8 Hot balloon held in place by force F

Solution Select a control volume as in Example 3.1 (see Figure 3.8). Let us examine the right-hand side of Equation (3.24) using the sign convention that vertically up is positive. First, we approximate

$$\frac{\mathrm{d}}{\mathrm{d}t}\iiint \rho v_x \,\mathrm{d}V \approx 0$$

because the velocity of the air in the balloon is stationary except near the shrinking membrane and at the exit plane. If the membrane had a significant mass we might need to reconsider this approximation. However, if the balloon were to retain a spherical shape while shrinking, the spherical symmetry would nearly cancel the vertical momentum between the upper and lower hemispheres. Since we expect v_0 to be more significant, the second term becomes

$$\iint \rho v_x (\boldsymbol{v} - \boldsymbol{w}) \cdot \boldsymbol{n} \,\mathrm{d}S = \rho(-v_0)(+v_0)\, S_0$$

Let us now consider the sum of the forces upwards. F is designated as the force, downwards, needed to hold the balloon in place. Another force is due to the pressure distribution of the cold atmosphere on the balloon. Assuming the atmosphere is motionless, fluid statistics gives the pressure variation with height x as

$$\frac{\partial p}{\partial x} = -\rho_\infty g$$

The pressure acts normal everywhere on the membrane, but opposite to $\boldsymbol{n}$. This force can therefore be expressed as

$$\iint_{\mathrm{CS}} (-p\boldsymbol{n})\,\mathrm{d}S \cdot \boldsymbol{i}$$

where $\boldsymbol{i}$ is the unit vector upwards. We can struggle with this surface integration or we can resort to the divergence theorem, which allows us to express

$$\iint_{\mathrm{CS}} (p\boldsymbol{i}) \cdot \boldsymbol{n}\, \mathrm{d}S = \iiint \nabla \cdot (p\boldsymbol{i})\, \mathrm{d}V = \iiint (\nabla p \cdot \boldsymbol{i})\, \mathrm{d}V$$

since the unit vector $\boldsymbol{i}$ is constant in magnitude, i.e. $\nabla \cdot \boldsymbol{i} = 0$. However,

$$\nabla p \cdot \boldsymbol{i} = \frac{\partial p}{\partial x} = -\rho_\infty g$$

Therefore,

$$\iint_{\mathrm{CS}} (-p\boldsymbol{n})\, \mathrm{d}S \cdot \boldsymbol{i} = -\iiint_{\mathrm{CV}} (-\rho_\infty g)\, \mathrm{d}V = +\rho_\infty g\, V_{\mathrm{B}}$$

This is a very important result which is significant for problems involving buoyant fluids. Finally, the last force to consider is the weight for the balloon consisting of the massless membrane and the hot air: $\rho g\, V_{\mathrm{B}}$.

Combining all the terms into Equation (3.24) gives

$$-F - \rho g\, V_{\mathrm{B}} + \rho_\infty g\, V_{\mathrm{B}} = -\rho v_0^2 S_0$$

or solving for F, the needed force,

$$F = (\rho_\infty - \rho) g\, V_{\mathrm{B}} + \rho v_0^2 S_0$$

The first term on the right-hand side is known as the buoyant force, the second is known as thrust. If this were just a puff of hot air without the balloon exhaust, we would only have the buoyant force acting. In this case we could not ignore $(\mathrm{d}/\mathrm{d}t) \iiint \rho v_x\, \mathrm{d}V$ since the puff would rise $(v_x\ +)$ solely due to its buoyancy, with viscous effects retarding it. Buoyancy generated flow is an important controlling mechanism in many fire problems.

3.7 Conservation of Energy for a Control Volume

As before, we ignore kinetic energy and consider Equation (2.18) as a rate equation for the system separating out the rate of work due to shear and shafts, $\dot{W}_{\mathrm{s}}$, and pressure, $\dot{W}_{\mathrm{p}}$. Then

$$\left(\frac{\mathrm{d}U}{\mathrm{d}t}\right)_{\mathrm{System}} = \dot{Q} - \dot{W}_{\mathrm{s}} - \dot{W}_{\mathrm{p}} \tag{3.25}$$

The internal energy of the system is composed of the contributions from each species. Therefore it is appropriate to represent U as

$$U = \sum_{i=1}^{N} \iiint_{V(t)} \rho_i u_i \, \mathrm{d}V \tag{3.26}$$

A subtle point is that the species always occupy the same volume $V(t)$ as the mixture. We can interchange the operations of the sum and the integration. Then, taking f as $\rho_i u_i$ for a system volume always enclosing species i, we need to use $\boldsymbol{v}_i$ as our velocity, as was done in Equation (3.16a), to obtain

$$\left(\frac{\mathrm{d}U}{\mathrm{d}t}\right)_{\text{System}} = \frac{\mathrm{d}}{\mathrm{d}t} \iiint_{\mathrm{CV}} \sum_{i=1}^{N} \rho_i u_i \, \mathrm{d}V + \iint_{\mathrm{CS}} \sum_{i=1}^{n} \rho_i u_i (\boldsymbol{v} + \boldsymbol{V}_i - \boldsymbol{w}) \cdot \boldsymbol{n} \, \mathrm{d}\boldsymbol{S} \tag{3.27}$$

It is sometimes useful to write the rate of heat added to the control volume in terms of a heat flux vector, $\dot{\boldsymbol{q}}''$, as

$$\dot{Q} = - \iint_{\mathrm{CS}} \dot{\boldsymbol{q}}'' \cdot \boldsymbol{n} \, \mathrm{d}S \tag{3.28}$$

where $\dot{\boldsymbol{q}}''$ is taken as directed out of the control volume or system. Note that when we apply the transformation to the Q and W terms there is no change since the system coincides with the control volume at time t. As with the mass flux of species in Equation (3.17), we can express the heat flux for pure conduction as

$$\dot{\boldsymbol{q}}'' = -k \nabla \boldsymbol{T} \tag{3.29}$$

where k is the thermal conductivity. In addition, $\dot{\boldsymbol{q}}''$ would have a radiative component.

The pressure work term is also made up of all of the species. However, let us first consider it for a single species and not include any subscripts. Since $\dot{W}$ has the convention of 'work done by the system on the surroundings' and the definition of work is the product of force times distance moved in the direction of the force, we can show from Figure 3.9 that

$$\Delta \dot{W}_{\mathrm{p}} = p\boldsymbol{n} \cdot \boldsymbol{v}(\Delta S) \tag{3.30}$$

for element ΔS that is displaced. For all of the species within the control volume, the total pressure work is

$$\dot{W}_{\mathrm{p}} = \iint_{\mathrm{CS}} \sum_{i=1}^{N} p_i \boldsymbol{n} \cdot \boldsymbol{v}_i \, \mathrm{d}S \tag{3.31}$$

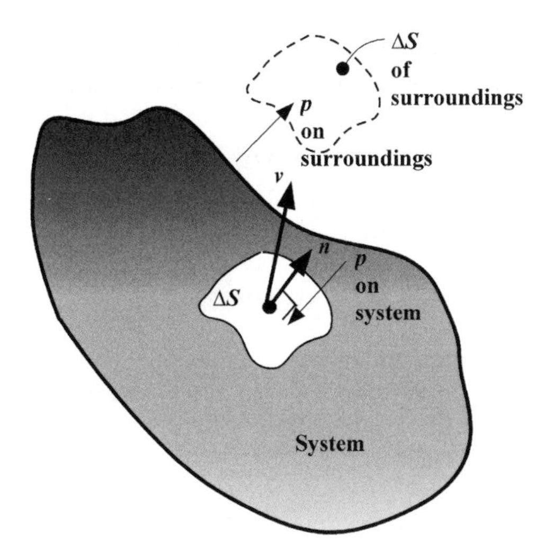

Figure 3.9 Work due to pressure on the surroundings

where p_i is the partial pressure of species i. Combining Eqations (3.27), (3.28) and (3.31) into Equation (3.25) gives

$$\frac{\mathrm{d}}{\mathrm{d}t}\iiint_{\mathrm{CV}}\left(\sum_{i=1}^{N}\rho_i u_i\right)\mathrm{d}V+\iint_{\mathrm{CS}}\sum_{i=1}^{N}\rho_i u_i(\boldsymbol{v}+\boldsymbol{V}_i-\boldsymbol{w})\cdot\boldsymbol{n}\,\mathrm{d}S$$
$$=-\iint_{\mathrm{CS}}\dot{\boldsymbol{q}}''\cdot\boldsymbol{n}\,\mathrm{d}S-\dot{\boldsymbol{W}}_{\mathrm{s}}-\iint_{\mathrm{CS}}\sum_{i=1}^{N}p_i(\boldsymbol{v}+\boldsymbol{V}_i)\cdot\boldsymbol{n}\,\mathrm{d}S \tag{3.32}$$

Introducing the species enthalpy as $h_i = u_i + p_i/\rho_i$ and rearranging, we obtain

$$\frac{\mathrm{d}}{\mathrm{d}t}\iiint_{\mathrm{CV}}\sum_{i=1}^{N}(\rho_i h_i-p_i)\,\mathrm{d}V+\iint_{\mathrm{CS}}\left(\sum_{i=1}^{N}\rho_i h_i\right)(\boldsymbol{v}-\boldsymbol{w})\cdot\boldsymbol{n}\,\mathrm{d}S$$
$$=-\iint_{\mathrm{CS}}\left(\dot{\boldsymbol{q}}''+\sum_{i=1}^{N}\rho_i h_i\boldsymbol{V}_i\right)\cdot\boldsymbol{n}\,\mathrm{d}\boldsymbol{S}-\dot{\boldsymbol{W}}_{\mathrm{s}}-\iint_{\mathrm{CS}}\left(\sum_{i=1}^{N}p_i\right)\boldsymbol{w}\cdot\boldsymbol{n}\,\mathrm{d}S \tag{3.33}$$

By Dalton's law, Equation (2.9), the mixture pressure, p, is $\sum_{i=1}^{N}p_i$. The term $\sum_{i=1}^{N}\rho_i\boldsymbol{V}_i h_i$ is sometimes considered to be a heat flow rate due to the transport of enthalpy by the species. (This is not the same as $\dot{\boldsymbol{q}}''$ arising from ∇Y_i, which is called the Dufour effect and is generally negligible in combustion.) With the exception of the enthalpy diffusion term, all the sums can be represented in mixture properties since $\rho h = \sum_{i=1}^{N}\rho_i h_i$. However, it is convenient to express the enthalpies in terms of the heat of formation and specific heat terms, and then to separate these two parts.

From Equation (2.24) we substitute for h_i. Without a loss in generality, but to improve the clarity of our discussion, we consider a three-component reaction

$$\text{Fuel (F)} + \text{oxygen } (O_2) \rightarrow \text{products (P)}$$

In terms of the molar stoichiometric coefficients

$$\nu_F F + \nu_{O_2} O_2 \rightarrow \nu_P P$$

Let us represent the mass consumption rate of the fuel per unit volume (of mixture) to be $\left(+\dot{m}'''_{F,r}\right)$. This is opposite to the sign convention adopted in Equation (3.18) in which species produced are positive in sign. However, it is more natural to treat $\dot{m}'''_{F,r}$ as inherently positive in sign. Then

$$\dot{m}'''_{O_2,r} = \left(\frac{\nu_{O_2}}{\nu_F}\right)\left(\frac{M_{O_2}}{M_F}\right)\dot{m}'''_{F,r} \tag{3.34a}$$

and

$$\dot{m}'''_{P,r} = -\left(\frac{\nu_P}{\nu_F}\right)\left(\frac{M_P}{M_F}\right)\dot{m}'''_{F,r} \tag{3.34b}$$

Substituting into Equation (3.33) gives

$$\begin{aligned} &\frac{d}{dt}\iiint_{CV}\left[\sum_{i=1}^{3}\rho_i\left(\Delta h^\circ_{f,i} + \int_{25\,^\circ C}^{T} c_{p_i}\,dT\right) + p\right]dV \\ &+ \iint_{CS}\sum_{i=1}^{3}\rho_i\left(\Delta h^\circ_{f,i} + \int_{25\,^\circ C}^{T} c_{p_i}\,dt\right)(\boldsymbol{v} + \boldsymbol{V}_i - \boldsymbol{w})\cdot\boldsymbol{n}\,dS \\ &= -\iint_{CS}\dot{\boldsymbol{q}}''\cdot\boldsymbol{n}\,dS - \dot{W}_s - \iint_{CS} p\boldsymbol{w}\cdot\boldsymbol{n}\,dS \end{aligned} \tag{3.35}$$

Since $\Delta h^\circ_{f,i}$ is a constant for each species, we can multiply it by each term in its respective species conservation equation, given by Equation (3.18). Summing over $i = 1$ to 3 for all species and using Equations (3.19) and (3.34), we obtain

$$\begin{aligned} &\frac{d}{dt}\iiint_{CV}\left(\sum_{i=1}^{3}\rho_i\Delta h^\circ_{f,i}\right)dV + \iint_{CS}\sum_{i=1}^{3}\rho_i\Delta h^\circ_{f,i}(\boldsymbol{v} + \boldsymbol{V}_i - \boldsymbol{w})\cdot\boldsymbol{n}\,dS \\ &= \iiint_{CV}\left(-\dot{m}'''_F\right)\left[\Delta h^\circ_{f,F} + \left(\frac{\nu_{O_2}}{\nu_F}\right)\left(\frac{M_{O_2}}{M_F}\right)\Delta h^\circ_{f,O_2} - \left(\frac{\nu_P}{\nu_F}\right)\left(\frac{M_P}{M_F}\right)\Delta h^\circ_{f,P}\right]dV \end{aligned} \tag{3.36}$$

The quantity in the brackets is exactly the heat combustion in mass terms, Δh_c, which follows from Equation (2.25). Now subtracting Equation (3.36) from Equation (3.35), we obtain (for $i = 1, N$)

$$\begin{aligned}
&\frac{\mathrm{d}}{\mathrm{d}t}\iiint_{\mathrm{CV}}\left[\left(\sum_{i=1}^{N}\rho_i\int_{25\,^\circ C}^{T}c_{p_i}\,\mathrm{d}T\right)-p\right]\mathrm{d}V \\
&+\iint_{\mathrm{CS}}\left(\sum_{i=1}^{N}\rho_i\int_{25\,^\circ\mathrm{C}}^{T}c_{p_i}\,\mathrm{d}t\right)(\boldsymbol{v}+\boldsymbol{V}_i-\boldsymbol{w})\cdot\boldsymbol{n}\,\mathrm{d}S \\
&=-\iint_{\mathrm{CS}}\dot{\boldsymbol{q}}''\cdot\boldsymbol{n}\,\mathrm{d}S+\Delta h_c\iiint_{\mathrm{CV}}\dot{m}'''_{\mathrm{F}}\,\mathrm{d}V-\dot{W}_{\mathrm{s}}-\iint_{\mathrm{CS}}p\boldsymbol{w}\cdot\boldsymbol{n}\,\mathrm{d}S
\end{aligned} \tag{3.37}$$

This is a general form of the energy equation for a control volume useful in combustion problems. The terms can be literally expressed as

$$\begin{aligned}
&\begin{bmatrix}\text{Rate of change for}\\ \text{the internal energy}\\ \text{of the mixture in}\\ \text{the CV}\end{bmatrix}+\begin{bmatrix}\text{net rate of enthalpy}\\ \text{flow due to bulk}\\ \text{flow and diffusion flow}\\ \text{from the CV}\end{bmatrix} \\
&=\begin{bmatrix}\text{net rate of}\\ \text{heat added}\\ \text{to the CV}\end{bmatrix}+\begin{bmatrix}\text{rate of chemical}\\ \text{energy released}\\ \text{within the CV}\end{bmatrix} \\
&-\begin{bmatrix}\text{net rate of shaft}\\ \text{and shear work done}\\ \text{by the fluid in the CV}\end{bmatrix}-\begin{bmatrix}\text{rate of pressure}\\ \text{work due to}\\ \text{a moving CV}\end{bmatrix}
\end{aligned}$$

A series of approximations will make Equation (3.37) more practical for engineering applications. These approximations are described below, and the resulting transformed equations are presented as each approximation is sequentially applied.

1. Equal specific heats, $c_{p_i} = c_p(T)$. This assumption allows the diffusion term involving $\rho_i \boldsymbol{V}_i$ to vanish since $\sum_{i=1}^{N} \rho_i \boldsymbol{V}_i = 0$ from Equation (3.16).
2. Equal and constant specific heats, $c_p =$ constant. From the Reynolds transport theorem, Equation (3.9), the pressure terms can be combined as

$$\frac{\mathrm{d}}{\mathrm{d}t}\iiint_{\mathrm{CV}} p\,\mathrm{d}V = \iiint_{\mathrm{CV}}\left(\frac{\partial p}{\partial t}\right)\mathrm{d}V + \iint_{\mathrm{CS}} p\boldsymbol{w}\cdot\boldsymbol{n}\,\mathrm{d}S \tag{3.38}$$

Also for all specifics heats equal and independent of temperature, i.e. $c_{p_i} = c_p =$ constant, Equation (3.37) becomes

$$\begin{aligned}
&c_p\frac{\mathrm{d}}{\mathrm{d}t}\iiint_{\mathrm{CV}}\rho(T-\ 25\,^\circ\mathrm{C})\mathrm{d}V-\iiint_{\mathrm{CV}}\frac{\partial p}{\partial t}\mathrm{d}V+c_p\iint_{\mathrm{CS}}\rho(T-25\,^\circ\mathrm{C})(\boldsymbol{v}-\boldsymbol{w})\cdot\boldsymbol{n}\,\mathrm{d}S \\
&=\Delta h_{\mathrm{c}}\iiint_{\mathrm{CV}}\dot{m}'''_{\mathrm{F,r}}\,\mathrm{d}V-\iint_{\mathrm{CS}}\dot{\boldsymbol{q}}''\cdot\boldsymbol{n}\,\mathrm{d}S-\dot{W}_{\mathrm{shaft}}
\end{aligned} \tag{3.39}$$

where the diffusion term has been eliminated by Equation (3.16a) as $\sum \rho_i \boldsymbol{V}_i = 0$. Finally, by multiplying the conservation of mass, Equation (3.13), by 25 °C and subtracting from Equation (3.39), it follows that

$$c_p \frac{\mathrm{d}}{\mathrm{d}t} \iiint_{\mathrm{CV}} \rho T \,\mathrm{d}V - \iiint_{\mathrm{CV}} \frac{\partial p}{\partial t} \mathrm{d}V + \iint_{\mathrm{CS}} \rho c_p T (\boldsymbol{v} - \boldsymbol{w}) \cdot \boldsymbol{n} \,\mathrm{d}S$$
$$= \iiint_{\mathrm{CV}} \dot{m}_{\mathrm{F}}''' \,\Delta h_{\mathrm{c}} \,\mathrm{d}V - \iint_{\mathrm{CS}} \dot{\boldsymbol{q}}'' \cdot \boldsymbol{n} \,\mathrm{d}S - \dot{W}_{\mathrm{shaft}} \tag{3.40}$$

3. Uniform fluxes and properties. For the case of discrete fluxes of mass and heat at the surface of the control volume with uniform boundary temperatures at these regions, the equation can be expressed as

$$c_p \frac{\mathrm{d}}{\mathrm{d}t} \iiint_{\mathrm{CV}} \rho T \,\mathrm{d}V - \iiint_{\mathrm{CV}} \frac{\partial p}{\partial t} \mathrm{d}V + \sum_{j,\mathrm{net,out}} \dot{m}_j c_p T_j$$
$$= \iiint_{\mathrm{CV}} \dot{m}_{\mathrm{F}}''' \Delta h_{\mathrm{c}} \,\mathrm{d}V + \dot{Q}_{\mathrm{added}} - \dot{W}_{\mathrm{shaft}} \tag{3.41}$$

For uniform properties within the control volume and discrete j-flows at the control surface

$$c_p \frac{\mathrm{d}}{\mathrm{d}t}(mT) - V \frac{\mathrm{d}p}{\mathrm{d}t} + \sum_{j,\mathrm{net,out}} \dot{m}_j c_p T_j = \dot{m}_{\mathrm{F,reacted}} \Delta h_{\mathrm{c}} + \dot{Q} - \dot{W}_{\mathrm{shaft}} \tag{3.42}$$

If the fluid mixture is a perfect gas following

$$pV = RmT \tag{3.43}$$

where R the gas constant is given as

$$R = c_p - c_v \tag{3.44}$$

we can write

$$\boxed{c_v \frac{\mathrm{d}}{\mathrm{d}t}(mT) + p \frac{\mathrm{d}V}{\mathrm{d}t} + \sum_{j,\mathrm{net,out}} \dot{m}_j c_p T_j = \dot{m}_{\mathrm{F,reacted}} \Delta h_{\mathrm{c}} + \dot{Q} - \dot{W}_{\mathrm{shaft}}} \tag{3.45}$$

Alternatively, for a perfect gas, we can use Equation (3.42) in a modified form. The first term on the left can be expanded as

$$c_p \frac{\mathrm{d}}{\mathrm{d}t}(mT) = c_p m \frac{\mathrm{d}T}{\mathrm{d}t} + c_p T \frac{\mathrm{d}m}{\mathrm{d}t} \tag{3.46}$$

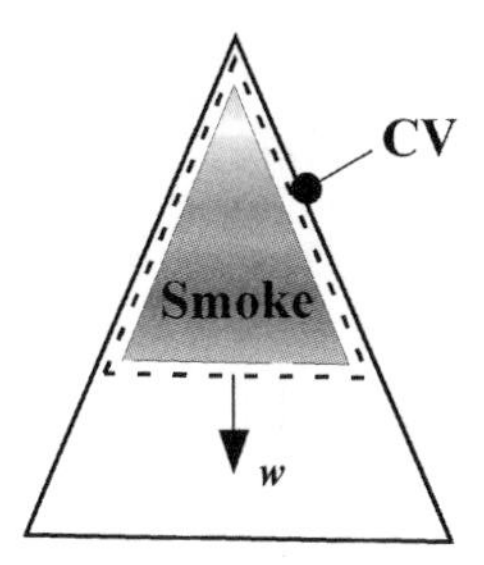

Figure 3.10 A control volume encasing an expanding smoke layer in a compartment

The conservation of mass is multiplied by c_pT to give

$$c_pT\frac{\mathrm{d}m}{\mathrm{d}t} + \sum_{j,\text{net,out}} \dot{m}_j c_p T = 0 \tag{3.47}$$

Subtracting Equation (3.47) from Equation (3.42) gives, for a perfect gas mixture with uniform properties in the control volume,

$$\boxed{c_p m\frac{\mathrm{d}T}{\mathrm{d}t} - V\frac{\mathrm{d}p}{\mathrm{d}t} + \sum_{j,\text{in only}} \dot{m}_j c_p (T - T_j) = \dot{m}_{\text{F,reacted}}\,\Delta h_c + \dot{Q} - \dot{W}_{\text{shaft}}} \tag{3.48}$$

Here T is the uniform temperature in the CV. Equations (3.45) and (3.48) are all equivalent under the three approximations, and either could be useful in problems. The development of governing equations for the zone model in compartment fires is based on these approximations. The properties of the smoke layer in a compartment have been described by selecting a control volume around the smoke. The control volume surface at the bottom of the smoke layer moves with the velocity of the fluid there. This is illustrated in Figure 3.10.

Example 3.4 A turbulent fire plume is experimentally found to burn with 10 times the required stoichiometric air up to the tip of the flame. It is also measured that 20 % of the chemical energy is radiated to the surroundings from the flame. The fuel is methane, which is supplied at 25 °C and burns in air which is also at 25 °C. Calculate the average temperature of the gases leaving the flame tip. Assume constant and equal specific heats and steady state.

Solution A fixed control volume is selected as a cylindrical shape with its bases at the burner and flame tip respectively, as shown in Figure 3.11. The radius of the CV cylinder is just large enough to contain all of the flow and thermal effects of the plume. This is illustrated in Figure 3.11 with the ambient edge of the temperature and velocity distributions included in the CV.

From the conservation of mass, Equation (3.15),

$$\dot{m} - \dot{m}_A - \dot{m}_{CH_4} = 0$$

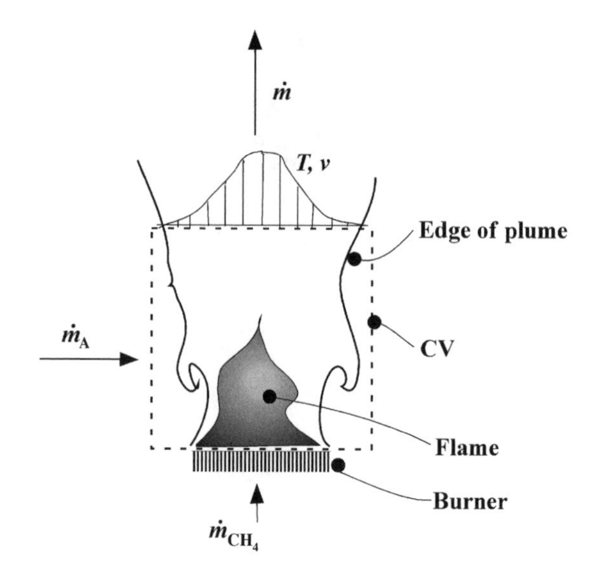

Figure 3.11 Fire plume problem

Apply the conservation of energy, Equation (3.40). Since the control volume is fixed the pressure work term does not apply. The shear work ($\boldsymbol{v} \times$ shear force) is zero because (a) the radius of the control volume was selected so that the velocity and its gradient are zero on the cylindrical face; and (b) at the base faces, the velocity is normal to any shear surface force. Similarly, no heat is conducted at the cylindrical surface because the radial temperature gradient is zero, and conduction is ignored at the bases since we assume the axial temperature gradients are small. However, heat is lost by radiation as

$$\dot{Q} = -(0.20)(-\dot{m}_{\mathrm{F,r}})\,\Delta h_{\mathrm{c}}$$

Here $\dot{m}_{\mathrm{F,r}}$ is defined as the production rate of fuel.

The heat of combustion of the methane is computed from Equation (2.25) or its mass counterpart in Equation (3.36). It can be computed from the stoichiometric reaction involving oxygen, or air, or 10 times stoichiometric air without change since any diluent or species not taking part in the reaction will not contribute to this energy. Therefore, we use

$$\mathrm{CH_4 + 2\ O_2 \rightarrow CO_2 + 2\ H_2O}$$

and obtain, using Table 2.2,

$$\Delta h_{\mathrm{c}} = \left[\frac{(-74.9\,\mathrm{kJ/mole\,CH_4})}{(16\,\mathrm{g/mole\,CH_4})} + 0\right] - \left[\left(\frac{1\ \mathrm{mole\,CO_2}}{1\ \mathrm{mole\,CH_4}}\right)\left(\frac{44\,\mathrm{g/mole\,CO_2}}{16\,\mathrm{g/mole\,CH_4}}\right)\right.$$
$$\times\left(\frac{-393.5\,\mathrm{KJ/mole\,CO_2}}{44\,\mathrm{g/mole\,CO_2}}\right) + \left(\frac{2\,\mathrm{moles\,H_2O}}{1\,\mathrm{mole\,CH_4}}\right)\left(\frac{18\,\mathrm{g/mole\,H_2O}}{16\,\mathrm{g/mole\,CH_4}}\right)$$
$$\left.\times\left(\frac{-241.8\,\mathrm{KJ/mole\,H_2O}}{18\,\mathrm{g/mole\,H_2O}}\right)\right]\Delta h_{\mathrm{c}} = 50.1\,\mathrm{kJ/g\,CH_4}$$

Alternatively, we could have taken this value from data in the literature as given by Table 2.3.

Since we have 2 moles of stoichiometric oxygen, we must have 2(4.76) moles of stoichiometric air. For a molecular weight of air of 28.84 g/mole air, we compute

$$\left[(2)(4.76)\text{moles of air}\right](28.84\ \text{g/mole of air})$$

for every mole of methane consumed. The mass rate of air consumed is

$$\begin{aligned}(\dot{m}_{\mathrm{A}})_{\mathrm{st}} &= \frac{(2)(4.76)(28.84)\text{g of air}}{(1\ \text{mole CH}_4)(16\text{g/mole CH}_4)}\left(\dot{m}_{\mathrm{CH_4,r}}\right)\\ &= 17.16\,\dot{m}_{\mathrm{CH_4,r}}\end{aligned}$$

The supply rate of air is 10 times stoichiometric , or

$$\dot{m}_{\mathrm{A}} = 171.6\left(-\dot{m}_{\mathrm{CH_4,r}}\right)$$

Since $\left(-\dot{m}_{\mathrm{CH_4,r}}\right)$ is equal to the fuel supply rate $(\dot{m}_{\mathrm{CH_4}})$ by definition for a stoichiometric reaction, then the exit mass flow rate is

$$\dot{m} = \dot{m}_{\mathrm{A}} + \dot{m}_{\mathrm{CH_4}} = (171.6 + 1)\left(-\dot{m}_{\mathrm{CH_4,r}}\right)$$

The energy equation can now be written from Equation (3.40) as

$$\dot{m}c_pT - \dot{m}_{\mathrm{A}}c_pT_{\mathrm{A}} - \dot{m}_{\mathrm{CH_4}}c_pT_{\mathrm{CH_4}} = -(0.20)(-\dot{m}_{\mathrm{CH_4,r}})\Delta h_{\mathrm{c}} + \left(-\dot{m}_{\mathrm{CH_4,r}}\right)\Delta h_{\mathrm{c}}$$

We need to select a suitable value for c_p. Since air contains mostly nitrogen and so does the heavily diluted exit stream, a value based on nitrogen can be a good choice. From Table 2.1 at 600 K

$$c_p = \left(\frac{30.1\,\text{J/mole K}}{28\,\text{g/mole}}\right) = 1.08\,\text{J/g K}$$

For methane at 25 °C or 298 K, $c_p = 2.24\,\text{J/g K}$. Note that on substituting for the temperatures in this steady state example it makes no difference whether K or °C units are used. This follows from the conservation of mass. However, for unsteady applications of Equation (3.40), since we have used the perfect gas law in which T is in K, we should be consistent and use it through the equations. When in doubt, use K without error. Substituting:

$$\begin{aligned}&(171.6 + 1)(-\dot{m}_{\mathrm{CH_4,r}})(1.08 \times 10^{-3}\text{kJ/g K})T\\ &\quad - 171.6(-_{\mathrm{CH_4,r}})(1.08 \times 10^{-3}\,\text{kJ/g K}) \quad (298\,\text{K})\\ &\quad - (-_{\mathrm{CH_4,r}})(1.08 \times 10^{-3}\,\text{kJ/g K}) \quad (298\,\text{K})\\ &= (0.8)(-_{\mathrm{CH_4,r}})(50.1\,\text{kJ/g})\end{aligned}$$

or

$$T = \frac{(0.8)(50.1) + (172.6)(1.08 \times 10^{-3})(298)}{(172.6)(1.08 \times 10^{-3})}$$

$$T = 513.0\,\text{K} = 240\,^{\circ}\text{C}$$

This is a realistic approximate average temperature at the flame tip for most fire plumes.

Problems

3.1 Water flows into a tank at a uniform velocity of 10 cm/s through a nozzle of cross-sectional area 4 cm^2. The density of water is 1 kg/m^3. There is a leak at the bottom of the tank that causes 0.5 g/s of water to be lost during filling. The tank is open at the top, has a cross-sectional area of 150 cm^2 and is 20 cm high. How long does it take for the tank to fill completely? For a control volume enclosing the water in the tank, is the process steady?

3.2 Propane and air are supplied to a combustion chamber so that 20 g/s of propane reacts. The reaction forms H_2O, CO_2 and CO where the molar ratio of CO to CO_2 is 0.1. The exhaust gases flow at a rate of 360 g/s. Assuming the process is steady and conditions are uniform at the exit, compute the exit mass fraction of the CO.

3.3 Methane is supplied to a burner in a room for an experiment at 10 g/s. The measurements indicate that air flow into the room doorway is 800 g/s and the door is the only opening. The exhaust leaves the room through the upper part of the doorway at a uniform temperature of 400 °C. Assume that the methane burns completely to CO_2 and H_2O and that steady conditions prevail in the room.

(a) Calculate the rate of heat loss to the surroundings.

(b) Calculate the mass fraction of CO_2 leaving the room.

3.4 Flashover occurs in a room causing 200 g/s of fuel to be generated. It is empirically known that the 'yield' of CO is 0.08. Yield is defined as the mass of product generated per mass of fuel supplied. The room has a volume of 30 m^3 and is connected to a closed corridor that has a volume of 200 m^3. The temperatures of the room and corridor are assumed uniform and constant at 800 and 80 °C respectively. An equation of state for these gases can be taken as $\rho T = 360$ kg K/m^3. Steady mass flow rates prevail at the window of the room (to the outside air) and at the doorway to the corridor. These flow rates are 600 and 900 g/s respectively. Derive expressions for the mass fraction of CO in the room and corridor as a function of time assuming uniform concentrations in each region.

3.5 Carbon monoxide is released in a room and burns in a pure oxygen environment; e.g. perhaps an accident on a space station. The combustion is given by the chemical equation

$$\text{CO} + \frac{1}{2}\text{O}_2 \rightarrow \text{CO}_2$$

(a) What is the stoichiometric fuel (CO) to oxygen (O_2) mass ratio, r, for this reaction?

(b) If the flow rate of oxygen supplied to the room is 500 g/s and the release rate of CO is 200 g/s, what is the rate of combustion of the CO?

(c) If the flow rate of oxygen remains as given in (b) but the release rate of CO suddenly changes to 1000 g/s, what is the mass fraction of CO in the room?

(d) For the case (b), what is the mass fraction of CO_2 leaving the room?

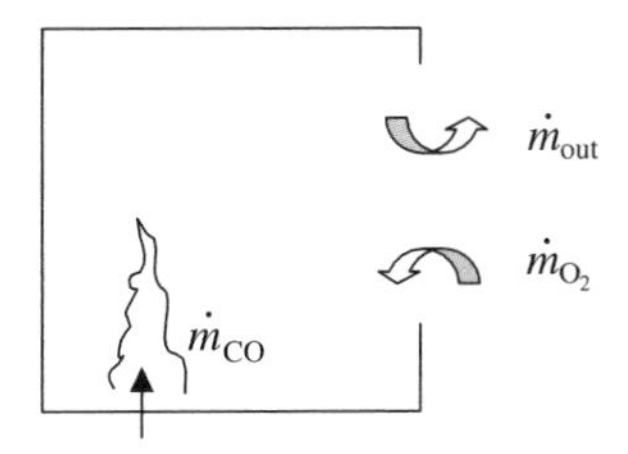

3.6 C_3H_8 is burned with 10 times the stoichiometric air in a steady flow process. The reaction is complete, forming CO_2 and H_2O. The fuel and air are mixed at 400 °C before entering the combustor. The combustor is adiabatic. Specific heats are all constant, $c_p = 1$ J/g.

(a) Calculate the stoichiometric air to fuel mass ratio.

(b) Calculate the stoichiometric oxygen to fuel mass ratio

(c) If the flow rate of the C_3H_8 is 10 g/s, calculate the exit flow rate.

(d) Calculate the oxygen mass fraction at the exit.

(e) Calculate the C_3H_8 mass fraction at the inlet.

(f) Calculate the exit temperature.

(g) Calculate the enthalpy per unit mass for the H_2O(g) in the exit stream (with respect to the 25 °C reference state).

3.7 An oxygen consumption calorimeter is shown below in which a chair is burned with all of the combustion products captured by the exhaust system.

The flow rate of the gases measured in the stack at station A is 2.0 m^3/s, and the temperature of the mass fraction and O_2 are 327 °C and 0.19 respectively. Assume steady state conditions. The surrounding air for combustion is at 20 °C. The specific heat of air is 1 kJ/kg K and its density is 1.1 kg/m^3. State clearly any other necessary assumptions in working the problem.

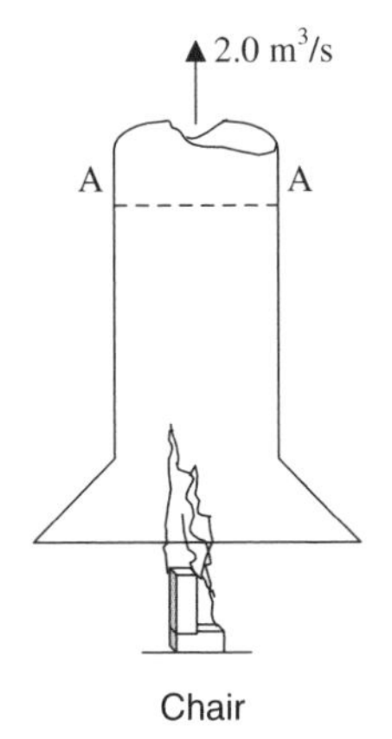

Compute:

(a) the mass flow rate in the exhaust stack;

(b) the oxygen consumption rate;

(c) the rate of chemical energy released by the chair fire (use the approximation that 13 kJ are released for every gram of O_2 consumed);

(d) the rate of heat lost from the flame and exhaust gases up to station A.

3.8 A fire occurs in a space station at 200 kW. The walls can be considered adiabatic and of negligible heat capacity. The initial and fuel temperatures are at 25 °C. Assume the station atmosphere has uniform properties with constant specific heats as given. Assume that the constant and equal specific heats of constant pressure and volume are 1.2 and 1.0 kJ/kg K respectively. Conduct your analysis for the control volume (CV) consisting of the station uniform atmosphere, excluding all solids and the fuel in its solid state.

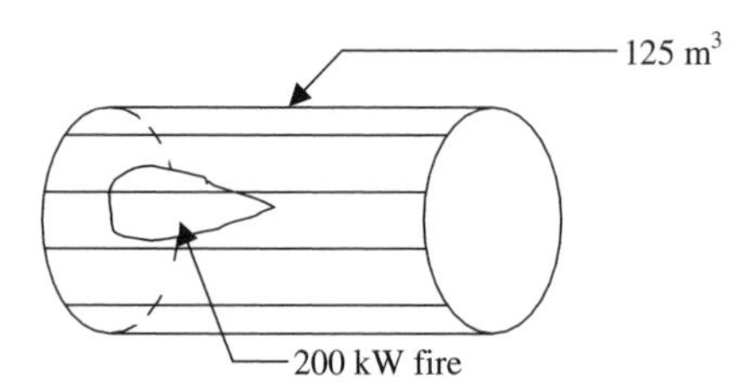

Note that $p = \rho RT$, where p is pressure.

(a) Indicate all mass exchanges for the CV shown and their magnitudes.

(b) Apply conservation principles to the CV and determine an equation for the atmosphere temperature in the CV. Identify your terms clearly.

(c) How can the pressure be determined? Explain. Write an equation.

3.9 At the instant of time shown, the entrainment of air into the fire plume is 300 g/s, the outflow of smoke through the door is 295 g/s and the liquid fuel evaporation rate of a spill burning on the floor is 10 g/s. What is the mass rate of smoke accumulation within the room?

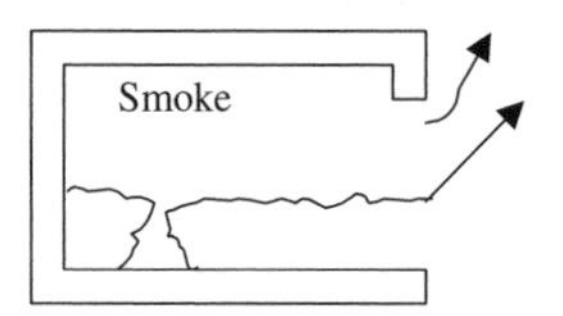

3.10 A chair burns in a room, releasing 500 kW, and the heat of combustion for the chair is estimated as 20 kJ/g. Fire and room conditions can be considered steady. The air flow rates into a door and through a window are 200 and 100 g/s respectively. Assume the room is adiabatic. The gases have constant and equal specific heats =1.5 kJ/kg K. The ambient air is 25 °C and the chair fuel gasifies at 350 °C. Compute the average temperature of the room gases. Account for all terms and show all work.

3.11 A furnace used preheated air to improve its efficiency. Determine the adiabatic flame temperature (in K) when the furnace is operating at a mass air to mass fuel ratio of 16. Air enters at 600 K and the fuel enters at 298 K. Use the following approximate properties:

All molecular weights are constant at 29 g/g mol.

All specific heats are constant at 1200 J/kg K.

Enthalpy of formation: zero for air and all products, 4×10^7 J/kg for the fuel.

3.12 Methane gas burns with air in a closed but leaky room. The leak in the room ensures that the room pressure stays equal to the surrounding atmosphere, which is at 10^5 Pa and 20 °C. The room is adiabatic and no heat is lost to the boundary surfaces. When the methane burns the flow is out of the room only. Assume that the gas in the room is well mixed. Its initial temperature is 20 °C and its volume is 30 m^3. The power of the fire during this combustion is $\dot{Q} = 30$ kW. The specific heat of the mixture is constant and equal to 1 J/g K. (Hint: $pV = mRT$, where $R = (8.314$ J/mole K)/(29 g/mole), 1 N/m^2 = 1 Pa, 1 N = 1 kg m/s^2 and 1 J = 1 N m.)

(a) Show that the mass flow rate out is $\dot{Q}/c_pT$, where T is the gas temperature in the room.

(b) Compute the temperature in the room after 2 seconds of combustion.

3.13 A room has mass flow rates of fuel, air and fire products at an instant of time in the fire, as shown in the sketch.

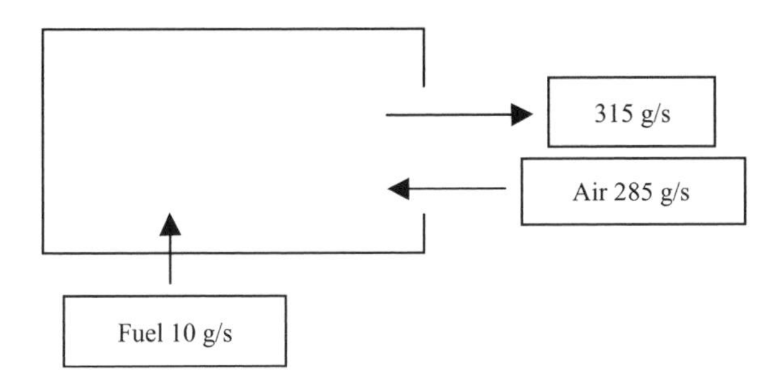

Data:

Room volume	28.8 m^3
Ambient temperature	35 °C
Ambient density	1.18 kg/m^3
Heat of combustion	24 kJ/g
Fuel gas entering temperature	300 °C
Exiting gas temperature	275 °C
Exiting oxygen mass fraction	0.11
Specific heat at constant pressure	1.3 J/g K (same for all species)

Assume constant pressure throughout and a control volume of the room gas can be considered to enclose uniform properties.

(a) Compute the time rate of change of the room gas density.

(b) Compute the rate of heat transfer from the room gases and state whether it is a loss or a gain.

3.14 Heptane spills on a very hot metal plate causing its hot vapor to mix with air. At some point the mixture of heptane vapor and air reaches a condition that can cause autoignition.

In the sketch, T is the temperature of the mixture and Y_F is the mass fraction of heptane vapor.

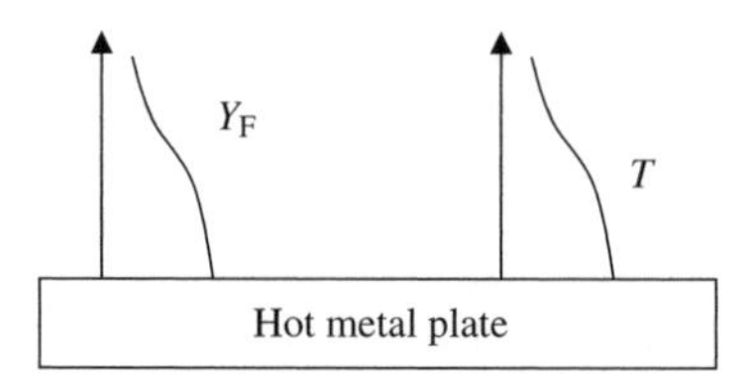

Autoignition is to be investigated by examining local conditions at a small volume in the mixture to see what temperature and fuel mass fraction are needed in the immediate surroundings (T_∞ and $Y_{F,\infty}$) to just allow ignition. Here we will assume that the small volume of fluid enclosed in a control volume has uniform properties: ρ, density; T, temperature; Y_F, fuel mass fraction; and c_p the specific heat and constant for all species. In addition, the pressure is constant throughout, and heat and fuel mass diffusion occurs between the reacting fluid in the volume and its surroundings at fixed conditions (T_∞ and $Y_{F,\infty}$).

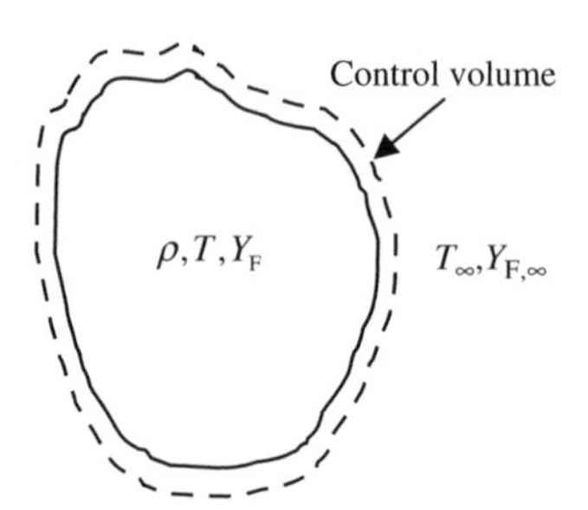

For the control volume, the heat flux at the boundary is given as $\dot{q}'' = h_c(T - T_\infty)$. The diffusion mass flux supplying the reaction is given as $\dot{m}_F'' = h_m(Y_{F,\infty} - Y_F)$, where from heat and mass transfer principles $h_m = h_c/c_p$. Let V and S be the volume and surface area of the control volume. The reaction rate per unit volume is given as $\dot{m}_F''' = AY_F e^{-E/(RT)}$ for the fuel in this problem.

Perform the following:

(a) Write the conservation of mass, fuel species and energy for this control volume. Show your development.

(b) Show, for fuel mass diffusion as negligible in the mass and energy conservation equations (only), that the conservation of species (fuel) and energy can be written as

$$\rho \frac{dY_F}{dt} = -AY_F e^{-E/(RT)} + \left(\frac{S}{V}\right)\left(\frac{h_c}{c_p}\right)(Y_{F,\infty} - Y_F)$$

$$\rho c_p \frac{dT}{dt} = \Delta h_c A Y_F e^{-E/(RT)} - \left(\frac{S}{V}\right) h_c (T - T_\infty)$$

(c) Show under steady conditions that

$$Y_F = Y_{F,\infty} - \frac{c_p}{\Delta h_c}(T - T_\infty)$$

(*Hint*: use part (b) here.)

(d) Consider $Y_{F,\infty}$ to be at its stoichiometric value for a mixture with air. For heptane, C_7H_{16}, going to complete combustion, calculate $Y_{F,\infty}$ from the stoichiometric equation.

(e) Using a graphical, computational analysis for heptane, compute the following (use the result of part (b) and the data below):

(i) T_∞, the minimum temperature to allow autoignition;

(ii) Y_F, the fuel mass fraction ignited;

(iii) identify the points of steady burning and compute the flame temperature for the surroundings maintained at T_∞ of part (e)(i).

Assume a constant volume sphere that reacts having a radius (r) of 1 cm ($V = \frac{4}{3}\pi r^3$, $S = 4\pi r^2$) using the following data:

Preexponential, $A = 9.0 \times 10^9$ g/m^3 s

Activation energy, $E = 160$ kJ/mol ($R = 8.314$ J/mol K)

Heat of combustion, $\Delta h_c = 41.2$ kJ/g

Specific heat, $c_p = 1.1$ J/g K

Conductivity, $k = 0.08$ W/m K

Heat transfer coefficient, $h_c \approx k/r$

Surrounding fuel mass fraction, $Y_{F,\infty} = 0.069$

4

Premixed Flames

4.1 Introduction

A premixed flame is a chemical reaction in which the fuel and the oxygen (oxidizer) are mixed before they burn. In a diffusion flame the fuel and oxygen are separated and must come together before they burn. In both cases we must have the fuel and the oxidizer together in order to burn. Thermodynamics can tell what happens when they burn, provided we know the chemical equation. We can place a gaseous fuel and oxygen together, yet nothing may occur. Even the requirements of chemical equilibrium from thermodynamics are likely not to ensure the occurrence of a chemical reaction. To answer the question of whether a reaction will occur, we must understand chemical kinetics, the subject related to the rate of the reaction and its dependent factors.

We shall examine several aspects of premixed flames:

1. What determines whether a reaction occurs?
2. What is the speed of the reaction in the fuel–oxidizer mixture?
3. What determines extinction of the reaction?

We shall discuss only gas mixtures, but the results can apply to fuel–oxidizer mixtures in liquid and solid form. The fuel in an aerosol state can also apply. At high speeds, compressibility effects will be important. At reaction front speeds greater than the speed of sound, a shock front will precede the reaction front. Such a process is known as a detonation, compared to reaction front speeds below sonic velocity which are called deflagrations. We shall only investigate the starting speed under laminar conditions. Beyond this point, one might regard the subject as moving from fire to explosion. The subject of explosion is excluded here. However, the means to prevent a potential explosion hazard are not excluded.

Fundamentals of Fire Phenomena James G. Quintiere

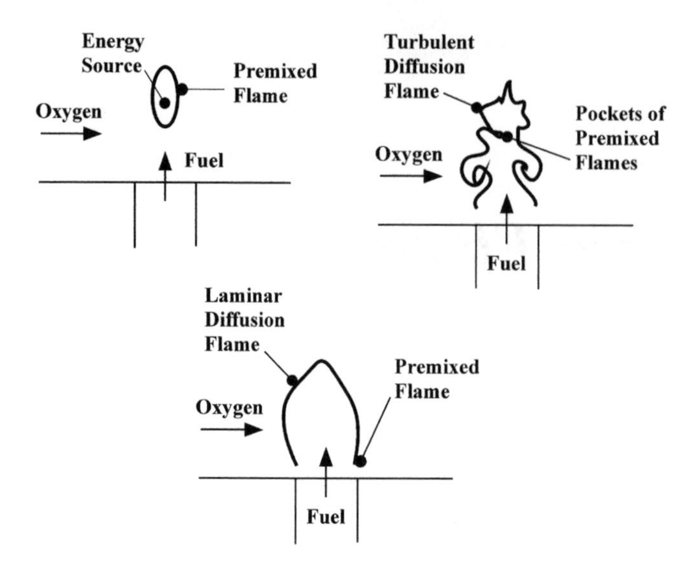

Figure 4.1 Premixed and diffusion flames

Although fire mainly involves the study and consequences of diffusion flames, premixed flames are important precursors. In order to initiate a diffusion flame, we must first have a premixed flame. In regions where a diffusion flame is near a cold wall, we are likely to have an intermediary premixed flame. Even in a turbulent diffusion flame, some state of a premixed flame must exist (see Figure 4.1).

4.2 Reaction Rate

There are many molecular and atomic mechanisms for the promotion of a chemical reaction. A catalyst is a substance that does not take part in the reaction, but promotes it. A radical is an unstable species seeking out electrons, as molecules are colliding and bonding in a complex path to the end of the reaction. The temperature is directly related to the kinetic energy of the molecules and their collision rate. This microscopic description is very approximately described by an empirical global kinetic rate equation. The basis for it is attributed to Suante Arrhenius [1], who showed that the thermal behavior for the rate is proportional to $e^{-E/(RT)}$, where R is the universal gas constant and E is called the activation energy having units of kJ/mole. $E/(RT)$ is dimensionless. The significance of E is that the lower its value, the easier it is for a chemical reaction to be thermally initiated. Despite the apparent restriction by R to the perfect gas law, the relationship is also used to describe chemical rates in solids and liquids. Intuitively, the rate for a chemical reaction must depend on the concentrations of fuel and oxidizer in the mixture. An overall reaction rate expression, which is often used to match the effect of the complex reaction steps needed, is given by

$$\dot{m}'''_{\mathrm{F,r}} = \rho A_0 Y_{\mathrm{F}}^n Y_{\mathrm{O_2}}^m \, \mathrm{e}^{-E/(RT)} \tag{4.1a}$$

Table 4.1 Overall reaction rates for fuels burning in air (from Westbrook and Dryer [2])[a]

Fuel	Formula	B	E(kcal/mole)	n	m
Methane	CH_4	8.3×10^5	30.0	−0.3	1.3
Ethane	C_2H_6	1.1×10^{12}	30.0	0.1	1.65
Propane	C_3H_8	8.6×10^{11}	30.0	0.1	1.65
n-Hexane	C_6H_{14}	5.7×10^{11}	30.0	0.25	1.5
n-Heptane	C_7H_{16}	5.1×10^{11}	30.0	0.25	1.5
Octane	C_8H_{18}	4.6×10^{11}	30.0	0.25	1.5
Methanol	CH_3OH	3.2×10^{12}	30.0	0.25	1.5
Ethanol	C_2H_5OH	1.5×10^{12}	30.0	0.15	1.6

[a] Here the rate constant is given by

$$\frac{\mathrm{d}[\mathrm{F}]}{\mathrm{d}t} = k[\mathrm{F}]^n[\mathrm{O_2}]^m$$

where [ ] implies molar concentration of the species in mole/volume,

$$k \equiv B\,\mathrm{e}^{-E/(RT)}$$

This compares to Equation (4.1a) with

$$A_0 = \frac{B\rho^{n+m-1}}{M_\mathrm{F}^{n-1}M_{\mathrm{O_2}}^m}$$

or

$$\dot{m}'''_{\mathrm{F,r}} = A\,\mathrm{e}^{-E/(RT)} \tag{4.1b}$$

This is commonly called the Arrhenius equation. Table 4.1 gives typical values for fuels in terms of the specific rate constant, k. In Equation (4.1), $\dot{m}'''_{\mathrm{F,r}}$ is taken as positive for the mass rate of fuel consumed per unit volume. Henceforth, in the text we will adopt this new sign convention to avoid the minus sign we were carrying in Chapter 3. The quantity A is called the pre-exponential factor and must have appropriate units to give the correct units to $\dot{m}'''_{\mathrm{F,r}}$. The exponents n and m as well as A must be arrived at by experimental means. The sum $(n + m)$ is called the order of the reaction. Often a zeroth-order reaction is considered, and it will suffice for our tutorial purposes.

Illustrative values for a zeroth-order reaction are given as follows:

$$A = 10^{13}\ \mathrm{g/m^3\,s}$$

and

$$E = 160\ \mathrm{kJ/mole}$$

(This value for A is somewhat arbitrary since it could range from about 10^5 to 10^{15}.) Table 4.2 accordingly gives a range of values for $\dot{m}'''_{\mathrm{F,r}}$. The tabulation shows a strong sensitivity to temperature. It is clear that between 298 and 1200 K there is a significant change. In energy units, for a typical heat of combustion of a gaseous fuel of 45 kJ/g, we have about 50×10^6 or $50\ \mathrm{kJ/cm^3\,s}$ at 1200 K. One calorie or 4.184 J can raise 1 $\mathrm{cm^3}$ of

Table 4.2 Typical fuel production for a zeroth-order arrhenius rate

T(K)	$E/(RT)$	$\exp[-E/(RT)]$	$\dot{m}'''_{F,r}$ (g/m^3 s)	Energy release (W/cm^3)
298	64.6	9.0×10^{-30}	9.0×10^{-17}	0.40×10^{-17}
600	32.1	1.18×10^{-14}	1.2×10^{-1}	0.53×10^{-2}
1200	16.0	1.08×10^{-7}	1.1×10^{6}	0.53×10^{5}
1600	12.0	5.98×10^{-6}	6.0×10^{7}	0.27×10^{7}
2000	9.6	6.62×10^{-5}	6.6×10^{8}	0.30×10^{8}
2500	7.7	4.54×10^{-4}	4.5×10^{9}	0.20×10^{9}

water 1 K. Here we have 50 000 J released within 1 cm^3 for each second. At 298 K this is imperceptable (0.4×10^{-17} J/cm^3 s) and at 600 K it is still not perceptible (0.54×10^{-2} J/cm^3 s).

Combustion is sometimes described as a chemical reaction giving off significant energy in the form of heat and light. It is easy to see that for this Arrhenius reaction, representative of gaseous fuels, the occurrence of combustion by this definition might be defined for some critical temperature between 600 and 1200 K.

4.3 Autoignition

Raising a mixture of fuel and oxidizer to a given temperature might result in a combustion reaction according to the Arrhenius rate equation, Equation (4.1). This will depend on the ability to sustain a 'critical' temperature and on the concentration of fuel and oxidizer. As the reaction proceeds, we use up both fuel and oxidizer, so the rate will slow down according to Arrhenius. Consequently, at some point, combustion will cease. Let us ignore the effect of concentration, i.e. we will take a zeroth-order reaction, and examine the concept of a 'critical' temperature for combustion. We follow an approach due to Semenov [3].

A gaseous mixture of fuel and oxidizer is placed in a closed vessel of fixed volume. Regard the vessel as a perfect conductor of heat with negligible thickness. The properties of the gaseous mixture are uniform in the vessel. Of course, as the vessel becomes large in size, motion of the fluid could lead to nonuniformities of the properties. We will ignore these effects. Assuming constant and equal specific heats, we adopt Equation (3.45) for a control volume enclosing the constant volume, fixed mass vessel (Figure 4.2):

$$mc_v \frac{dT}{dt} = \dot{Q} + (\dot{m}'''_{F,r})(V)(\Delta h_c) \tag{4.2}$$

The heat added term is given in terms of a convective heat transfer coefficient, h, and surface area, S,

$$\dot{Q} = hS(T_\infty - T) \tag{4.3}$$

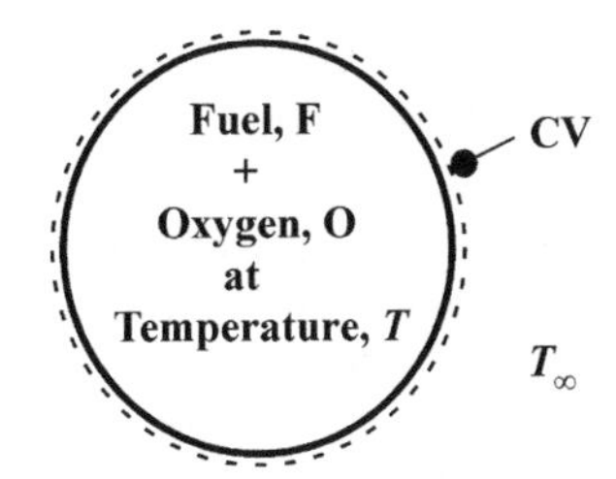

Figure 4.2 Autoignition in a closed vessel

Actually, we expect $T > T_\infty$ due to the release of chemical energy, so this is a heat loss. We write this loss as

$$\dot{Q}_\mathrm{L} = hS(T - T_\infty) \tag{4.4}$$

The chemical energy release rate is written as

$$\dot{Q}_\mathrm{R} = \dot{m}'''_\mathrm{F,r} V \, \Delta h_\mathrm{c} \tag{4.5}$$

Then we can write

$$mc_v \frac{\mathrm{d}T}{\mathrm{d}t} = \dot{Q}_\mathrm{R} - \dot{Q}_\mathrm{L} \tag{4.6}$$

Figure 4.3 tries to show the behavior of these terms in which $\dot{Q}_\mathrm{R}$ is only approximately represented for the Arrhenius dependence in temperature. For a given fuel and its associated kinetics, $\dot{Q}_\mathrm{R}$ is a unique function of temperature. However, the heat loss term depends on the surface area of the vessel. In Figure 4.3, we see the curves for increasing

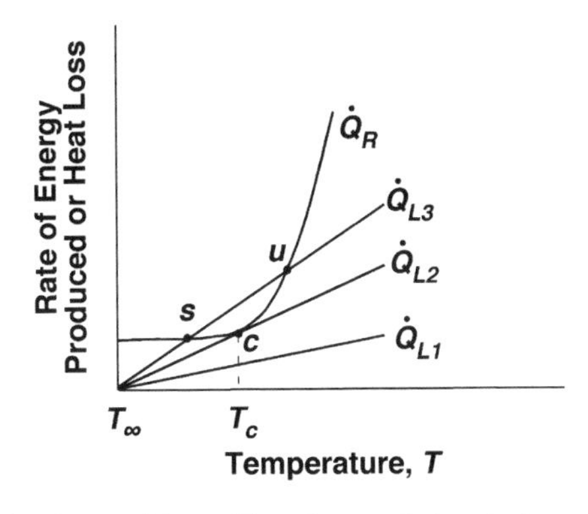

Figure 4.3 Competition of heat loss and chemical energy release

size vessels (S_1, S_2, S_3) for a constant h. Where we have no intersection of $\dot{Q}_L$ and $\dot{Q}_R$, $\dot{Q}_R > \dot{Q}_L$ and by Equation (4.6) $dT/dt > 0$. This means that the temperature will continue to rise at an increasing rate by the difference between the curves. This is usually called thermal runaway or a thermal explosion. On the other hand, when the curves intersect, we must have $dT/dt = 0$ by Equation (4.6) or a steady state solution. Two intersections are possible for the zeroth-order reaction. The one labeled s is stable, while that labeled u is unstable. This can be seen by imposing small changes in temperature about u or s. For example, a small positive temperature perturbation at s gives $\dot{Q}_{L_3} > \dot{Q}_R$ so by Equation (4.6) $dT/dt < 0$, and the imposed temperature will be forced back to that of s. Likewise small changes of temperature about point u will either cause a runaway $(+\Delta T)$ or a drop to s $(-\Delta T)$.

The critical condition is given by point c with T_c being the critical temperature. It should be noted that the critical heat loss rate, $\dot{Q}_{L_2}$, depends not just on the vessel size but also on the surrounding temperature, T_∞. Hence, the slope and the T intersection of $\dot{Q}_L$ in Equation (4.4), as tangent to $\dot{Q}_R$, will give a different T_c. We shall see that the Arrhenius character of the reaction rate will lead to $T_c \approx T_\infty$. This is called the autoignition temperature. The mathematical analysis is due to Semenov [3].

An approximate mathematical analysis is considered to estimate the functional dependence of the gas properties and vessel size on the autoignition temperature. We anticipate that T_c will be close to T_∞, so we write

$$\begin{aligned} e^{-E/(RT)} &= \exp\left[\left(\frac{-E}{RT_\infty}\right)\left(\frac{T_\infty}{T - T_\infty + T_\infty}\right)\right] \\ &= \exp\left[\left(\frac{-E}{RT_\infty}\right)\left(\frac{T - T_\infty}{T_\infty} + 1\right)^{-1}\right] \end{aligned} \tag{4.7}$$

Since $(T - T_\infty)/T_\infty$ is expected to be small, we approximate

$$\left(\frac{T - T_\infty}{T_\infty} + 1\right)^{-1} \approx 1 - \frac{T - T_\infty}{T_\infty} \tag{4.8}$$

where these are the first two terms of an infinite power series. Then, defining

$$\theta \equiv \left(\frac{T - T_\infty}{T_\infty}\right)\left(\frac{E}{RT_\infty}\right) \tag{4.9}$$

$$e^{-E/(RT)} \approx e^{\theta} e^{-E/(RT_\infty)} \tag{4.10}$$

from Equations (4.7) to (4.9). Since θ is dimensionless, it is useful to make Equation (4.2) dimensionless. By examining the dimensions of

$$\rho V c_v \frac{dT}{dt} \sim hS(T - T_\infty)$$

a reference time can be defined as

$$t_c = \frac{\rho V c_v}{hS} \tag{4.11}$$

Physically, t_c is proportional to the time required for the gases in the vessel to lose their stored energy to the surroundings by convection. Let us define a dimensionless time as

$$\tau = \frac{t}{t_c} \tag{4.12}$$

Now Equation (4.2) can be written with the approximation for $e^{-E/(RT)}$ as

$$\frac{d\theta}{d\tau} = \delta e^{\theta} - \theta \tag{4.13}$$

where

$$\boxed{\delta = \left(\frac{\Delta h_c V A e^{-E/(RT_\infty)}}{hST_\infty}\right)\left(\frac{E}{RT_\infty}\right)} \tag{4.14}$$

is a dimensionless quantity. The equation can be solved with $\theta = 0$ at $\tau = 0$ to give the temperature rise for a given δ. The increase in θ from its initial value is directly related to δ, a Damkohlar number.

We are only interested here in whether combustion will occur. From Figure 4.3, we need to examine point c. In terms of Equation (4.13), we have the dimensionless forms of $\dot{Q}_R$ and $\dot{Q}_L$ on the right-hand side. We plot two quantities, θ and e^{θ}, against θ in Figure 4.4. The critical condition c is denoted by the tangency of the two quantities. Mathematically, this requires that each quantity and their slopes must be equal at c. Hence,

$$\delta_c e^{\theta_c} = \theta_c \tag{4.15}$$

and

$$\delta_c e^{\theta_c} = 1 \tag{4.16}$$

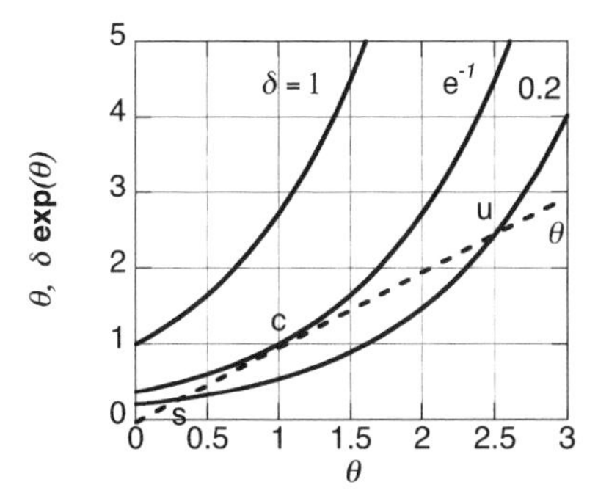

Figure 4.4 Conditions for autoignition

Dividing Equation (4.15) by Equation (4.16) gives

$$\theta_c = 1 \tag{4.17}$$

and then it follows that

$$\delta_c = e^{-1} \tag{4.18}$$

For a given vessel size, Equations (4.14) and (4.18) give the minimum value of T_∞ to cause autoignition. This value, $T_{\infty,c}$, is the minimum temperature needed, and is called the autoignition temperature. The dependence on the chemical parameters, A, E and Δh_c, as well as size are clearly contained in the parameter, δ. Also, from Equations (4.9) and (4.17),

$$\frac{T_c - T_\infty}{T_\infty} = \frac{RT_\infty}{E} \tag{4.19}$$

which according to Table 4.1 suggests a value for T_c of roughly 1/30 above a T_∞ of 600 K, i.e.

$$\frac{T_c - T_\infty}{T_\infty} = \frac{(8.314\,\text{J/mole K})(600\,\text{K})}{(160 \times 10^3\,\text{J/mole})} = \frac{1}{32.1} \tag{4.20}$$

or $T_c = 618.7\,\text{K}$.

As we can see, the value T_c is very close to the value $T_{\infty,c}$ needed to initiate the combustion. Practical devices, representative of the system in Figure 4.2, can regulate T_∞ until ignition is recognized, e.g. ASTM E 659 [4]. Table 4.3 gives some typical measured autoignition temperatures (AIT). There is a tendency for the AIT to drop as the pressure is increased.

Table 4.3 Autoignition temperatures (at 25 °C and 1 atm, usually at stoichiometric conditions in air) (from Zabetakis [5])

Fuel	AIT (°C)
Methane, CH_4	540
Ethane, C_2H_6	515
Propane, C_3H_8	450
n-Hexane, C_6H_{14}	225
n-Heptane, C_7H_{16}	215
Methanol, CH_3OH	385
Ethanol, C_2H_5OH	365
Kerosene	210
Gasoline	~450
JP-4	240
Hydrogen, H_2	400

4.4 Piloted Ignition

Often combustion is initiated in a mixture of fuel and oxidizer by a localized source of energy. This source might be an electric arc (or spark) (moving charged particles in a fluid or a plasma) or a small flame itself. Because the spark or small flame would locally raise the temperature of the mixture (as T_∞ did in the autoignition case), this case is defined as piloted ignition. The bulk of the mixture remains at T_∞, well below the AIT.

To illustrate the process of piloted ignition, we consider a cylindrical spark of radius, r, and length, ℓ, discharged in a stoichiometric mixture of fuel and air. A control volume enclosing the spark is shown in Figure 4.5 where it can expand on heating to match the ambient pressure, p_∞. The governing energy equation for this isobaric system is taken from Equation (3.45), where $p = p_\infty$, $\dot{m}_j = 0$ and

$$\rho V c_p \frac{\mathrm{d}T}{\mathrm{d}t} = \dot{m}_{\mathrm{F}}''' \Delta h_{\mathrm{c}} V + \dot{q}''' V - hS(T - T_\infty) \tag{4.21}$$

where $\dot{q}'''$ is the spark rate of energy supplied per unit volume and h is an effective conductance, k/r, with k as the mixture thermal conductivity. This is an approximate analysis as in Equations (4.2) to (4.5), except that we have added the spark energy and we approximate a conduction rather than a convective heat loss. By introducing the same dimensionless variables, it can be shown that

$$\frac{\mathrm{d}\theta}{\mathrm{d}\tau} = \delta\, \mathrm{e}^{\theta} - \theta + \left(\frac{\dot{q}'''\, V}{hST_\infty}\right)\left(\frac{E}{RT}\right) \tag{4.22}$$

where here $t_{\mathrm{c}} = \rho V c_p/(hS) = r^2/(2\alpha)$ with α, the thermal diffusivity, $k/\rho c_p$, and $V = \pi r^2 \ell$ and $S = 2\pi r \ell$ as approximations. By examining the definition of δ, Equation (4.14), it can be recognized as the ratio of two times

$$\delta = \frac{t_{\mathrm{c}}}{t_{\mathrm{R}}} \tag{4.23}$$

where t_{c} is the time needed for the conduction loss, or thermal diffusion time, and t_{R} is the chemical reaction time:

$$\boxed{t_{\mathrm{R}} \equiv \frac{\rho c_p T_\infty}{[E/(RT_\infty)]\Delta h_{\mathrm{c}} A\, \mathrm{e}^{-E/(RT_\infty)}}} \tag{4.24}$$

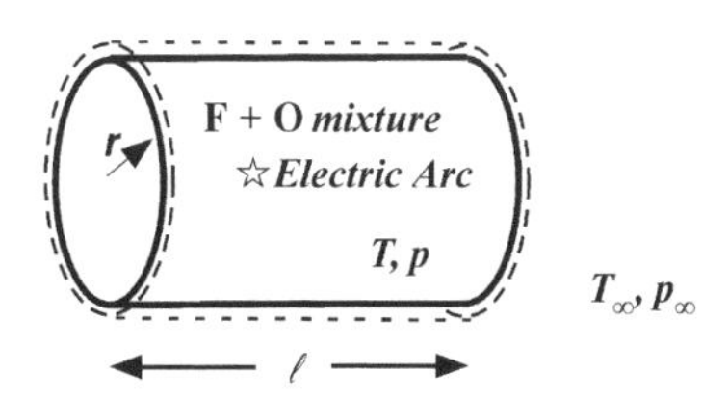

Figure 4.5 Spark ignition

Table 4.4 Minimum spark energy to ignite stoichiometric mixtures in air (at 25 °C, 1 atm) (from various sources)

Fuel	Energy (μJ)
Methane	300
n-Hexane	290
Hydrogen	17

The critical solution follows as in Equations (4.15) to (4.18):

$$\theta_c = 1 + \left(\frac{\dot{q}''' V}{hST_\infty}\right)\left(\frac{E}{RT}\right) = 1 + \left(\frac{\dot{q}''' r^2}{2kT_\infty}\right)\left(\frac{E}{RT}\right) \tag{4.25a}$$

and

$$\delta_c = e^{-\theta_c} = e^{-1} e^{-(\dot{q}''' r^2/(2kT_\infty))(E/(RT))} \tag{4.25b}$$

Hence the minimum value of δ to cause ignition is smaller for a system under the heating of a 'pilot' compared to autoignition, e.g. Equation (4.18).

Analogous to Figure 4.4, the point of tangency, c, corresponds to a minimum spark energy. This has usually been measured in terms of energy $(\int \dot{q}''' V \, dt)$ for the spark duration. Table 4.4 gives some typical spark energies. While the minimum electric spark energy is generally found to range from 10 to 1000 μJ, the power density of a spark in these experiments is greater than 1 MW/cm^3. Figure 4.6 shows the minimum spark energy needed to ignite methane over a range of flammable mixtures. Within the

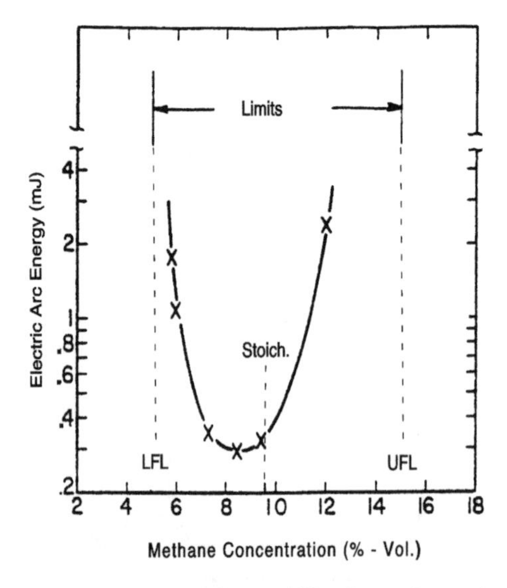

Figure 4.6 Ignitability curve and limits of flammability for methane-air mixtures at atmospheric pressure and 26 °C (taken from Zabetakis [5])

U-shaped curve, we have mixtures that can be ignited for a sufficiently high spark energy. From Equation (4.25) and the dependence of the kinetics on both temperatures and reactant concentrations, it is possible to see why the experimental curve may have this shape. The lowest spark energy occurs near the stoichiometric mixture of $X_{CH_4} = 9.5\,\%$. In principle, it should be possible to use Equation (4.25) and data from Table 4.1 to compute these ignitability limits, but the complexities of temperature gradients and induced flows due to buoyancy tend to make such analysis only qualitative. From the theory described, it is possible to illustrate the process as a quasi-steady state $(\mathrm{d}T/\mathrm{d}t = 0)$. From Equation (4.21) the energy release term represented as

$$\dot{Q}_{\mathrm{R}} = \dot{m}_{\mathrm{F}}''' \Delta h_{\mathrm{c}} V \tag{4.26}$$

and the net loss term as

$$\dot{Q}_{\mathrm{L}} = hS(T - T_{\infty}) - \dot{q}''' V \tag{4.27}$$

can be sketched as a function of the reacting mixture temperature. In this portrayal, we are ignoring spatial gradients except for their effect in the heat loss part of $\dot{Q}_{\mathrm{L}}$. As shown in Figure 4.7, $\dot{Q}_{\mathrm{R}}$ will first sharply increase due to the Arrhenius temperature dependence, $\mathrm{e}^{-E/(RT)}$, but later it will diminish due to the reduction of the reactants. The net loss term can be approximated as linear in T, with T_{∞} representative of an ambient starting

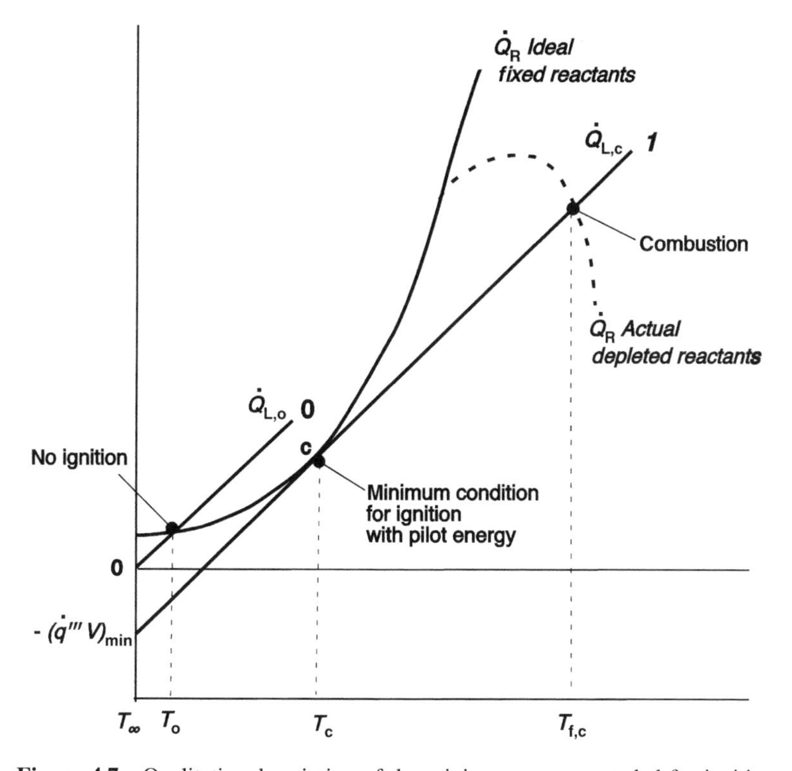

Figure 4.7 Qualitative description of the minimum energy needed for ignition

temperature (e.g. 26 °C and 1 atm as in Figure 4.6). The net loss curve labeled 1 has a stable intersection corresponding to T_f, a flame temperature. For this protrayal, the reactants would have to be continuously supplied at a concentration corresponding to this intersection. If they were not replenished, the fire would go out as $\dot{Q}_R$ goes to zero. Curve 1 corresponds to an ignition (spark) energy greater than the minimum value. Curve 1 gives the critical ignition result where two solution temperatures, T_c and $T_{f,c}$ are possible. In the dynamic problem dT/dt is always greater than zero because $\dot{Q}_R > \dot{Q}_L$ at the T_c neighborhood. Hence, the steady physically plausible solution will be the fire condition at $T_{f,c}$. Any lower value of $(\dot{q}'''V)$ will result in an intersection corresponding to T_0, a very small and imperceptible difference from T_∞ (see curve 0). This intersection is stable and not a state of combustion. Here, it should be emphasized that the coordinate scales in Figure 4.7 are nonlinear (e.g. logarithmic), and quantitative results can only be seen by making computations.

In dimensionless terms, there is a critical value for δ (Damkohler number) that makes ignition possible. From Equation (4.23), this qualitatively means that the reaction time must be smaller than the time needed for the diffusion of heat. The pulse of the spark energy must at least be longer than the reaction time. Also, the time for autoignition at a given temperature T_∞ is directly related to the reaction time according to Semenov (as reported in Reference [5]) by

$$\log(t_{\text{auto}}) = \kappa\left(\frac{E}{RT_\infty}\right) + b \tag{4.28}$$

where $\kappa/R = 0.22$ cal/mole K and b is an empirical constant for the specific mixture. Theoretically, b would have some small dependency on T_∞, i.e. $\log(T_\infty)$, as would follow from Equation (4.24). Thus, $t_{\text{auto}} \approx t_R$ and $t_{\text{spark}} \geq t_R$ for piloted ignition.

4.5 Flame Speed, S_u

Flammability is defined for a mixture of fuel and oxidizer when a sustained propagation occurs after ignition. This result for a given mixture depends on temperature, pressure, heat losses and flow effects, namely due to gravity. The speed of this propagation also depends on whether the flow remains laminar or becomes turbulent, and whether it exceeds the speed of sound. For speeds below the speed of sound for the medium, the process is called a deflagration. The pressure rise across the burning region is relatively low for a deflagration. However, for a detonation – speeds greater than the speed of sound – the pressure rise is considerable. This is the result of a shock wave that must develop before the combustion region in order to accommodate the conservation laws. We will not discuss detonations in any detail. However, we will discuss the initiation of a deflagration process in a gas mixture that can ideally be regarded to occur without heat losses. This ideal flame speed is called the burning velocity and is designated by S_u. Specifically, S_u is defined as the normal speed of an adiabatic plane (laminar) combustion region measured with respect to the unburned gas mixture. S_u, in this idealization, can be treated as a property of the fuel–oxidizer mixture, but it can only be approximately measured. Typical measured values of S_u are shown in Figure 4.8 which peak for a slightly fuel-rich mixture. Typical maximum values of S_u (25 °C, 1 atm) in air

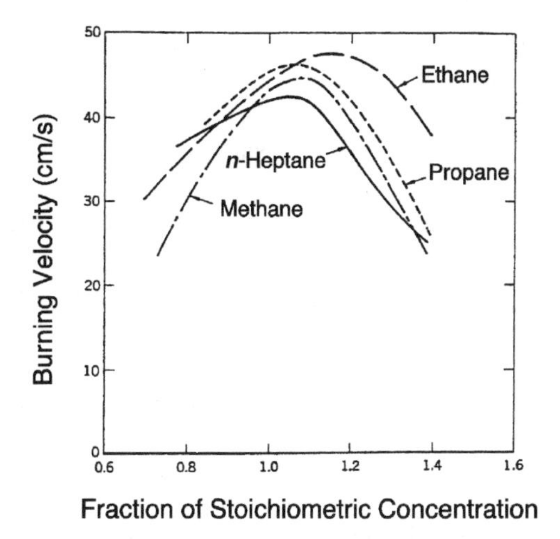

Figure 4.8 Typical burning velocities (taken from Zabetakis [5])

are slightly less than 1 m/s with acetylene at 1.55 m/s and hydrogen at 2.9 m/s (which actually peaks at an equivalence ratio of 1.8). In contrast detonation speeds can range from 1.5 to 2.8 km/s. Turbulence and pressure effects can accelerate a flame to a detonation.

4.5.1 Measurement techniques

Figures 4.9(a) and (b) show two measurement strategies for measuring S_u. One is a Bunsen burner and the other is a burner described by Botha and Spalding [6]. In the

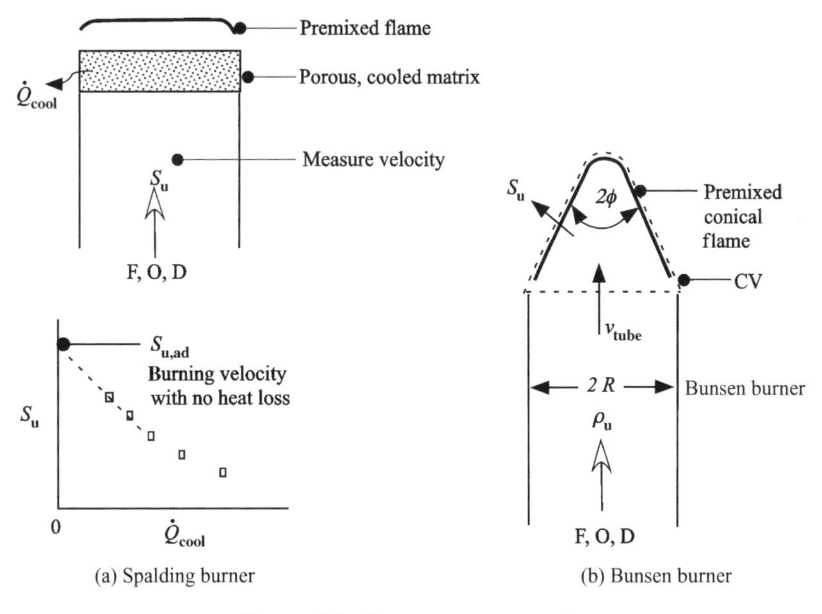

Figure 4.9 Burners to measure S_u

Bunsen burner the conical flame angle (2ϕ) and the mean unburned mixture velocity (v_u) in the tube can approximately give S_u:

$$S_u = v_u \sin\phi \tag{4.29}$$

In the Spalding burner, a flat plane flame stands off the surface of a cooled porous matrix. Data are compiled to give S_u as measured by the speed in the burner supply, to maintain a stable flame, for a given measured cooling rate. By plotting these data so as to extrapolate to a zero cooling condition yields S_u under nearly adiabatic conditions.

In both of these experimental arrangements, for a given mixture, there is a unique duct velocity (v_u) that matches the burning velocity. In the Spalding burner, this is the adiabatic burning velocity (or the true S_u). If $v_u > S_u$ the condition is not stable and the flame will blow off or move away from the exit of the duct until a reduced upstream velocity matches S_u. If $v_u < S_u$, the flame will propagate into the duct at a speed where the flame velocity is $S_u - v_u$. This phenomenon of upstream propagation is known as flashback. It is not a desirable effect since the flame may propagate to a larger chamber of flammable gases and the larger energy release could result in a destructive pressure rise. An appropriate safety design must avoid such a result.

4.5.2 *Approximate theory*

The ability to predict S_u is limited by the same factors used to predict the autoignition or flammability limits. However, an approximate analysis first considered by Mallard and Le Chatelier in 1883 [7] can be useful for quantitative estimates.

Consider a planar combustion region in an adiabatic duct that is fixed in space and is steadily supplied with a mixture of fuel (F), oxidizer (O) and diluent (D) at the velocity S_u. This condition defines S_u and is depicted in Figure 4.10. The process is divided into two stages:

I. A preheat region in which the heat transfer from the flame brings the unburned mixture to its critical temperature for ignition, T_{ig}. This is much like what occurred in describing auto and piloted ignition, except that the the heat is supplied from the flame itself.

II. The second stage is the region where a significant chemical energy is released. This is perceived as a flame.

The boundary between these two regions is not sharp, but it can be recognized.

We apply the conservation laws to two control volumes enclosing these regions. Since there is no change in area, conservation of mass (Equation (3.15)) gives, for the unburned mixture (u) and burned product (b),

$$\rho_u S_u = \rho_b v_b = \dot{m}'' \tag{4.30}$$

or the mass flux, where $\dot{m}''$ is constant.

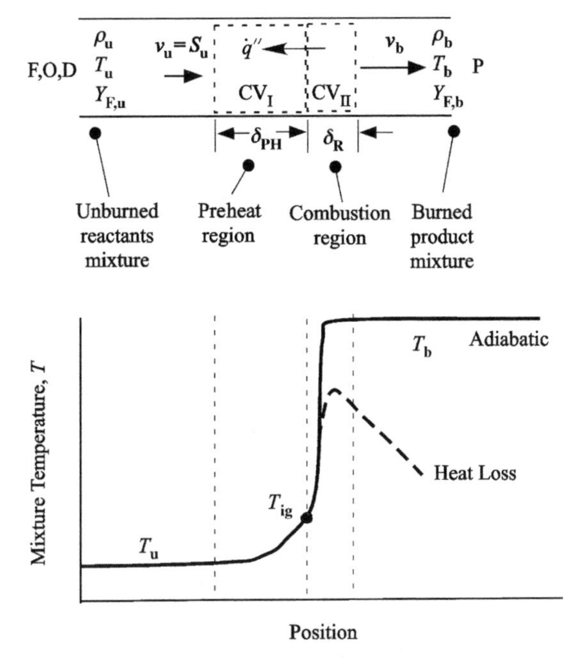

Figure 4.10 Burning velocity theory

The conservation of energy applied to CV_I requires a knowledge of the 'ignition' temperature, T_{ig} and the heat transferred to the preheat region, $\dot{q}''$. We assume that radiation effects are negligible and approximate this heat flux as

$$\dot{q}'' = -k\frac{dT}{dx} \approx k\left(\frac{T_b - T_{ig}}{\delta_R}\right) \tag{4.31}$$

We are only seeking an approximate theory since a more precise simple analytical result is not possible. We only seek an order of magnitude estimation and insight into the important variables. With this approximation the energy equation for CV_I is

$$\rho_u S_u c_p (T_{ig} - T_u) \approx k\left(\frac{T_b - T_{ig}}{\delta_R}\right) \tag{4.32}$$

Regarding properties as constants, evaluated at the unburned mixture condition, Equations (4.30) and (4.32) combine to give

$$\boxed{S_u \approx \left(\frac{T_b - T_{ig}}{T_{ig} - T_u}\right)\frac{\alpha}{\delta_R}} \tag{4.33}$$

where the thermal diffusivity, α, is $k/\rho c_p$. As an order of magnitude estimate, we can regard T_{ig} as a typical autoignition temperature $\sim$525 °C, $T_u \sim 25$ °C and T_b as an adiabatic flame temperature ~2000 °C. Therefore,

$$\frac{T_b - T_{ig}}{T_{ig} - T_u} \approx \frac{1500}{500} = 3$$

However, S_u is typically 0.5 m/s and α is typically $2 \times 10^{-5}\,\mathrm{m^2/s}$. Hence, the thickness of the flame is estimated as

$$\delta_R \sim (3)\frac{(2 \times 10^{-5}\,\mathrm{m^2/s})}{(0.5\,\mathrm{m/s})} = 12 \times 10^{-5}\,\mathrm{m} \sim 10^{-4}\,\mathrm{m}$$

or the flame is roughly of the order of 0.1 mm thick. This would be consistent with the laminar flame of a Bunsen burner or oxyacetylene torch.

To gain further insight into the effect of kinetics on the flame speed, we write the energy equation for the control volume $\mathrm{CV_{II}}$ as

$$\rho_u S_u c_p (T_b - T_{ig}) = \dot{m}'''_F \Delta h_c \delta_R - \dot{q}'' \tag{4.34}$$

We can make the same approximation for $\dot{q}''$, as in Equation (4.31), and substitute for δ_R using Equation (4.33). It can be shown that

$$S_u = \left\{ \frac{[(T_b - T_{ig})/(T_{ig} - T_u)]\alpha \dot{m}'''_F \Delta h_c}{\rho_u c_p (T_b - T_u)} \right\}^{1/2} \tag{4.35}$$

As before, since T_{ig} is not precisely known, we approximate

$$\frac{T_b - T_{ig}}{T_{ig} - T_u} \approx \frac{1500}{500} = 3$$

and substitute to estimate the flame speed as

$$\boxed{S_u \approx \left[\frac{3\alpha \Delta h_c \dot{m}'''_F}{\rho_u c_p (T_b - T_u)} \right]^{1/2}} \tag{4.36}$$

The constant 3 is an order of magnitude estimate which depends on the approximate analysis for the model of Figure (4.10).* In addition, the burning rate $\dot{m}'''_F$ and the properties α and c_p should be evaluated at some appropriate mean temperature. For example, in Equation (4.34), the more correct expression for $\dot{m}'''_F \delta_R$ is

$$\int_0^{\delta_R} \dot{m}'''_F \,\mathrm{d}x = \int_{T_{ig}}^{T_b} \frac{\dot{m}'''_F \,\mathrm{d}T}{\mathrm{d}T/\mathrm{d}x}$$

*An asymptotic analysis for a large activation energy or small reaction zone obtains 2 instead of 3 for the constant [8].

or

$$\dot{m}_F''' \delta_R \approx \frac{\int_{T_{ig}}^{T_b} \dot{m}_F''' \, dT}{T_b - T_{ig}} \delta_R$$

4.5.3 Fuel lean results

For our estimations and the adiabatic control volume in Figure 4.10, T_b should be the adiabatic flame temperature. Consider a fuel-lean case in which no excess fuel leaves the control volume. All the fuel is burned. Then by the conservation of species,

$$\rho_u S_u Y_{F,u} = \dot{m}_F''' \delta_R \tag{4.37}$$

In addition, conservation of energy for the entire control volume (I + II) gives

$$\rho_u S_u c_p (T_b - T_u) = \dot{m}_F''' \delta_R \Delta h_c \tag{4.38}$$

Combining these equations gives

$$c_p (T_b - T_u) = Y_{F,u} \Delta h_c \tag{4.39}$$

for the fuel-lean adiabatic case. Then Equation (4.36) can be written as

$$\boxed{S_u = \left(\frac{3\alpha \dot{m}_F'''}{\rho_u Y_{F,u}} \right)^{1/2}} \tag{4.40}$$

where this applies to fuel-lean conditions and T_b is the adiabatic flame temperature given by Equation (4.39). These results show that the flame speed depends on kinetics and thermal diffusion. Any turbulent effects will enhance the thermal diffusion and the surface area so as to increase the flame speed. Any heat losses will decrease the temperature and therefore the burning rate and flame speed as well. Indeed, if the heat losses are sufficient, a solution to Equation (4.38) may not be possible. Let us investigate this possibility.

4.5.4 Heat loss effects and extinction

Consider the entire control volume in Figure 4.10, but now heat is lost to the surroundings at the temperature of the unburned mixture, T_u. Such an analysis was described by Meyer [9]. The conservation of energy for the entire control volume is

$$\rho_u S_u c_p (T_b - T_u) = \dot{m}_F''' \delta_R \Delta h_c - \dot{q}'' \tag{4.41}$$

Consider the fuel-lean case, in which Equation (4.37) applies. Substituting gives

$$\dot{m}_F''' \delta_R \left[\Delta h_c - \frac{c_p (T_b - T_u)}{Y_{F,u}} \right] = \dot{q}'' \tag{4.42}$$

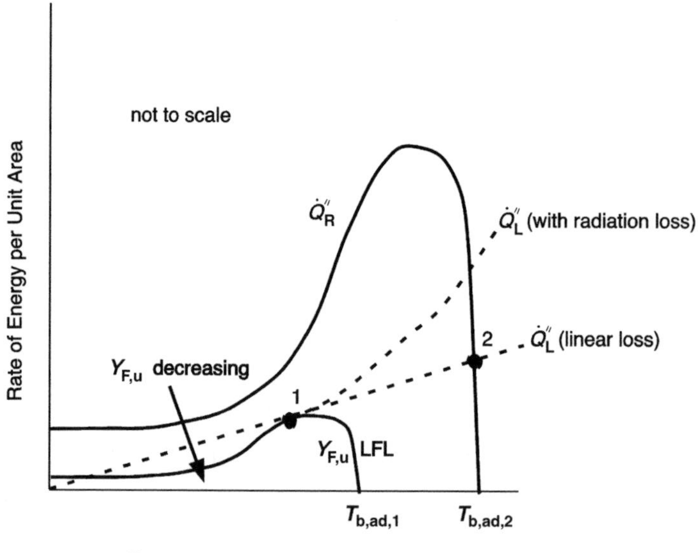

Figure 4.11 Solution for steady flame speed

Note that when $\dot{q}'' = 0$, T_b is the adiabatic flame temperature. We can regard this equation as a balance between the net energy released and energy lost:

$$\dot{Q}_R'' = \dot{m}_F''' \delta_R \left[\Delta h_c - \frac{c_p(T_b - T_u)}{Y_{F,u}} \right] \tag{4.43a}$$

and

$$\dot{Q}_L'' \approx \frac{k(T_b - T_u)}{\delta_R} \quad \text{or} \quad \sim (T_b - T_u) \tag{4.43b}$$

which approximates the heat loss from a linear decrease in the product temperature.* The flame thickness can be approximated from Equation (4.33) as

$$\delta_R \approx \frac{3\alpha}{S_u} \tag{4.44}$$

and S_u taken from Equation (4.36). Then the steady flame temperature can be formed by equating Equations (4.43a) and (4.43b).

Figure 4.11 contains a qualitative sketch of the $\dot{Q}''_R$ and $\dot{Q}''_L$ curves as a function of T_b. In the fuel-lean range, $Y_{F,u}$ is greatest at the stoichiometric mixture. For such a case, an intersection of the curves gives a steady propagation solution, depicted as 1. The

*The heat losss from the combustion products in Figure 4.10 is due to conduction and radiation.

temperature is below the adiabatic flame temperature for the stoichiometric mixture. As $Y_{F,u}$ is decreased, a point of tangency will occur for some value. This is the lowest fuel concentration for which a steady propagation is possible. Any further decrease in $Y_{F,u}$ will not allow propagation even if ignition is achieved. The critical tangency condition value for this fuel concentration is called the lower flammable limit (LFL). This condition is depicted as 2 in Figure 4.11, and the intersection with the zero energy horizontal axis is the adiabatic flame temperature at the lower limit. This analysis is analogous to the autoignition problem defined by Equation (4.5), in which here

$$m''c_p\frac{\mathrm{d}T}{\mathrm{d}t} = \dot{Q}''_{\mathrm{R}} - \dot{Q}''_{\mathrm{L}} \tag{4.45}$$

and at any time $\dot{Q}''_{\mathrm{R}} < \dot{Q}''_{\mathrm{L}}$ and the temperature decreases. At the lower limit, any perturbation to increase heat losses will lead to extinction. Hence, we see that there is a lower limit concentration, on the lean side, below which steady propagation is not possible. As the fuel concentration is increased above stoichiometric on the rich side, the incompleteness of combustion will reduce Δh_c. Hence this will decrease T and the burning rate, causing a similar effect and leading to extinction. This critical fuel-rich concentration is called the upper flammable limit (UFL).

4.6 Quenching Diameter

In the previous analysis, the heat loss was taken to be from the flame to the surrounding gas. If a solid is present, it will introduce some heat loss. This will be examined in terms of flame propagation in a duct. For a large-diameter duct, a moving flame may only be affected at its edges where heat loss can occur to the duct wall. This affected region is called the quenching distance. For a small-diameter duct, the duct diameter that will not allow a flame to propagate is called the quenching diameter, D_{Q}.

The process of steady flame propagation into a premixed system is depicted in Figure 4.12 for a moving control volume bounding the combustion region δ_{R}. The heat loss in this case is only considered to the duct wall. With h as the convection heat transfer coefficient, the loss rate can be written as

$$\dot{Q}_{\mathrm{L}} = h\pi D\delta_{\mathrm{R}}(T_{\mathrm{b}} - T_{\mathrm{w}}) \tag{4.46}$$

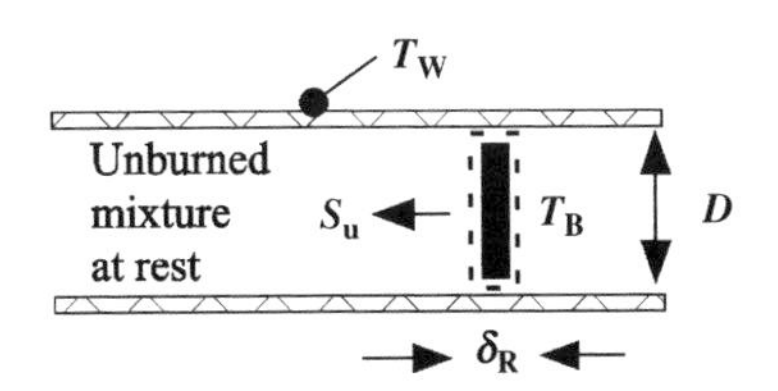

Figure 4.12 Quenching distance analysis

where T_b is approximated as uniform over δ_R and T_w is the wall temperature. For fully developed laminar flow in a duct of diameter D, the Nusselt number is a constant of

$$\frac{hD}{k} = 3.66 \tag{4.47}$$

according to the heat transfer literature [10]. (Note that in fully developed flow the velocity profile is parabolic and zero at the wall, so viscous effects will also have a bearing on the flame speed. We are ignoring this profile effect in our one-dimensional model.)

The energy balance for this problem can be expressed by Equation (4.45) with $\dot{Q}''_R$ given in Equation (4.43a) as before and $\dot{Q}''_L$ given from Equations (4.47) and (4.48) as

$$\begin{aligned}\dot{Q}''_L &= \left(\frac{3.66\,k}{D}\right)\left(\frac{\pi D \delta_R}{\pi D^2/4}\right)(T_b - T_w) \\ &= 14.64\,k\frac{\delta_R(T_b - T_w)}{D^2}\end{aligned} \tag{4.48}$$

Again δ_R follows from Equation (4.44) and S_u from Equation (4.36). The equilibrium solutions for steady propagation are sketched in Figure 4.13. Only the loss curves depend on D. For $D > D_Q$, two intersections are possible at s and u. The former, s, is a stable solution since any perturbation in temperature, T_b, will cause the state to revert to s. For example, if $T_b > T_{b,s}$, $\dot{Q}''_L > \dot{Q}''_R$ and $\mathrm{d}T/\mathrm{d}t < 0$ from Equation (4.45). Then T_b will decrease until $T_{b,s}$ is reached. The intersection at u can be shown to be unstable, and therefore physically unrealistic. The point Q corresponds to the smallest diameter that will still give a stable solution. For $D < D_Q$, $\mathrm{d}T/\mathrm{d}t < 0$ always, and any ignition will not

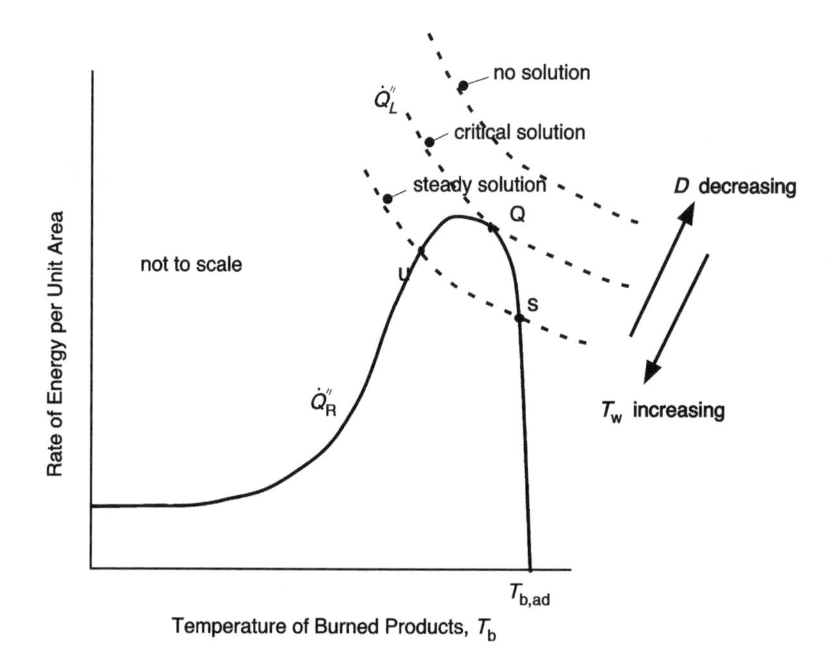

Figure 4.13 Quenching distance solution

Table 4.5 Flame speed and quenching diameter

Fuel	S_u, Maximum (in air) (m/s)	D_Q, Stoichiometric (in air) (mm)
Hydrogen	2.9	0.64
Acetylene	1.5	2.3
Methane	0.52	2.5

lead to an appropriate temperature for propagation. Any increase in wall temperature, $T_w > T_u$, will have a tendency to enable propagation.

By equating $\dot{Q}''_R$ and $\dot{Q}''_L$, with the following approximations a relationship between S_u and D_Q is suggested. From the sketch in Figure 4.13, the positive term in $\dot{Q}''_R$ is more dominant near the solution, Q, so let $T_b \approx T_{f,ad}$. Therefore

$$\dot{Q}''_R \approx \dot{m}'''_F \delta_R \Delta h_c$$

and for $T_w = T_u$,

$$\dot{Q}''_L \approx 14.64\, k \frac{(T_{f,ad} - T_u)\delta_R}{D_Q^2}$$

Then

$$D_Q \approx \left[\frac{14.64\, k(T_{f,ad} - T_u)}{\dot{m}'''_F \Delta h_c} \right]^{1/2} \tag{4.49}$$

From Equation (4.36),

$$D_Q \approx \frac{6.6\, \alpha}{S_u} \tag{4.50}$$

This seems to qualitatively follow data in the literature, as illustrated in Table 4.5.

The quenching distance is of practical importance in preventing flashback by means of a flame arrestor. This safety device is simply a porous matrix whose pores are below the quenching distance in size. However, care must be taken to maintain the flame arrestor cool, since by Equation (4.49) an increase in its temperature will reduce the quenching distance.

The quenching distance is also related to the flame stand-off distance depicted in Figure 4.14. This is the closest distance that a premixed flame can come to a surface.

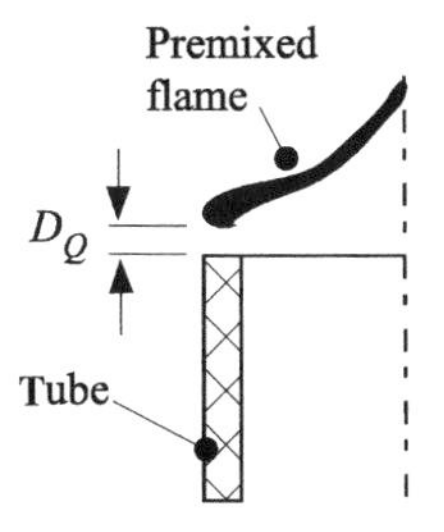

Figure 4.14 Stand-off distance

Even in a diffusion flame the region near a surface becomes a premixed flame since fuel and oxygen can come together in this 'quenched' region.

4.7 Flammability Limits

In Figure 4.6 there are two apparent vertical asymptotics where no ignitable mixture is possible regardless of the size of the ignition energy source. These limits given in concentration of fuel are defined as the lower (X_L) and upper (X_U) flammability limits. They are typically given in terms of molar (or volume) concentrations and depend for a given mixture (e.g. fuel in air or fuel in oxygen and an additive species) on temperature and pressure. This demarcation is also dependent on the nature of the test apparatus and conditions of the test in practice. Generally, data are reported from a standard test developed by the US Bureau of Mines [11], which consists of a spark igniter at the base of an open 2 inch diameter glass tube. Flammability is defined for a uniformly distributed mixture if a sustained flame propagates vertically up the tube. Outside the flammable regions, a spark may succeed in producing a flame, but by the definition of flammability, it will not continue to propagate in the mixture. Since flammability, as ignitability, can depend on flow and heat transfer effects, it cannot be considered to be a unique property of the mixture. It must be regarded as somewhat dependent on test conditions. Nevertheless, reported values for upper and lower flammability limits have a useful value in fire safety design considerations and can be used with some generality.

A useful diagram to show the effect of fuel phase and mixture temperature on flammability is shown in Figure 4.15. The region of autoignition is characterized as a combustion reaction throughout the mixture at those temperatures and concentrations indicated. Outside that temperature domain only piloted ignition is possible, and flammability is manifested by a propagating combustion reaction through the mixture. At temperatures low enough, the gaseous fuel in the mixture could exist as a liquid. The saturation states describe the gaseous fuel concentration in equilibrium with its liquid. In other words, this is the vapor concentration at the surface of the liquid fuel, say in air, at the given temperature. Recall that this saturation (or mixture partial pressure) is only a function of temperature for a pure substance, such as single-component liquid fuels. The lowest flammable temperature for a liquid fuel in air is known as the flashpoint (FP).

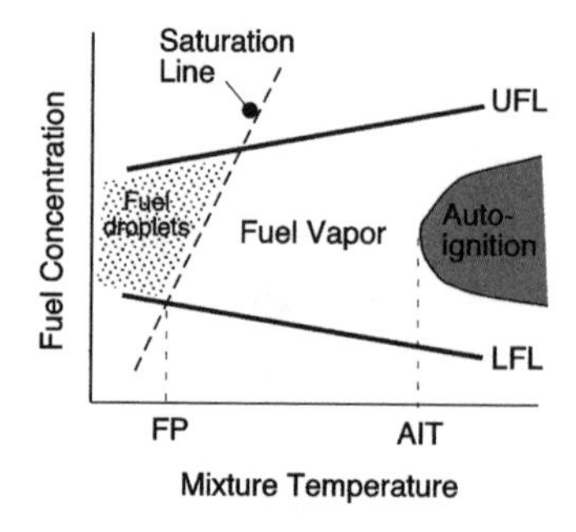

Figure 4.15 Effect of temperature on flammability for a given pressure (taken from Zabetakis [5])

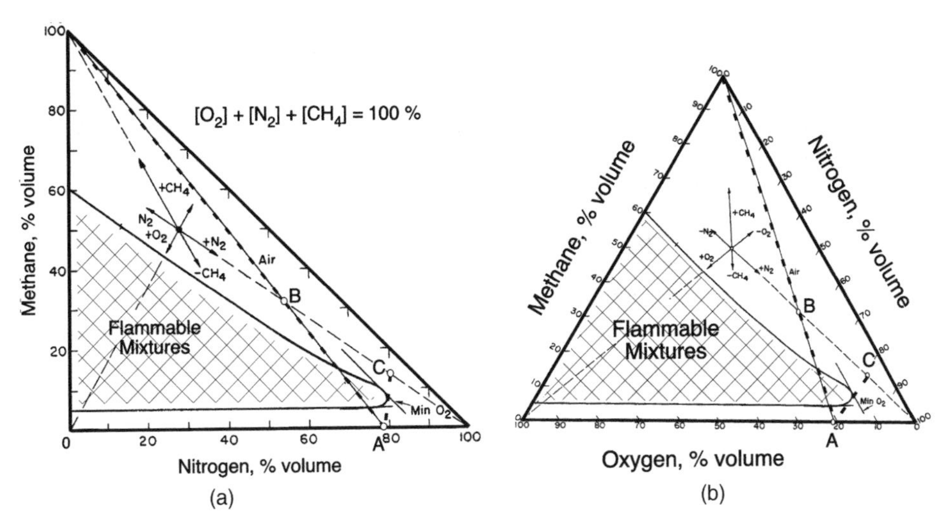

Figure 4.16 Flammability diagram for three-component systems at 26 °C and 1 atm (taken from Zabetakis [5])

However, if the fuel is in the form of droplets in air, it can be flammable at temperatures below the flashpoint. Droplet size will also play a role in this aerosol region.

Zabetakis [5] describes two types of flammability diagrams useful to analyze three component mixtures consisting of (1) fuel, (2) oxidizer and (3) inert (diluent) or fire retardant. For example, an inert species could be nitrogen or carbon dioxide which can serve to dilute the fuel upon its addition to the mixture. A fire retardant third body is, for example, carbon tetrafluoride (CF_4). A fire retardant dilutes as well as inhibits the combustion reaction by playing a chemical role. Either a rectangular or triangular format can be used to represent the flammability of such a three-component system. Figures 4.16(a) and (b) illustrate CH_4—O_2—N_2. The characteristics of each diagram are:

1. Each point must add to 100 % in molar concentration: $X_{CH_4} + X_{O_2} + X_{N_2} = 100\,\%$.
2. The concentration of each species at a state on the diagram is read parallel to its zero concentration locus.
3. A species is added by moving towards its vertex at 100 % concentration.
4. The air-line always has O_2 and N_2 concentrations equal to that of air.

Example 4.1 We take an example from *The SFPE Handbook* described by Beyler [12].

A methane leak fills a 200 m^3 room until the methane concentration is 30 % by volume. Calculate how much nitrogen must be added to the room before air can be allowed in the space.

Solution Assume the process takes place at constant temperature and pressure and Figure 4.16 applies. The constant pressure assumption requires that as the methane leaks into the room, the pressure is not increased, as would occur in a constant volume process, i.e. $p = mRT/V$. Hence, as air is added, the volume of the gases in the room must increase; therefore, these gases leak out of the room (B).

We further assume that the mixture in the room is well mixed. In general, this will not be true since methane is lighter than air and will then layer at the ceiling. Under this unmixed condition, different states of flammability will exist in the room. The mixed state is given by B in Figure 4.16(b):

$$\text{B}: \quad X_{CH_4} = 30\,\%, \quad X_{O_2} = 0.21(70)\,\%, \quad X_{N_2} = 0.79(70)\,\%$$

We add N_2 uniformly to the mixture, moving along the B–C line towards 100 % N_2. The least amount of N_2 before air can safely be added is found by locating state C just when the addition of air no longer intersects the flammable mixture region. This is shown by the process B → C and C → A in Figure 4.16(b):

$$\text{C}: \quad X_{CH_4} = 13\,\%, \quad X_{O_2} = 5\,\%, \quad X_{N_2} = 82\,\%$$

Hence state C is found, but conservation of mass for species N_2 must be used to find the quantity of N_2 added.

Consider the control volume defined by the dashed lines in C. Employ Equation (3.23):

$$m\frac{dY_{N_2}}{dt} + \dot{m}_{N_2}(Y_{N_2} - 1) = 0$$

where Y_{N_2} is the mass fraction of the N_2 in the mixture of the room, m is the mass of the room and $\dot{m}_{N_2}$ is the mass flow rate of N_2 added. There is no chemical reaction involving N_2 so the right-hand side of Equation (3.23) is zero. Also, diffusion of N_2 has been neglected in the inlet and exit streams because it is reasonable to assume negligible gradients of nitrogen concentration in the flow directions. The mass of the mixture may change during the process B → C. Expressing it in terms of density (ρ) and volume of the room ($V = 200\,\text{m}^3$),

$$m = \rho V$$

By the perfect gas law for the mixture:

$$\rho = \frac{p}{RT}M$$

or ρ depends only on the molecular weight of the mixture,

$$M = \sum_{i=1}^{3} X_i M_i$$

For state B:

$$M_B = (0.30)(16) + (0.147)(32) + (0.543)(28) = 24.7$$

and for state C:

$$M_C = (0.13)(16) + (0.05)(32) + (0.82)(28) = 26.6$$

Hence, it is reasonable to approximate the density as constant. With this approximation we can integrate the differential equation. Let $y = Y_{N_2} - 1$ and $Y_{N_2} = X_{N_2}(M_{N_2}/M)$. Then

$$\frac{dy}{dt} = -\frac{\dot{m}_{N_2}}{\rho V}y$$

where (B) $t = 0$, $Y_{N_2} = (0.543)(28/24.7) = 0.616$ and (C) t_{final}, $Y_{N_2} = (0.82)(28/26.6) = 0.863$. Thus

$$\ln\left(\frac{1-0.863}{1-0.616}\right) = \frac{-\int_0^{t_{final}} \dot{m}_{N_2}\,dt}{\rho V}$$

If we represent $\dot{m}_{N_2}$ as a volumetric flow rate of N_2, then the volume of nitrogen added (at $p = 1$ atm, T = 26 °C) is

$$\rho_{N_2} V_{N_2} = \int_0^{t_{final}} \dot{m}_{N_2}\,dt$$

or

$$V_{N_2} = \left(\frac{\rho}{\rho_{N_2}}\right) V \ln(2.80) \approx (1)(200\,\text{m}^3)(1.031) = 206\,\text{m}^3$$

It is interesting to realize that as the N_2 or air is being added, CH_4 is always part of the gas mixture leaking from the room. This exhausted methane can have a flammable concentration as it now mixes with the external air. The flammable states for the exhaust stream in exterior air can be described by a succession of lines from the B–C locus to A.

Figure 4.17 shows the effect of elevated pressure on the flammability of natural gas in mixtures with nitrogen and air. We see the upper limit significantly increase with

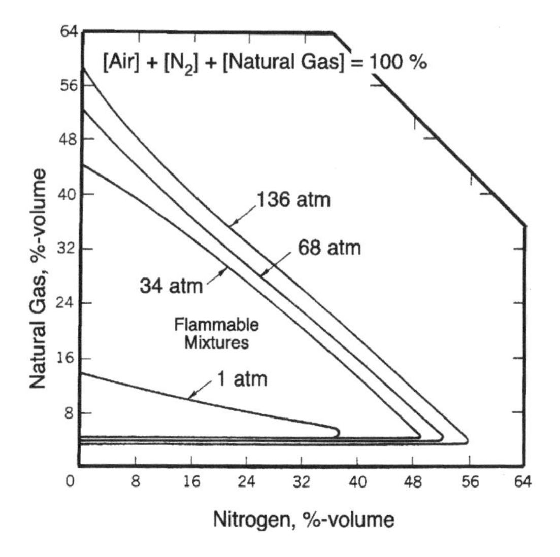

Figure 4.17 Effect of pressure on limits of flammability of natural gas–nitrogen–air mixtures at 26 °C (taken from Zabetakis [5])

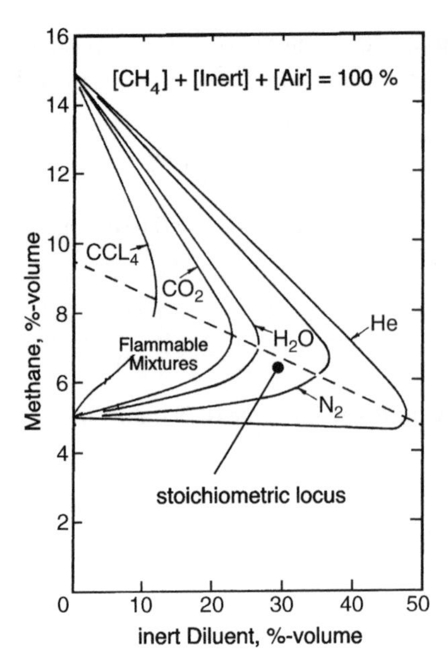

Figure 4.18 Limits of flammability of various methane–inert gas–air mixtures at 25 °C and atmospheric pressure (taken from Zabetakis) [5]

pressure. Figure 4.18 shows the effect of other diluents on the flammability of CH_4, including the halogenated retardant, carbon tetrachloride (CCl_4). The optimization of a specific diluent in terms of weight or toxicity can be assessed from such a figure. It should be pointed out that such diagrams are not only used to assess the potential flammability of a mixture given an ignition source but also to assess the suppression or extinguishment of a fire (say CH_4 in air) by the addition of a diluent (inert or halogenated). In Figure 4.18, the lower concentration of the (CCl_4) needed to achieve the nonflammability of a stoichiometric mixture is less than the inert diluents because of the chemical inhibition effect of chloride ions on the fuel species free radicals. The free radicals are reduced and the reaction rate is effectively slowed.

4.8 Empirical Relationships for the Lower Flammability Limit

Properties of the paraffin hydrocarbons (C_nH_{2n+2}) exhibit some approximate relationships that are sometimes used to describe common fuels generally. Table 4.6 show some data [5].

The lower flammability limit (LFL) is found to be approximately proportional to the fuel stoichiometric concentration:

$$X_L = 0.55 X_{st} \tag{4.51}$$

Table 4.6 Flammability properties of paraffin hydrocarbons (at 25 °C, 1 atm where relevant)

Fuel	M	ρ/ρ_{air}	Δh_c (kJ/mole)	X_L (%)	X_{st} (%)	X_U (%)
Methane, CH_4	16.0	0.55	802	5.0	9.5	15.0
Ethane, C_2H_6	30.1	1.04	1430	3.0	5.6	12.4
Propane, C_3H_8	44.1	1.52	2030	2.1	4.0	9.5
n-Butane, C_4H_{10}	58.1	2.01	2640	1.8	3.1	8.4
n-Pentane, C_5H_{12}	72.2	2.49	3270	1.4	2.5	7.8
n-Hexane, C_6H_{14}	86.2	2.98	3870	1.2	2.2	7.4
n-Heptane, C_7H_{16}	100.	3.46	4460	1.1	1.9	6.7

for 25 °C and 1 atm conditions. The upper limit concentration follows:

$$X_U = 6.5\sqrt{X_L} \tag{4.52}$$

It is further found that the adiabatic flame temperature is approximately 1300 °C for mixtures involving inert diluents at the lower flammable limit concentration. The accuracy of this approximation is illustrated in Figure 4.19 for propane in air. This approximate relationship allows us to estimate the lower limit under a variety of conditions. Consider the resultant temperature due to combustion of a given mixture. The adiabatic flame temperature ($T_{f,ad}$), given by Equation (2.22) for a mixture of fuel (X_F), oxygen (X_{O_2}) and inert diluent (X_D) originally at T_u, where all of the fuel is consumed, is

$$\sum_{\substack{i=1\\ \text{products}}}^{N} \int_{25\,^\circ\mathrm{C}}^{T_{f,ad}} X_i \left(\frac{n_p}{n_u}\right) \tilde{c}_{p_i} \mathrm{d}T - \sum_{\substack{i=1\\ \text{reactants}}}^{N} \int_{25\,^\circ\mathrm{C}}^{T_u} X_i \tilde{c}_{p_i} \, \mathrm{d}T = X_{F,u} \widetilde{\Delta h_c} \tag{4.53}$$

Basically, the chemical energy released in burning all of the fuel in the original (unburned) mixture ($X_{F,u}$) is equal to the sensible energy stored in raising the temperature

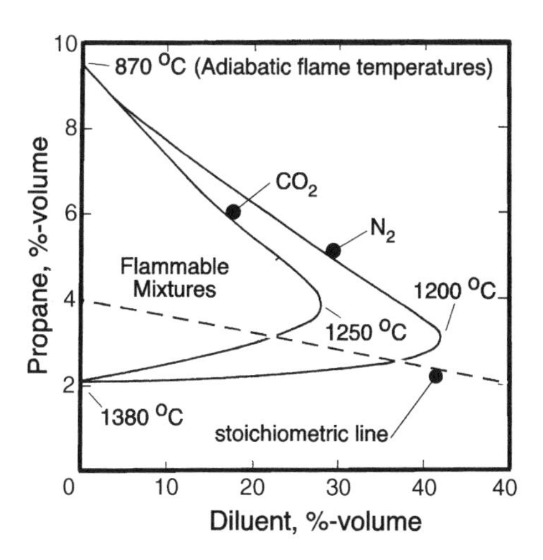

Figure 4.19 Adiabatic flame temperature and mixture concentration for propane [5,13]

of the products of combustion. For an approximate constant specific heat of the product and reactant mixtures, realizing for air at typical lower limits the specific heat is approximately that of nitrogen. Then Equation (4.53) becomes

$$T_{\mathrm{f,ad}} = T_{\mathrm{u}} + \frac{X_{\mathrm{F,u}}\,\widetilde{\Delta h}_{\mathrm{c}}}{\tilde{c}_{p,\mathrm{u}}} \tag{4.54}$$

where $X_{\mathrm{F,u}}$ is the unburned fuel mole fraction and $\widetilde{\Delta h}_{\mathrm{c}}$ is the heat of combustion in molar units. Applying the constant adiabatic temperature approximation of 1300 °C to mixtures at different initial temperatures (T_{u}) gives a relationship for the lower limit as a function of temperature:

$$\boxed{X_{\mathrm{L}}(T_{\mathrm{u}}) = \frac{\tilde{c}_{p,\mathrm{u}}(1300\,^{\circ}\mathrm{C} - T_{\mathrm{u}})}{\widetilde{\Delta h}_{\mathrm{c}}}} \tag{4.55}$$

Notice that at $T_{\mathrm{u}} = 25\,^{\circ}\mathrm{C}$ and for $\tilde{c}_{p,\mathrm{u}} = 0.030$ kJ/mole K of the original unburned mixture

$$X_{\mathrm{L}}\widetilde{\Delta h}_{\mathrm{c}} \approx 38\,\mathrm{kJ/mole} \tag{4.56}$$

Similar analyses can be applied to mixtures of paraffin hydrocarbons where we define X_i to be the ith fuel concentration of the fuel mixture, X_{L_i} to be the lower limit of only the single ith fuel in air and X_{L} to be the lower limit of the fuel mixture in air. For a mixture of N fuels, the heat of combustion with respect to the fuel mixture (e.g. 10 % CH_4, 90 % C_3H_8) is

$$\widetilde{\Delta h}_{\mathrm{c}} = \sum_{i=1}^{N} X_i \widetilde{\Delta h}_{\mathrm{c}_i} \tag{4.57}$$

where $\widetilde{\Delta h}_{\mathrm{c}_i}$ is the heat of combustion for the ith fuel. Since we can approximate from Equation (4.55),

$$X_{\mathrm{L}_i}\,\widetilde{\Delta h}_{\mathrm{c}_i} = X_{\mathrm{L}}\,\widetilde{\Delta h}_{\mathrm{c}} = \tilde{c}_{p,\mathrm{u}}(1300\,^{\circ}\mathrm{C} - 25\,^{\circ}\mathrm{C}) \tag{4.58}$$

Combining,

$$\tilde{c}_{p,\mathrm{u}}(1300\,^{\circ}\mathrm{C} - 25\,^{\circ}\mathrm{C}) = X_{\mathrm{L}} \sum_{i=1}^{N} X_i \left[\frac{\tilde{c}_{p,\mathrm{u}}(1300\,^{\circ}\mathrm{C} - 25\,^{\circ}\mathrm{C})}{X_{\mathrm{L}_i}}\right]$$

or

$$\boxed{\frac{1}{X_{\mathrm{L}}} = \sum_{i=1}^{N\,\mathrm{fuels}} \left(\frac{X_i}{X_{\mathrm{L}_i}}\right)} \tag{4.59}$$

This is sometimes known as Le Chatelier's law, which now gives the lower limit of the mixture of *N* fuels in terms of fuel concentration (X_i) and the respective individual lower limits $(X_{L,i})$.

4.9 A Quantitative Analysis of Ignition, Propagation and Extinction

The ignition, propagation and extinction of premixed combustion systems depends on detailed chemical kinetics and highly nonlinear temperature effects. Although we have qualitatively protrayed these phenomena through a series of graphs involving the energy production and heat loss terms as a function of temperature, it is difficult to quantitatively appreciate the magnitude of these phenomena. Unfortunately, it is not possible to make accurate computations without a serious effort. However, the approximate equations for flame speed and burning rate for a zeroth-order reaction offer a way to obtain order of magnitude results. We shall pursue this course.

Quantitative results will depend very much on the property data we select. The following properties have been found to yield realistic results and are representative of air and fuel mixtures.

$$\begin{aligned}
\rho_u &= 1.1\,\text{kg/m}^3 \\
c_p &= 1.0\,\text{kJ/kg K} \\
k &= 0.026\,\text{W/m K} \\
\alpha &= \text{k}/\rho_u c_p = 2.4 \times 10^{-5}\text{m}^2/\text{s} \\
\Delta h_c &= 40\,\text{kJ/g} \\
E &= 160\,\text{kJ/mole} \\
A &= 10^7\,\text{kg/m}^3\,\text{s}
\end{aligned}$$

Also, representative lean fuel mass fractions between the lower limit for propagation and stoichiometric conditions, range from about $Y_{F,u} = 0.03$ to 0.05. This corresponds to adiabatic flame temperatures, from

$$c_p(T_{f,ad} - T_u) = Y_{F,u}\,\Delta h_c$$

of 1225 °C to 2025 °C for $T_u = 25\,°C$.

Using Equation (4.40), the corresponding flame speeds are computed as 0.24 and 1.7 m/s, respectively. These are realistic values according to Figure 4.8.

4.9.1 Autoignition Calculations

Consider a spherical flammable mixture of radius $r = 1\,\text{cm}$, surrounded by air at temperature T_∞. The heat loss coefficient is approximated as pure conduction,

$$h \approx k/r = 0.026/0.01 = 2.2\,\text{W/m}^2\,\text{K}$$

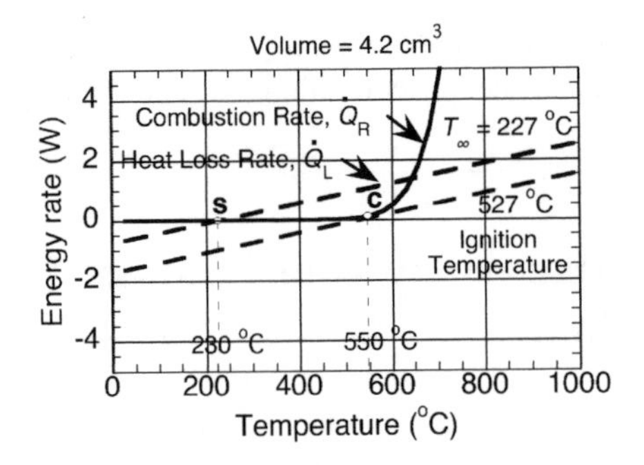

Figure 4.20 Autoignition calculations

From Equations (4.3) and (4.4), a solution for the minimum T_∞ needed to cause ignition can be found. We plot $\dot{Q}_L$ and $\dot{Q}_R$ as functions of T, the flammable mixture average temperature:

$$\dot{Q}_L = h(4\pi r^2)(T - T_\infty) = 0.00327(T - T_\infty)\ \text{W}$$
$$\dot{Q}_R = \dot{m}_F''' \Delta h_c (4/3)\pi r^3 = 1.67 \times 10^6\, e^{-1.92\times 10^4/T}\ \text{W}$$

The results are plotted in Figure 4.20. For $T_\infty < 527\,^\circ\text{C}$ ignition does not occur, and only stable solutions result, e.g. $T_\infty = 227\,^\circ\text{C}$, $T = 230\,^\circ\text{C}$. The critical tangency condition, c, is the minimum condition to initiate ignition with $T_\infty = 527\,^\circ\text{C}$ and $T = 550\,^\circ\text{C}$. The time to reach this ignition condition can be estimated from Equations (4.11) and (4.24) and is

$$t_{ig} \approx t_c + t_R$$

where

$$\begin{aligned} t_c &= \frac{\rho c_p (4/3)\pi r^3}{(k/r)4\pi r^2} \\ &= \frac{r^2}{k/\rho c_p} \\ &= \frac{(0.01\ \text{m})^2}{2.4 \times 10^{-5}\ \text{m}^2/\text{s}} \\ &= 4.2\ \text{s} \end{aligned}$$

and

$$\begin{aligned} t_R &= \frac{\rho c_p T_\infty}{[E/(RT_\infty)]\Delta h_c\, \dot{m}_F'''} \\ &= \frac{(1.1\ \text{kg/m}^3)(1.0\ \text{kJ/kg K})(800\ \text{K})}{\{160 \times 10^3/(8.314)(800)\}(40 \times 10^3\ \text{kJ/kg})10^7\ \text{kg/m}^3\ \text{s}\ e^{-1.92/800}} \\ &= 2.4\ \text{s} \end{aligned}$$

Hence, the time ignition is about 6 s.

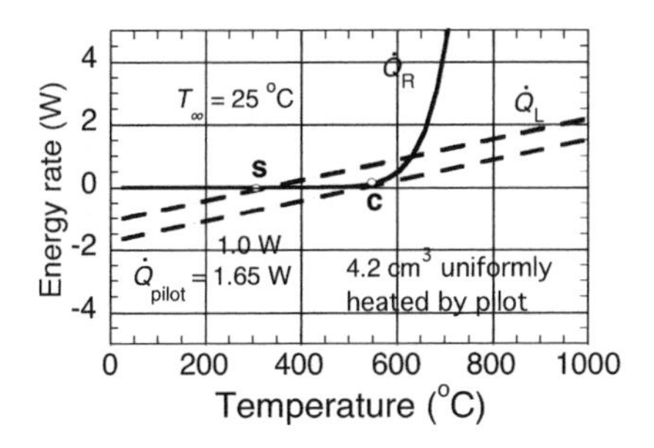

Figure 4.21 Piloted ignition calculations

4.9.2 Piloted ignition calculations

Consider the same spherical flammable mixure of 1 cm radius with a uniformly distributed energy source. This is an idealization of a more realistic electric arc ignition source which could be roughly 10 000 °C, or a flame of roughly 2000 °C, distributed over less than 1 mm. For our idealization, we plot Equations (4.26) and (4.27) as a function of the mixture temperature. Here $T_\infty = 25\,°C$, and we determine the energy supply rate needed from the pilot. The plot in Figure 4.21 is similar to that of Figure 4.20 except that here the needed energy comes from the pilot, not the surroundings. The critical condition for ignition corresponds to a supply energy rate of 1.65 W, with anything less leading to a stable solution at an elevated temperature for as long as the fuel and oxidizer exist. At the critical condition, the mixture temperature is at 550 °C, which was also found for the autoignition case. Since the heating is uniform and from within, we might estimate the ignition time as the reaction time estimated earlier at 527 °C, or 2.4 s. For the uniform constant pilot at 1.65 W, this would require a minimum energy of about 4 J. Spark or electric energies are typically about 1 mJ as a result of their very high temperature over a much smaller region. The uniform pilot assumption leads to the discrepancy. Nevertheless, a relatively small amount of energy, but sufficiently high enough temperature, will induce ignition of a flammable mixture.

4.9.3 Flame propagation and extinction calculations

Once ignition has occurred in a mixture of fuel and oxidizer, propagation will continue, provided the concentrations are sufficient and no disturbance results in excessive cooling. The zeroth-order rate model is assumed to represent the lean case. Substituting the selected properties into Equation (4.43), the net release and loss curves are plotted in Figure 4.22 as a function of the flame temperature. The initial temperature of the mixture is 25 °C and fuel mass fractions are 0.05 and 0.03, representative of stoichiometric and the lower limit respectively. At this lower limit, we should see that a steady solution is not possible, and the calculations should bear this out. The burning rate is evaluated at the flame temperature, and δ_R is found from Equation (4.44) with S_u at the flame temperature

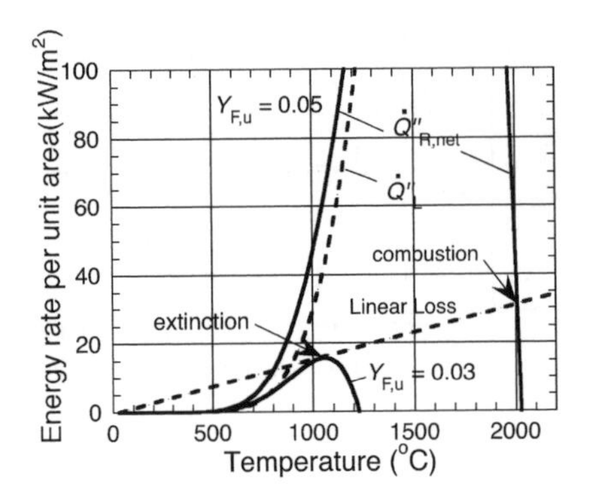

Figure 4.22 Flame propagation calculations

also:

$$\dot{Q}''_{\mathrm{R}} = \dot{m}'''_{\mathrm{F}}\delta_{\mathrm{R}}[40 \times 10^3\,\mathrm{kJ/g} - 1\,\mathrm{kJ/kg\;K}(T - 298\,\mathrm{K})/Y_{\mathrm{F,u}}] \qquad \mathrm{kW/m^2}$$

$$\dot{Q}''_{\mathrm{L}} = (0.026 \times 10^{-3}\,\mathrm{kW/m\,K})(T - 298\,\mathrm{K})/\delta_{\mathrm{R}} \qquad \mathrm{kW/m^2}$$

$$\dot{m}'''_{\mathrm{F}} = 10^7\,\mathrm{kg/m^3\,s}\exp(-1.92 \times 10^4/T) \qquad \mathrm{kg/m^3\,s}$$

$$\delta_{\mathrm{R}} = 3(2.4 \times 10^{-5}\,\mathrm{m^2\,s})/S_{\mathrm{u}} \qquad \mathrm{m}$$

$$S_{\mathrm{u}} = \left[\frac{3(2.4 \times 10^{-5}\,\mathrm{m^2\,s})(40 \times 10^3\,\mathrm{kJ/kg})\dot{m}'''_{\mathrm{F}}}{(1.1\,\mathrm{kg/m^3})(1\,\mathrm{kJ/kg\,K})(T - 298)}\right]^{1/2} \qquad \mathrm{m/s.}$$

Because the results for $\dot{Q}''_{\mathrm{L}}$ were bound to be somewhat too large due to underestimating δ_{R}, an alternative linear heat loss relationship in shown in Figure 4.22. The linear loss considers a constant reaction thickness of 1.65 mm. This shows the sensitivity of the properties and kinetic modeling parameters. Let us use the linear heat loss as it gives more sensible results. For $Y_{\mathrm{F,u}} = 0.05$, a distinct stable solution is found at the flame temperature of about 2000 °C with a corresponding adiabatic flame temperature of 2025 °C. At $Y_{\mathrm{F,u}} = 0.03$, extinction is suggested with a tangent intersection at about 1050 °C with a corresponding adiabatic flame temperature of 1225 °C.

4.9.4 *Quenching diameter calculations*

In this case, the loss is considered solely to the duct wall at 25 °C. The net energy release is computed as above (Equation (4.43)) and the loss to the wall per unit duct cross-sectional area is

$$\dot{Q}''_{\mathrm{L}} = 14.64(0.02 \times 10^{-3}\,\mathrm{kW/m\,K})(T - 298\,\mathrm{K})\delta_{\mathrm{R}}/D^2 \qquad \mathrm{kW/m^2}$$

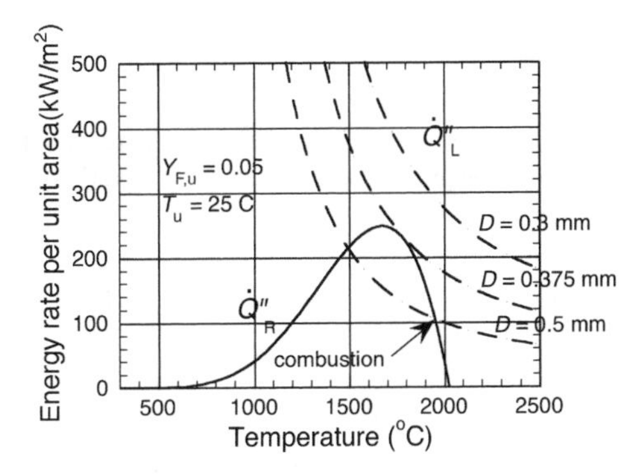

Figure 4.23 Quenching diameter calculations

where δ_R is evaluated as before. Figure 4.23 shows computed results for $Y_{F,u} = 0.05$ and $T_u = 25\,°C$. The diameter is varied to show the solution possibilities. The critical condition shows the quenching diameter at 0.375 mm. Combustion or stable propagation can occur at greater diameters.

References

1. Arrhenius, S., On the reaction velocity of the inversion of cane sugar by acids, *Z. Phys. Chem.*, 1889, **4**, 226–48.
2. Westbrook, C.K. and Dryer, F.L., Chemical kinetic modeling of hydrocarbon combustion, *Prog. Energy Combust. Sci.*, 1984, **10**, 1–57.
3. Semenov, N.N., Zur theorie des verbrennungsprozesses, *Z. Phys. Chem.*, 1928, **48**, 571.
4. ASTM E 659, *Standard Test Method for Autoignition of Liquid Chemicals*, American Society for Testing and Materials, Philadelphia, Pennsylvania, 1978.
5. Zabetakis, M.G., Flammability characteristics of combustible gases and vapors, Bureau of Mines Bulletin 627, 1965.
6. Botha, J.P. and Spalding, D.B., The laminar flame speed of propane–air mixtures with heat extraction from the flame, *Proc. Royal Soc. London, Ser. A.*, 1954, **225**, 71–96.
7. Mallard, E. and Le Chatelier, H., Recherches experimentales et theoriques sur la combustion des melanges gaseoux explosifs, *Ann. Mines*, 1883, **4**, 379.
8. Zeldovich, Ya. B., Barenblatt, G.I., Librovich, V.B. and Makviladze, G.M., *The Mathematical Theory of Combustion and Explosions*, Consultants Bureau, New York, 1985, p. 268.
9. Meyer, E., A theory of flame proposition limits due to heat loss, *Combustion and Flame*, 1957, **1**, 438–52.
10. Incroprera, F.P. and DeWitt, D.P., *Fundamentals of Heat and Mass Transfer*, 4th edn., John Wiley & Sons, New York, 1996, p. 444.
11. Coward, H.F. and Jones, G.W., Limits of flammability of gases and vapors, Bureau of Mines Bulletin 627, 1965.
12. Beyler, C., Flammability limits of premixed and diffusion flames, in *The SFPE Handbook of Fire Protection Engineering*, 2nd edn (eds P.J. Di Nenno *et al.*), Section 2, Chapter 9, National Fire Protection Association, Quiney, Massachusetts, 1995, p. **2**–154.
13. Lewis, B. and von Elbe, G., *Combustion, Flames and Explosions of Gases*, 2nd edn., Academic Press, New York, 1961, p. 30.

Problems

4.1 The energy conservation is written for a control volume surrounding a moving premixed flame at velocity S_u into a fuel–air mixture at rest. The equation is given below:

$$c_p \frac{d}{dt} \iiint_{CV} \rho T \, dV = \dot{m}_F''' \delta_R \Delta h_c - \rho_u S_u c_p (T_b - T_u) - k(T_b - T_u)/\delta_R$$

Identify in words each of the energy terms.

4.2 The burning velocity, S_u, can be determined by a technique that measures the speed of a spherical flame in a soap bubble. This process is shown below.

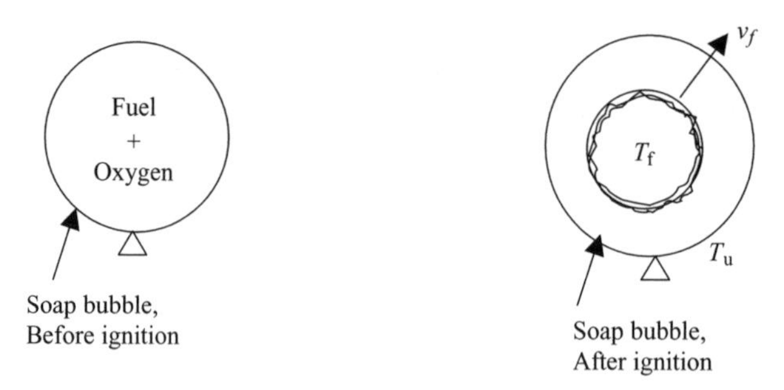

The flame front moves at a speed $v_f = 6\,m/s$.

The temperature in the burnt region is $T_f = 2400\,K$.

The temperature in the unburnt region is $T_u = 300\,K$.

The burned gas is at rest.

Determine the burning velocity, S_u. (*Hint*: draw a control volume around the flame and write the conservation of mass.)

4.3 A mixture of methane and air is supplied to a Bunsen burner at 25 °C. The diameter of the tube is 1 cm and the mixture has an average velocity of 0.6 m/s.

(a) Calculate the minimum and maximum mass flow rates of CH_4 that will permit a flame for the burner.

(b) For a stoichiometric mixture, the flame forms a stable conical shape at the tube exit. The laminar burning velocity for CH_4 in air is 0.37 m/s. Calculate the cone half-angle, ϕ.

Surface area of cone $A_s = \frac{\pi R^2}{\sin \phi}$

(c) What do we expect to happen if we increase the mixture flow rate while keeping the mixture at the stoichiometric condition?

(d) The mixture flow velocity is reduced to 0.1 m/s while maintaining stoichiometric conditions. A flame begins to propagate back into the tube as a plane wave. Assume the propagation to be steady and then calculate the flame propagation velocity (v_f).

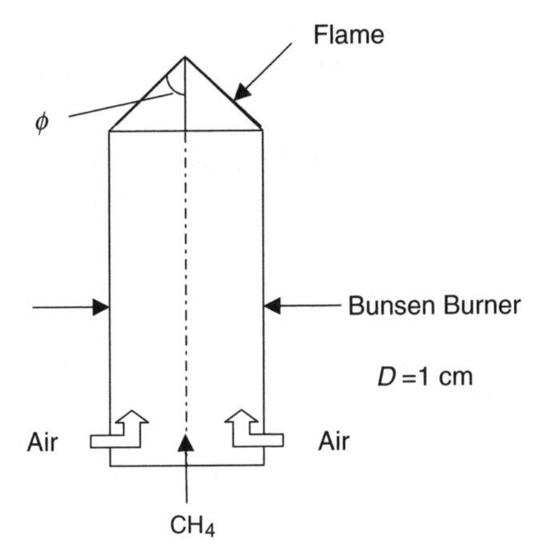

4.4 Consider the ignition of a flammable mixture at the closed end of a horizontal circular tube of cross-sectional area A. The tube has a smaller opening at the other end of area A_o. The process is depicted in the figure below. The flame velocity in the tube is designated as v_f. It is distinct from the ideal burning velocity, S_u, which is constant in this adiabatic case. The following assumptions apply:

- The flame spread process in adiabatic. There is no heat transfer to the tube wall and the temperature of the burned products, T_b, and the unburned mixture, T_u, are therefore constant.
- Bernoulli's equation applies at the tube exit.
- The contraction area ratio to the vena contracta of the exiting jet is C_o.
- The perfect gas law applies.
- The pressure in the tube varies with time, but is uniform in the tube.
- Specific heats are constant.
- The reaction rate, $\dot{m}'''_F$, and flame thickness, δ_R, are related to the ideal burning speed by

$$\rho_u c_p S_u (T_b - T_u) = \dot{m}'''_F \delta_R \Delta h_c$$

$$S_u \approx \frac{3k}{\rho_u c_p \delta_R}$$

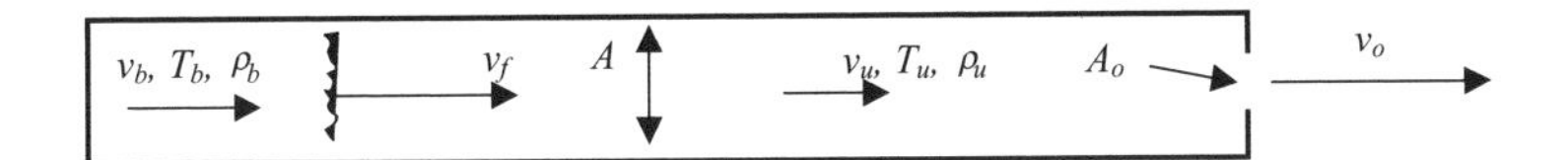

(a) Express the velocity of the flame in terms of the ideal burning speed and the velocity of the unburned mixture. (*Hint*: use the definition of the ideal burning speed.)

(b) Express the velocity of the unburned mixture in terms of the pressure differences between the tube gas and the outside environment, p–p_o. (*Hint*: use Bernoulli.)

(c) Derive the differential equation for the pressure in the tube. Use the nomenclature here and in the text for any other variables that enter. (*Hint*: use the energy equation, Equation (3.45).)

Solve the energy equation and discuss the behavior of the flame velocity in the tube. (There is a limit to the speed since the exit velocity cannot exceed the speed of sound, i.e. choked flow.)

4.5 A flammable gas mixture is at a uniform concentration and at rest in a room at 25 °C and 1.06×10^5 Pa. The properties of the gas mixture is that it has a burning velocity $S_u = 0.5$ m/s and its adiabatic flame temperature is 2130 °C in air. The room is totally open (3 m × 3 m)at one end where an observer turns on a light switch. This triggers a plane-wave adiabatic laminar flame propagation in the mixture that propagates over 15 m. The room processes are also adiabatic so that the unburned and burned gas temperatures are fixed.

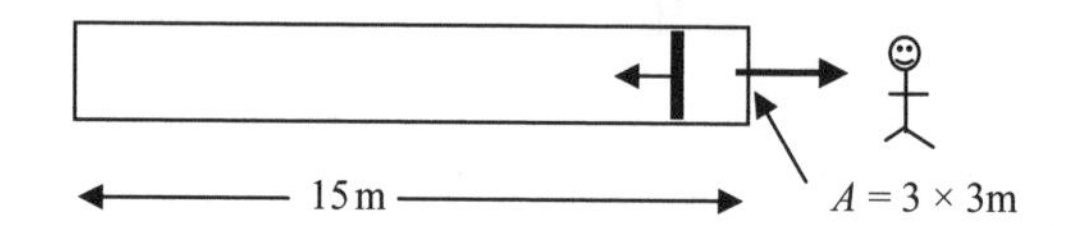

(a) What is the velocity of the flame propagation as seen by the observer?

(b) What is the velocity of the burned gases as they exit the room as seen by the observer? Assume the pressure is uniform in the room, but can change with time.

(c) Compute the pressure rise in the compartment after 0.5 seconds.

Other properties: mixture specific heats, $c_p = 1.0$ kJ/kg K, $c_v = 0.71$ kJ/kg K, $p = R\rho T$, $1\,\text{N/m}^2 = 1$ Pa, $1\,\text{N} = 1\,\text{kg m/s}^2$ and $1\,\text{J} = 1\,\text{N m}$.

4.6 Given that the lower flammability limit of *n*-butane (*n*-C_4H_{10}) in air is 1.8 % by volume, calculate the adiabatic flame temperature at the limit. Assume the initial temperature to be 25 °C. Use Table 4.5.

4.7 What is the LFL in air of a mixture of 80 % (molar) methane and 20 % propane by Le Chatelier's rule? What is the adiabatic flame temperature of this system?

4.8 Felix Weinberg, a British combustion scientist, has shown a mixture that is 1 % by volume of methane in air and can burn if preheated to 1270 K. Although this temperature is higher than the autoignition temperature for methane in air (~550 °C), the mixture burned by pilot ('spark') ignition before the autoignition effects could develop. Make your own estimation of the LFL at the initial mixture temperature of 1270 K. Does it agree with the experimental results?

4.9 Calculate the adiabatic flame temperature, at constant pressure, for ethane (C_2H_6) in air at 25°C:

(a) at the lower flammability limit (X_L);

(b) at the stoichiometric mixture condition (X_{st}).

Use Tables 2.1 and 4.6.

4.10 It is reported that the adiabatic flame temperature for H_2 at the lower flammability limit (LFL) in air is 700 °C. From this information, estimate the LFL, in % by volume, for the hydrogen–air mixture at 25 °C. Assume water is in its vapor phase within the products.

4.11 A gaseous mixture of 2 % by volume of acetone and 4 % ethanol in air is at 25 °C and pressure of 1 atm.

Data:

Acetone (C_3H_6O), heat of combustion: $\Delta h_c = 1786$ kJ/g mol

Ethanol (C_2H_5OH), heat of combustion: $\Delta h_c = 1232$ kJ/g mol

Atomic weights: H = 1, C = 12, O = 16 and N = 14

Specific heat, c_p, I = 1 kJ/kg K, constant for each species.

Find:

(a) For a constant pressure reaction, calculate the partial pressure of the oxygen in the product mixture.

(b) Determine the adiabatic flame temperature of this mixture.

(c) Do you think this mixture is above or below the lower flammable limit? Why?

(d) If this mixture was initially at 400 °C, what will the resultant adiabatic flame temperature be?

4.12 Compute the lower flammable limit in % mole of acetonitrile in air at 25 °C using the 1600 K adiabatic temperature rule. Acetonitrile ($C_2H_3N(g)$) burns to form hydrogen cyanide ($HCN(g)$), carbon dioxide and water vapor.

Heat of formation in kcal/g mol

Hydrogen cyanide:	32.2
Acetonitrile:	21.0
Water vapor:	−57.8
Carbon dioxide:	−94.1
Oxygen:	0.0

Assume constant and equal specific heats of constant pressure and constant volume of 1.2 and 1.0 kJ/kg K, respectively.

4.13 An electrical fault causes the heating of wire insulation producing fuel gases that mix instantly with air. The wire is in a narrow shaft in which air enters at a uniform velocity of 0.5 m/s at 25 °C (and density 1.18 kg/m^3). The cross-sectional area of the shaft is 4 cm^2. The gaseous fuel generated is at 300 °C and its heat of combustion is 25 kJ/g. Assume steady state conditions and constant specific heat at 1.0 J/g K.

(a) What is the mass flow rate of the air into the shaft?

(b) The wire begins to arc above the point of fuel generation. Draw a control volume that would allow you to compute the conditions just before ignition, and state the minimum temperature and boundary conditions needed for ignition.

(c) Find the minimum mass flow rate of fuel necessary for ignition by the arc. Show the control volume used in your analysis.

(d) Find the minimum fuel mass fraction needed for ignition by the arc.

Show the control volume used in your analysis.

4.14 After combustion in a closed room, the temperature of the well-mixed gases achieves 350 °C. There is gaseous fuel in the room left over at a mass fraction of 0.015. There is ample air left to burn all of this fuel. Do you consider this mixture flammable, as defined as capable of propagating a flame from an energy source? Quantitatively explain your answer. You may assume a specific heat of 1 J/g K and a heat of combustion from the fuel of 42 kJ/g.

4.15 The mole (volume) fraction of acetone (C_3H_6O) as a vapor in air is 6 %. This condition is uniform throughout a chamber of volume 20 m^3. Air is supplied at 0.5 m^3/s and the resulting mixture is withdrawn at the same rate. Assume that the mixture always has a uniform concentration of acetone for the chamber. Assume the gas density of the mixture is nearly that of pure air, 1.2 kg/m^3. How long does it take to reduce the acetone mole fraction to 2 %?

4.16 A methane leak in a closed room is assumed to mix uniformly with air in the room. The room is $(4 \times 4 \times 2.5)$m high. Take the air density as 1.1 kg/m^3 with an average molecular weight of 29 g/g mole. How many grams of methane must be added to make the room gases flammable? The lower and upper flammability limits of methane are 5 and 15 % by volume respectively.

4.17 How much N_2 would we have to add to a mixture of 50 % by volume of methane, 40 % oxygen and 10 % nitrogen to make it nonflammable? Use Figure 4.16 and determine graphically.

4.18 Use the flammability diagram below to answer the following:

(a) Is pure hydrazine flammable? Label this state A on the diagram.

(b) Locate a mixture B on the diagram that is 50 % hydrazine and 45 % air.

(c) If pure heptane vapor is added to mixture B find the mixture C that just makes the new mixture nonflammable. Locate C in the figure.

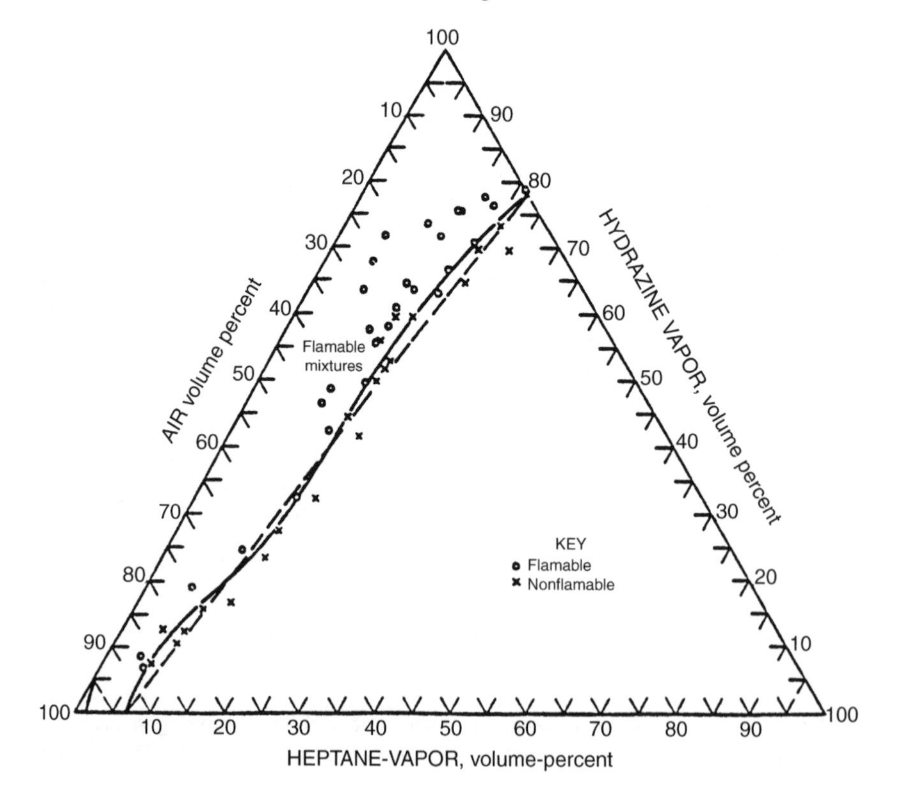

4.19 Using the methane–oxygen–nitrogen flammability diagram (Figure 4.16(b)), complete the following and show your work on the diagram:

(a) Locate the mixture B of 20 % methane, 50 % oxygen and the remainder, nitrogen. Is this in the flammable range?

(b) If we move methane from this mixture, find the composition just on the boundary of the flammable range.

4.20 Methane leaks from a tank in a 50 m^3 sealed room. Its concentration is found to be 30 % by volume, as recorded by a combustible gas detector. The watchman runs to open the door of the room. The lighter mixture of the room gases flows out to the door at a steady rate of 50 g/s. The flammable limits are 5 and 15 % by volume for the methane in air. Assume a constant temperature at 25 °C and well-mixed conditions in the room. The mixture of the room gases can be approximated at a constant molecular weight and density of 25 g/mol and 1.05 kg/m^3 respectively. After the door is opened, when will the mixture in the room become flammable?

4.21 Indicate which of the following are true or false:

(a) The lower flammability limit is usually about one-half of the stoichiometric fuel concentration. T___F___

(b) The lower flammability limit does not depend on the mixture temperature and pressure. T___F___

(c) The burning velocity of a natural gas flame on a stovetop burner is zero because the flame is stationary. T___F___

(d) The quenching diameter is the minimum water droplet diameter needed to extinguish a flame. T___F___

(e) The heat of combustion is related to the enthalpy of formation of all the reactant and product species. T___F___

5

Spontaneous Ignition

5.1 Introduction

Spontaneous ignition or spontaneous combustion are terms applied to the process of autoignition arising from exothermiscity of the material itself. The term self-heating is also used to describe the exothermic process leading to the ignition event. As discussed in the theory of ignition, the ignition event is a criticality that results in a dramatic temperature change leading to a state of combustion. For solid materials, spontaneous ignition can manifest itself as smoldering combustion at temperatures as low as 300 °C without perceptible glowing, or is visible flaming combustion with gas temperatures in excess of 1300 °C. It is probably more common to apply the term spontaneous ignition to the event of flaming combustion due to self-heating.

Factors that can contribute to the spontaneous ignition of solids are many:

(a) bulk size: contributes to the storage of energy and interior temperature rise;

(b) porosity: contributes to the diffusion of air to promote oxidation and metabolic energy release of biological organisms;

(c) moisture: contributes to the growth of bacteria and other micro-organisms;

(d) contamination: additives, such as unsaturated oils, can promote exothermiscity;

(e) process defects: storage of heated materials, insufficient additives (such as antioxidants in synthetic polymers) or terminating a process before its chemical reactions are completed.

These processes listed here are just some of the factors that can cause and promote spontaneous ignition.

Common examples of spontaneous ignition have included moist haystacks, oily cotton rags and mounds of low-grade coal from mining operations. Many have experienced the

Fundamentals of Fire Phenomena James G. Quintiere

warm interior of a mulch pile used in gardening. As with haystacks, moist woody matter will experience a biological process that produces energy. This energy can raise the temperature of the wood matter where it can oxidize or degrade and chemically produce more energy. Thus, self-heating occurs. The chemical process, in general, can be exothermic due to oxidation or due to thermal decomposition, as in ammonium potassium nitrate. Bowes [1] presents an extensive theoretical and practical review of the subject of spontaneous ignition. An excerpt is presented from this book on the propensity of spontaneous ignition and self-heating in oily rags:

> *The amounts of material involved in the initiation of fires can be quite small. For example Taradoire [1925], investigating the self-ignition of cleaning rags impregnated with painting materials, stated that experience indicated the occurrence of self ignition in as little as 25 g of rags but, in a series of experiments, found that the best results (sic) were obtained with 75 g of cotton rags impregnated with an equal weight of mixtures of linseed oil, turpentine and liquid 'driers' (containing manganese resinate) and exposed to the air in cylindrical containers made of large-mesh wire gauze (dimensions not stated). Times to ignition generally lay between 1 h and 6 h, but for reasons which were not discovered, some samples, which had not ignited within these times and were considered safe, ignited after several days. Other examples giving an indication of scale have been described by Kissling [1895], who obtained self-heating to ignition, from an initial temperature of 23.5 °C, in 50 g of cotton wool soaked with 100 g of linseed oil and by Gamble [1941] who reports two experiments with cotton waste soaked in boiled linseed oil (quantities not stated) and lightly packed into boxes; with the smaller of the two, a cardboard box of dimensions 10 cm × 10 cm × 15 cm, the temperature rose from 21 °C to 226 °C in $6\frac{1}{4}$ h and the cotton was charred.*

We have had similar experiences in examining cotton rags impregnated with linseed oil and loosely packed in a slightly closed cardboard box about 30 cm on a side. Ignition could occur in 2 to 4 hours. It was proceeded by a noticeable odor early, then visible white 'smokey' vapor. It could result in a totally charred cotton residue – indicating smoldering – or into flaming ignition. On some occasions, no noticeable ignition resulted from the self-heating. This simple demonstration was utilized by a surprised newscaster on Philadelphia TV when the contents of a cardboard box with only rags burst into flames. He was exploring the alleged cause of the fire at One Meridian Plaza (1988) which led to multistory fire spread, an unusable 38 story building, three firefighter fatalities and damage in civil liabilities of up to $400 million.

A less expensive accident attributed to wood fiberboard in 1950 brought the subject of spontaneous ignition to the forefront of study. This is vividly described in the opening of an article by Mitchelle [2] from the *NFPA Quarterly*, 'New light on self-ignition':

> *One carload of wood fiberboard burned in transit and eight carloads burned in a warehouse 15 days after shipment, causing a $2,609,000 loss. Tests to determine the susceptibility of wood fiberboard to self-heating revealed that the ambient temperature to cause ignition varied inversely with the size of the specimen, ranging from 603 °F for a $\frac{1}{8}$-inch cube to 252 °F for an octagonal 12-in prism. Pronounced self-heating was evident when the surrounding air was held at 147 °F.*
>
> *An untoward event sometimes points the way to useful discoveries. A fire in an Army Warehouse, June 1950, pointed the way to the subject of this discussion. The story briefly is: on June 2, nine carloads of insulation fiberboard were shipped from a factory in the South, bound for upstate New York. During the afternoon of June 9, in the course of switching, one carload was discovered to be afire and left in the railyard. The other eight cars were sent on a few miles farther to their destination, where they were unloaded into a warehouse during the period June 12 to 15. This lot of eight carloads of fiberboard was stacked in one pile*

Table 5.1 Ignition temperatures for wood fiberboard [2]

Characteristic dimension (in)	Ignition temperature (°F)
1/8	603
12	252
22	228

measuring more than 24,000 cubic feet. In the forenoon of June 17, two days afterward, fire was discovered which destroyed the warehouse and its contents. The monetary loss was reported to have been $2,609,000.

It is evident from the excerpt of Mitchell's article that the temperature necessary to cause ignition of the fiberboard depends on the size of the wood. Table 5.1 gives data from Mitchell [2]. The implication is that a stack of the wood of 24 000 ft^3 (or about 29 ft as a characteristic length) led to ignition in the warehouse after two days in a New York June environment at roughly 75 °F. Such events are not uncommon, and have occurred for many types of stored processed materials.

5.2 Theory of Spontaneous Ignition

Although many factors (chemical, biological, physical) can be important variables, we shall restrict our attention to the effects of size and heating conditions for the problem. The chemical or biological effects will be accounted for by a single Arrhenius zeroth-order rate relationship. The material or medium considered will be isotropic and undergo no change in phase with respect to its thermal properties, taken as constant. Thus, no direct accounting for the role and transport of oxygen and fuel gases will be explicit in the following theory. If we could include these effects it would make our analysis more complex, but we aim to remain simple and recognize that the thermal effects are most relevant.

Since size is very important; we will consider a one-dimensional slab subject to convective heating. The geometry and heating condition can be modified resulting in a new problem to solve, but the strategy will not change. Hence, one case will be sufficient to illustrate the process of using a theory to assess the prospect of spontaneous ignition. Later we will generalize it to other cases.

We follow the analysis of Frank-Kamenetskii [3] of a slab of half-thickness, r_o, heated by convection with a constant convective heat transfer coefficient, h, from an ambient of T_∞. The initial temperature is $T_i \leq T_\infty$; however, we consider no solution over time. We only examine the steady state solution, and look for conditions where it is not valid. If we return to the analysis for autoignition, under a uniform temperature state (see the Semenov model in Section 4.3) we saw that a critical state exists that was just on the fringe of valid steady solutions. Physically, this means that as the self-heating proceeds, there is a state of relatively low temperature where a steady condition is sustained. This is like the warm bag of mulch where the interior is a slightly higher temperature than the ambient. The exothermiscity is exactly balanced by the heat conducted away from the interior. However, under some critical condition of size (r_o) or ambient heating (h and T_∞), we might leave the content world of steady state and a dynamic condition will

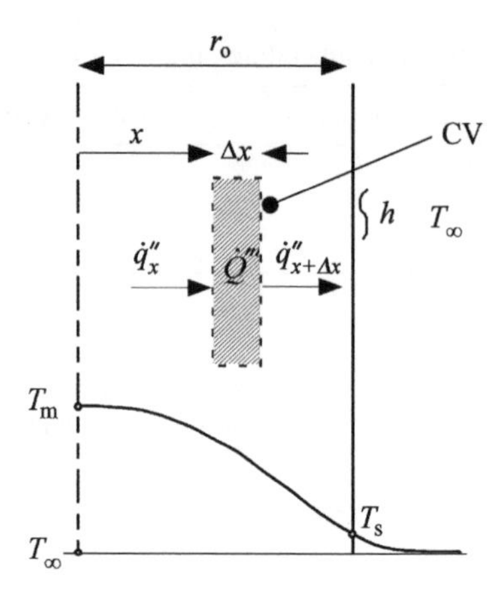

Figure 5.1 Theoretical model

ensue. This is the phenomenon of thermal runaway or thermal explosion, which means that we will move to a state of combustion – smoldering or flaming.

Let us derive the governing equation for steady state conduction in a one-dimensional slab of conductivity, k, with internal heating described by an energy release rate per unit volume

$$\dot{Q}''' = (A\Delta h_c)e^{-E/(RT)} \tag{5.1}$$

where $(A\Delta h_c)$ and E are empirical properties. Consider a small element, Δx, of the slab (Figure 5.1) by the conservation of energy for the control volume surrounding the element

$$0 = \dot{Q}'''\Delta x + \dot{q}''_x - \dot{q}''_{x+\Delta x} \tag{5.2}$$

By Fourier's law of conduction,

$$\dot{q}''_x = -k\frac{dT}{dx} \tag{5.3}$$

and by Taylor's expansion theorem,

$$\dot{q}''_{x+\Delta x} = \dot{q}''_x + \frac{d}{dx}(\dot{q}''_x)\Delta x + O(\Delta x)^2 \tag{5.4}$$

By substitution in Equation (5.2), dividing by Δx and letting $\Delta x \to 0$, it follows that

$$0 = \dot{Q}''' - \frac{d}{dx}(\dot{q}''_x) - O(\Delta x)$$

or

$$\frac{d}{dx}\left(k\frac{dT}{dx}\right) + (A\Delta h_c)e^{-E/(RT)} = 0 \tag{5.5}$$

The boundary conditions are

$$x = 0, \qquad \frac{\mathrm{d}T}{\mathrm{d}x} = 0 \qquad \text{by symmetry} \tag{5.6}$$

and

$$x = r_o, \qquad -k\frac{\mathrm{d}T}{\mathrm{d}x} = h(T - T_\infty) \tag{5.7}$$

by applying the conservation of energy to the surface.

It is useful to rearrange these equations into dimensionless form as was done in Section 4.3, Equation (4.9):

$$\boxed{\theta = \left(\frac{T - T_\infty}{T_\infty}\right)\frac{E}{RT_\infty}} \tag{5.8}$$

where again we approximate

$$\mathrm{e}^{-E/(RT)} \approx \mathrm{e}^{\theta}\mathrm{e}^{-E/(RT_\infty)} \tag{5.9}$$

anticipating that the critical temperature will be close to T_∞. With

$$\xi = x/r_o \tag{5.10}$$

the problem becomes

$$\frac{\mathrm{d}^2\theta}{\mathrm{d}\xi^2} + \delta \mathrm{e}^{\theta} = 0 \tag{5.11}$$

with $\mathrm{d}\theta/\mathrm{d}\xi = 0$ at $\xi = 0$ and $\mathrm{d}\theta/\mathrm{d}\xi + Bi\ \theta = 0$ at $\xi = 1$. The dimensionless parameters are:

(a) the Damkohler number, δ:

$$\boxed{\delta = \left(\frac{E}{RT_\infty}\right)\left[\frac{r_o^2(A\Delta h_c)\mathrm{e}^{-E/(RT_\infty)}}{kT_\infty}\right]} \tag{5.12}$$

which represents the ratio of chemical energy to thermal conduction, or the ratio of thermal diffusion (r_o^2/α) to chemical reaction time from Equations (4.23) and (4.24), and
(b) the Biot number, Bi:

$$Bi = \frac{hr_o}{k} \tag{5.13}$$

the ratio of convection to conduction heat transfer in the solid. If h is very large for high flow conditions, $Bi \rightarrow \infty$. Mathematically, this reduces to the special case of $T(r_o) = T_\infty$ or $\theta(1) = 0$.

As in the solution to the simpler Semenov problem (see section 4.3), we expect that as δ increases (increasing the exothermiscity over the material's ability to lose heat) a critical δ_c is reached. At this point, an unsteady solution must hold and we have no valid solution to Equation (5.11). For a given geometric shape of the solid and a given heating condition, there will be a unique δ_c that allows this 'ignition' event to occur. For the slab problem described, we see that since Bi is an independent parameter, then δ_c must be a function of Bi.

For the special case of a constant temperature boundary ($\xi = 1, \theta = 0$), Frank-Kamenetskii [3] gives the solution to Equation (5.11) as

$$e^{\theta} = \frac{a}{\cosh^2\left[b \pm \sqrt{a\delta/2}\xi\right]} \tag{5.14}$$

where a and b are constants of integration. From Equation (5.6), it can be shown that $b = 0$, and from $\theta(1) = 0$, it follows that a must satisfy

$$a = \cosh^2\left(\frac{a\delta}{2}\right) \tag{5.15}$$

The maximum temperature at the center of the slab can be expressed from Equation (5.14) as

$$e^{\theta_m} = a \tag{5.16a}$$

or

$$\theta_m = \ln(a) = f(\delta) \tag{5.16b}$$

From Equation (5.15), a is an implicit function of δ and θ_m is a function only of δ. A solution is displayed in Figure 5.2. Steady physical solutions are given on the lower

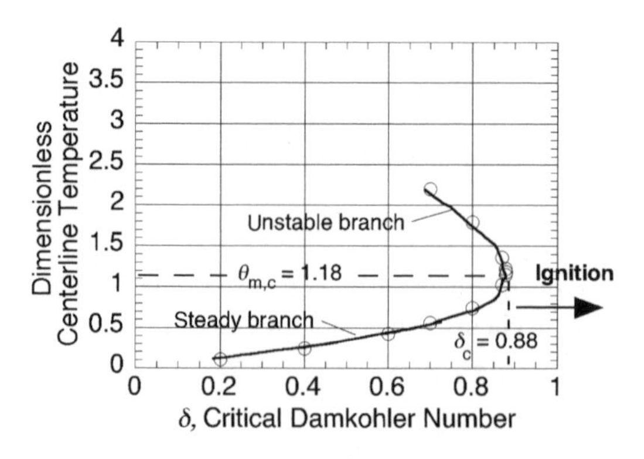

Figure 5.2 Solution for a slab at a constant surface temperature

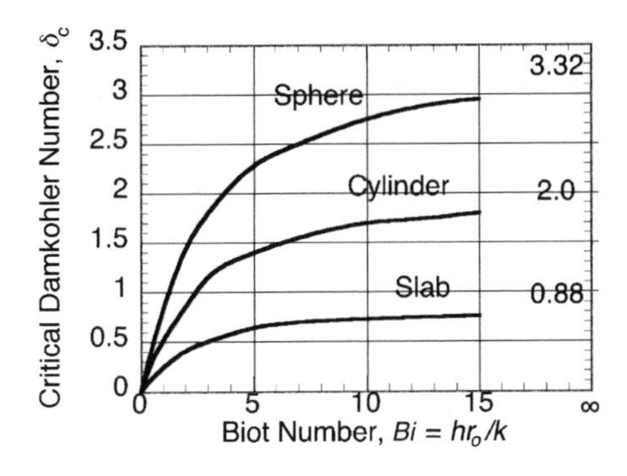

Figure 5.3 δ_c As a function of *Bi* for a sphere, cylinder and slab (from Bowes [1])

branch; i.e. for small values of $\delta < \delta_c = 0.88$, the slab is able to maintain a state in which the chemical energy is conducted to the surface at T_∞. From Equation (5.12), δ can increase as T_∞ increases or as r_o increases. If δ exceeds 0.88 for any reason, a steady solution is not possible, and a thermal runaway occurs leading to ignition, and finally to another possible steady state at a new temperature corresponding to combustion. The upper branch of the curve has mathematically possible steady solutions but is believed to be unstable, and thus not relevant. Therefore, to prevent spontaneous ignition of this slab, we must be assured that its Damkohler number is less than 0.88. Another configuration or heating condition will have another solution and a different critical Damkohler number, δ_c.

The critical dimensionless temperature corresponding to $\delta_c = 0.88$ is approximately $\theta_{m,c} = 1.18$. As found for Equation (4.19),

$$\frac{T_{m,c} - T_\infty}{T_\infty} = (1.18)\frac{RT_\infty}{E} \sim \frac{T_\infty}{10^4}$$

as a typical result for $E \sim 30$ kcal/mole. This confirms that $T_{m,c}$ at ignition is a small departure from T_∞, and T_∞ can be regarded as the ignition temperature corresponding to a given r_o. It is not a unique property of the material!

Figure 5.3 and Table 5.2 give solutions for δ_c for convective heating of a sphere, cylinder and slab. As the degrees of freedom for heat conduction increase, it is less likely to achieve ignition as δ_c increases from a slab (1), to a cylinder (2), to a sphere (3). Also as $Bi \to 0$, the material becomes perfectly adiabatic, and only an unsteady solution is

Table 5.2 Critical Damkohler numbers for $Bi \to \infty$ (medium)

	δ_c
Semenov, $r_o = V/S$	$e^{-1} = 0.368$
Slab, r_o is half-width	0.88
Infinite cylinder, r_o is radius	2.0
Sphere, r_o is radius	3.32
Cube, r_o is half-side	2.52

possible, $\delta_c \to 0$. For $Bi \to \infty$, the constant surface temperature case, δ_c for a cube of half side, r_o, is 2.52, and for a rectangular box of half dimensions r_o, ℓ_o, and s_o is

$$\delta_c = 0.88\left[1 + \left(\frac{r_o}{\ell_o}\right)^2 + \left(\frac{r_o}{s_o}\right)^2\right]$$

5.3 Experimental Methods

In order to assess spontaneous ignition propensity for a material practically, data must be developed. According to the model presented, these data would consist of (E/R) and $(A\Delta h_c/k)$ from Equation (5.12) for δ. If Bi is finite, we must also know k separately. To simplify our discussion, let us consider only the constant surface temperature case ($Bi \to \infty$). Experiments are conducted in a convective oven with good circulation (to make h large) and at a controlled and uniform oven temperature, T_∞ (Figure 5.4). Data of cubical configurations of sawdust, with and without an oil additive, are shown in Table 5.3. The temperatures presented have been determined by trial and error to find a sufficient T_∞ ($T_{\infty,2}$ Figure 5.4) to cause ignition. Ignition should be observed to determine whether it is flaming or smoldering. The center temperature, T_m, is used as the indicator of thermal runaway. A surface temperature should also be recorded to examine whether it is consistent with our supposition that $T_s = T_\infty$ for the convective oven. If this is not true, then h must be determined for the oven. While the experiment is simple, care must be taken to properly secure the thermocouples to the material. Also, patience is needed since it can take many hours before ignition occurs. If a sample changes and melts, the data can be invalid. This is particularly true since small oven-size samples require relatively high temperatures for ignition, and such temperatures are likely to exceed the melting point of common polymers. Hence, the test conditions must be as close as possible to the conditions of investigation; however, this may not be fully possible to achieve. Figure 5.5 gives data from Hill [4] for a cotton material used in the buffing of aluminum for aircraft surfaces. This was tested because fires were periodically occurring in a hopper where the waste cotton was discarded. It can be seen from these data that only a small oven temperature change is needed when we are near the critical conditions, i.e. $\delta \approx \delta_c$.

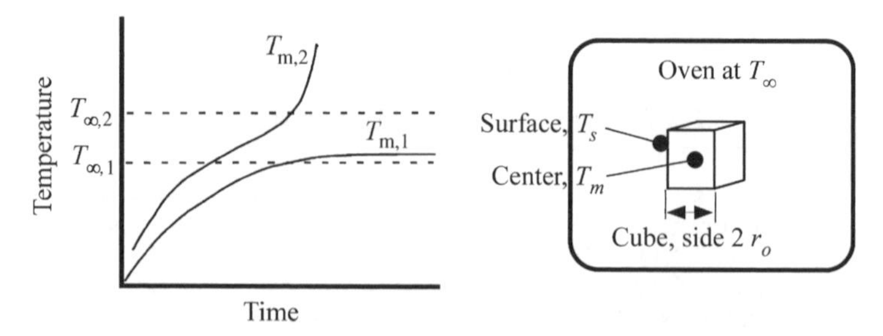

Figure 5.4 Experimental method

Table 5.3 Critical oven temperatures for cubes of sawdust with and without oil (from Bowes [1])

Cube size $2r_0$ (mm)	Oil content (%)	Ignition temperature (°C)
25.4	0	212
25.4	11.1	208
51	0	185
51	11.1	167
76	0	173
76	11.1	146
152	0	152[a]
152	11.1	116
303	0	135
303	11.1	99
910	0	109[a]
910	11.1	65

[a] Calculated.

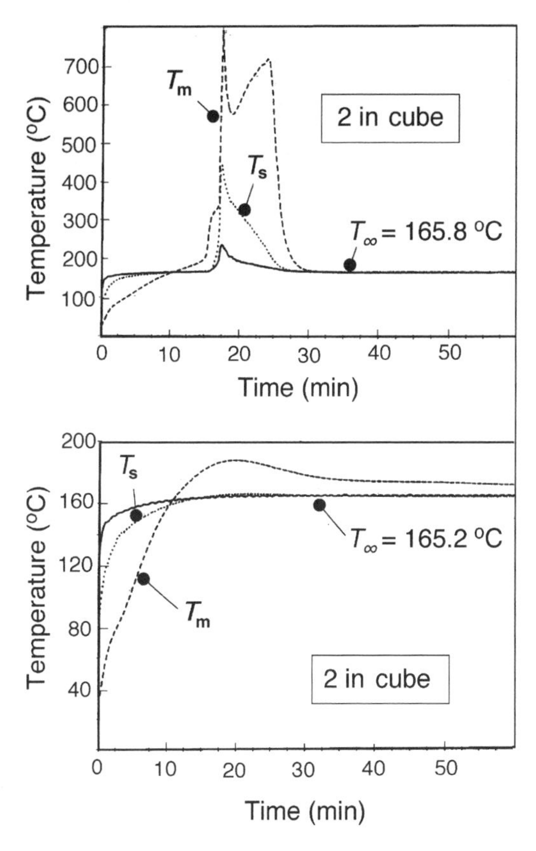

Figure 5.5 Example of a thermal response for a cotton-based material used in an aluminum polishing process (from Hill [4])

The needed properties can be derived from data illustrated in Table 5.3. To derive these data, it is convenient to rearrange Equation (5.12) as

$$\ln\left(\frac{\delta_c T_\infty^2}{r_o^2}\right) = P - \left(\frac{E}{R}\right)\left(\frac{1}{T_\infty}\right) \tag{5.17}$$

where

$$P \equiv \ln\left[\left(\frac{E}{R}\right)\left(\frac{A\Delta h_c}{k}\right)\right]$$

Since the data are for cubes, $\delta_c = 2.52$ and we plot

$$\ln\left[\delta_c\left(\frac{T_\infty(\mathrm{K})}{r_o(\mathrm{m})}\right)^2\right] \quad \text{versus} \quad \frac{1}{T_\infty(\mathrm{K})}$$

in Figure 5.6. This gives information on the bulk kinetic properties of the sawdust. The fact that the data are linear in $(1/T_\infty)$ suggests that the theory we are using can, at least, explain these experimental results. From the slope of the lines, E/R is 9.02×10^3 K and 13.8×10^3 K for oil and no added oil to the sawdust respectively. For $R = 8.314$ J/mole K, we obtain the activation energies of

$$E_{sd} = 115 \text{ kJ/mole and } E_{sd,oil} = 75 \text{ kJ/mole}$$

showing that the oil-impregnated sawdust can react more easily. If we approximate, for illustrative purposes, a thermal conductivity of the sawdust as 0.1 W/m K, then

$$\ln\left[\left(\frac{E}{R}\right)\left(\frac{A\Delta h_c}{k}\right)\right] = \begin{cases} 39.5, & \text{with oil} \\ 49.3, & \text{no oil} \end{cases}$$

or

$$A\Delta h_c = \begin{cases} 1.6 \times 10^{12}\ \mathrm{W/m^3}, & \text{with oil} \\ 1.9 \times 10^{16}\ \mathrm{W/m^3}, & \text{no oil} \end{cases}$$

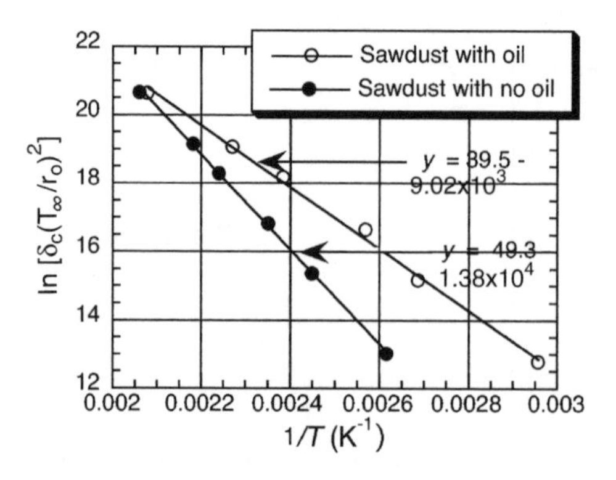

Figure 5.6 Bulk kinetics of sawdust

These results give some appreciation for the kinetic properties. They are 'effective' properties since the actual mechanism of the exothermic reaction is more complex than a zeroth-order Arrhenius model. However, it is seen to be quite satisfactory for these data. The practical question is the accuracy of extrapolating these results to large cubes, or other sawdust configurations and conditions. Nevertheless, the procedure described does give a powerful analytical tool, when combined with good data, to make a quantitative judgment about the propensity of a product to reach spontaneous ignition. Let us consider a practical application.

Example 5.1 Animal feedstock, a distillery byproduct, is to be stored in a silo, a rectangular facility 12 m high and 3 m × 3 m square. For an ambient temperature of 20 °C do we expect a problem? What would be the critical ambient temperature sufficient to cause spontaneous ignition in the silo? The following oven data was obtained for cubes of side $s(= 2r_o)$ from feedstock samples.

s(mm)	Critical oven temperature (°C ± 3 °C)
50	158
75	140
150	114
300	85

Solution Using the procedure just outlined with $\delta_c = 2.52$ for a cube, we can show that $P = 39.88$ in units of r_o in m and $E/R = 8404$ K. Substituting these values into Equation (5.12) yields

$$\delta = 2.081 \times 10^{17} \frac{r_o^2(\mathrm{m})}{T_\infty^2(\mathrm{K})} \exp\left[-\frac{8404}{T_\infty(\mathrm{K})}\right]$$

as a general equation for the Damkohler number corresponding to any r_o and T_∞ for the animal feedstock. This can now be applied to the silo configuration.

For the rectangular silo, we compute the corresponding critical Damkohler number, $\delta_c(r_o)$:

$$\delta_c = 0.88\left[1 + \left(\frac{1.5}{6}\right)^2 + \left(\frac{1.5}{1.5}\right)^2\right] = 1.80$$

Note that this δ_c is defined in terms of $r_o = 1.5$ m. Calculating δ for $r_o = 1.5$ m and $T_\infty = 20\,°\mathrm{C}$ from our feedstock equation gives $\delta = 1.9$. Since the δ for this condition is greater than the critical value of 1.8, we will have spontaneous ignition.

For $\delta = \delta_c = 1.80$ and $r_o = 1.5$ m, we solve for T_∞ to find that the critical ambient temperature is 19.5 °C. Any value T_∞ below 19.5 °C will be safe.

5.4 Time for Spontaneous Ignition

The prospect of successfully estimating the likelihood of spontaneous ignition depends on our ability to know δ_c for our configuration and heating condition, and the bulk kinetic properties (E/R) and $(A\Delta h_c/k)$ of the material. Since the kinetics are very sensitive to temperature, any uncertainties can be magnified by its nonlinear effects. Therefore, it may be pushing our hopes to consider an estimate of the time for spontaneous condition. However, since the self-heating period can be hours or even days before spontaneous ignition occurs, it is useful to examine its dependence. The Frank-Kamenetskii model for a given geometry is too complex to achieve analytical results, so we return to the Semenov model of Section 4.3, Equation (4.13):

$$\frac{d\theta}{d\tau} = \delta e^{\theta} - \theta \tag{5.18}$$

This deals with purely convective heating, but defining δ by Equation (5.12), we take $h = k/r_o$ and $r_o = V/S$ of the Semenov model. Equation (5.18) approximates the constant surface temperature case. Also, here the initial temperature is taken as T_∞. If we allow no convective cooling, the adiabatic equation becomes

$$\frac{d\theta}{d\tau} = \delta e^{\theta} \tag{5.19}$$

Equation (5.19) can also be obtained from the unsteady counterpart of Equation (5.5). If internal conduction is taken as zero in Equation (5.5), this adiabatic equation arises. Equation (5.19) also becomes the basis of an alternative method for determining the bulk kinetic properties. The method is called the 'adiabatic furnace' used by Gross and Robertson [5]. The furnace is controlled to make its temperature equal to the surface temperature of the material, thus producing an adiabatic boundary condition. This method relies on the measurement of the time for ignition by varying either r_o or T_∞, the initial temperature.

Now let us return to determining an estimate for the ignition time with the sample initially at T_∞. (There may be a thermal conduction time needed to bring the sample to T_∞; this is ignored.) The time to ignite under adiabatic conditions will give a lower bound on our estimate. Equation (5.19) can be integrated to give

$$\delta \int_0^{\tau} d\tau = \int_0^{\theta} e^{-\theta} d\theta$$

or

$$\delta\tau = 1 - e^{-\theta} \tag{5.20}$$

where τ is the time (dimensionless) to achieve θ. We assume that the dimensionless critical temperature for ignition is always a constant value. For the nongeometry-dependent Semenov case (from Equation (4.17)),

$$\theta_c = 1$$

and from the constant surface temperature slab case,

$$\theta_{m,c} = 1.18$$

Hence, selecting $\theta_c \approx 1$ appears to be a reasonable universal estimate. Then from Equation (5.20),

$$\delta \tau_{ig} = 1 - e^{-1} = 0.632$$

or

$$\tau_{ig} \approx 0.632/\delta \tag{5.21a}$$

In dimensional terms, this adiabatic ignition time is

$$t_{ig,\ ad} \approx \frac{0.632 r_o^2}{\alpha \delta} \tag{5.21b}$$

If we substitute Equation (5.12) for δ,

$$\boxed{t_{ig,\ ad} \approx 0.632 t_R} \tag{5.21c}$$

where t_R is the chemical reaction time given in Equation (4.24). The characteristic time chosen for making time dimensionless was

$$t_c \equiv \frac{r_o^2}{\alpha}$$

which is the time for thermal diffusion through the medium of the effective radius, r_o. Again, this shows that δ is the ratio of these times:

$$\delta = \frac{t_c}{t_R}$$

Applying Equations (5.21) to the adiabatic time corresponding to the critical Damkohler number, and realizing for a three-dimensional pile of effective radius, $r_o, \delta_c \approx 3$ (e.g. $\delta_c = 3.32$ for a sphere for $Bi \to \infty$), then we estimate a typical ignition time at $\delta = \delta_c \approx 3$ of

$$t_{ig,\ c} = 0.21 r_o^2/\alpha \tag{5.22}$$

To estimate the time to ignite when $\delta > \delta_c$, from Equations (5.21) and (5.22),

$$t_{ig} \approx t_{ig,\ c}\left(\frac{\delta_c}{\delta}\right) = 0.21\left(\frac{r_o^2}{\alpha}\right)\left(\frac{\delta_c}{\delta}\right) \tag{5.23}$$

Equation (5.23) is still based on the adiabatic system. A better, but still approximate, result can be produced from the nonadiabatic Semenov model where

$$\frac{d\theta}{d\tau} = \delta e^{\theta} - \theta \tag{5.24}$$

from Equation (4.13). We know at ignition that

$$\theta \approx 1 \text{ and } \delta_c = e^{-1}, \text{ or } e^{\theta} \approx \delta_c^{-1}$$

Then Equation (5.24), moving toward ignition, might be approximated as

$$\frac{d\theta}{d\tau} \approx \frac{\delta}{\delta_c} - 1 \tag{5.25}$$

Integrating from the initial state to ignition,

$$\int_o^1 d\theta \approx \left(\frac{\delta}{\delta_c} - 1\right)\tau_{ig} \tag{5.26a}$$

or

$$t_{ig} \approx \frac{(r_o^2/\alpha)}{(\delta/\delta_c - 1)} \tag{5.26b}$$

Beever [6] reports an approximation of Boddington *et al.* for the nonadiabatic case, as

$$t_{ig} \approx M \frac{(r_o^2/\alpha)}{\left[\delta(\delta/\delta_c - 1)^{1/2}\right]} \tag{5.27}$$

where M is approximately 1.6. Good accuracy is claimed for $(\delta/\delta_c) < 3$. For a given size, r_o, where the ambient temperature is greater than the critical temperature, the ratio of the Damkohler numbers is

$$\frac{\delta}{\delta_c} = \frac{e^{(T_\infty/T_{\infty,c}-1)[E/(RT_\infty)]}}{(T_\infty/T_{\infty,\,c})^2} \tag{5.28}$$

A typical value of $E/(RT_\infty)$ is about 30, and with an ambient temperature of 10 % above the critical temperature, $\delta/\delta_c = 16.5$. Thus, the time for ignition is greatly reduced as the temperature exceeds its critical value.

References

1. Bowes, P.C., *Self-Heating: Evaluating and Controlling the Hazards*, Elsevier, The Netherlands, 1984.
2. Mitchelle, N.D., New light on self-ignition, *NFPA Quarterly*, October 1951, 139.
3. Frank-Kamenetskii, D.A., *Diffusion and Heat in Chemical Kinetics*, 2nd edn., Plenum Press, New York, 1969.
4. Hill, S.M., Investigating materials from fire using a test method for spontaneous ignition: case studies, MS Thesis, Department of Fire and Protection Engineering, University of Maryland, College Park, Maryland, 1997.

5. Gross, D. and Robertson, A.F., Self-ignition temperatures of materials from kinetic-reaction data, *J. Research National Bureau of Standards*, 1958, **61**, 413–7.
6. Beever, P.F., Self-heating and spontaneous combustion, in *The SFPE Handbook of Fire Protection Engineering*, 2nd edn (eds P.J. DiNenno *et al.*), Section 2, National Fire Protection Association, Quincy, Massachusetts, 1995, p. **2**-186.
7. Gray, B.F. and Halliburton, B., The thermal decomposition of hydrated calcium hypochlorite (UN 2880), *Fire Safety J.*, 2000, **35**, 233–239.

Problems

5.1 In June 1950 an Army Warehouse was destroyed by fire with a \$2 609 000 loss. It occurred two days after a shipment of fiberboard arrived in which a freight car of the shipment had experienced a fire. The remaining fiberboard was stored in a cubical pile of 24 000 cubic feet. Spontaneous ignition was suspected. Data for the fiberboard was developed (Gross and Robertson [5]):

$$\rho = 0.25 \text{ g/cm}^3$$

$$c_p = 0.33 \text{ cal/g K}$$

$$k = 0.000\,12 \text{ cal/s cm K}$$

$$E = 25.7 \text{ kcal/mol}$$

$$A\Delta h_c = 1.97 \times 10^9 \text{ cal/s cm}^3$$

Determine whether an ambient temperature of 50 °C could cause the fire. If not consider 100 °C.

5.2 A cubical stack of natural foam rubber pillows was stored in a basement near a furnace. That part of the basement could achieve 60 °C. A fire occurs in the furnace area and destroyed the building. The night watchman is held for arson. However, an enterprising investigator suspects spontaneous ignition. The stack was 4 meters on a side. The investigator obtains the following data for the foam:

$$\rho = 0.108 \text{ g/cm}^3$$

$$c_p = 0.50 \text{ cal/g K}$$

$$k = 0.000\,096 \text{ cal/s cm K}$$

$$E = 27.6 \text{ kcal/mol}$$

$$A\Delta h_c = 7.48 \times 10^{10} \text{ cal/s cm}^3$$

Does the watchman have a plausible defense? Show calculations to support your view.

5.3 Calculate the radius of a spherical pile of cotton gauze saturated with cottonseed oil to cause ignition in an environment with an air temperature (T_a) of 35 °C and 100 °C. Assume perfect heat transfer between the gauze surface and the air. The gauze was found to follow the Frank-Kamenetskii ignition model, i.e.

$$k\frac{d^2T}{dx^2} = A\Delta h_c e^{-E/(RT)}$$

Small-scale data yield:

$$k = 0.000\,11 \text{ cal/s cm K}$$

$$E = 24\,100 \text{ cal/mol}$$

$$A\Delta h_c = 2.42 \times 10^{11} \text{ cal/s cm}^3$$

Note: $R = 8.31$ J/mol K

5.4 A fire disaster costing \$67 million occurred in Texas City, Texas, on the *SS Grandcamp* (16 April 1987) due to spontaneous ignition of stored fertilizer in the ship's hold. A release of steam from an engine leak caused the atmosphere of the ammonium nitrate fertilizer to be exposed to temperatures of 100 °C. Ammonium nitrate (NH_4NO_3) decomposes exothermically releasing 378 kJ/g mol. Its rate of decomposition can be described by the Arrhenius equation:

$$\dot{m}''' = \rho(6 \times 10^{13})e^{-170\,\text{kJ/g mol}/(RT)} \text{ in kg/m}^3\text{s}$$

where

T is temperature in K

Density, $\rho = 1750$ kg/m^3

Thermal conductivity, $k = 0.1256$ W/m K and

Universal gas constant, $R = 8.3144$ J/g mol K

(a) What is the minimum cubical size of the fertilizer under this steam atmosphere to achieve self-ignition?

(b) Explain what happens to the remaining stored fertilizer after self-ignition of this cube.

5.5 A flat material of thickness ℓ is placed on a hot plate of controlled temperature T_b. The material is energetic and exothermic with a heat of combustion of Δh_c and its reaction is governed by zeroth-order kinetics, $Ae^{-E/(RT)}$, the mass loss rate per unit volume. Notation is as used in the text. The differential equation governing the process to ignition is given as

$$\rho c \frac{\partial T}{\partial t} = Ae^{-E/(RT)} \Delta h_c + k \frac{\partial^2 T}{\partial x^2}$$

where x is the coordinate measured from the hot plate's surface. Let $\xi = x/\ell$ and $\phi = T/T_b$. Convection occurs at the free boundary with a heat transfer coefficient h_c and the surrounding temperature at T_∞. Invoke the approximation that $\phi - 1$ is small (see Equation (4.8)), and use the new dimensionless temperature variable $\theta = (E/(RT_\infty)(\phi - 1)$. Write the governing equation and the boundary and initial conditions in terms of the dimensionless variables : θ, ξ and τ,

5.6 A polishing cloth residue is bundled into a cube of 1 m on the half-side dimension. It is stored in an environment of 35 °C and the convective heat transfer is very good between the air and the bundle. Experiments show that the cloth residue has the following properties:

Thermal conductivity, $k = 0.036$ W/m K

Density, $\rho = 68.8$ kg/m^3

Specific heat, $c = 670$ J/kg K

Heat of combustion, $\Delta h_c = 15$ kJ/g

Kinetic properties:

$A = 9.98 \times 10^4$ kg/m^3 s

$E/R = 8202$ K

(a) Show that it is possible for this bundle to spontaneously ignite.

(b) Compute an estimate for the time for ignition to occur.

5.7 A material is put on a hot plate. It has a cross-sectional area, S, and thickness, r_o. The power output of the plate is P. The plate and edges of the material are perfectly insulated.

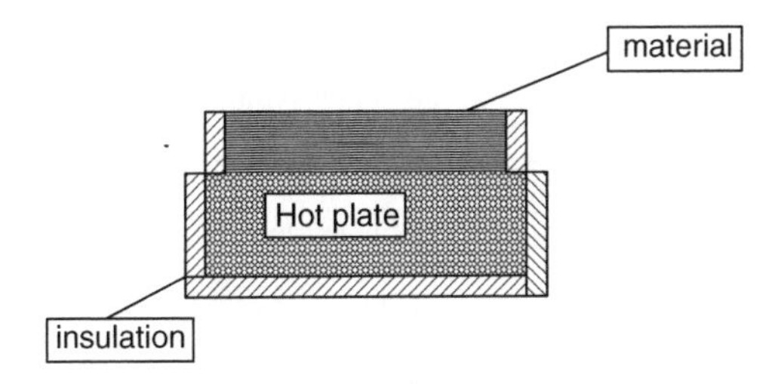

The material can be chemically exothermic, and follows the theory:

$$\frac{d^2\theta}{d\xi^2} + \delta e^{\theta} = 0$$

where

$\xi = x/r_o$, $\quad \theta = \left(\frac{T-T_\infty}{T_\infty}\right)\left(\frac{E}{RT_\infty}\right)$, $\quad \delta$ is the Damkohler number

Boundary conditions for the material:

At the hot plate: $-k\frac{dT}{dx} = P/S$

At the ambient: $-k\frac{dT}{dx} = h_c(T - T_\infty)$

with T_∞ the ambient temperature and h_c the convective coefficient.

From the boundary conditions, show the dimensionless parameters that the critical Damkohler number will depend on.

5.8 Hydrated calcium hypochlorite exothermically decomposes to release oxygen. The heat of decomposition is relatively small, but the resulting limited increase in temperature coupled with the increase in oxygen concentration can cause the ignition of adjacent items. Ignition has occurred in its storage within polyethylene (PE) kegs. Consider the keg to be an equicylinder – a cylinder of radius r_o and height $2r_o$. The critical Damkohler number and critical temperature are found for this geometry to be $\delta_c = 2.76$ and $\theta_c = 1.78$ for a large *Bi* number. Data collected by Gray and Halliburton [7] are given below for an equicylinder geometry.

Radius, r_o (m)	Critical temperature (°C)
0.00925	149.5, 154.2
0.01875	135.5, 141.5
0.02875	126.5, 128.5
0.055	120.5, 123.0
0.075	102.5, 101.0
0.105	87.5, 90.6
0.175	64.0
0.175	60.1 (in PE keg)

(a) Plot the data as in Figure 5.6. Show that there are two reactions. Find the Arrhenius parameters and the temperature range over which they apply. The thermal conductivity can be taken as 0.147 W/m K.

(b) For an equicylinder exposed to an ambient temperature of 45 °C, determine the critical radius for ignition, and the center temperature at the onset of ignition.

5.9 The parameter δ entered into the solutions for spontaneous ignition.

(a) What is the name associated with this dimensionless parameter?

(b) Explain its significance in the Semenov and Frank-Kamenetskii models with respect to whether ignition is possible.

(c) For a given medium, spontaneous ignition depends on ambient temperature as well as other factors. Name one more.

(d) Once conditions are sufficient for the occurrence of spontaneous ignition, it will always occur very quickly. True or false?

5.10 The time for spontaneous ignition to occur will decrease as the Damkohler number _____increases or _____decreases with respect to its critical value.

5.11 List three properties or parameters that play a significant role in causing spontaneous ignition of solid commodities.

6

Ignition of Liquids

6.1 Introduction

The ignition of a liquid fuel is no different from the processes described for autoignition and piloted ignition of a mixture of gaseous fuel, oxidizer and a diluent, with one exception. The liquid fuel must first be evaporated to a sufficient extent to permit ignition of the gaseous fuel. In this regard, the fuel concentration must at least be at the lower flammable limit for piloted ignition, and within specific limits for autoignition. The liquid temperature corresponding to the temperatures at which a sufficient amount of fuel is evaporated for either case is dependent on the measurement apparatus. For piloted ignition, the position of the ignitor (spark) and the mixing of the evaporated liquid fuel with air is critical. For autoignition, both mixing, chamber size and heat loss can make a difference (see Figure 6.1). Temperature and fuel concentration play a role, as illustrated in Figure 6.2 for piloted ignition in air. The fuel can only be ignited in a concentration range of X_L to X_U, and a sufficient temperature must exist in this region, corresponding to piloted and autoignition respectively. These profiles of X_F and T depend on the thermodynamic, transport properties and flow conditions. We will see that the heat of vaporization (h_{fg}) is an important thermodynamic property that controls the evaporation rate and hence the local concentration. The smallest heating condition to allow for piloted ignition would be with a spark at the surface when $X_{F,s} = X_L$, the lower flammable limit.

6.2 Flashpoint

The liquid temperature (T_L) corresponding to X_L is measured for practical purposes in two apparatuses known as either the 'closed' or 'open' cup flashpoint test, e.g. ASTM D56 and D1310. These are illustrated in Figure 6.3. The surface concentration (X_s) will be shown to be a unique function of temperature for a pure liquid fuel. This temperature is known as the saturation temperature, denoting the state of thermodynamic equilibrium

Fundamentals of Fire Phenomena James G. Quintiere

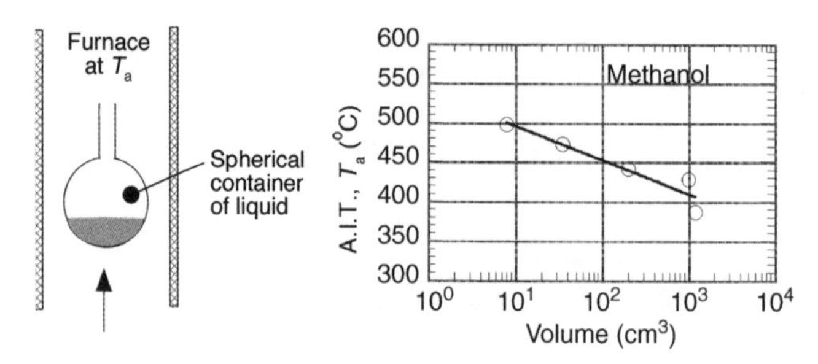

Figure 6.1 Autoignition of an evaporated fuel in a spherical vessel of heated air (taken from Setchkin [1])

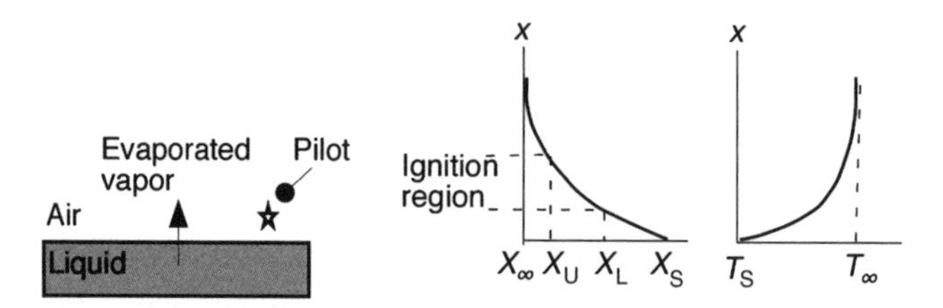

Figure 6.2 Dynamics of piloted ignition for a liquid fuel in air under convective heating from the atmosphere

between a liquid and its vapor. Theoretically, the saturation temperature corresponding to X_L is the lower limit temperature, T_L. Because the spark location is above the liquid surface and the two configurations affect the profile shape of X_F above the liquid, the ignition temperature corresponding to the measurement in the bulk liquid can vary between the open and closed cup tests. Typically, the closed cup might insure a lower T at ignition than the open cup. The minimum liquid temperature at which a flame is seen at the spark is called the flashpoint. This temperature may need to be raised to achieved sustained burning for the liquid, and this increased temperature is called the firepoint. The flashpoint corresponds to the propagation of a premixed flame, while the firepoint corresponds to the sustaining of a diffusion flame (Figure 6.4). In these test apparatuses, the liquid fuel is heated or cooled to a fixed temperature and the spark is applied until the

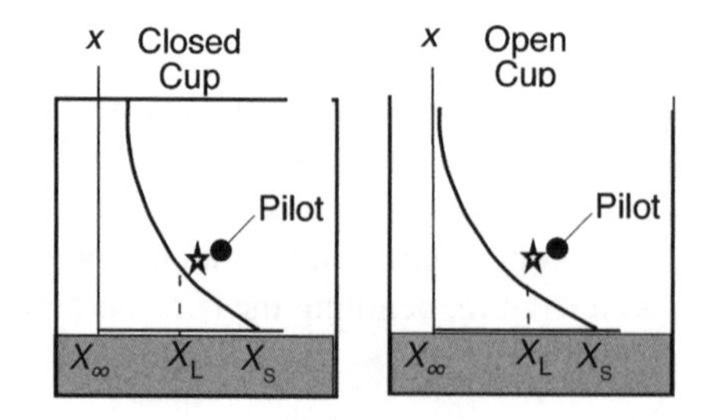

Figure 6.3 Fuel concentration differences in the closed and open cup flashpoint tests

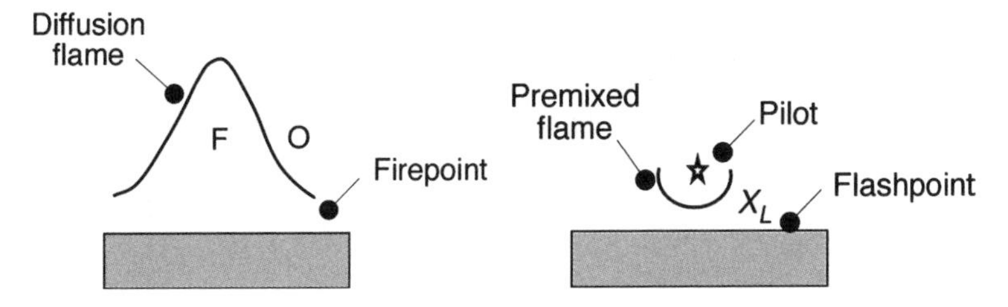

Figure 6.4 Flashpoint and firepoint, liquid fuel at a uniform temperature

lowest temperature for ignition is found. These 'flashpoints' are commonly reported for the surrounding air at 25 °C and 1 atm pressure. They will vary for atmospheres other than air since X_L changes. However, X_L tends to range from about 1 to 8 % for many common conditions. Thus, a relatively small amount of fuel must be evaporated to permit ignition.

It is useful to put in perspective the range of temperature conditions during the combustion of a liquid fuel. In increasing order:

1. T_L = flashpoint or the saturation temperature corresponding to the lower flammable limit.
2. T_b = boiling point corresponding to the saturation pressure of 1 atm.
3. T_a = autoignition temperature corresponding to the temperature that a mixture of fuel and air can self-ignite. Usually this is measured at or near a stoichiometric mixture.
4. $T_{f,ad}$ = adiabatic flame temperature corresponding to the maximum achievable temperature after combustion. This is usually reported for a stoichiometric mixture of fuel in air.

Table 6.1 gives these temperature data along with the heat of vaporization and heat of combustion, two significant variables. The former controls the evaporation process and the latter is significant for sustained burning. To have sustained burning it is necessary that $\Delta h_c > h_{fg}$; h_{fg} is usually a weak function of temperature and is usually taken as constant for a fuel at temperatures relevant to the combustion process. Table 6.2 gives some additional relevant properties.

6.3 Dynamics of Evaporation

Let us examine methanol. Its flashpoint temperature is 12 to 16 °C (285–289 K) or, say, 15 °C. If this is in an open cup, then the concentration near the surface is $X_L = 6.7\,\%$. Performed under normal room temperatures of, say, 25 °C, the temperature profile would be as in Figure 6.2. This must be the case because heat must be added from the air to cause this evaporated fuel vapor at the surface. This decrease in temperature of an evaporating surface below its environment is sometimes referred to as evaporative cooling. If the convective heat transfer coefficient, typical of natural convection, is,

Table 6.1 Approximate combustion properties of liquid fuels in air from various sources [2,3]

Fuel	Formula	T_L(K)		T_b	T_a	$T_{f,ad}$[a]	X_L	h_{fg}	Δh_c^b
		Closed	Open	(K)	(K)	(K)	(%)	(kJ/g)	(kJ/g)
Methane	CH_4	—	—	111	910	2226	5.3	0.59	50.2
Propane	C_3H_8	—	169	231	723	2334	2.2	0.43	46.4
n-Butane	C_4H_{10}	—	213	273	561	2270	1.9	0.39	45.9
n-Hexane	C_6H_{14}	251	247	342	498	2273	1.2	0.35	45.1
n-Heptane	C_7H_{16}	269	—	371	—	2274	1.2	0.32	44.9
n-Octane	C_8H_{18}	286	—	398	479	2275	0.8	0.30	44.8
n-Decane	$C_{10}H_{22}$	317	—	447	474	2277	0.6	0.28	44.6
Kerosene	$\sim C_{14}H_{30}$	322	—	505	533	—	0.6	0.29	44.0
Benzene	C_6H_6	262	—	353	771	2342	1.2	0.39	40.6
Toluene	C_7H_8	277	280	383	753	2344	1.3	0.36	41.0
Naphthalene	$C_{10}H_8$	352	361	491	799	—	0.9	0.32	40.3
Methanol	CH_3OH	285	289	337	658	—	6.7	1.10	20.8
Ethanol	C_2H_5OH	286	295	351	636	—	3.3	0.84	27.8
n-Butanol	C_4H_9OH	302	316	390	616	—	11.3	0.62	36.1
Formaldehyde	CH_2O	366	—	370	703	—	7.0	0.83	18.7
Acetone	C_3H_6O	255	264	329	738	2121	2.6	0.52	29.1
Gasoline	—	228	—	306	644	—	1.4	0.34	44.1

[a] Based on stoichiometric combustion in air.
[b] Based on water and fuel in the gaseous state.

say, 5 W/m^2 K, then the heat transfer is

$$\dot{q}'' = (5\ W/m^2\ K)(25 - 15)\ K = 50\ W/m^2$$

A corresponding mass flux of fuel leaving the surface and moving by diffusion to the pure air with $Y_{F,\infty} = 0$ would be given approximately as

$$\dot{m}''_F = h_m(Y_F(0) - Y_{F,\infty}) \tag{6.1}$$

Table 6.2 Approximate physical properties of liquid fuels from various sources [4]

Fuel	Formula	Liquid properties[a]		Vapor Properties	
		$\rho(kg/m^3)$	c_p(J/g K)	$\rho(kg/m^3)$	c_p(J/g K)
Methane	CH_4	—	—	0.72(0 °C)	2.2
Propane	C_3H_8	—	2.4	2.0(0 °C)	1.7
n-Butane	C_4H_{10}	—	2.3	2.62(0 °C)	1.7
n-Hexane	C_6H_{14}	527	2.2	—	1.7
n-Heptane	C_7H_{16}	676	2.2	—	1.7
n-Octane	C_8H_{18}	715	2.2	—	1.7
Benzene	C_6H_6	894	1.7	—	1.1
Toluene	C_7H_8	896	1.7	—	1.5
Naphthalene	$C_{10}H_8$	1160	1.2	—	1.4
Methanol	CH_3OH	790	2.5	—	1.4
Ethanol	C_2H_5OH	790	2.4	1.5 (100 °C)	1.7
n-Butanol	C_4H_9OH	810	2.3	—	1.7
Acetone	C_3H_6O	790	2.2	—	1.3

[a] At normal atmospheric temperature and pressure.

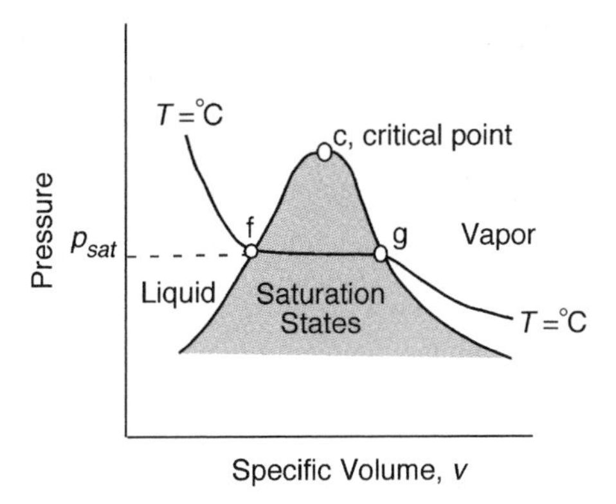

Figure 6.5 p–v Liquid–vapor diagram

where h_{m} is a convective mass transfer coefficient. $Y_{\mathrm{F}}(0)$ theoretically corresponds to the lower flammable limit, or

$$Y_{\mathrm{F}}(0) = (0.067)\left(\frac{M_{\mathrm{meth}}}{M_{\mathrm{mix}}}\right) = 0.067\left(\frac{32}{29}\right) = 0.073$$

Up to now we have presented this example without any regard for consistency, i.e. satisfying thermodynamic and conservation principles. This fuel mass flux must exactly equal the mass flux evaporated, which must depend on $\dot{q}''$ and h_{fg}. Furthermore, the concentration at the surface where fuel vapor and liquid coexist must satisfy thermodynamic equilibrium of the saturated state. This latter fact is consistent with the overall approximation that local thermodynamic equilibrium applies during this evaporation process.

The conditions that apply for the saturated liquid–vapor states can be illustrated with a typical p–v, or $(1/\rho)$, diagram for the liquid–vapor phase of a pure substance, as shown in Figure 6.5. The saturated liquid states and vapor states are given by the locus of the f and g curves respectively, with the critical point at the peak. A line of constant temperature T is sketched, and shows that the saturation temperature is a function of pressure only, $T_{\mathrm{sat}}(p)$ or $p_{\mathrm{sat}}(T)$. In the vapor regime, at near normal atmospheric pressures the perfect gas laws can be used as an acceptable approximation, $pv = (R/M)T$, where R/M is the specific gas constant for the gas of molecular weight M. Furthermore, for a mixture of perfect gases in equilibrium with the liquid fuel, the following holds for the partial pressure of the fuel vapor in the mixture:

$$\boxed{p_{\mathrm{F}} = p_{\mathrm{sat}}(T) = X_{\mathrm{F}}p} \tag{6.2}$$

from Dalton's law, where p is the mixture pressure. Since theoretically $X_{\mathrm{F}} = X_{\mathrm{L}}$ at the condition of piloted ignition, the surface temperature T must satisfy the thermodynamic state equation $p_{\mathrm{F}} = p_{\mathrm{sat}}(T)$ for the fuel. We will derive an approximate relationship for this state equation in Section 6.4.

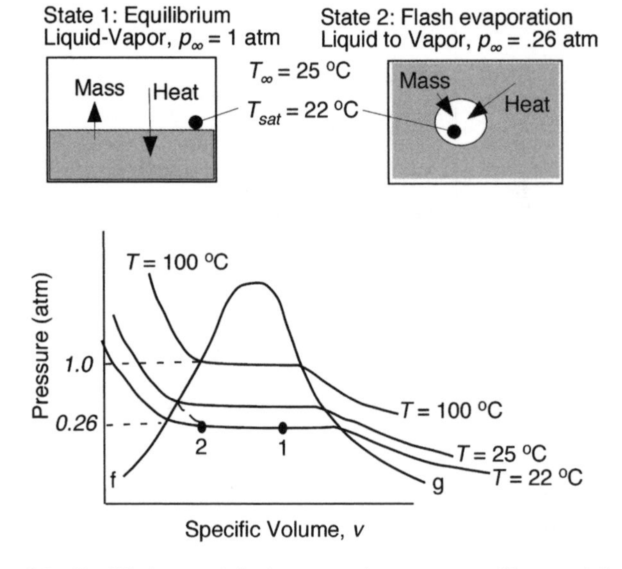

Figure 6.6 Equilibrium and flash evaporation processes illustrated for water

Here we are talking about evaporation under thermodynamic equilibrium. We can also have evaporation under nonequilibrium conditions. For example, if the pressure of a liquid is suddenly dropped below its saturation pressure, flash evaporation will occur. The resulting vapor will be at the boiling point or saturation temperature corresponding to the new pressure, but the bulk of the original liquid will remain (out of equilibrium) at the former higher temperature. Eventually, all of the liquid will become vapor at the lower pressure. The distinction between flash evaporation and equilibrium evaporation is illustrated in Figure 6.6 for water.

An interesting relevant accident scenario is the 'explosion' of a high-pressure tank of fuel under heating conditions. This is illustrated in Figure 6.7 for a propane rail tank car whose escaping gases from its safety valve have been ignited. The overturned tank is heated, and its liquid and vapor segregate due to gravity, as shown. The tank metal shell adjacent to the vapor region cannot be cooled as easily compared to that by boiling in the

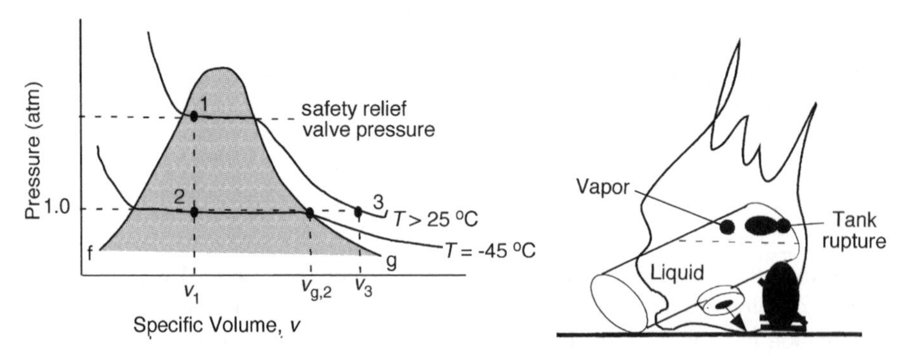

Figure 6.7 BLEVE Illustration

liquid region. Eventually the heated metal's structural properties weaken due to a temperature rise, and a localized rupture occurs in the vapor region. At this instant (before vapor expansion can occur), we have a constant volume (and constant mass) process for the liquid–vapor propane mixture. This is illustrated by 1 to 2 on Figure 6.7. Since this is a sudden event, the contents do not have a chance to cool to the new equilibrium state. The contents of the tank now seek the external pressure, $p_2 = 1$ atm. By the perfect gas law we have a considerable increase in vapor volume,

$$v_3 = \left(\frac{R}{M}\right)\left(\frac{T_1}{p_2}\right) > v_{g,2} > v_{g,1} \tag{6.3}$$

which in turn causes a more dramatic rupture of the tank. In this case, lack of thermodynamic equilibrium for the gas phase in achieving v_3, instead of the smaller $v_{g,2}$, exacerbates the problem. If the leakage at the rupture cannot accommodate this change in vapor volume, a pressure rise will cause a more catastrophic rupture. This phenomenon is commonly called a BLEVE (boiling liquid expanding vapor explosion). Indeed, the liquid–vapor system will expand to meet the atmospheric pressure, $p_2 = 1$ atm, at the temperature T_1 and all of the mass (liquid and vapor) will reach state 3. The boiling point of the liquid propane is 231 K (−42 °C), which is well below normal atmospheric temperatures. Not only will the liquid receive significant heat transfer, but the new equilibrium state 3 does not allow any liquid to exist. This constant pressure process 2 → 3 is referred to as a flash evaporation of the liquid. The liquid state cannot be maintained for atmospheric conditions of approximately 25 °C and 1 atm since it is above the boiling point of −42 °C. The actual process from 1 to 3 is not in two steps as shown, since it depends on the rupture process. The new volume based on v_3 and the mass of the fuel can be considerable. This sudden release of a 'ball' of gaseous fuel leads to a fireball culminating in a mushroom shaped fire that has a relatively short lifetime.

6.4 Clausius–Clapeyron Equation

The Clausius–Clapeyron equation provides a relationship between the thermodynamic properties for the relationship $p_{sat} = p_{sat}(T)$ for a pure substance involving two-phase equilibrium. In its derivation it incorporates the Gibbs function (G), named after the nineteenth century scientist, Willard Gibbs. The Gibbs function per unit mass is defined as

$$G/m = g \equiv h - Ts \tag{6.4}$$

where s is entropy per unit mass.

For a pure substance at rest with no electrical or magnetic effects, the first law of thermodynamics for a closed system is expressed for a differential change as

$$\mathrm{d}u = \delta q - \delta w \tag{6.5}$$

where both q and w are incremental heat added and work done by the substance per unit mass. For this simple system (see Section 2.4.1)

$$\delta w = p\,\mathrm{d}v \tag{6.6}$$

if no viscous or frictional effects are present, and

$$\delta q = T\,\mathrm{d}s \tag{6.7}$$

if the process is reversible according to the second law of thermodynamics. Combining Equations (6.5) to (6.7) gives a relationship only between thermodynamic (equilibrium) properties for this (simple) substance,

$$\mathrm{d}u = T\,\mathrm{d}s - p\,\mathrm{d}v \tag{6.8}$$

Equation (6.8) is a general state equation for the pure substance. From the definition of enthalpy, $h = u + pv$, Equation (6.7) can be expressed in an alternative form:

$$\mathrm{d}h = \mathrm{d}u + p\,\mathrm{d}v + v\,\mathrm{d}p \tag{6.8a}$$

$$\mathrm{d}h = T\,\mathrm{d}s + v\,\mathrm{d}p$$

$$\text{then}\quad \mathrm{d}g = \mathrm{d}h - T\,\mathrm{d}s - s\,\mathrm{d}T \tag{6.8b}$$

$$\text{or}\quad \mathrm{d}g = v\,\mathrm{d}p - s\,\mathrm{d}T$$

Let us apply Equation (6.8) to the two-phase liquid–vapor equilibrium requirement for a pure substance, namely $p = p(T)$ only. This applies to the mixed-phase region under the 'dome' in Figure 6.5. In that region along a p-constant line, we must also have T constant. Then for all state changes along this horizontal line, under the p–v dome, $\mathrm{d}g = 0$ from Equation (6.8b). The pure end states must then have equal Gibbs functions:

$$g_\mathrm{f}(T) = g_\mathrm{g}(T) \tag{6.9}$$

Consequently, it is permissible to equate the derivatives, as these are continuous functions,

$$\frac{\mathrm{d}g_\mathrm{f}}{\mathrm{d}T} = \frac{\mathrm{d}g_\mathrm{g}}{\mathrm{d}T} \tag{6.10}$$

From Equation (6.8b), by division with $\mathrm{d}T$ or by a formal limit operation, we recognize that in this two-phase region,

$$\frac{\mathrm{d}g}{\mathrm{d}T} = v\frac{\mathrm{d}p}{\mathrm{d}T} - s \tag{6.11}$$

From Equation (6.8a) in this two-phase region along a $p =$ constant line,

$$T\mathrm{d}s = \mathrm{d}h$$

However, T is also constant here, so integrating between the two limit extremes gives

$$T(s_\mathrm{g} - s_\mathrm{f}) = h_\mathrm{g} - h_\mathrm{f} \tag{6.12}$$

Substitution of Equation (6.11) into Equation (6.10) gives

$$v_f \frac{dp}{dT} - s_f = v_g \frac{dp}{dT} - s_g \tag{6.13}$$

By combining Equations (6.12) and (6.13), it follows that

$$\boxed{\frac{dp}{dT} = \frac{s_g - s_f}{v_g - v_f} = \left(\frac{1}{T}\right)\left(\frac{h_g - h_f}{v_g - v_f}\right)} \tag{6.14}$$

This is known as the Clausius–Clapeyron equation. It is a state relationship that allows the determination of the saturation condition $p = p(T)$ at which the vapor and liquid are in equilibrium at a pressure corresponding to a given temperature.

An important extension of this relationship is its application to the evaporation of liquids into a given atmosphere, such as air. Consider the evaporation of water into air ($X_{O_2} = 0.21$ and $X_{N_2} = 0.79$). Suppose the air is at 21 °C. If the water is in thermal equilibrium with the air, also at 21 °C, its vapor at the surface must have a vapor pressure of 0.0247 atm (from standard Steam Tables). However, if the water and the air are at the same temperature, no further heat transfer can occur. Therefore no evaporation can take place. We know this cannot be true. As discussed, in the phenomenon of 'evaporative cooling', the surface of the liquid water will have to drop in temperature until a new equilibrium $p(T)$ can satisfy the conservation laws, i.e.

$$\text{Mass evaporation rate} = \text{diffusion transport rate to the air}$$
$$\text{Heat transfer rate} = \text{energy required to vaporize}$$

Let us say that this new surface equilibrium temperature of the water is 15.5 °C; then the vapor pressure is 0.0174 atm. At the surface of the water, not only does water vapor exist, but so must the original gas mixture components of air, O_2 and N_2. By Dalton's law the molar concentration of the water vapor is

$$X_{H_2O}(15.5\,^\circ\text{C}) = 0.0174$$

and the new values for O_2 and N_2 are

$$X_{O_2} = 0.21(1 - 0.0174) = 0.2063$$
$$X_{N_2} = 0.79(1 - 0.0174) = 0.776$$

For this value of the water vapor concentration at the surface, $X_{H_2O}(15.5\,^\circ\text{C}) = 0.0174$, the relative humidity (ϕ) at this point for the atmosphere at 21 °C is 0.0174/0.0247 or 70.4 %.

In general, relative humidity is given as

$$\boxed{\phi = \frac{p_{H_2O}(T_\infty)}{p_g(T_\infty)} \times 100\,\%} \tag{6.15}$$

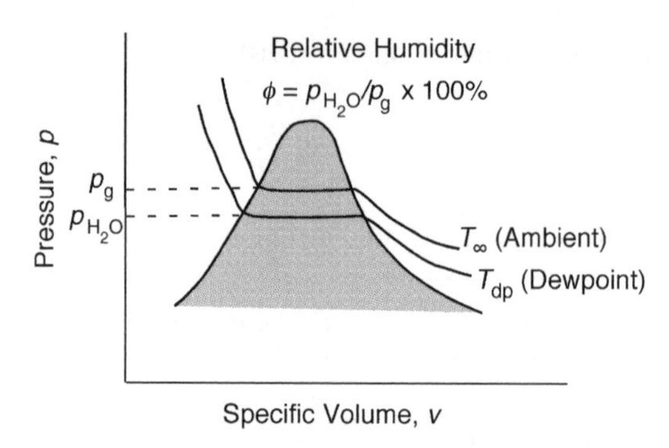

Figure 6.8 Relative humidity and dew point

where p_{H_2O} is the partial pressure of water vapor in the atmosphere and p_g is the water saturation pressure at the temperature of the atmosphere, T_∞. The dew point (T_{dp}) is the saturation temperature corresonding to $p_{H_2O}(T_\infty)$. The relationship of these temperatures are illustrated in Figure 6.8. In other words, if the atmosphere at a given relative humidity (RH) is suddenly reduced in temperature to T_{dp}, the new state becomes 100 % relative humidity. Any further cooling will cause the water vapor in the atmosphere to condense. This process causes the dew on the grass in early morning as the blades of grass have undergone radiative cooling to the black night sky and their temperature has dropped below the dew point of the 'moist' atmosphere.

By assuming that a perfect gas law applies for the vapor, including the saturation state, it can be seen that

$$\left(\frac{\phi}{100}\right) = \frac{p_{H_2O}}{p_g} = \frac{RT_\infty/v_{H_2O}M_{H_2O}}{RT_\infty/v_g M_{H_2O}} = \frac{v_g}{v_{H_2O}} = \frac{\rho_{H_2O}}{\rho_g} \tag{6.16}$$

Thus at 100 % relative humidity the air atmosphere contains the most mass of water vapor possible for the temperature T_∞. No further evaporation of liquid water is possible.

The introduction of the perfect gas law to the Clausius–Clapeyron equation (Equation (6.14)) allows us to obtain a more direct approximation to $p = p(T)$ in the saturation region. We use the following:

1. $pv_g = (R/M)T$.
2. $h_g - h_f \equiv h_{fg}$, the heat of vaporization at temperature T.
3. $v_g \gg v_f$, $v_g - v_f \approx v_g$, except near the critical point.

Substituting into Equation (6.14), we obtain

$$\frac{dp}{dT} = \frac{h_{fg}(T)}{T^2 R/(Mp)} \tag{6.17}$$

If we assume that $h_{fg}(T)$ is independent of T, reasonable under expected temperatures, then we can integrate this equation. The condition selected to evaluate the constant of integration is $p = 1$ atm, $T = T_b$, which is the definition of the boiling point usually reported (see Table 6.1). With M_g the molecular weight of the evaporated component,

$$\int_{p_\infty = 1\,\text{atm}}^{p} \frac{dp}{p} = \frac{h_{fg}}{R/M_g} \int_{T_b}^{T} \frac{dT}{T^2}$$

or

$$\ln\left(\frac{p}{p_\infty}\right) = -\frac{h_{fg} M_g}{RT}\Bigg|_{T=T_b}^{T=T}$$

or

$$\boxed{X_g = \frac{p}{p_\infty} = e^{-(h_{fg}M_g/R)(1/T - 1/T_b)}} \qquad (6.18)$$

where $p_\infty = 1$ atm for the normally defined boiling point, or in general is the mixture pressure and T_b is the saturation temperature where $p = p_\infty$. Notice that this equation has a similar character to the Arrhenius form, with $h_{fg}M_g$ playing the role of an activation energy (kJ/mole).

Example 6.1 Estimate the minimum piloted ignition temperature for methanol in air which is at 25 °C and 1 atm.

Solution Use the data in Table 6.1:

$T_b = 337 \text{ K} = 64\,°\text{C}$

$h_{fg} = 1.10 \text{ kJ/g}$

$M_g = 12 + 4 + 16 = 32 \text{ g/mole}$

$R = 8.315 \text{ J/mole K}$
$\hat{h}_{fg} = (1.1)(32) = 35.2 \text{ kJ/mole}$

The minimum temperature for piloted ignition is given by $X_L = X_g(T_L)$. From Equation (6.18),

$$0.067 = \exp\left[-\frac{35.2 \text{ kJ/mole}}{8.315 \times 10^{-3} \text{ kJ/mole K}}\left(\frac{1}{T_L} - \frac{1}{337 \text{ K}}\right)\right]$$

Solving gives

$$T_L = 277 \text{ K} = 4\,°\text{C}$$

This should be compared to the measured open or closed cups of 285 and 289 K. We expect the computed value to be less, since the ignitor is located above the surface and mixing affects the ignitor concentration of fuel. Thus, a higher liquid surface temperature is needed to achieve X_L at the ignitor.

6.5 Evaporation Rates

Let us consider a case of steady evaporation. We will assume a one-dimensional transport of heat in the liquid whose bulk temperature is maintained at the atmospheric temperature, T_∞. This would apply to a deep pool of liquid with no edge or container effects. The process is shown in Figure 6.9. We select a differential control volume between x and $x + dx$, moving with a surface velocity $(-(dx_0/dt)\,\boldsymbol{i})$. Our coordinate system is selected with respect to the moving, regressing, evaporating liquid surface. Although the control volume moves, the liquid velocity is zero, with respect to a stationary observer, since no circulation is considered in the contained liquid.

We apply the conservation of mass to the control volume:

$$\frac{d}{dt}(\bar{\rho}dx) + \rho(x+dx)\left[0 - \left(\frac{dx_0}{dt}\,\boldsymbol{i}\right)\right]\cdot(+\boldsymbol{i}) + \rho(x)\left[0 - \left(\frac{dx_0}{dt}\,\boldsymbol{i}\right)\right]\cdot(-\boldsymbol{i}) = 0$$

where $\bar{\rho}$ is the mean density between x and $x + dx$. As we formerly divide by dx and take the limit as $dx \to 0$, with $\bar{\rho}(x + dx)$ expressed in terms of x by Taylor's expansion theorem

$$\rho(x+dx) \approx \rho(x) + \frac{\partial \rho}{\partial x}dx$$

we obtain

$$\left(\frac{\partial \rho}{\partial t}\right)_x - v_0\left(\frac{\partial \rho}{\partial x}\right)_t = 0 \tag{6.19}$$

where $v_0 = dx_0/dt$. Since the density is constant in the liquid, this equation is automatically satisfied.

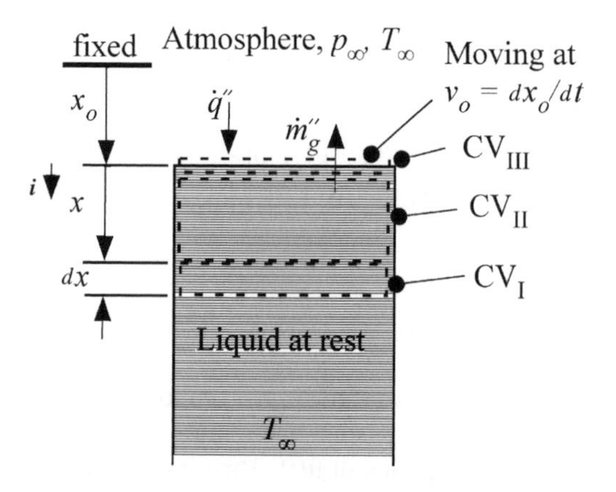

Figure 6.9 Liquid evaporation

It is more significant to take a control volume between $x = 0$ (in the gas) and any x in the liquid. For this control volume (CV_{II}),

$$\frac{d}{dt}\int_0^x \rho \, dx + \rho(x)[0 - (v_0 \boldsymbol{i})] \cdot (+\boldsymbol{i}) + \dot{m}_g'' = 0$$

where $\dot{m}_g''$ is the mean flux of vapor evaporated at the surface. Since ρ is constant

$$\boxed{\rho v_0 = \dot{m}_g''} \tag{6.20}$$

The energy equation can be derived in a similar fashion for CV_I. The liquid is considered to be at the same pressure as the atmosphere and therefore and Equation (3.48) applies:

$$\rho c_p \frac{d}{dt}(\bar{T}\, dx) + \rho c_p v_0 [T(x) - T(x + dx)] = -\left[\dot{q}_k''\right]\Big|_x^{x+dx}$$

where $\dot{q}_k''$ is the heat flux in the x direction.

Expanding and taking the limit as $dx \to 0$,

$$\rho c_p \frac{\partial T}{\partial t} - \rho c_p v_0 \frac{\partial T}{\partial x} = -\frac{\partial \dot{q}_k''}{\partial x} \tag{6.21}$$

Considering Fourier's law for heat conduction,

$$\dot{q}_k'' = -k\frac{\partial T}{\partial x} \tag{6.22}$$

Then we combine Equations (6.21) and (6.22) to obtain

$$\boxed{\rho c_p \frac{\partial T}{\partial t} - \rho c_p v_0 \frac{\partial T}{\partial x} = k\frac{\partial^2 T}{\partial x^2}} \tag{6.23}$$

for k = constant.

At $x = 0$, we can employ a control volume (CV_{III}) just surrounding the surface. For this control volume, the conservation of mass is Equation (6.20). The conservation of energy becomes

$$\text{at } x = 0: \qquad \dot{m}_g'' h_g - \rho v_0 h_f = \dot{q}'' - \dot{q}_k'' \tag{6.24}$$

where $\dot{q}''$ is the net incident heat flux to the surface. Note that there is no volumetric term in the equation since the volume is zero for this 'thin' control volume. From Equations (6.20) and (6.22), we obtain

$$\text{at } x = 0: \qquad \dot{m}_g'' h_{fg} = \dot{q}'' + k\left(\frac{\partial T}{\partial x}\right)_{x=0} \tag{6.25}$$

We also know at $x = 0$, $T = T_s$ that the saturation temperature at $p_g(0)$ or $X_g(0) \equiv X_s = p_g(0)/p_\infty$, in terms of molar concentrations. At $x \to \infty$, $T = T_\infty$ for a deep pool of liquid. In summary, for steady evaporation, Equation (6.23) becomes

$$-v_0 \frac{dT}{dx} = \alpha \frac{d^2T}{dx^2} \tag{6.26}$$

The thermal diffusivity $\alpha = k/\rho c_p$. We have a second-order differential equation requiring two boundary conditions:

$$x = 0, \; T = T_s \tag{6.27a}$$

$$x \to \infty, \; T = T_\infty \tag{6.27b}$$

The evaporation rate is also unknown. However, for pure convective heating, we can write

$$\dot{q}'' = h(T_\infty - T_s) \tag{6.28}$$

The convective heat transfer coefficient may be approximated as that due to heat transfer without the presence of mass transfer. This assumption is acceptable when the evaporation rate is small, such as drying in normal air, and for conditions of piloted ignition, since X_L is typically small. Mass transfer due to diffusion is still present and can be approximated by

$$\dot{m}''_g = \left(\frac{h}{c_g}\right)(Y_s - Y_\infty) \tag{6.29}$$

where the convective mass transfer coefficient is given as h/c_g (c_g is the specific heat at constant pressure for the gas mixture) with Y_∞ as the mass fraction of the transferred phase (liquid $\to$ gas) in the atmosphere. For the evaporation of water into dry air, $Y_\infty = 0$.

If we consider the unknowns in the above equations, we count four:

$$T_s, \; Y_s, \; \dot{m}_g \;\; \text{and} \;\; \dot{q}''$$

The differential equation can be solved to give $T = T(x, T_s, T_\infty, v_0/\alpha)$. The four unknowns require four equations: Equations (6.26), (6.28) and (6.29), together with the Clausius–Clapeyron equation in the form of

$$Y_s(T_s) = \frac{M_g}{M}\, e^{-(M_g h_{fg}/R)(1/T_s - 1/T_b)} \tag{6.30}$$

where M is the molecular weight of the gas mixture ($M \approx 29$ for air mixtures). Furthermore, it can be shown that where the mass transfer rate is small, T_s will be only several tens of degrees, at most, below T_∞. Hence we might avoid solving explicity for T_s by approximating

$$Y_s(T_s) \approx Y_s(T_\infty) \tag{6.31}$$

Let us complete the problem by solving Equation (6.26). Let

$$\psi = \frac{dT}{dx}$$

Therefore,

$$\frac{d\psi}{dx} = -\frac{v_0}{\alpha}\psi$$

Integrating once gives

$$\psi = c_1 e^{-(v_0/\alpha)x}$$

and integrating again gives

$$T = c_1\left(\frac{\alpha}{v_0}\right)e^{-(v_0/\alpha)x} + c_2 \tag{6.32}$$

By the boundary conditions of Equation (6.27),

$$c_2 = T_\infty$$

and

$$T_s - T_\infty = c_1\left(\frac{\alpha}{v_0}\right)$$

Then, the solution can be expressed as

$$\boxed{T - T_\infty = (T_s - T_\infty)e^{-(v_0/\alpha)x}} \tag{6.33}$$

Substituting into Equation (6.25) gives

$$\begin{aligned} k\left(\frac{dT}{dx}\right)_{x=0} &= -k(T_s - T_\infty)\left(\frac{v_0}{\alpha}\right) \\ &= -\rho c_p v_0 (T_s - T_\infty) \\ &= -\dot{m}_g'' c_p (T_s - T_\infty) \end{aligned}$$

(c_p is the specific heat of the liquid) or

$$\boxed{\dot{q}'' = \dot{m}''_{g}\left[h_{fg} + c_p(T_s - T_\infty)\right]} \tag{6.34}$$

where $h_{fg} + c_p(T_s - T_\infty)$ is called the heat of gasification since it is the total energy needed to evaporate a unit mass of liquid originally at temperature T_∞. It is usually given by the symbol L and is composed of the heat of vaporization plus the 'sensible energy' needed to bring the liquid from its original temperature T_∞ to its evaporation temperature T_s.

The special case of a dry atmosphere, under the one-dimensional steady assumptions with a low mass transfer flux, can be written as (from Equations (6.29) and (6.30))

$$\dot{m}''_{g} = \left(\frac{h}{c_g}\right)\left(\frac{M_g}{M}\right) e^{M_g h_{fg}/(RT_b)} e^{-M_g h_{fg}/(RT_s)} \tag{6.35}$$

This shows that surface evaporation into a dry atmosphere has the form of an Arrhenius rule for a zeroth-order chemical reaction.

Example 6.2 Estimate the mass flux evaporated for methanol in dry air at 25 °C and 1 atm. Assume natural convection conditions apply at the liquid–vapor surface with $h = 8\ \text{W/m}^2\ \text{K}$.

Solution By Equation (6.35) and Example 6.1, we first assume $T_s \approx 25\ ^\circ\text{C}$:

$$\dot{m}''_{g} = \left(\frac{8\ \text{W/m}^2\ \text{K}}{1.0\ \text{J/g K}}\right)\left(\frac{32}{29}\right) \times e^{-\left[(32)(1.10\ \text{kJ/g})/(8.315\times 10^{-3}\ \text{kJ/mole K})\right](1/298\ \text{K} - 1/337\ \text{K})}$$

$$\dot{m}''_{g} = 1.7\ \text{g/m}^2\ \text{s}$$

where we have approximated the specific heat of the gas mixture as 1.0 J/g K. This is a typical fuel mass flux at ignition, and at extinction also; normally 1–5 g/m^2 s. The approximate heat flux needed to make this happen (for $T_s \approx T_\infty = 25\ ^\circ\text{C}$) is

$$\dot{q}'' = \dot{m}''_{g} h_{fg} = (1.7\ \text{g/m}^2\ \text{s})(1.10\ \text{kJ/g}) = 1.88\ \text{kW/m}^2$$

From Equation (6.27), we can now estimate T_s as

$$T_s = T_\infty - \frac{\dot{q}''}{h} = 25\ ^\circ\text{C} - \frac{1.88\ \text{kW/m}^2}{8\times 10^{-3}\ \text{kW/m}^2\ \text{K}}$$

$$T_s = 25\ ^\circ\text{C} - 235 = -210\ ^\circ\text{C}$$

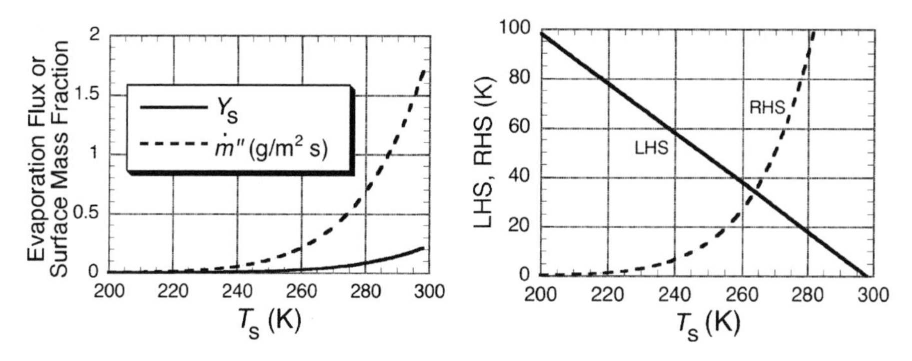

Figure 6.10 Example 6.2, solution for T_s

which obviously does not satisfy our assumption that $T_s = T_\infty = 25\,^\circ\mathrm{C}$!

This calls for a more accurate solution. We do this as follows: equate Equations (6.28) and (6.34) as $\dot{q}''/h$; and substitute Equation (6.35):

$$T_\infty - T_s = \frac{M_g}{c_g M}\, e^{-(M_g h_{fg}/R)(1/T_s - 1/T_b)} \left[h_{fg} - c_p(T_\infty - T_s)\right]$$

This equation requires an iterative solution for T_s. We plot the LHS and RHS as functions of T_s for the following substituted equations:

$$\begin{aligned} 298\ \mathrm{K} - T_s = & \frac{(32\ \mathrm{g/mole})/(29\ \mathrm{g/mole})}{10^{-3}\ \mathrm{kJ/gK}} \\ & \times\ e^{-(32\ \mathrm{g/mole})(1.10\ \mathrm{kJ/g})/(8.315\times 10^{-3}\ \mathrm{kJ/mole})(1/T_s - 1/337\ \mathrm{K})} \\ & \times \left[1.10\ \mathrm{kJ/g} - (2.37\times 10^{-3}\ \mathrm{kJ/gK})\ (298\ \mathrm{K} - \mathrm{T_s})\right] \end{aligned}$$

giving an intersection (see Figure 6.10) for

$$T_s \approx 264\ \mathrm{K} = -9\,^\circ\mathrm{C}$$

Equation (6.18) could now be used to assess whether the methanol vapor exceeds its LFL. At this temperature, by Equations (6.28) and (6.34), and with $c_p = 2.37\ \mathrm{J/g\,K}$ for liquid methanol,

$$\begin{aligned} \dot{m}_g'' &= \frac{h(T_\infty - T_s)}{h_{fg} - c_p(T_\infty - T_s)} \\ &= \frac{8\times 10^{-3}\ \mathrm{kW/m^2\,K}\ (298 - 264)\ \mathrm{K}}{[1.10 - (2.37\times 10^{-3})(298 - 264)]\ \mathrm{kJ/g}} \\ \dot{m}_g'' &= 0.26\ \mathrm{g/m^2\,s} \end{aligned}$$

Hence this accurate solution gives a big change in the surface temperature of the methanol, but the order of magnitude of the mass flux at ignition is about the same, $\sim O(1\ \mathrm{g/m^2\,s})$.

Let us examine some more details of this methanol problem. The regression rate of the surface is, for a liquid density of 0.79 g/cm^3,

$$v_0 = \frac{\dot{m}''_g}{\rho} = \frac{0.26 \text{ g/m}^2 \text{ s}}{0.79 \text{ g/cm}^3 \times 10^6 \text{ cm}^3/\text{m}^3}$$

or

$$v_0 = 3.3 \times 10^{-7} \text{ m/s}$$

From Equation (6.33), the temperature profile in the liquid pool is

$$\frac{T - 25\,°\text{C}}{(-9 - 25)\,°\text{C}} = \text{e}^{-(v_0/\alpha)x}$$

where, using $k = 0.5$ W/m K for the liquid,

$$\frac{v_0}{\alpha} = \rho v_0 c_p/k = \dot{m}''_g c_p/k = \frac{(0.26 \text{ g/m}^2 \text{ s})(2.37 \text{ J/g K})}{0.5 \text{ W/m K}} = 1.23 \text{ m}^{-1}$$

The thermal penetration depth of the cooled liquid can be approximated as

$$\text{e}^{-(1.23)\delta_\text{p}} = 0.01$$

for $T(\delta_\text{p}) = 24.7\,°\text{C}$ or $\delta_\text{p} = 3.7$ m. Hence this pool of liquid must maintain a depth of roughly 4 m while evaporating in order for our semi-infinite model to apply. If it is a thin pool then the temperature at the bottom of the container or ground will have a significant effect on the result.

Suppose the bottom temperature of the liquid is maintained at 25 °C for a thin pool. Let us consider this case where the bottom of the pool is maintained at 25 °C. For the pool case, the temperature is higher in the liquid methanol as depth increases. This is likely to create a recirculating flow due to buoyancy. This flow was ignored in developing Equation (6.33); only pure conduction was considered. For a finite thickness pool with its back face maintained at a higher temperature than the surface, recirculation is likely. Let us treat this as an effective heat transfer coefficient, h_L, between the pool bottom and surface temperatures. For purely convective heating, conservation of energy at the liquid surface is

$$\dot{m}''_g h_\text{fg} = h(T_\infty - T_\text{s}) + h_\text{L}(T_\text{bot} - T_\text{s}). \tag{6.36}$$

Substituting Equation (6.35) and taking $T_\text{bot} = T_\infty$ gives an implicit equation for T_s:

$$T_\infty - T_\text{s} = \left(\frac{h}{h + h_\text{L}}\right) \frac{h_\text{fg}}{c_\text{g}} \frac{M_\text{g}}{M} \text{e}^{-(h_\text{fg} M_\text{g}/R)(1/T_\text{s} - 1/T_\text{b})} \tag{6.37}$$

This gives the surface temperature for a liquid evaporating pool convectively heated from above and below.

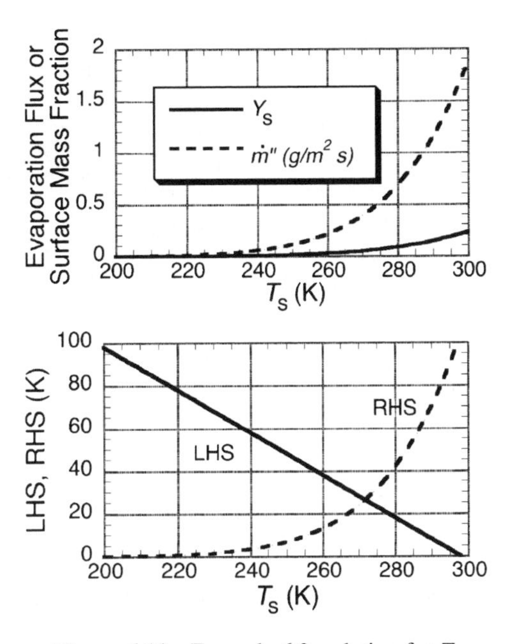

Figure 6.11 Example 6.3, solution for T_s

Example 6.3 Consider Example 6.2 for a shallow pool of methanol with its bottom surface maintained at 25 °C. Assume that natural convection occurs in the liquid with an effective convective heat transfer coefficient in the liquid taken as 10 W/m^2 K. Find the surface temperature, surface vapor mass fraction and the evaporation flux for this pool.

Solution From Equation (6.37), we substitute, for known quantities,

$$298 \text{ K} - T_s = \left(\frac{8}{8+10}\right)\left(\frac{32}{29}\right)\left(\frac{1100 \text{ J/g}}{1 \text{ J/g K}}\right)e^{-4240.9(1/T_s - 1/337 \text{ K})}$$

The LHS and RHS are plotted in Figure 6.11, yielding $T_s = 272$ K or -1 °C. From Equation (6.30),

$$Y_s = \left(\frac{32}{29}\right)e^{-4240.9(1/272 - 1/337)} = 0.054$$

The mass flux evaporated into dry air is, by Equation (6.29),

$$\dot{m}''_g = \frac{h}{c_g}Y_s = \frac{8 \text{ W/m}^2 \text{ K}}{1 \text{ J/g K}}(0.054)$$

or

$$\dot{m}''_g = 0.44 \text{ g/m}^2 \text{ s}$$

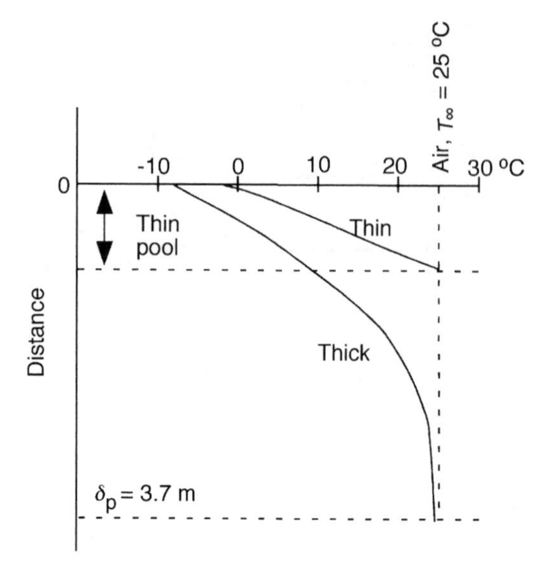

Figure 6.12 Methanol evaporation into air at 25 °C under natural convection (not drawn to scale)

These results are close to the theoretical flammable limit conditions at the surface of $T_s = 277$ K, $Y_s = 0.074$ and $\dot{m}''_g = 0.58$ g/m^2 s (see Example 6.1).

The thermal dynamics of Examples 6.2 and 6.3 for methanol are sketched in Figure 6.12.

References

1. Setchkin, N.P., Self-ignition temperature of combustible liquids, *J. Research, National Bureau of Standards*, 1954, **53**, 49–66.
2. Kanury, A.M., Ignition of liquid fuels, in *The SFPE Handbook of Fire Protection Engineering*, 2nd edn, (eds P.J. DiNenno *et al.*), Section 2, *National Fire Protection Association*, Quincy, Massachusetts, 1995, pp. **2**-164 to **2**-165.
3. Turns, S.R., *An Introduction to Combustion*, 2nd edn, McGraw-Hill, Boston, Massachusetts, 2000.
4. Bolz, R.E. and Tuve, G.L. (eds), *Handbook of Tables for Applied Engineering Science*, 2nd edn, CRC Press, Cleveland, Ohio, 1973.

Problems

6.1 A runner is in a long race on a cloudy day and is perspiring. Will the joggerss' skin temperature be _____higher or _____lower than the air temperature? Check one.

6.2 Why is the vapor pressure of a liquid fuel important in assessing the cause of the ignition of the fuel?

6.3 Under purely convective evaporation from gas phase heating, the liquid surface temperature is ______less than or____ greater than the ambient. Check the blank that fits.

6.4 At its boiling point, what is the entropy change of benzene from a liquid to vapor per gram? Use Table 6.1.

6.5 Calculate the vapor pressure of the following pure liquids at 0 °C. Use the Clausius–Claperyon equation and Table 6.1.

(a) *n*-Octane

(b) Methanol

(c) Acetone

6.6 Calculate the range of temperatures within which the vapor–air mixture above the liquid surface in a can of *n*-hexane at atmospheric pressure will be flammable. Data are found in Table 4.5. Calculate the range of ambient pressures within which the vapor/air mixture above the liquid surface in a can of *n*-decane ($n\text{-}C_{10}H_{22}$) will be flammable at 25 °C.

6.7 Using data in Table 6.1, calculate the closed cup flashpoint of *n*-octane. Compare your results with the value given in Table 6.1.

6.8 Calculate the temperature at which the vapor pressure of *n*-decane corresponds to a stoichiometric vapor–air mixture. Compare your result with the value quoted for the firepoint of *n*-decane in Table 6.1.

6.9 A liquid spill of benzene, C_6H_6, falls on a sun-heated roadway, and is heated to 40 °C. Safety crews are concerned about sparks from vehicles that occur at the surface of the benzene. Show by calculation that the concern of the crew is either correct or not. Use the following data as needed:

$T_{boil} = 353$ K $\quad M = 78 \quad h_{fg} = 432$ kJ/kg

$\Delta h_c = 40.7$ kJ/g $\quad X_{lower} = 1.2\,\% \quad X_{upper} = 7.1\,\%$

$R = 8.314$ J/mole K

6.10 A bundle of used cleaning rags and its flames uniformly impinge on a vat of naphthalene ($C_{10}H_8$). The liquid naphthalene is filled to the brim in a hemispherical vat of 1 m radius. The flame heats the thin metal vat with a uniform temperature of 500 °C and a convective heat transfer coefficient of 20 W/m^2 K, and the liquid in the vat circulates so that it can be taken at a uniform temperature. Heat transfer from the surface of the liquid can be neglected, and the metal vat has negligible mass. A faulty switch at the top surface of the vat continually admits an electric arc. Using the properties for the naphthalene, determine the time when the liquid will ignite. In this process you need to compute the temperature of the liquid that will support ignition. The initial temperature is 35 °C.

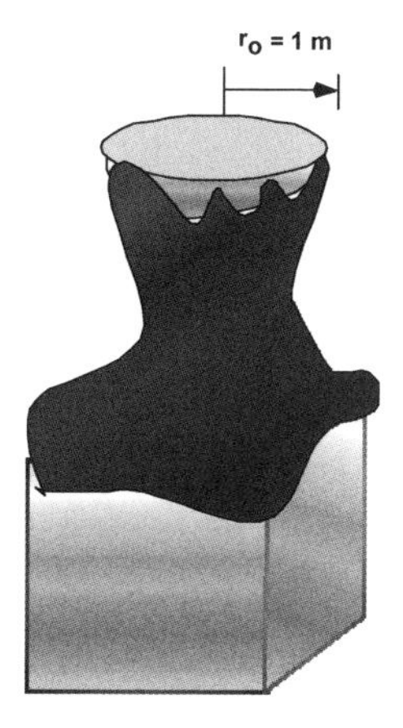

Properties:

Liquid density, $\rho = 1160\ kg/m^3$

Liquid specific heat, $c = 1.2\ J/g\ K$

Heat of vaporization, $h_{fg} = 0.32\ kJ/g$

Lower flammability limit, $X_L = 0.9\ \%$

Boiling point, $T_b = 491\ K$

Gas constant, $R = 8.315\ J/mole\ K$

Surface area of the sphere $= 4\pi r^2$

Volume of the sphere $= (4/3)\pi r^3$

(a) Find the temperature required for ignition.

(b) Find the time for ignition.

6.11 *n*-Octane spills on a hot pavement during a summer day. The pavement is 40 °C and heats the octane to this temperature. The wind temperature is 33 °C and the pressure is 1 atm. Use Table 6.1.

Data:

Ambient conditions are at a temperature of 33 °C and pressure of 760 mmHg. Density of air is $1.2\ kg/m^3$ and has a specific heat of 1.04 J/g K.

Octane properties:

Boiling temperature is 125.6 °C

Heat of vaporization is 0.305 kJ/g

Specific heat (liquid) is 2.20 J/g K

Specific heat (gas) is 1.67 J/g K

Density (liquid) is $705\ kg/m^3$

Lower flammability limit in air is 0.95 %

Upper flammability limit in air is 3.20 %

(a) Determine the molar concentration of the octane vapor at the surface of the liquid spill.

(b) If a spark is placed just at the surface, will the spill ignite? Explain the reason for your answer.

(c) Assume a linear distribution of octane vapor over the boundary layer where it is 4 cm thick. Determine the vertical region over the 4 cm thick boundary layer where the mixture is flammable.

6.12 This problem could be boring or it could be exciting. You are to determine, by theoretical calculations, the drying characteristics of a wet washcloth.

A cloth is first saturated with water, then weighed and hung to dry for a 40 minutes, after which it is weighed again. The room temperature and humidity are recorded. Theoretically, compute the following, stating any assumptions and specifications in your analysis:

(a) The heat and mass transfer coefficients for the cloth suspended in the vertical orientation.

(b) The average evaporation rate per unit area and the drying time, assuming it continues to hang in the same position.

(c) The surface temperature and the mass fraction of water vapor at the surface of the cloth.

Data:

Room temperature: 23 °C	Room humidity: 29.5%
Washcloth mass: 55.48 g	Dimensions: 33 cm × 32.5 cm × 3 cm
Initial wet washcloth mass: 135.8 g	Time: 11:25 a.m.
Final mass: 122.6 g	Time: 12:05 p.m.

6.13 The US Attorney General authorizes the use of 'nonpyrotechnic tear gas' to force cult members from a building under siege. She is assured the tear gas is not flammable. The teargas (CS) is delivered in a droplet of methylene chloride (dichloromethane, CH_2Cl_2) as an aerosol. Twenty canisters of the tear gas are simultaneously delivered, each containing 0.5 liters of methylene chloride. The properties of the CH_2Cl_2 are listed below:

Properties of CH_2Cl_2:

Liquid specific heat = 2 J/g K

Boiling point = 40 °C

Liquid density = 1.33 g/cm^3

Heat of vaporization = 7527 cal/g mole

Lower flammability limit = 12 % (molar)

(a) Compute the flashpoint of the liquid CH_2Cl_2.

(b) Compute the evaporation rate per unit area of the CH_2Cl_2 once released by the breaking canisters in the building. The air temperature is 28 °C, and no previous tear gas was introduced. Assume the spill can be considered as a deep pool, and the heat transfer coefficient between the floor and air is 10 W/m^2 K.

(c) What is the maximum concentration of CH_2Cl_2 gas in the building assuming it is well mixed. The building is 60 ft × 30 ft × 8 ft high, and has eight broken windows 3 ft × 5 ft high. Each canister makes a 1 m diameter spill, and the air flow rate into the building can be taken as $0.05\ A_0H_0^{1/2}$ in kg/s, where A_0 is the total area of the windows and H_0 is the height of a window.

6.14 A burglar in the night enters the back room of a jewelry store where the valuables are kept. The room is at 32 °C and is 3.5 m × 3.5 m × 2.5 m high. He enters through an open window 0.8 m × 0.8 m high. The colder outside temperature will allow air to flow into the room through the window based on the formula

$$\dot{m}_{\text{air}}(\text{g/s}) = 110 A_0 \sqrt{H_0}$$

where A_0 is the area of the window in m^2 and H_0 is its height in m.

The burglar trips and breaks his flashlight and also a large bottle of methanol that spills into a holding tank of 1.8 m diameter. He then uses a cigarette lighter to see. Use property data from the tables in Chapter 6; the gas specific heat is constant at 1.2 J/g K, the density for the gas in the room can be considered constant at 1.18 kg/m^3 and the heat transfer coefficient at the methanol surface is 15 W/m^2 K.

(a) If the burglar *immediately* drops the lit lighter into the methanol tank, explain by calculations what happens. You need to compute the concentration of the methanol.

(b) What is the *initial* evaporation rate of the methanol in g/s?

(c) If the burglar does not drop his lit lighter, but continues to use it to see, will an ignition occur and when? Explain your results by calculations. Assume the composition of the evaporated methanol is uniformly mixed into the room, and further assume that the methanol surface temperature remains constant and can be taken for this estimate at the temperature when no methanol vapor was in the room.

6.15 Tung oil is fully absorbed into a cotton cloth, 0.5 cm thick, and suspended in dry air at a uniform temperature. The properties of the tung oil and wet cloth are given below:

Heat of combustion	40 kJ/g
Heat of vaporization	43.0 kJ/mole tung oil
Lower flammability limit of tung oil	1 % (by volume)
Boiling point of tung oil	159 °C
Thermal conductivity of wet cloth	0.15 W/m K
Activation energy, E	75 kJ/mole
Pre-exponential factor, A	0.38×10^8 g/m^3 s
Gas constant, R	8.315 J/mole K

(a) Ignition occurs when a small flame comes close to the cloth. What is the minimum cloth surface temperature to allow flaming ignition?

(b) If the cloth were exposed to air at this minimum temperature for (a), could it have led to ignition without the flame coming close to the cloth? Explain in quantitative terms.

7

Ignition of Solids

7.1 Introduction

Up to now we have considered ignition phenomena associated with gas mixtures for auto and piloted cases (Chapter 4), the conditions needed for spontaneous ignition of bulk products (Chapter 5) and the conditions needed for the ignition of liquids. The time for ignition to occur has not been carefully studied. In fire scenarios, the role of ignition for typical solid commodities and furnishings plays an important role in fire growth. In particular, the time for materials to ignite can help to explain or anticipate the time sequence of events in a fire (time-line). Indeed, we will see (Chapter 8) that the time to ignite for solids is inversely related to the speed of flame spread. In this chapter, we will confine our discussion to the ignition of solids, but liquids can be included as well provided they are reasonably stagnant. A hybrid case of a liquid fuel absorbed by a porous solid might combine the properties of both the liquid and the solid in controlling the process.

To understand the factors controlling the ignition of a solid fuel, we consider several steps. These steps are depicted in Figure 7.1.

Step 1 The solid is heated, raising its temperature to produce pyrolysis products that contain gaseous fuel. This decomposition to produce gaseous fuel can be idealized by an Arrhenius type reaction

$$\dot{m}_{\mathrm{F}}''' = A_{\mathrm{s}}\,\mathrm{e}^{-E_{\mathrm{s}}/(RT)} \tag{7.1}$$

where A_s and E_s are the properties of the solid. As with pure liquid evaporation (Equation (6.35)) this is a very nonlinear function of temperature in which a critical temperature, T_{py}, might exist that allows a significant fuel vapor concentration to make ignition

Fundamentals of Fire Phenomena James G. Quintiere

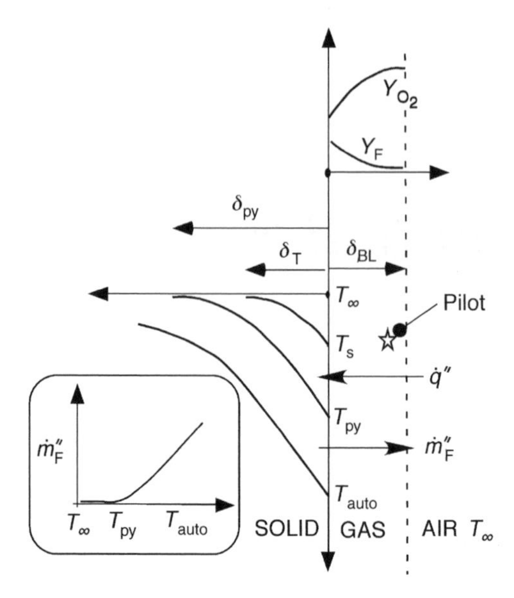

Figure 7.1 Factors involved in ignition of a solid

possible. For a flat solid heated in depth, the fuel mass flux leaving the surface could be described by

$$\dot{m}_F'' = \int_0^{\delta_{py}} A_s \, e^{-E_s/(RT)} \, dx \tag{7.2}$$

where δ_{py} is a critical depth heated. This process is illustrated in Figure 7.1. We imply that before the surface temperature reaches T_{py} there is no significant evolution of fuel vapor. At T_{py}, the value $\dot{m}_F''$ is sufficient to allow piloted ignition.

Step 2 The evolved fuel vapor must be transported through the fluid boundary layer where the ambient air is mixed with the fuel. For piloted ignition, a sufficient energy source such as an electrical spark or small flame must be located where the mixed fuel is flammable, $X_F \geq X_L$. For autoignition, we must not only achieve a near stoichiometric mixture but a sufficient volume of the vapor mixture must achieve its autoignition temperature (~300–500 °C). If the solid is radiatively heated, then either heat transfer from the surface or radiation absorption in the boundary layer must raise the temperature of the mixture to its autoignition value. This will be substantially higher than the typical surface temperature, T_{py}, needed for piloted ignition. For a charring material, surface oxidation can occur before ignition in the gas phase results. This would be especially possible for the autoignition case or for low heat fluxes. For example, Spearpoint [1] saw thick wood ignite as low as 8 kW/m^2 by radiant heat in the presence of a pilot. However, this flaming piloted ignition was preceded by surface glowing of the wood. Hence the energy of reaction for this oxidation augmented the radiant heating. Even without a pilot, glowing could induce flaming ignition. We shall ignore such glowing ignition and its effects in our continuing discussion, but it can be an important mechanism in the ignition of charring materials.

Let us just consider the piloted ignition case. Then, at T_{py} a sufficient fuel mass flux is released at the surface. Under typical fire conditions, the fuel vapor will diffuse by turbulent natural convection to meet incoming air within the boundary layer. This will take some increment of time to reach the pilot, whereby the surface temperature has continued to rise.

Step 3 Once the flammable mixture is at the pilot, there will be another moment in time for the chemical reaction to proceed to 'thermal runaway' or a flaming condition.

This three-step process can allow us to express the time for ignition as the sum of all three steps [2]:

$$t_{ig} = t_{py} + t_{mix} + t_{chem} \tag{7.3}$$

where

t_{py} = conduction heating time for the solid to achieve T_{py}
t_{mix} = diffusion or transport time needed for the flammable fuel concentration and oxygen to reach the pilot
t_{chem} = time needed for the flammable mixture to proceed to combustion once at the pilot

7.2 Estimate of Ignition Time Components

The processes described by Equation (7.3) are complex and require an elaborate analysis to make precise determinations of t_{ig}. By making appropriate estimates of each of the time components, we can considerably simplify practical ignition analyses for most typical fire applications. Again, we are just considering piloted-flaming ignition but autoignition can be similarly described.

7.2.1 Chemical time

Consider the chemical time component, t_{chem}. From Equation (5.22) we estimated the time for ignition of a flammable mixture for, say, a spherical volume of radius r_o as

$$t_{chem} = 0.21\ r_o^2/\alpha \tag{7.4}$$

where α is the thermal diffusivity of the gas mixture. For a runaway thermal reaction, the critical Damkohler number must be achieved for this sphere. Therefore, substituting for r_o,

$$\delta_c = \left(\frac{E}{RT_\infty}\right)\frac{r_o^2 A \Delta h_c\, e^{-E/(RT_\infty)}}{kT_\infty} = 3.32 \tag{7.5}$$

Combining,

$$t_{chem} = \frac{(0.21)(3.3)k\ T_\infty}{\alpha[E/(RT_\infty)]A\ \Delta h_c\, e^{-E/(RT_\infty)}} \tag{7.6}$$

We estimate this time in the following manner (using Table 4.2), selecting

$$\begin{aligned} E &= 160 \text{ kJ/mole} \\ A &= 10^{13} \text{ g/m}^3\text{ s} \\ k &= 0.10 \text{ W/m K} \\ \alpha &= 16 \times 10^{-6}\text{m}^2/\text{s} \\ \Delta h_c &= 45 \text{ kJ/g} \end{aligned}$$

as typical flammable mixtures in air, and we select $T_\infty = 1600$ K($\sim$1300 °C) as the lowest adiabatic flame temperature for a fuel mixture in air to achieve flaming combustion. In general, we would have to investigate the energy release rate,

$$\dot{Q}_{\text{chem}}(t) = A\,\Delta h_c e^{-E/(RT)}\left(\tfrac{4}{3}\pi r_o^3\right) \tag{7.7}$$

and find $t = t_{\text{chem}}$ when a perceptible amount of energy and light is observed at the pilot spark after the flammable mixture arrives. This is too cumbersome for illustrative purposes, so instead we select $T_\infty = 1600$ K as a representative combustion temperature. Substituting these values, with $E/(RT_\infty) = 12.0$,

$$\begin{aligned} t_{\text{chem}} &= \frac{(0.21)(3.32)(0.10 \text{ W/m K})(1600\text{ K})}{(16\times10^{-6}\text{ m}^2/\text{s})(12.0)(10^{13}\text{ g/m}^3\text{ s})(45\text{ kJ/g})e^{-12.0}} \\ &= 2.1\times10^{-4}\text{ s.} \end{aligned}$$

Thus, typical chemical reaction times are very fast. Note, if a chemical gas phase retardant is present or if the oxygen concentration is reduced in the ambient, A would be affected and reduced. Thus the chemical time could become longer, or combustion might not be possible at all.

7.2.2 *Mixing time*

Now let us consider the mixing time, t_{mix}. This will be estimated by an order of magnitude estimate for diffusion to occur across the boundary layer thickness, δ_{BL}. If we have turbulent natural conditions, it is common to represent the heat transfer in terms of the Nusselt number for a vertical plate of height, ℓ, as

$$\frac{h_c \ell}{k} = 0.021(Gr\,Pr)^{2/5} \text{ for } Gr\,Pr > 10^9 \tag{7.8}$$

where Gr is the Grashof number, $g[(T_s - T_\infty)/T_\infty](\ell^3/\nu^2)$ for a gas, and Pr is the Prandtl number, ν/α, with ν the kinematic viscosity of the gas. The heat transfer coefficient, h_c, can be related to the gas conductivity k, approximately as

$$h_c \approx \frac{k}{\delta_{\text{BL}}} \tag{7.9}$$

Then for a heated surface of $\ell = 0.5$ m, combining Equations (7.8) and (7.9), for typical air values at $T_s = 325\,^\circ$C and $T_\infty = 25\,^\circ$C ($\bar{T} = 175\,^\circ$C)

$$Pr = 0.69$$
$$\alpha = 45 \times 10^{-6}\ \mathrm{m^2/s}$$
$$\nu = 31 \times 10^{-6}\ \mathrm{m^2/s}$$
$$k = 35 \times 10^{-3}\ \mathrm{W/m-K}$$
$$Gr = (9.81\ \mathrm{m/s^2})\frac{300\ \mathrm{K}}{298\ \mathrm{K}}\frac{(0.5\ \mathrm{m})^3}{(31 \times 10^{-6}\ \mathrm{m^2/s})^2} = 1.28 \times 10^9$$
$$\frac{\delta_{\mathrm{BL}}}{\ell} = \frac{1}{(0.021)[(1.28 \times 10^9)(0.69)]^{2/5}} = 0.0126$$

or

$$\delta_{\mathrm{BL}} = (0.0126)(0.5\ \mathrm{m}) = 6.29 \times 10^{-3}\ \mathrm{m}$$

The time for diffusion to occur across δ_{BL} is estimated by

$$\delta_{\mathrm{BL}} \approx \sqrt{Dt_{\mathrm{mix}}} \tag{7.10}$$

where D is the diffusion coefficient, say $1.02 \times 10^{-5}\ \mathrm{m^2/s}$ for the laminar diffusion of ethanol in air. For turbulent diffusion, the effective D is larger; however, the laminar value will yield a conservatively, larger time. Equation (7.10) gives a diffusion penetration depth in time t_{mix}. Therefore, we estimate

$$t_{\mathrm{mix}} \approx \frac{(6.29 \times 10^{-3}\ \mathrm{m})^2}{1.02 \times 10^{-5}\ \mathrm{m^2/s}} = 3.9\ \mathrm{s}$$

Hence, we see that this mixing time is also very small. However, this analysis implicity assumed that the pilot is located where the fuel vapor evolved. A pilot downstream would introduce an additional transport time in the process.

7.2.3 Pyrolysis

Finally, we estimate the order of magnitude of the time of the heated solid to achieve T_{py} at the surface. This is primarily a problem in heat conduction provided the decomposition and gasification of the solid (or condensed phase) is negligible. We know that typically low fuel concentrations are required for piloted ignition ($X_{\mathrm{L}} \sim 0.01$–0.10) and by low mass flux ($\dot{m}''_{\mathrm{F}} \sim 1$–$5\ \mathrm{g/m^2\ s}$) accordingly. Thus, a pure conduction approximation is satisfactory. A thermal penetration depth for heat conduction can be estimated as

$$\delta_{\mathrm{T}} = \sqrt{\alpha t} \tag{7.11}$$

where α is the thermal diffusivity of the condensed phase. We can estimate the pyrolysis time by determining a typical δ_{T}.

The net incident heat flux at the surface is given by $\dot{q}''$. An energy conservation for the surface, ignoring any phase change or decomposition energies, gives

$$\dot{q}'' = -\left(k\frac{\partial T}{\partial x}\right)_{x=0} \tag{7.12}$$

At the pyrolysis time, we approximate the derivative as

$$-\left(k\frac{\partial T}{\partial x}\right)_{x=0} \approx \frac{k(T_{py} - T_\infty)}{\delta_T} \tag{7.13}$$

which allows an estimate of the pyrolysis time from Equation (7.11) as

$$t_{py} \approx k\rho c\left(\frac{T_{py} - T_\infty}{\dot{q}''}\right)^2 \tag{7.14}$$

The property $k\rho c$ is the thermal inertia; the higher it is, the more difficult it is to raise the temperature of the solid. Let us estimate t_{py} for typical conditions for a wood product

$k\rho c = 0.5\ (\text{kW/m}^2\text{K})^2\text{s}$
$T_{py} = 325\ ^\circ\text{C}$
$\dot{q}'' = 20\ \text{kW/m}^2$ (a relatively low value for ignition conditions)

Then from Equation (7.14),

$$t_{py} \approx (0.5)\left(\frac{325 - 25}{20}\right)^2 = 113\ \text{s}$$

Hence, the conduction time is the most significant of all, and it controls ignition.

In general, thermal conductivity is directly related to density for solids, so

$$k\rho c \sim \rho^2$$

which emphasizes why lightweight solids, such as foamed plastics, can ignite so quickly. Also k and c usually increase as temperature increases in a solid. Hence the values selected for use in $k\rho c$ at elevated temperatures should be higher than generally given for normal ambient conditions.

7.3 Pure Conduction Model for Ignition

Therefore, for piloted ignition we can say

$$t_{ig} \approx t_{py}$$

and select a reasonable surface temperature for ignition. Although decomposition is going to occur, it mostly occurs at the end of the conduction time and is usually of short duration, especially for piloted ignition. The ignition temperature can be selected as the surface temperature corresponding to pure conductive heating as interpreted from experimental measurements. This is illustrated in Figure 7.2, and would apply to either

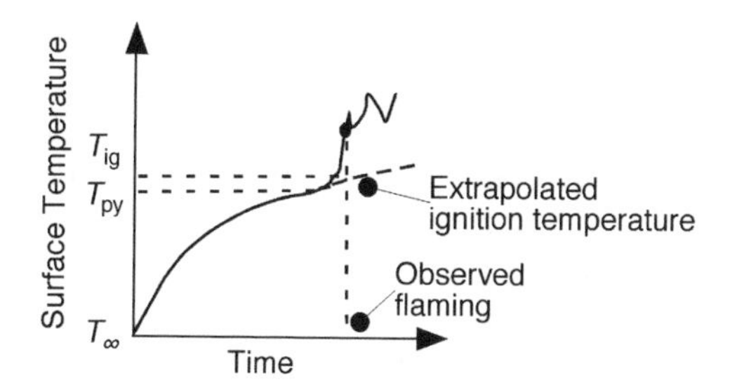

Figure 7.2 Surface temperature response for ignition

a piloted or autoignition case. A higher T_{ig} would result for autoignition with more distortion of a pure conductive heating curve near ignition. Nevertheless, a pure conduction model for the ignition of solids is a satisfactory approximation.

Of course, information for T_{ig} and thermal properties k, ρ and c must be available. They can be regarded as 'properties' and could be derived as 'fitting constants' with respect to thermal conduction theory. In this regard, ignition temperature is not a fundamental thermodynamic property of the liquid flashpoint, but more likened to modeling parameters as eddy viscosity for turbulent flow, or for that matter kinematic viscosity (ν) for a fluid approximating Newtonian flow behavior. The concept of using the ignition temperature is very powerful in implementing solutions, not only for ignition problems but for flame spread as well. Since the determination of T_{ig} by a conduction model buries the processes of degradation and phase change under various possible heating conditions, we cannot expect a precisely measured T_{ig} to be the same for all cases. Figure 7.3 shows results for the temperature of a wood particle board measured by K.

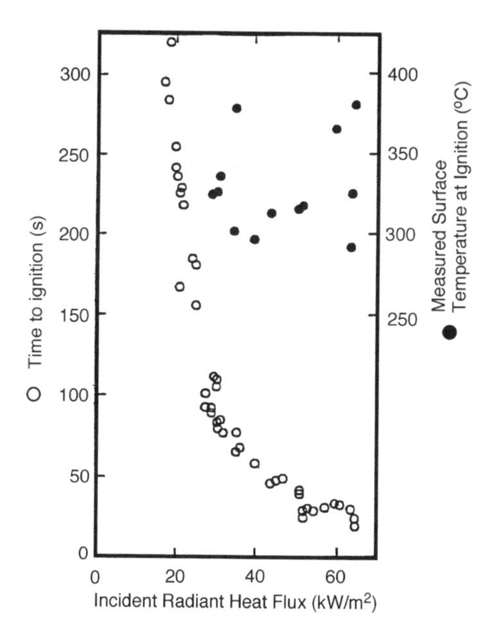

Figure 7.3 Piloted ignition of wood particle board (vertical) by radiant heating [3]

Saito with a fine wire thermocouple embedded at the surface [3]. The scatter in the results are most likely due to the decomposition variables and the accuracy of this difficult measurement. (Note that the surface temperature here is being measured with a thermocouple bead of finite size and having properties dissimilar to wood.) Likewise the properties k, ρ and c cannot be expected to be equal to values found in the literature for generic common materials since temperature variations in the least will make them change. We expect k and c to increase with temperature, and c to effectively increase due to decomposition, phase change and the evaporation of absorbed water. While we are not modeling all of these effects, we can still use the effective properties of T_{ig}, k, ρ and c to explain the ignition behavior. For example,

t_{ig} increases as T_{ig} increases (addition of retardants, more fire resistance polymers);
t_{ig} increases as k, ρ increase;
t_{ig} increases as c increases (phase change, decomposition evaporation of water).

7.4 Heat Flux in Fire

It would be useful to pause for a moment in our discussion of solids to put into perspective the heat fluxes encountered under fire conditions. It is these heat fluxes that promote ignition, flame spread and burning rate – the three components of fire growth.

7.4.1 Typical heat flux levels

Table 7.1 lists heat flux levels commonly encountered in fire and contrasts them with perceptible levels. It is typically found for common materials that the lowest heat fluxes to cause piloted ignition are about 10 kW/m^2 for thin materials and 20 kW/m^2 for thick materials. The time for ignition at these critical fluxes is theoretically infinite, but practically can be O(1 min) (order of magnitude of a minute). Flashover, or more precisely the onset to a fully involved compartment fire, is sometimes associated with a heat flux of 20 kW/m^2 to the floor. This flow heat flux can be associated with typical

Table 7.1 Common heat flux levels

Source	kW/m^2
Irradiance of the sun on the Earth's surface	≤ 1
Minimum for pain to skin (relative short exposure)[4]	~ 1
Minimum for burn injury (relative short exposure)[4]	~ 4
Usually necessary to ignite thin items	≥ 10
Usually necessary to ignite common furnishings	≥ 20
Surface heating by a small laminar flame[5]	50–70
Surface heating by a turbulent wall flame[6–8]	20–40
ISO 9705 room–corner test burner to wall	
(0.17 m square propane burner) at 100 kW	40–60
(0.17 m square propane burner) at 300 kW	60–80
Within a fully involved room fire (800–1000°C)	75–150
Within a large pool fire (800–1200°C)	75–267

room smoke layer temperatures of 500 °C (20 kW/m^2 to ceiling) and 600 °C (20 kW/m^2 to the floor, as estimated by an effective smoke layer emissivity and configuration factor product of 0.6).

The highest possible heat flux from the sun is not possible of igniting common solids. However, magnification of the sun's rays through a fish bowl was found to have ignited thin draperies in an accidental fire.

7.4.2 *Radiation properties of surfaces in fire*

For radiant heating it is not sufficient to know just the magnitude of the incident heat flux to determine the temperature rise. We must also know the spectral characteristics of the source of radiation and the spectral properties (absorptivity, α_λ, transmittance, τ_λ, and reflectance, ρ_λ) of the material. Recall that

$$\alpha_\lambda + \rho_\lambda + \tau_\lambda = 1 \tag{7.15}$$

and by equilibrium at a given surface temperature for a material, the emissivity is

$$\epsilon_\lambda = \alpha_\lambda \tag{7.16}$$

Figures 7.4(a), (b) and (c) give the spectral characteristics of three sources: (a) the sun at the earth's surface, (b) a JP-4 pool fire and (c) blackbody sources at typical room fire conditions (800–1100 K). The solar irradiance is mostly contained in about 0.3–2.4 μm while fire conditions span about 1–10 μm.

Figures 7.5 and 7.6 give the measured spectral reflectances and transmittances of fabrics. It is clear from Figure 7.5 that color (6,white; 7,black; 1,yellow) has a significant effect in reflecting solar irradiance, and also we see why these colors can be discriminated in the visible spectral region of 0.6 μm. However, in the spectral range relevant to fire conditions, color has less of an effect. Also, the reflectance of dirty (5a) or wet (5b) fabrics drop to ≤ 0.1. Hence, for practical purposes in fire analyses, where no other information is available, it is reasonable to take the reflectance to be zero, or the absorptivity as equal to 1. This is allowable since only thin fabrics (Figure 7.6) show transmittance levels of 0.2 or less and decrease to near zero after 2 μm.

7.4.3 *Convective heating in fire*

Convective heating in fire conditions is principally under natural convection conditions where for turbulent flow, a heat transfer coefficient of about 10 W/m^2 K is typical. Therefore, under typical turbulent average flame temperatures of 800 °C, we expect convective heat fluxes of about 8 kW/m^2. Consequently, under turbulent conditions, radiative heat transfer becomes more important to fire growth. This is one reason why fire growth is not easy to predict.

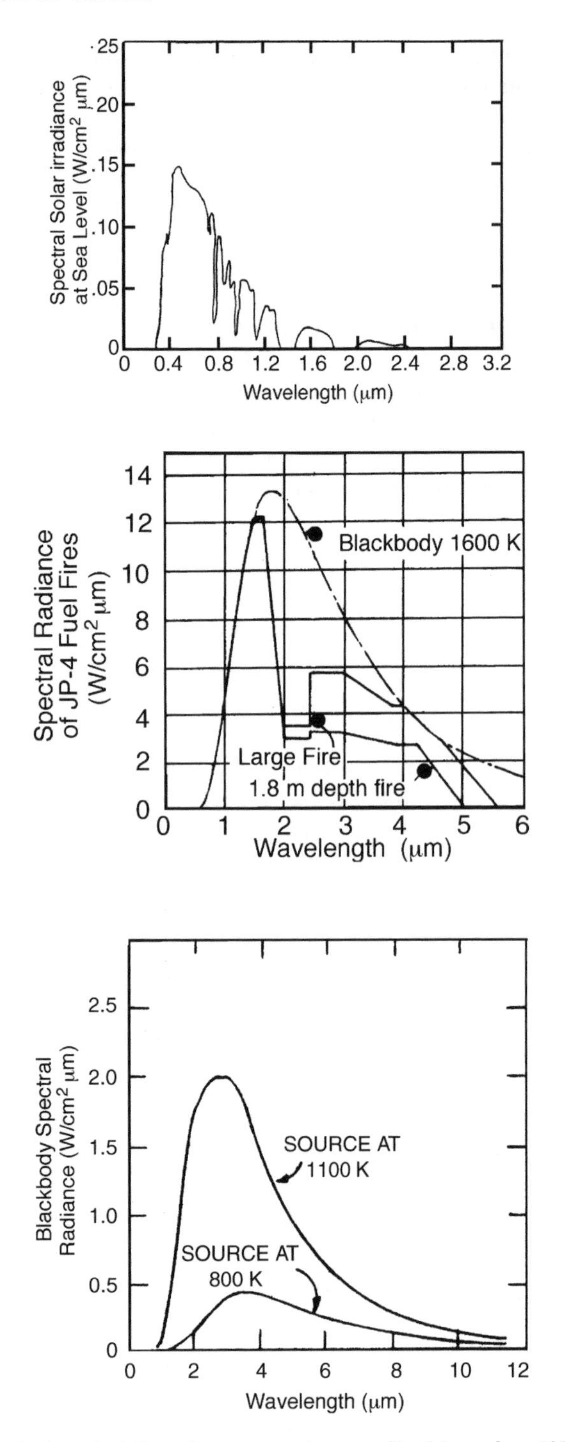

Figure 7.4 Spectral characteristics of sources: (a) sun at Earth's surface [9], (b) JP-4 fuel fire [10], (c) blackbodies [11]

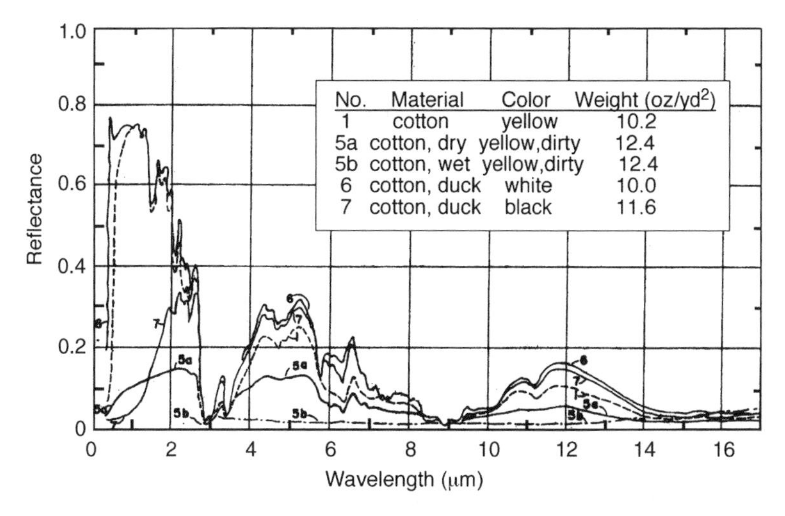

Figure 7.5 Spectral reflectance of cotton fabrics [11]

7.4.4 *Flame radiation*

Radiation from flames and combustion products involve complex processes, and its determination depends on knowing the temporal and spatial distributions of temperature, soot size distribution and concentration, and emitting and absorbing gas species concentrations. While, in principle, it is possible to compute radiative heat transfer if

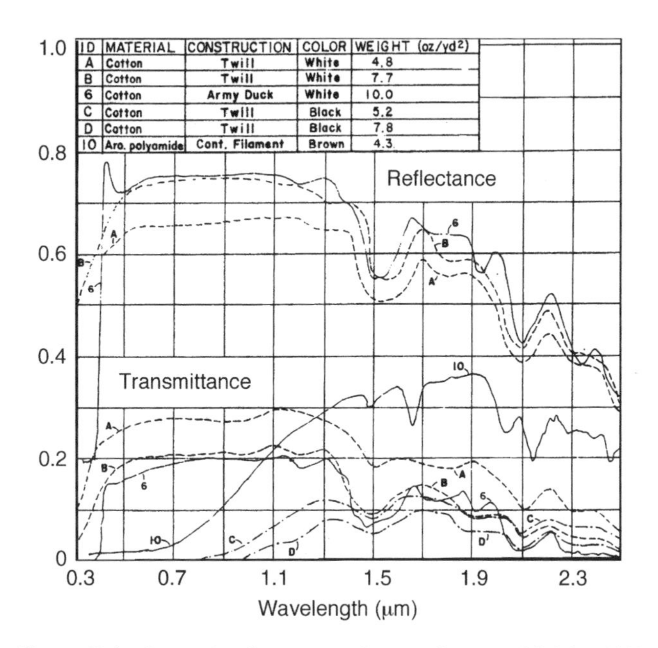

Figure 7.6 Spectral reflectance and transmittance of fabrics [11]

these quantities are fully known, it has not become fully practical for the spectrum of realistic fire scenarios. Hence, only rational estimates, empirical correlations or measurements must be used. The difficulties can be appreciated by portraying the mean time heat flux emitted from a flame by

$$\overline{\dot{q}''} = \overline{(\bar{\epsilon}_g + \epsilon_g')\sigma(\bar{T} + T')^4} \tag{7.17}$$

where the overbars denote a time average and the primes the fluctuating turbulent components. Spatial variations, not considered here, would compound the complexity. In addition, the emissivity depends on fuel properties and flame shape. The flame or gas emissivity can be expressed in a simplified form as

$$\epsilon_g = 1 - e^{-\kappa_g \ell} \tag{7.18}$$

where κ_g is the absorption coefficient, which may typically be $O(1)\ \mathrm{m}^{-1}$ for typical fuels in fire, and ℓ is a mean beam length or characteristic length scale for the fire [12]. Flames having length scales of 1–2 m would approach a blackbody emitter, $\epsilon_g \approx 1$.

7.4.5 *Heat flux measurements*

Heat fluxes in fire conditions have commonly been measured by steady state (fast time response) devices: namely a Schmidt–Boelter heat flux meter or a Gordon heat flux meter. The former uses a thermopile over a thin film of known conductivity, with a controlled back-face temperature; the latter uses a suspended foil with a fixed edge temperature. The temperature difference between the center of the foil and its edge is directly proportional to an imposed uniform heat flux. Because the Gordon meter does not have a uniform temperature over its surface, convective heat flux may not be accurately measured.

For a foil of radius, R, thickness, d, and conductivity, k, it can be shown that the Gordon gage temperature difference is

$$\begin{aligned} T(0) - T(R) &= \left[\frac{\dot{q}_r'' + h_c(T_\infty - T(R))}{h_c}\right][1 - 1/I_o(mR)] \\ &\approx \left\{\frac{\dot{q}_r''}{h_c} + [T_\infty - T(R)]\right\}\left(\frac{R^2}{4kd}\right) \end{aligned}$$

for small $mR, m = \sqrt{h_c/kd}$, I_o is a Bessel function and $\dot{q}_r''$ and $h_c(T_\infty - T(R))$ are the net radiative and convective heat fluxes to the cooled meter at $T(R)$ [6].

7.4.6 *Heat flux boundary conditions*

Heat flux is an important variable in fire growth and its determination is necessary for many problems. In general, it depends on scale (laminar or turbulent, beam length ℓ), material (soot, combustion products) and flow features (geometric, natural or forced). We will make appropriate estimates for the heat flux as needed in quantitative discussions and problems, but its routine determination must await further scientific and engineering progress in fire. We will represent the general net fire heat flux condition as

$$\begin{aligned}\dot{q}'' &= \left(-k\frac{\partial T}{\partial x}\right)_{x=0} \\ &= \dot{q}''_{\mathrm{e}} + \dot{q}''_{\mathrm{f}} + h_{\mathrm{c}}(T_\infty - T(0,t)) - \epsilon\sigma(T^4(0,t) - T^4_\infty)\end{aligned} \tag{7.19}$$

which gives the general surface boundary condition. With

$\dot{q}''_{\mathrm{e}}$ = external incident radiative heat flux
$\dot{q}''_{\mathrm{f}}$ = total (radiative + convective) flame heat flux
$h_{\mathrm{c}}(T_\infty - T(0,t))$ = convective heating flux when no flame is present
$\epsilon\sigma(T^4(0,t) - T^4_\infty)$ = net reradiative heat flux to a large surrounding at T_∞

Note that if this net flux is for a heat flux meter cooled at T_∞, the ambient temperature, the gage directly measures the external incident radiative and flame heat fluxes.

7.5 Ignition in Thermally Thin Solids

Let us return to our discussion of the prediction of ignition time by thermal conduction models. The problem reduces to the prediction of a heat conduction problem for which many have been analytically solved (e.g. see Reference [13]). Therefore, we will not dwell on these multitudinous solutions, especially since more can be generated by finite difference analysis using digital computers and available software. Instead, we will illustrate the basic theory to relatively simple problems to show the exact nature of their solution and its applicability to data.

7.5.1 Criterion for thermally thin

First, we will consider thin objects – more specifically, those that can be approximated as having no spatial, internal temperature gradients. This class of problem is called thermally thin. Its domain can be estimated from Equations (7.11) to (7.12), in which we say the physical thickness, d, must be less than the thermal penetration depth. This is illustrated in Figure 7.7. For the temperature gradient to be small over region d, we require

$$d \ll \delta_{\mathrm{T}} \approx \sqrt{\alpha t} \approx \frac{k(T_{\mathrm{s}} - T_{\mathrm{o}})}{\dot{q}''} \tag{7.20a}$$

or

$$Bi \equiv \frac{dh_{\mathrm{c}}}{k} \ll \frac{h_{\mathrm{c}}(T_{\mathrm{s}} - T_{\mathrm{o}})}{\dot{q}''} \tag{7.20b}$$

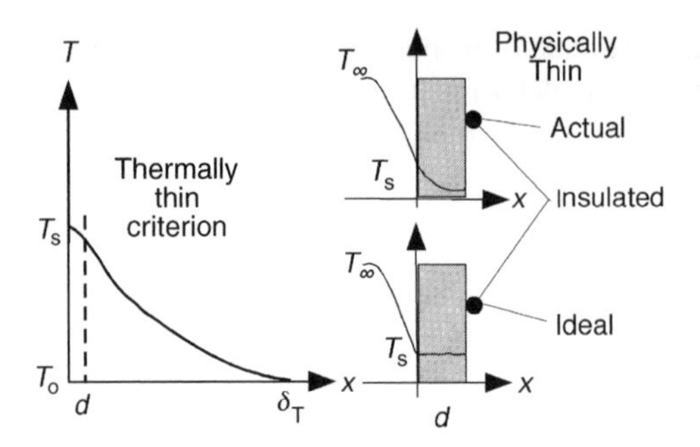

Figure 7.7 Thermally thin approximation

where Bi is the Biot number. If $\dot{q}''$ is based just on convective heating alone, i.e. $h_c(T_\infty - T_s)$, then Equation (7.20b) becomes

$$Bi \ll \frac{T_s - T_o}{T_\infty - T_s} \quad \text{(pure convection)} \tag{7.21}$$

which says that for a thermally thin case – where d is small enough – the internal temperature difference must be much smaller than the difference across the boundary layer. Under typical ignition conditions for a solid fuel,

$$k \sim 0.2 \text{ W/m K}$$
$$T_s = T_{ig} \sim 325 \text{ °C}$$
$$T_o = T_\infty = 25\text{°C}$$
$$\dot{q}'' = 20 \text{ kW/m}^3$$

then

$$d \ll (0.2 \times 10^{-3})(325 - 25)/20 = 3 \text{ mm}$$

Typically items with a thickness of less than about 1 mm can be treated as thermally thin. This constitutes single sheets of paper, fabrics, plastic films, etc. It does not apply to thin coatings or their laminates on noninsulating substances, as the conduction effects of the substrate could make the laminate act as a thick material. The paper covering on fiberglass insulation batting would be thin; pages in a closed book would not.

7.5.2 *Thin theory*

The theory of thermally thin ignition is straightforward and can apply to (a) a material of thickness d insulated on one side or (b) a material of thickness $2d$ heated symmetrically. The boundary conditions are given as

$$\dot{q}'' = \left(-k\frac{\partial T}{\partial x}\right)_{x_s} \quad \text{(surface)}$$

and

$$\left(\frac{\partial T}{\partial x}\right)_{x_i} = 0 \quad \text{(center or insulated face)}$$

However, by the thermally thin approximation, $T(x,t) \approx T(t)$ only. A control volume surrounding the thin material with the conservation of energy applied, Equation (3.45), gives (for a solid at equal pressure with its surroundings)

$$\rho c_p d \frac{\mathrm{d}T}{\mathrm{d}t} = \dot{q}'' = \dot{q}''_\mathrm{e} - h_\mathrm{c}(T - T_\infty) - \epsilon(T^4 - T_\infty^4) \tag{7.22}$$

The net heat flux is taken here to represent radiative heating in an environment at T_∞ with an initial temperature T_∞ as well. From Equation (7.20) a more general form can apply if the flame heat flux is taken as constant. This nonlinear problem cannot yield an analytical solution. To circumvent this difficulty, the radiative loss term is approximated by a linearized relationship using an effective coefficient, h_r:

$$h_\mathrm{r} \equiv \frac{\epsilon\sigma(T^4 - T_\infty^4)}{T - T_\infty} \approx 4\epsilon\sigma T_\infty^3 \tag{7.23}$$

for $T - T_\infty$ small. Hence, with $h_\mathrm{t} \equiv h_\mathrm{c} + h_\mathrm{r}$, we have

$$\rho c_p d \frac{\mathrm{d}T}{\mathrm{d}t} = \dot{q}''_\mathrm{e} - h_\mathrm{t}(T - T_\infty) \tag{7.24}$$

Let

$$\theta = \frac{T - T_\infty}{\dot{q}''_\mathrm{e}/h_\mathrm{t}} \quad \text{and} \quad \tau \equiv \frac{h_\mathrm{t} t}{\rho c_p d} \tag{7.25}$$

Then

$$\frac{\mathrm{d}\theta}{\mathrm{d}\tau} = 1 - \theta \quad \text{with} \quad \theta(0) = O \tag{7.26}$$

which can be solved to give

$$\boxed{T - T_\infty = \frac{\dot{q}''_\mathrm{e}}{h_\mathrm{t}}\left[1 - \exp\left(-\frac{h_\mathrm{t} t}{\rho c_p d}\right)\right]} \tag{7.27}$$

for an insulated solid of thickness d heated by an incident radiative heat flux $\dot{q}''_\mathrm{e}$. In general, the flame heat flux can be included here. From Equation (7.27) as $t \rightarrow \infty$, we reach a steady state, and as $t \rightarrow 0, \mathrm{e}^{-\tau} \approx 1 - \tau$. This allows simple expressions for the ignition time, t_ig, which occurs when $T = T_\mathrm{ig}$.

For small ignition times, since $1 - e^{-\tau} \approx \tau$,

$$t_{ig} \approx \frac{\rho c_p d(T_{ig} - T_\infty)}{\dot{q}''_e} \tag{7.28}$$

Note that the effect of cooling has been totally eliminated since we have implicitly considered the high heating case where $\dot{q}''_e \gg h_t(T_{ig} - T_\infty)$. Consider

$$\begin{aligned} h_c &\sim 0.010\ \text{kW/m}^2\,\text{K} \\ h_r &\sim 4(1)\left(5.671 \times 10^{-11}\ \text{kW/m}^2\,\text{K}^4\right)(298\ \text{K})^3 \\ &= 0.006\ \text{kW/m}^2\ \text{K} \\ T_{ig} &= 325\ ^\circ\text{C} \end{aligned}$$

Then, the heat flux due to surface cooling is

$$\begin{aligned} h_t(T_{ig} - T_\infty) &= (0.010 + 0.006)(325 - 25) \\ &= 4.8\ \text{kW/m}^2 \end{aligned}$$

Therefore this approximation might prove valid for $\dot{q}''_e > 5$ kW/m^2 at least.

For long heating times, eventually at $t \to \infty$, the temperature just reaches T_{ig}. Thus for any heat flux below this critical heat flux for ignition, $\dot{q}''_{ig,crit}$, no ignition is possible by the conduction model. The critical flux is given by the steady state condition for Equation (7.27),

$$\dot{q}''_{ig,crit} = h_t(T_{ig} - T_\infty) \tag{7.29a}$$

or a more exact form (with $\epsilon = 1$),

$$\dot{q}''_{ig,crit} = h_c(T_{ig} - T_\infty) + \sigma(T_{ig}^4 - T_\infty^4) \tag{7.29b}$$

7.5.3 *Measurements for thin materials*

Figures 7.8(a) and (b) display piloted ignition results for a metalized polyvinyl fluoride (MPVF) film of 0.2 mm thickness over a 25 mm fiberglass batting [14]. The MPVF film was bonded to a shear glass scrim (with no adhesive or significant thickness increase) to prevent it from stretching and ripping. The unbounded MPVF film was also tested. This shows several features that confirm the theory and also indicate issues.

The ignition and rip-time data both yield critical heat fluxes of about 25 and 11 kW/m^2 respectively. Clearly, both events possess critical heat fluxes that would indicate the

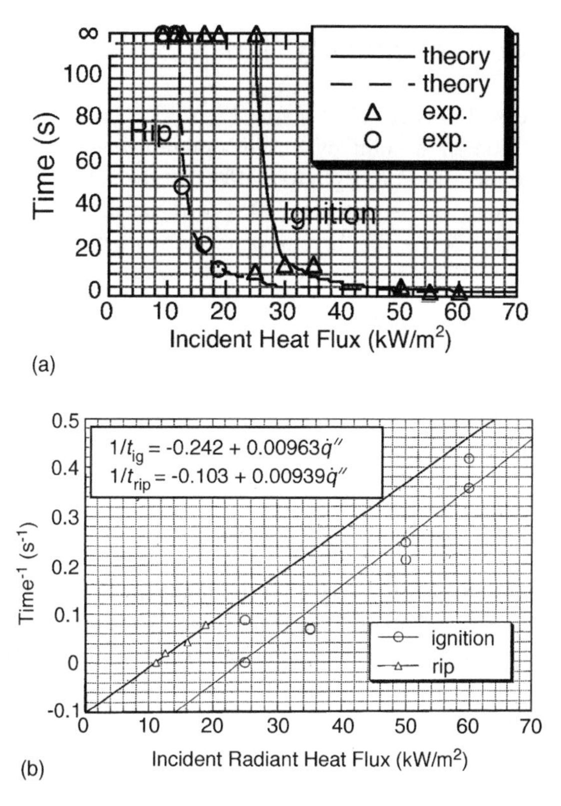

Figure 7.8 (a) Piloted ignition of glass-bonded MPVF and ripping times for unbonded MPVF (d = 0.2 mm) due to radiant heating and laid on fiberglass batting. (b) Reciprocal time results for ignition and ripping of thin MPVF ($d = 0.2$ mm)

minimum needed to ignite or rip the film. These results allow the determination of T_{ig} or T_{rip}, critical temperatures, from Equation (7.29b). Only h_{c} must be known for these determinations.

In Figure 7.8(b), the data are plotted as 1/t versus $\dot{q}''_{\mathrm{e}}$, and display a linear behavior as suggested by Equation (7.28). The linear results from the small time theory appear to hold for even 'long times'; the measured times close to the critical heat fluxes are less than 60 s. Hence, this small time theory appears to hold over a practical range of ignition times. The applicability of Equation (7.28) over a wide time range is found to be generally true in practice. Moreover, the slope of these lines gives $[\rho c d(T_{\mathrm{ig}} - T_{\infty})]^{-1}$, and allows a direct determination for c (ρd is the face weight per unit area and can easily be measured).

If this film (without bonding) is exposed to radiant heating, it would rip first. Depending on the final disposition of the ripped–stretched film, it may not ignite. Issues like this prevail in small-scale flame ignition tests, giving potentially misleading results. Thoughtful consideration of the fire scenario and its relevance to the test must be given.

7.6 Ignition of a Thermally Thick Solid

The thermally thin case holds for d of about 1 mm. Let us examine when we might approximate the ignition of a solid by a semi-infinite medium. In other words, the backface boundary condition has a negligible effect on the solution. This case is termed thermally thick. To obtain an estimate of values of d that hold for this case we would want the ignition to occur before the thermal penetration depth, δ_T reaches $x = d$. Let us estimate this by

$$d \geq \delta_T \approx \sqrt{\alpha t_{ig}} \tag{7.30}$$

From experiments on common fuels in fire, the times for ignition usually range well below 5 minutes (300 s). Taking a representative $\alpha = 2 \times 10^{-5}$ m^2/s, we calculate

$$d \geq \sqrt{(2 \times 10^{-5}\ \text{m}^2/\text{s})(300\,\text{s})} = 0.078\ \text{m}$$

Hence, we might expect solids to behave as thermally thick during ignition and to be about 8 cm for $t_{ig} = 300$ s and about 2.5 cm for 30 s. Therefore, a semi-infinite solution might have practical utility, and reduce the need for more tedious finite-thickness solutions. However, where thickness and other geometric effects are important, such solutions must be addressed for more accuracy.

7.6.1 *Thick theory*

Let us consider the semi-infinite (thermally thick) conduction problem for a constant temperature at the surface. The governing partial differential equation comes from the conservation of energy, and is described in standard heat transfer texts (e.g. Reference [13]):

$$\frac{\partial T}{\partial t} = \alpha \frac{\partial^2 T}{\partial x^2} \tag{7.31}$$

with boundary conditions:

$$x = 0, \quad T = T_s, \quad \text{constant}$$
$$x \to \infty, \quad T = T_\infty$$

and the initial condition:

$$t = 0, \quad T = T_\infty$$

We will explore a solution of the form:

$$\frac{T - T_\infty}{T_s - T_\infty} = \theta(\eta) \tag{7.32}$$

where

$$\eta = \frac{x}{2\sqrt{\alpha t}}$$

By substitution, it can be shown that

$$\left(\frac{\partial T}{\partial t}\right)_x = (T_s - T_\infty)\frac{d\theta}{d\eta}\left(-\frac{\eta}{2t}\right)$$

$$\frac{\partial}{\partial x}\left(\frac{\partial T}{\partial x}\right) = (T - T_\infty)\frac{d^2\theta}{d\eta^2}\left(\frac{1}{2\sqrt{\alpha t}}\right)^2$$

so that Equation (7.31) becomes

$$\frac{d^2\theta}{d\eta^2} + 2\eta\frac{d\theta}{d\eta} = 0 \tag{7.33}$$

with

$$\eta = 0, \quad \theta = 1$$
$$\eta \to \infty, \quad \theta = 0$$

This type of solution is called a similarity solution where the affected domain of the problem is proportional to $\sqrt{\alpha t}$ (the growing penetration depth), and the dimensionless temperature profile is similar in time and identical in the η variable. Fluid dynamic boundary layer problems have this same character.

We solve Equation (7.33) as follows. Let

$$\psi = \frac{d\theta}{d\eta}$$

Then

$$\frac{d\psi}{d\eta} = -2\eta\psi$$

Integrating,

$$\ln\psi = -\eta^2 + c_1$$

or

$$\psi = c_2\,e^{-\eta^2}$$

where we have changed the constant of integration. Integrating again gives

$$\theta - \theta(0) = c_2\int_0^\eta e^{-\eta^2}\,d\eta$$

or

$$\theta = 1 + c_2 \int_0^{\eta} e^{-\eta^2} d\eta$$

Applying the other boundary condition, $\eta \to \infty, \theta = 0$, gives

$$c_2 = \frac{-1}{\int_0^{\infty} e^{-\eta^2} d\eta}$$

The integral is defined in terms of the error function given by

$$\text{erf}(\eta) \equiv \frac{2}{\sqrt{\pi}} \int_0^{\eta} e^{-\eta^2} d\eta \tag{7.34}$$

where $\text{erf}(\infty) = 1$ and $\text{erf}(0) = 0$; c_2 is found as $(-2/\sqrt{\pi})$ and the solution is

$$\theta = 1 - \text{erf}(\eta) \tag{7.35}$$

A more accurate estimate of δ_T can now be found from this solution. We might define δ_T arbitrarily to be x when $\theta = 0.01$, i.e. $T = T_\infty + 0.01(T_s - T_\infty)$. This occurs where $\eta = 1.82$, or

$$\delta_T = 3.6\sqrt{\alpha t} \tag{7.36}$$

If we reduce our criterion for δ_T to x at $\theta = 0.5$, then $\delta_T \approx \sqrt{\alpha t}$, as we originally estimated.

Table 7.2 gives tabulated values of the error function and related functions in the solution of other semi-infinite conduction problems. For example, the more general boundary condition analogous to that of Equation (7.27), including a surface heat loss,

$$\left(-k\frac{\partial T}{\partial x}\right)_{x=0} = \dot{q}'' = \dot{q}''_e - h_t(T - T_\infty) \tag{7.37}$$

(with the initial condition, $t = 0, T = T_\infty$), gives the solution for the surface temperature (T_s)

$$\boxed{\frac{Ts - T_\infty}{(\dot{q}''_e/h_t)} = 1 - \exp(\gamma^2)\text{erfc}(\gamma)} \tag{7.38}$$

where

$$\gamma = h_t\sqrt{t/(k\rho c)}$$

Table 7.2 Error function [13]

η	$\text{erf}(\eta)$	$\text{erfc}(\eta)$	$\frac{2}{\sqrt{\pi}}e^{-\eta^2}$
0	0	1.00	1.128
0.05	0.0564	0.944	1.126
0.10	0.112	0.888	1.117
0.15	0.168	0.832	1.103
0.20	0.223	0.777	1.084
0.25	0.276	0.724	1.060
0.30	0.329	0.671	1.031
0.35	0.379	0.621	0.998
0.40	0.428	0.572	0.962
0.45	0.475	0.525	0.922
0.50	0.520	0.489	0.879
0.55	0.563	0.437	0.834
0.60	0.604	0.396	0.787
0.65	0.642	0.378	0.740
0.70	0.678	0.322	0.591
0.75	0.711	0.289	0.643
0.80	0.742	0.258	0.595
0.85	0.771	0.229	0.548
0.90	0.797	0.203	0.502
0.95	0.821	0.170	0.458
1.00	0.843	0.157	0.415
1.1	0.880	0.120	0.337
1.2	0.910	0.090	0.267
1.3	0.934	0.066	0.208
1.4	0.952	0.048	0.159
1.5	0.966	0.034	0.119
1.6	0.976	0.024	0.087
1.7	0.984	0.016	0.063
1.8	0.989	0.011	0.044
1.9	0.993	0.007	0.030
2.0	0.995	0.005	0.021

The quantity $\text{erfc}(\eta)$ is called the complementary error function, defined as

$$\text{erfc}(\eta) = 1 - \text{erf}(\eta) \tag{7.39}$$

Just as we examined the short ($t \to 0$) and long ($t \to \infty$) behavior of the thermally thin solution, we examine Equation (7.38) for $T_s = T_{ig}$ to find simpler results for t_{ig}. From the behavior of the functions by series expansions [13],

$$\gamma \to \text{large}, \quad \text{erfc}(\gamma) \approx \frac{1}{\sqrt{\pi}} \frac{\exp(-\gamma^2)}{\gamma} \tag{7.40}$$

so that Equation (7.38) becomes

$$\frac{T_s - T_\infty}{(\dot{q}''_e/h_t)} \approx 1 - \frac{\exp(\gamma^2)\exp(-\gamma^2)}{\sqrt{\pi}\gamma} = 1 - \frac{1}{\sqrt{\pi}\gamma} \tag{7.41}$$

For the ignition time, at long times

$$t_{\mathrm{ig}} \approx \frac{k\rho c}{\pi} \frac{(\dot{q}_{\mathrm{e}}''/h_{\mathrm{t}})^2}{[\dot{q}_{\mathrm{e}}'' - h_{\mathrm{t}}(T_{\mathrm{ig}} - T_\infty)]^2} \tag{7.42}$$

As in the thermally thin case, as $t_{\mathrm{ig}} \to \infty$, the critical flux for ignition is

$$\dot{q}_{\mathrm{ig,crit}}'' = h_{\mathrm{t}}(T_{\mathrm{ig}} - T_\infty) \tag{7.43}$$

Similarly, investigating small-time behavior, we use the approximations from series expansions [13]:

$$\gamma \text{ small}, \operatorname{erf}(\gamma) \approx \frac{2}{\sqrt{\pi}}\gamma$$
$$\exp(\gamma^2) \approx 1 + \gamma^2$$

Then

$$\begin{aligned}\frac{T_{\mathrm{s}} - T_\infty}{\dot{q}_{\mathrm{e}}''/h_{\mathrm{t}}} &\approx 1 - (1+\gamma^2)\left(1 - \frac{2}{\sqrt{\pi}}\gamma\right) \\ &= +\frac{2}{\sqrt{\pi}}\gamma - \gamma^2 + \frac{2}{\sqrt{\pi}}\gamma^3 \\ &\approx \frac{2}{\sqrt{\pi}}\gamma\end{aligned}$$

For the ignition time at small times,

$$\boxed{t_{\mathrm{ig}} \approx \frac{\pi}{4} k\rho c \left(\frac{T_{\mathrm{ig}} - T_\infty}{\dot{q}_{\mathrm{e}}''}\right)^2} \tag{7.44}$$

Again as in the thermally thin case, the heat loss term is not included, suggesting that this only holds where $\dot{q}_{\mathrm{e}}''$ is large.

7.6.2 *Measurements for thick materials*

Figures 7.9(a) and (b) show piloted ignition data taken for a Douglas fir (96 mm square by 50 mm thick) irradiated by Spearpoint [1]. There is a distinction between samples heated across or along the grain (see Figure 7.10). Experimental critical ignition fluxes were estimated as $12.0\,\mathrm{kW/m^2}$ (across) and $9.0\,\mathrm{kW/m^2}$ (along); however, despite waiting for nearly 1 hour for ignition, a flux at which no ignition could be truly identified was not determined. In Figure 7.9(b), the data are plotted as $1/\sqrt{t_{\mathrm{ig}}}$ against the incident radiation heat flux, as indicated by Equation (7.44). The 'high' heat flux data follow theory. The so-called 'low' heat flux data very likely contain surface oxidation, and are shifted to the left of the straight-line high heat flux data due to this extra heat addition. From the critical flux and the slope of the high heat flux line in Figure 7.9(b), T_{ig} and $k\rho c$

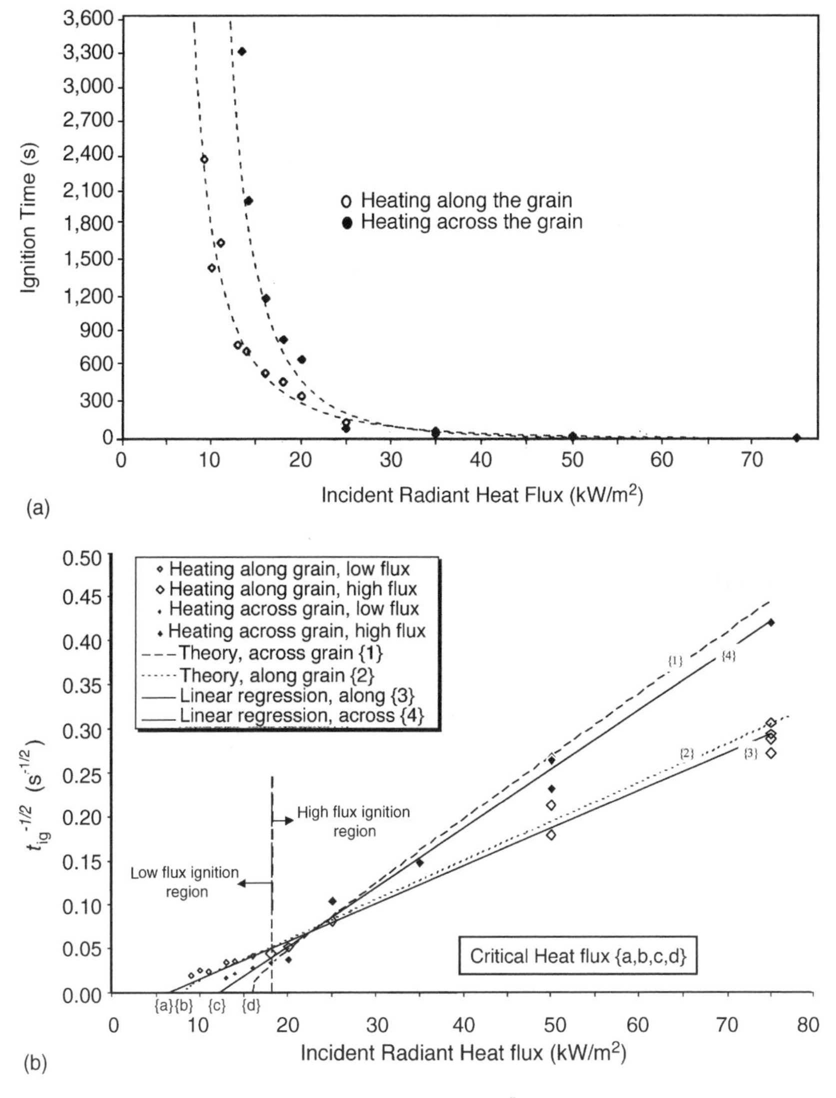

Figure 7.9 (a): Piloted ignition and (b) $1/\sqrt{t_{ig}}$ versus $\dot{q}''_e$ for a 50 mm thick Douglas fir (from Spearpoint [1])

could be determined. These are given in Table 7.3 for several wood species tested by Spearpoint [1] along with their measured densities.

In Figure 7.10 the anatomy of the virgin wood is depicted together with the definitions of heating 'along' and 'across' the grain used in Figure 7.9 and Table 7.3. For the 'along the grain' heating, it is easy for volatiles to escape the wood along the grain direction. This is likely to account for the lower T_{ig} in the 'along' orientation. It is known that the conductivity parallel to the grain (along) is about twice that perpendicular to the grain (across). This accounts for the differences in the $k\rho c$ values.

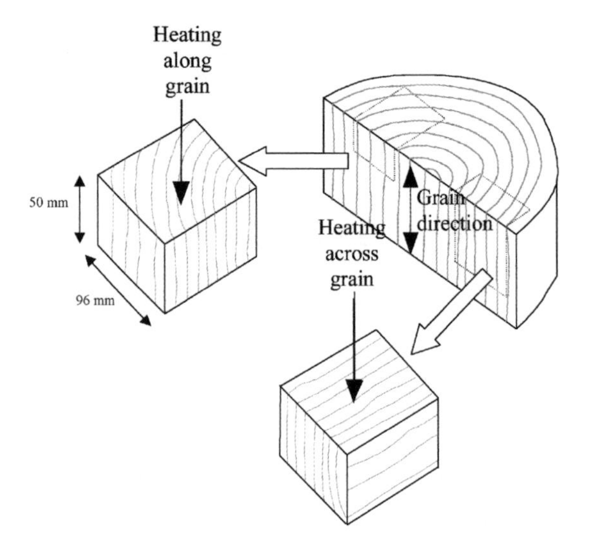

Figure 7.10 Wood grain orientation

7.6.3 Autoignition and surface ignition

Boonmee [15,16] investigated the effect of surface oxidation observed by Spearpoint [1] at low heat flux for wood. Autoignition and the onset of surface ignition ('glowing') was studied for redwood under radiative heating. Glowing ignition was not based on a perception of color for the surface of the redwood, but by a distinct departure from inert heating. This process is essentially ignition to smoldering in contrast to flaming. At a high heat flux, the smoldering or wood degradation is controlled by oxygen diffusion, while at a low heat flux, it is controlled by decomposition chemical kinetics. The onset of flaming autoignition showed a distinct reliance on the energy supplied by smoldering at heat fluxes below about 40 kW/m^2. These results are depicted in Figure 7.11 for the times for piloted ignition, autoignition and glowing ignition. There appears to be a critical flux for autoignition at 40 kW/m^2, yet below this value flaming ignition can still occur after considerable time following glowing ignition. This glowing effect on autoignition is

Table 7.3 Piloted ignition properties of wood species [1][a]

Species	Heating	Density (kg/m^3)	$\dot{q}''_{\rm ig,crit}$ measured (kW/m^2)	$\dot{q}''_{\rm ig,crit}$ derived[b] (kW/m^2)	$T_{\rm ig}$ (°C)[c]	$k\rho c\left(\frac{kW}{m^2\,K}\right)^2 s$
Redwood	Across	354	13.0	15.0	375	0.22
	Along	328	9.0	6.0	204	2.1
Douglas fir	Across	502	12.0	16.0	384	0.25
	Along	455	9.0	8.0	258	1.4
Red oak	Across	753	—	11.0	305	1.0
	Along	678	—	9.0	275	1.9
Maple	Across	741	12.0	14.0	354	0.67
	Along	742	8.0	4.0	150	11.0

[a] 50 mm thick samples, moisture content 5–10%.
[b] Implies theoretically derived from data.
[c] Based on theoretical values of $\dot{q}''_{\rm ig}$.

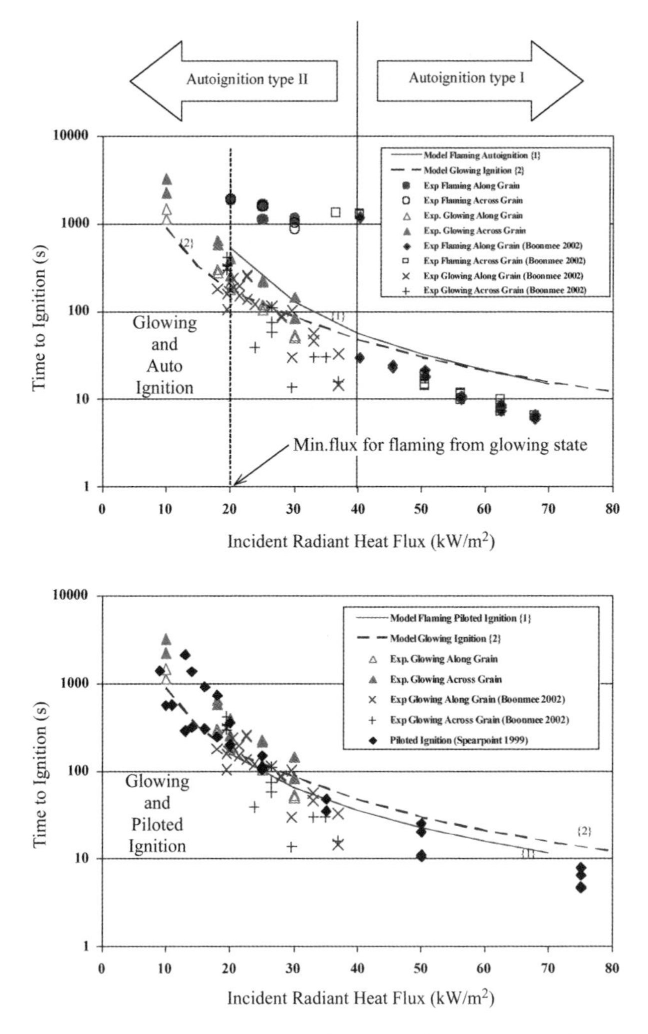

Figure 7.11 Ignition times as a function of incident heat flux [15]

present to a measured critical flux of 20 kW/m^2. The glowing transition to flaming cannot be predicted by a simple heat conduction model given by Equation (7.44). However, the remainder of the data follows conduction theory. Above 40 kW/m^2, no perceptible differences in ignition times are found within the accuracy of the measurements. Ignition temperatures in the conduction theory regions of data suggest values indicated in Table 7.4, as taken for Spearpoint and Boonmee's measurements given in Figure 7.12.

Table 7.4 Estimated ignition temperatures and critical flux for redwood

	Heating along grain (°C)	Heating across grain (°C)	Critical flux (kW/m^2)
Piloted ignition	204	375	9–13
Glowing ignition ($<40\ kW/m^2$)	400 ± 80	480 ± 80	10
Autoignition ($>40\ kW/m^2$)	350 ± 50	500 ± 50	20

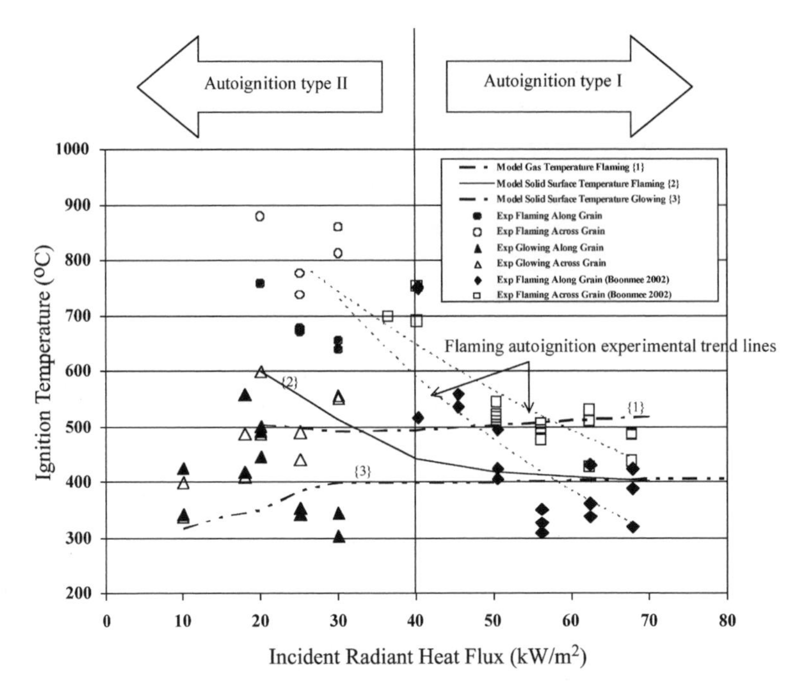

Figure 7.12 Flaming autoingition and glowing ignition temperatures as a function of incident heat flux [15]

Scatter in these data indicate accuracy, but they suggest that the earliest ignition time would be piloted, followed by glowing and then autoignition. This still may not preclude the effect of smoldering on piloted ignition at very low heat fluxes, as occurs with autoignition at 40 kW/m^2.

Table 7.4 clearly shows that the ignition temperature for autoignition is considerably higher than for piloted ignition. In conformance with this behavior, Boonmee [15] indicates that the corresponding fuel mass fractions needed are about 0.10 and 0.45 ± 0.15 for the pilot and autoignition of redwood respectively.

7.7 Ignition Properties of Common Materials

The values presented in Table 7.5 are typical of many common furnishings. A standard test method, ASTM E-1321 [17], has been developed to determine T_{ig} and $k\rho c$ for materials and products. These items are treated as thermally thick in the analysis since they are backed by an insulating board in the test. They combine the thickness effect of the material tested with the thickness of the inert substrate in the standard test. Figure 7.13 shows examples of ignition and lateral flame spread rate data taken from materials as thin as 1.27 mm. The solid curve through the ignition data (to the right) follows the thermally thick theory (Equation (7.44)). The velocity of spread data plotted here are taken at long irradiance heating times, and are seen to be asymptotically symmetrical to the ignition time data about the critical flux for ignition. This implies that surface flame velocity is infinite when the surface is at the ignition temperature. In actuality, flame spread occurs here in the gas phase at the lower flammable limit.

Table 7.5 Piloted ignition and flame spread properties from ASTM 1321 [18]

Material	T_{ig}(°C)	$k\rho c$ ((kW/m^{2}K)^2 s)	Φ ((kW)^2/m^3)	$T_{s,min}$ (°C)	$\Phi/(k\rho c)$ (m\K^2/s)
PMMA polycast, 1.59 mm	278	0.73	5.45	120	8
Polyurethane (535M)	280	—	—	105	82
Hardboard, 6.35 mm	298	1.87	4.51	170	2
Carpet (acrylic)	300	0.42	9.92	165	24
Fiberboard, low density (S119M)	330	—	—	90	42
Fiber insulation board	355	0.46	2.25	210	5
Hardboard, 3.175 mm	365	0.88	10.97	40	12
Hardboard (S159M)	372	—	—	80	18
PMMA Type G, 1.27 cm	378	1.02	14.43	90	14
Asphalt shingle	378	0.70	5.38	140	8
Douglas fir particle board, 1.27 cm	382	0.94	12.75	210	14
Wood panel (S178M)	385	—	—	155	43
Plywood, plain, 1.27 cm	390	0.54	12.91	120	24
Chipboard (S118M)	390	—	—	189	11
Plywood, plain, 0.635 cm	390	0.46	7.49	170	16
Foam, flexible, 2.54 cm	390	0.32	11.70	120	37
Glass/polyester, 2.24 mm	390	0.32	9.97	80	31
Mineral wool, textile paper (S160M)	400	—	—	105	34
Hardboard (gloss paint), 3.4 mm	400	1.22	3.58	320	3
Hardboard (nitrocellulose paint)	400	0.79	9.81	180	12
Glass/polyester, 1.14 mm	400	0.72	4.21	365	6
Particle board, 1.27 cm stock	412	0.93	4.27	275	5
Gypsum board, wall paper (S142M)	412	0.57	0.79	240	1
Carpet (nylon/wool blend)	412	0.68	11.12	265	16
Carpet #2 (wool, untreated)	435	0.25	7.32	335	30
Foam, rigid, 2.54 cm	435	0.03	4.09	215	141
Polyisocyanurate, 5.08 cm	445	0.02	4.94	275	201
Fiberglass shingle	445	0.50	9.08	415	18
Carpet #2 (wool, treated)	455	0.24	0.98	365	4
Carpet #1 (wool, stock)	465	0.11	1.83	450	17
Aircraft panel epoxy fiberite	505	0.24	—	505	—
Gypsum board, FR, 1.27 cm	510	0.40	9.25	300	23
Polycarbonate, 1.52 mm	528	1.16	14.74	455	13
Gypsum board, common, 1.27 mm	565	0.45	14.44	425	32
Plywood, FR, 1.27 cm	620	0.76	—	620	—
Polystyrene, 5.08 cm	630	0.38	—	630	—

Figure 7.14 shows the relationship of this standard test apparatus for Equation (7.29), for which $T_s = function(\dot{q}''_e)$:

$$T_s - T_\infty = \frac{\dot{q}''_e}{h_t} = \frac{\dot{q}''_e}{h_c} - \frac{\sigma(T_s^4 - T_\infty^4)}{h_c} \tag{7.45}$$

When $\dot{q}''_e = \dot{q}''_{ig,crit}$, then $T_s = T_{ig}$, and hence the material properties can be determined. We see these 'fitted' effective properties listed in Table 7.5 for a variety of materials. Table 7.6 gives generic data for materials at normal room conditions. It can be seen that for generic plywood at normal room temperature $k\rho c = 0.16(\text{kW/m}^2\ \text{K})^2\text{s}$ while under ignition conditions it is higher $\sim 0.5\ (\text{kW/m}^2\ \text{K})^2$ s. Increases in k and c with temperature can partly explain this difference. The data and property results were taken from Quintiere and Harkleroad [18].

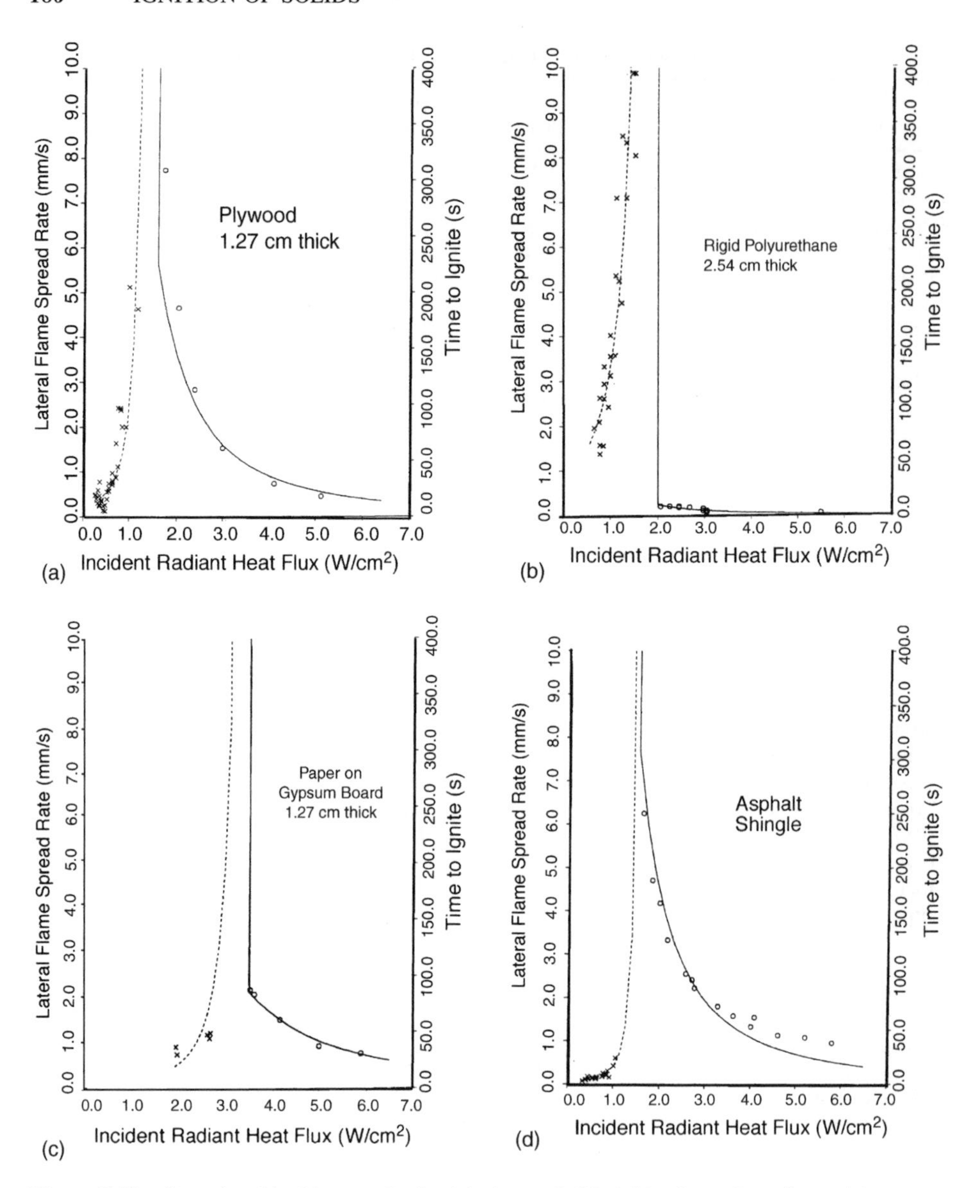

Figure 7.13 Spread and ignition results for (a) plywood, (b) rigid polyurethane foam, (c) gypsum board and (d) asphalt shingle [18]

An alternative form of an ignition correlation used in ASTM E-1321 follows from Equations (7.44) and (7.45):

$$\frac{\dot{q}''_{\text{ig,crit}}}{\dot{q}''_{\text{e}}} = \left[\frac{4}{\pi}\,\frac{h_{\text{t}}^2}{k\rho c}\,t_{\text{ig}}\right]^{1/2} \tag{7.46}$$

where $h_{\text{t}} = \dot{q}''_{\text{ig,crit}}/(T_{\text{ig}} - T_{\infty})$. A result for fiberboard is shown in Figure 7.15.

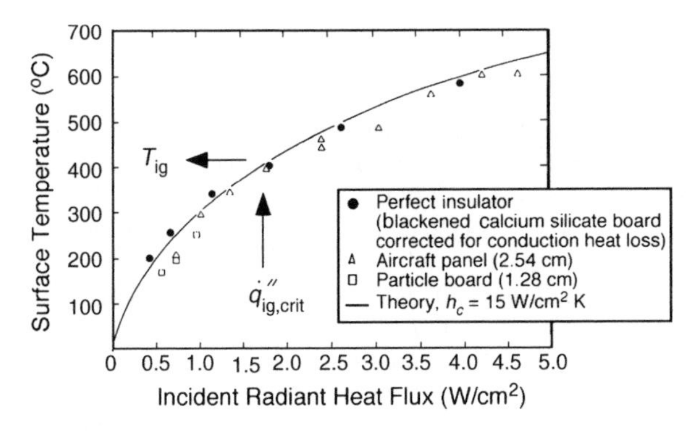

Figure 7.14 Equilibrium surface temperatures as a function of external radiant heating in the test apparatus [17]

Table 7.6 Thermal Properties of common materials at normal room temperature [a]

Material	Density, ρ (kg/m^3)	Conductivity, k (W/m K)	Specific heat, c (kJ/kg K)	$k\rho c$ ((kW/m^2 K)^2 s)
Concrete, stone	2200	1.7	0.75	2.8
Polymethyl methacrylate (PMMA)	1200	0.26	2.1	0.66
Gypsum board	950	0.17	1.1	0.18
Calcium silicate board	700	0.11	1.1	0.085
Particle board	650	0.11	2.0	0.14
Plywood	540	0.12	2.5	0.16
Cork board	200	0.040	1.9	0.015
Balsa wood	160	0.050	2.9	0.023
Polyurethane, rigid	32	0.020	1.3	0.0008
Polystyrene, expanded	20	0.034	1.5	0.0010

[a] Compiled from various sources.

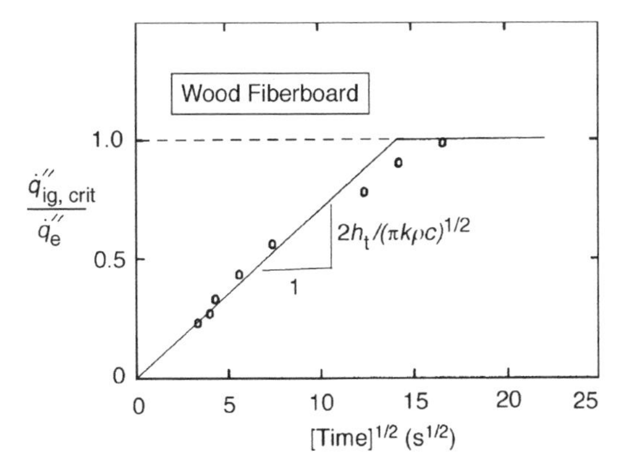

Figure 7.15 Correlation of ignition results for fiberboard [18]

References

1. Spearpoint, M.J., Predicting the ignition and burning rate of wood in the cone calorimeter using an integral model, MS Thesis, Department of Fire Protection Engineering, University of Maryland, College Park, Maryland 1999.
2. Fernandez-Pello, A.C., The solid phase, in *Combustion Fundamentals of Fire*, (ed. G. Cox), Academic Press, London, 1994.
3. Quintiere, J.G., The application of flame spread theory to predict material performance, *J. Research of the National Bureau of Standards*, 1988 **93**(1), 61.
4. Stoll, A.M. and Greene, L.C., Relationship between pain and tissue damage due to thermal radiation, *J. Applied Physiology*, 1959, **14**, 373–82.
5. Ito, A. and Kashiwagi, T., Characterization of flame spread over PMMA using holographic interferometry, sample orientation effects, *Combustion and Flame*, 1988, **71**, 189.
6. Quintiere, J.G., Harkleroad, M.F. and Hasemi, Y., Wall flames and implications for upward flame spread, *J. Combus. Sci. Technol.*, 1986, **48**(3/4), 191–222.
7. Ahmad, T. and Faeth, G.M., Turbulent wall fires, in 17th Symposium. (International) on *Combustion*, The Combustion Institute, Pittsburg, Pennsylvania, 1978, p. 1149.
8. Orloff, L., Modak, A.T. and Alpert, R.L., Turbulent wall fires, in 16th Symposium (International) on *Combustion*, The Combustion Institute, Pittsburg, Pennsylvania, 1979, pp. 1149–60.
9. Gast, P.R., *Handbook of Geophysics*, (ed. C.F. Campen), Macmillan Company, London, 1960, pp. 14–30.
10. Graves, K.W., *Fire Fighter Exposure Study*, Technical Report AGFSRS 71-1, Wright-Patterson Air Force Base, 1970.
11. Quintiere, J.G., Radiative characteristics of fire fighters' coat fabrics, *Fire Technol.*, 1974, **10**, 153.
12. Tien, C.L., Lee, K.Y. and Stratton, A.J., Radiation heat transfer, *The SFPE Handbook of Fire Protection Engineering*, 2nd edn (eds P.J. DiNenno *et al.*), Section 1, National Fire Protection Association, Quincy, Massachusetts, 1995, p. **1**-65.
13. Carslaw, H.S. and Jaeger, J.C., *Conduction of Heat in Solids*, 2nd edn, Oxford University Press, London, 1959, p. 485.
14. Quintiere, J.G., The effects of angular orientation on flame spread over thin materials, *Fire Safety J.*, 2001, **36**(3), 291–312.
15. Boonmee, N., Theoretical and experimental study of the auto-ignition of wood, PhD Dissertation, Department of Fire Protection Engineering, Univerity of Maryland, College Park, Maryland, 2004.
16. Boonmee, N. and Quintiere, J.G., Glowing and flaming auto-ignition of wood, in 29th Symposium (International) on *Combustion*, The Combustion Institute, Pittsburg, Pennsylvania, 2002, pp. 289–96.
17. ASTM E-1321, *Standard Test Method for Determining Material Ignition and Flame Spread Properties*, American Society for Testing and Materials Philadelphia, Pennsylvania, 1996.
18. Quintiere, J.G. and Harkleroad, M.F., New concepts for measuring flame spread properties, in *Fire Safety Science and Engineering*, ASTM STP882, (ed. T.-Z. Harmathy), American Society for Testing and Materials, Philadelphia, Pennsylvania, 1985, p. 239.

Problems

7.1 The time for ignition of solid fuels is *principally* controlled by

	Yes	*Not necessarily*	*No*
(a) The heating time of the solid	____	___	___
(b) The time for diffusion in air	____	___	___
(c) The reaction time in the gas	____	___	___

(d) Ignition temperature	____	___	___
(e) Density	____	___	___

7.2 Explain how the time to ignition of a solid can be based on conduction theory.

7.3 Kerosene as a thick, *motionless* fluid is in a large tank.

The overall heat transfer coefficient for convective and radiative cooling at the surface is 12 W/m^2 K.
The thermal conductivity of the kerosene is 0.14 W/m K.
Density of kerosene is 800 kg/m^3.
Specific heat of liquid kerosene is 1.2 J/g K.
Other properties are given in Table 6.1 as needed.
The ambient temperature is 25 °C.

If the incident heat flux is 10 kW/m^2, how long will it take the kerosene to ignite with a pilot flame present?

7.4 A flame impinges on a ceiling having a gas temperature of 800 °C and a radiation heat component of 2 W/cm^2 to the ceiling. The ceiling is oak. How long will it take to ignite? Assume $h = 30$ W/m K, $T_\infty = 20$ °C and use Tables 7.4 and 7.5.

7.5 Calculate the time to ignite (piloted) for the materials listed below if the irradiance is 30 kW/m^2 and the initial temperature is 25 °C. The materials are thick and the convective heat transfer coefficient is 15 W/m^2 K. Compute the critical flux for ignition as well.

Material	$T_{ig}(deg\ C)$	$k\rho c((kW/m^2K)^2)s$
PMMA	380	1.0
Fiberboard	330	0.46
Plywood	390	0.54
Wood carpet	435	0.25
Polyurethane foam	435	0.03
Paper on gypsum board	565	0.45

7.6 A thin layer of Masonite (wood veneer) is attached to an insulating layer of glass wool. The Masonite is 2 mm thick. It has the following properties:

$$k = 0.14\ \mathrm{W/m\ K}$$
$$\rho = 0.640\ \mathrm{g/m^3}$$
$$c_p = 2.85\ \mathrm{J/gK}$$
$$T_{ig} = 300\ ^\circ\mathrm{C}$$

The total radiative–convective heat transfer coefficient in still air at 25 °C is 30 W/m^2 K.

(a) Calculate the time to ignition for Masonite when subjected to a radiant heat flux of 50 kW/m^2.
(b) What is the critical or minimum heat flux for ignition?
(c) After ignition, what is the initial upward spread rate if the flame heat flux is uniform at 30 kW/m^2 and the flame extends 0.2 m beyond the ignited region of 0.1 m?

7.7 A flame radiates 40 % of its energy. The fuel supply is 100 g/s and its heat of combustion is 30 kJ/g. A thin drapery is 3 m from the flame. Assume piloted ignition. When will the drapery ignite? The ambient temperature is 20 °C and the heat transfer coefficient of the drapery is

$10\,\mathrm{W/m^2\,K}$. The drapery properties are listed below:

$$\rho = 40\ \mathrm{kg/m^3}$$
$$c_p = 1.4\ \mathrm{J/g}\ K$$
$$T_{ig} = 350\ ^\circ\mathrm{C}$$
$$\text{Thickness} = 2\mathrm{mm}\ \text{(assume thin)}$$

The incident flame heat flux can be estimated as $X_r\dot{Q}/(4\pi r^2)$, where X_r is the flame radiation fraction and r is the distance from the flame.

7.8 A vertical block of wood is heated to produce volatiles. The volatiles mix with entrained air in the boundary layer as shown below. A spark of sufficient energy is placed at the top edge of the wood within the boundary layer. The temperature can be assumed uniform across the boundary layer at the spark. The entrained air velocity (v_a), air density (ρ_a) and temperature (T_a) are given on the figure. All gases have fixed specific heat, c_p=1 J/g K. The wood is 10 cm wide by 10 cm high. The surface temperature (T_s) of the wood reaches 400 °C.

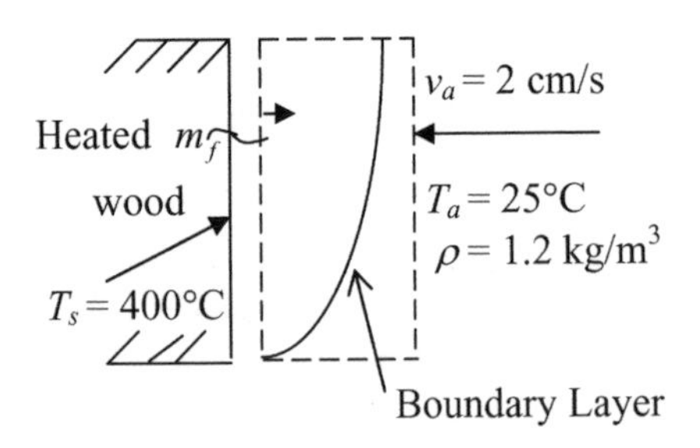

Using the assumption of a minimum flame temperature needed for ignition of the mixture, determine the minimum fuel mass loss rate per unit surface area ($\dot{m}_f''$) to cause flame propagation through the boundary layer. The heat of combustion that the volatile wood produces (Δh_c) is 15 kJ/g. (*Hint*: the adiabatic flame temperature at the lower flammable limit for the mixture in the boundary layer must be at least 1300 °C.)

7.9 The temperature at the corner edge of a thick wood element exposed to a uniform heat flux can be expressed, in terms of time, by the approximate equation given below for small time:

$$T_s - T_\infty \approx \frac{4.2}{\sqrt{\pi}} \dot{q}_e'' \sqrt{t/k\rho c}$$

The notation is consistent with that of the text. The properties of the wood are:

Thermal conductivity	0.12 W/m K
Density	510 kg/m^3
Specific heat	1380 J/kg K
Piloted ignition temperature	385 °C
Autoignition temperature	550 °C
Ambient temperature	25 °C

A radiant heat flux of 45 kW/m^2, *alone*, is applied to the edge of the wood. When will it ignite?

7.10 A thin piece of paper is placed in an oven and raised to 60 °C. When taken out of the oven, it is subjected to 30 kW/m^2 and ignites in 50 s. It is known to ignite in 130 s, when at 20 °C and similarly subjected to 30 kW/m^2. What is its ignition temperature?

8

Fire Spread on Surfaces and Through Solid Media

8.1 Introduction

In Chapter 4 we examined the spread rate of a premixed flame and found that its speed, S_u, depended on the rate of chemical energy release, $\dot{m}_F''' \Delta h_c$. Indeed, for a laminar flame, the idealized flame speed

$$S_u \approx \frac{\dot{m}_F''' \Delta h_c \delta_R}{\rho c_p (T_f - T_\infty)} \tag{8.1}$$

where

T_f = flame temperature
T_∞ = initial temperature of the mixture
δ_R = thickness of the chemical reaction zone

In words we can write

$$\text{Flame speed} = \frac{[\text{chemical energy release rate}]}{[\text{energy needed to raise the substance to its ignition temperature}]} \tag{8.2}$$

We substitute 'ignition temperature' for 'flame temperature (T_f)' because we are interested in the speed of the pyrolysis (or evaporation) front (v_p) just as it ignites. As we have seen from Chapter 7 on the ignition of solids, this critical surface temperature can be taken as the ignition temperature corresponding to the lower flammable limit. Indeed, we might use the same strategy of eliminating the complex chemistry of decomposition and phase changes to select an appropriate 'ignition temperature'. If we use a conduction-based model as in estimating the ignition of solids, then the same measurement techniques for identifying T_{ig} can prevail.

Fundamentals of Fire Phenomena James G. Quintiere

Although very fundamental theoretical and experimental studies have been done to examine the mechanisms of spread on surfaces (solid and liquids), a simple thermal model can be used to describe many fire spread phenomena. This was eloquently expressed in a paper by Forman Williams [1] in discussing flame spread on solids, liquids and smoldering through porous arrays. Excellent discussions on flame spread and fire growth can be found in reviews by Fernandez-Pello [2] and by Thomas [3]. Some noteworthy basic studies of surface flame spread against the wind include the first extensive solution by deRis [4], experimental correlations to show the effect of wind speed and ambient oxygen concentration by Fernandez-Pello, Ray and Glassman [5,6] and the effect of an induced gravitation field by Altenkirch, Eichhorn and Shang [7]. For surface flame spread in the same direction as the wind (or buoyant flow), laminar cases have been examined by Loh and Fernandez-Pello [8,9] with respect to forced flow speed and oxygen concentration, and by Sibulkin and Kim [10] for natural convection laminar upward flame spread. Many features have been investigated for flame spread. These include orientation, wind speed, gravitational variations, laminar and turbulent flows, solid, liquid or porous medium, flame size and radiation, and geometrical effects – corners, channels, ducts and forest terrain to urban terrains.

Figure 8.1 illustrates several modes of flame spread on a solid surface. We can have wind-aided (or concurrent) flame spread in which the pyrolysis front at x_p moves in the directions of the induced buoyant or pressure-driven ambient flow speed, u_∞. Consistent with the thermal model, we require, at $x = x_p$, T_s to be equal to the ignition temperature, T_{ig}. In general, $T_s(x)$ is the upstream or downstream surface temperature of the virgin material. A wind-directed flame on the floor will be different than one on the ceiling. In the floor case, the larger the fire becomes, the more its buoyant force grows and tries to make the flame become vertical despite the wind. This phenomenon will be profoundly different in a channel than on an open floor. In the ceiling case, the gravity vector helps to hold the flame to the ceiling, and even tries to suppress some of its turbulence.

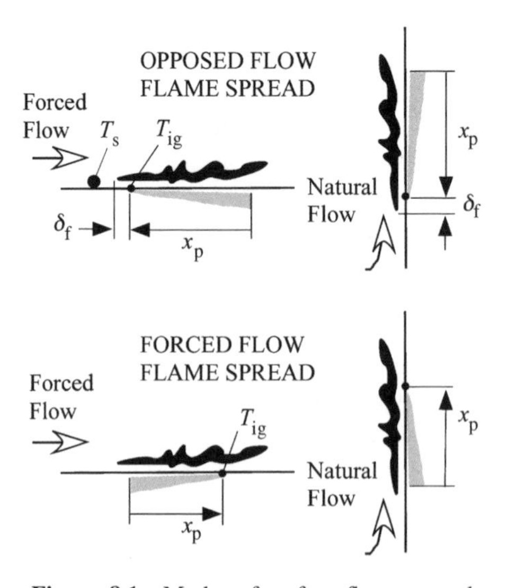

Figure 8.1 Modes of surface flame spread

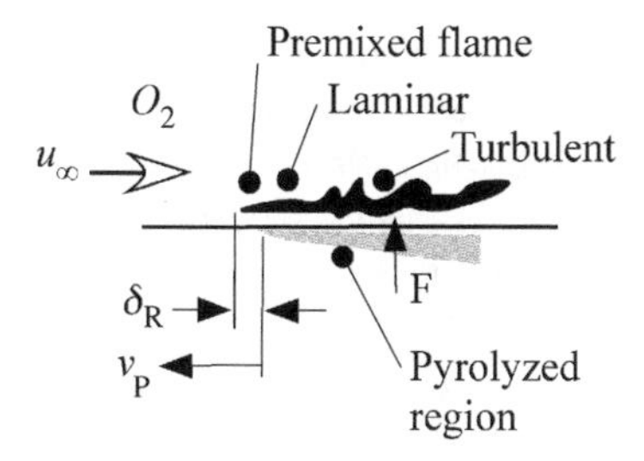

Figure 8.2 Dynamics of opposed flow surface flame spread

When the ambient flow is directed into the advancing flame, we call this case opposed flow flame spread. The flow can be both natural and artificially forced. We will see that opposed flow spread is much slower than wind-aided spread and tends to be steady. This can easily be seen by observing downward or lateral (horizontal) flame spread on a matchstick. However, invert the match to allow the flame to spread upward and we have an almost immeasurable response (provided we can hold the match!). In opposed flow spread, we know that the flame will be quenched near the surface since the typical range of ignition temperatures (250–450 °C) is not likely to support gas phase combustion. Hence, this gap will allow air and fuel to meet before burning, so a premixed flame is likely to lead the advancing diffusion flame, as illustrated in Figure 8.2. If the ambient speed $u_\infty = 0$, it will still appear to an observer on the moving flame (or the pyrolysis leading edge position, x_p) that an opposed flow is coming into the flame at the flame speed, v_p. Even under zero gravity conditions, this is true; therefore, opposed flow spread may play a significant role in spacecraft (microgravity) fires. As u_∞ increases, mixing of fuel and oxygen can be enhanced and we expect a higher $\dot{m}_F'''$ and therefore we expect v_p to increase. However, at high speeds the flow time of air or oxygen through the reaction zone (δ_R) may give a flow time $t_{flow} = \delta_R/u_\infty < t_{chem}$. Thus, the speed of oxygen through the leading premixed flame may be so great as to inhibit the chemical reaction rate. This can lead to a decrease in v_p. Similar arguments can describe the effect of ambient oxygen concentration on v_p; i.e. as $Y_{O_2,\infty}$ increases, $\dot{m}_F'''$ increases and so v_p increases. We shall examine some of these effects later in a more explicit thermal model for v_p.

Since accidental fire spread mostly occurs under natural convection conditions within buildings and enclosures, some examples of configurations leading to opposed or wind-aided types of spread are illustrated in Figure 8.3. Flame spread calculations are difficult

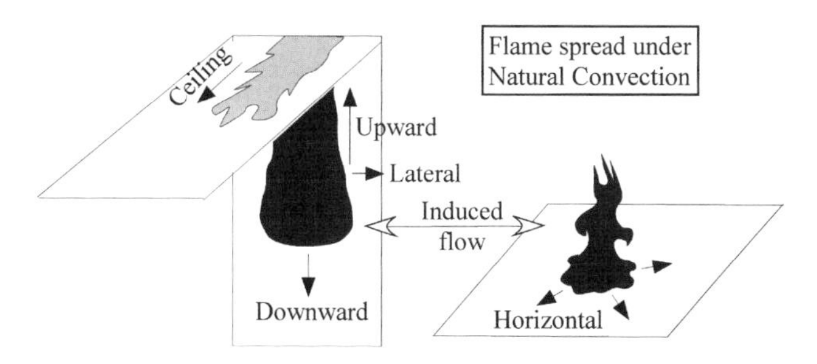

Figure 8.3 Examples of surface flame spread under natural convective conditions

because of the various flow conditions that can occur. Moreover, the size and scale of the fire introduce complexities of turbulence and radiation heat transfer. We will not be able to deal with all of these complexities directly. There are no practical and available solutions that exist for every case. In this chapter we will present flame spread theory in its simplest form. However, we will illustrate how practical engineering estimates can be made for materials based on data and empirical relationships. While we will not address forest fire spread, a similar practical engineering approach has proven useful in the defense of forest fires. It relies on forest vegetation data, humidity, wind, slope of terrain and empirical formulas [11]. We will not describe all these forest fire details, but we will address some aspects of flame spread through porous material, such as fields of grass, forest litter, etc. In this chapter we will also touch on flame spread over liquids and fire growth in multiroom dwellings. The scope of fire spread is very wide; this is why the issue of 'material flammability' is so complex.

8.2 Surface Flame Spread – The Thermally Thin Case

We will develop an analytical formulation of the statement in Equation (8.2). This will be done for surface flame spread on solids, but it can be used more generally [1]. As with the ignition of solids, it will be useful to consider the limiting cases of thermally thin and thermally thick solids. In practice, these solutions will be adequate for first-order approximations. However, the model will not consider any effects due to

(a) melting: involving dripping, heats of melting and vaporization;

(b) charring: causing an increase in surface temperature between pyrolysis and ignition;

(c) deformation: manifested by stretching, shrinking, ripping, curling, etc.;

(d) inhomogeneity: involving the effects of substrate, composites, porosity and transparency absorption to radiation.

Factors (a), (b) and (d) are all swept into the property parameters: T_{ig} and k, ρ and c. Deformation (c) is ignored, but may be significant in practice. A curling material could enhance spread, while ripping could retard it. Dripping in factor (a) could act in both ways, depending on the orientation of the sample and direction of spread. Any test results that lead to the thermal properties for ignition and those to be identified for spread must apply to the fire scenario for the material in its end use state and configuration. For example, a test of wallpaper by itself as a thin sheet would not suffice for evaluating its flame spread potential when adhered to a gypsum wall board in a building corridor subject to a fully involved room fire.

Figure 8.4 displays a thermally thin solid of thickness d, insulated at its back face and undergoing surface flame spread. Any of the modes shown in Figure 8.1 are applicable. We consider the case of steady flame spread, but this constraint can be relaxed. The surface flame spread speed is defined as

$$v_{\text{p}} = \frac{\text{d}x_{\text{p}}}{\text{d}t} \tag{8.3}$$

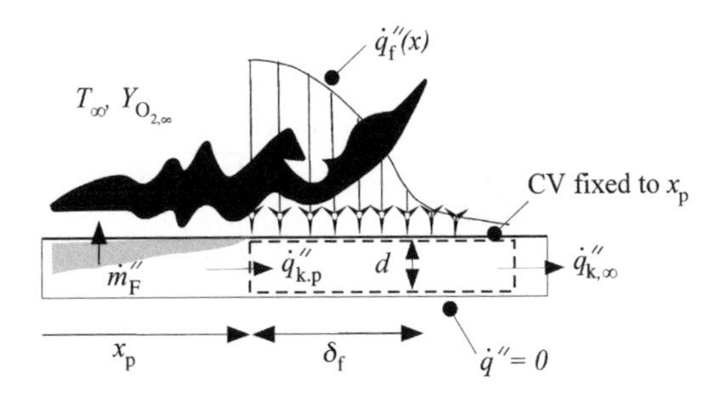

Figure 8.4 Thermal model for surface flame spread

In Figure 8.4 the control volume surrounds the solid and is fixed at $x = x_{\rm p}$ and a distance far enough ahead of the flame heat transfer affected region of the solid. The possible heat fluxes include:

$\dot{q}''_{\rm f}$ flame incident convective and radiative heat flux
$\dot{q}''_{k,\rm p}$ conduction from the pyrolyzed region
$\dot{q}''_{k,\infty}$, conduction from the CV to $T_{\rm s}$

Here $T_{\rm s}$ is taken to mean the surface temperature of the solid that is not affected by the flame heating. If $T_{\rm s}$ is constant or varies slowly with $x(\partial T/\partial x \approx 0)$, we can set $\dot{q}''_{k,\infty} = 0$. Much study has gone into assessing the importance of $\dot{q}''_{k,\rm p}$, and it can be significant. However, the dominant heat transfer, even under slow opposed spread, is $\dot{q}''_{\rm f}$. For concurrent spread this approximation is very accurate since the flame surface heating distance is very long and $\partial T/\partial x \approx 0$ near $x = x_{\rm p}$. We shall adopt the solid temperature necessary for piloted ignition as the ignition temperature. This is in keeping with the initiation of pyrolysis to cause ignition of the evolved fuel with a pilot being the advancing flame itself.

We apply Equation (3.45) for the CV fixed to $x = x_{\rm p}$. The solid with density ρ and specific heat $c_{\rm p}$ appears to enter and leave the CV with velocity $v_{\rm p}$, the flame speed. Thus,

$$\rho c_p \, d v_{\rm p}(T_{\rm ig} - T_{\rm s}) = \int_{x_{\rm p}}^{\infty} \left[\dot{q}''_{\rm f}(x) - \sigma(T^4(x) - T^4_\infty)\right] {\rm d}x \tag{8.4}$$

where we have approximated the surface as a blackbody and consider radiation to an ambient at T_∞. Since we expect $\dot{q}''_{\rm f} \gg \sigma(T^4_{\rm ig} - T^4_\infty)$ we will ignore this nonlinear temperature term. Furthermore, we define an average or characteristic $\dot{q}''_{\rm f}$ for an effective heating length $\delta_{\rm f}$ as

$$\dot{q}''_{\rm f}\delta_{\rm f} \equiv \int_{x_{\rm p}}^{\infty} \dot{q}''_{\rm f}(x) \, {\rm d}x \tag{8.5}$$

Physically $\delta_{\rm f}$ is the flame extension over the new material to ignite. This is illustrated for a match flame, spreading upward or downward in Figure 8.5 or as suggested in Figure 8.1. We shall see that the visible flame is a measure for $\delta_{\rm f}$ because at the flame tip in a turbulent flame the temperature can drop from 800 °C in the continuous luminous zone to

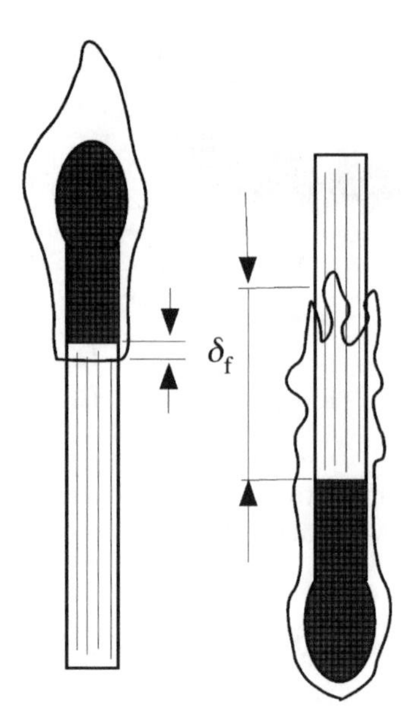

Figure 8.5 Flame heating extensions for a match flame

about 350 °C. Hence, a point between the continuous luminous zone and the fluctuating flame tip might constitute an approximation for δ_f. In any case, Equation (8.5) is mathematically acceptable for a continuous function, $\dot{q}''_f(x)$, according to the mean value theorem. Equations (8.4) and (8.5) yield a solution

$$v_p = \frac{\dot{q}''_f \delta_f}{\rho c_p d(T_{ig} - T_s)} \tag{8.6}$$

This can be put into a different form to reveal the relationship of flame spread to ignition:

$$v_p = \frac{\delta_f}{t_f} \tag{8.7a}$$

$$t_f = \frac{\rho c_p d(T_{ig} - T_s)}{\dot{q}''_f} \tag{8.7b}$$

This is identical with our result for the ignition time at a high heat flux in the thermally thin ignition of a solid. It can be generalized to say that flame speed – thermally driven – can be represented as the ratio of the flame extension length to the time needed to ignite this heated material, originally at T_s.

Equation (8.6) demonstrates that as the face weight, ρd, decreases the spread rate increases. Moreover, if a material undergoing spread is heated far away from the flame, such as would happen from smoke radiation in a room fire, T_s will increase over time. As $T_s \rightarrow T_{ig}$, an asymptotic infinite speed is suggested. This cannot physically happen. Instead, the surface temperature will reach a pyrolysis temperature sufficient to cause fuel vapor at the lower flammable concentration. Then the speed of the visible flame along the surface will equal the premixed speed. This speed in the gas phase starts at about 0.5 m/s

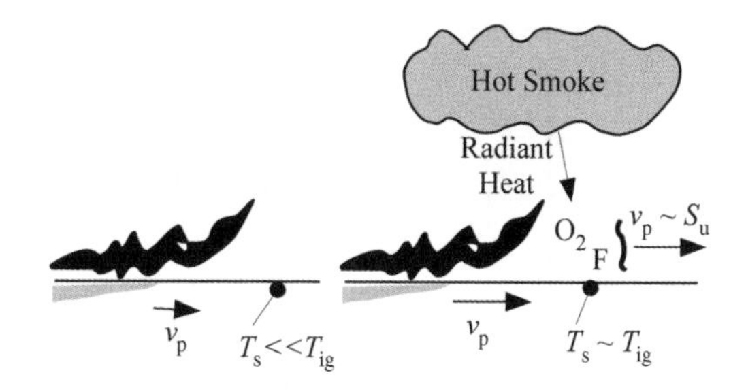

Figure 8.6 Enhanced flame spread due to preheating

and can accelerate due to turbulence (see Figure 8.6). We see under these conditions why smoke-induced radiant heat flux of about 20 kW/m^2 might provoke rapid flame spread across objects on a floor during a room fire. This mechanism can result in flashover – a rapid transition to a fully involved room fire.

It should be pointed out that Equation (8.6), and its counterpart for thermally thick materials, will hold only for $T_s > T_{s,min}$, a minimum surface temperature for spread. Even if we include the heat loss term in Equation (8.4) by a mean-value approximation for the integrand,

$$\frac{T_{ig}^4 + T_s^4}{2}$$

we obtain a physically valid solution still possible for all T_s as

$$v_p = \frac{\dot{q}_f'' \delta_f - \sigma\Big((T_{ig}^4 + T_s^4)/2\Big)\,\delta_f}{\rho c_p d(T_{ig} - T_s)} \tag{8.8}$$

as long as the numerator is positive. Usually at $T_s = T_{s,min}$, $v_p > 0$, as determined in the experiments. This is shown in Figures 8.7 and 8.8 for lateral spread and upward spread on a vertical surface (for thermally thick behavior of PMMA and wood respectively). The

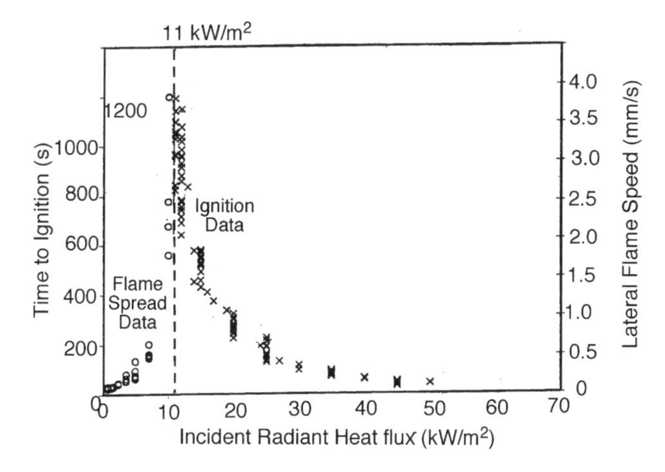

Figure 8.7 Natural steady flame speed and piloted ignition for PMMA (vertical) (from Long [12])

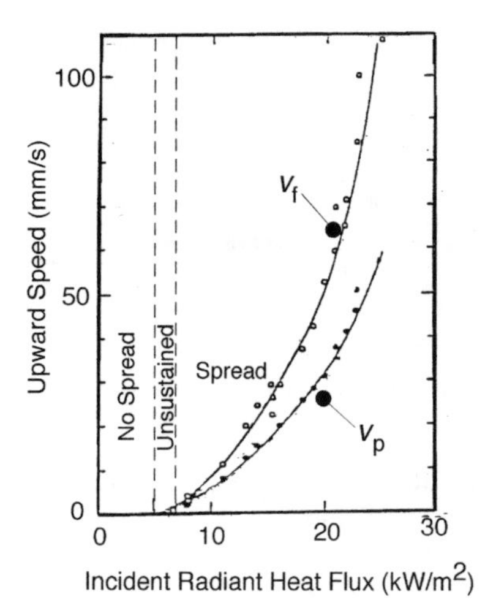

Figure 8.8 Pyrolysis-front propagation velocity v_p, and flame-tip propagation velocity v_f, as functions of the incident radiant energy flux, for a preheat time t_p of 2 min on a wood sample [13]

incident external radiant heat flux is a surrogate here for the steady state T_s at equilibrium. Thus, there is a critical heat flux for flame spread (as in ignition) given by

$$\dot{q}''_{s,crit} = \sigma(T^4_{s,min} - T^4_\infty) + h_c(T_{s,min} - T_\infty) \tag{8.9}$$

This follows by a steady state energy balance of the surface heated by $\dot{q}''_e$, outside the flame-heated region δ_f. It appears that a critical temperature exists for flame spread in both wind-aided and opposed flow modes for thin and thick materials. $T_{s,min}$ has not been shown to be a unique material property, but it appears to be constant for a given spread mode at least. Transient and chemical effects appear to be the cause of this flame spread limit exhibited by $T_{s,min}$. For example, at a slow enough speed, v_p, the time for the pyrolysis may be slower than the effective burning time:

$$t_f = \frac{\delta_f}{v_p} > t_b \tag{8.10}$$

and the inequality in this directly leads to extinction. For wind-aided spread the limit condition will depend on the flame extension and on the burning time; for opposed flow spread, the ability to achieve a flammable mixture at the flame front is key. In addition, transient effects will influence all of these factors.

8.3 Transient Effects

It is useful to consider the transient aspect of the flame spread problem. The ratio of characteristic times for the solid and the gas phase, based on d as a characteristic length scale, is

$$\frac{t_{gas}}{t_{solid}} = \frac{d^2/\alpha_{gas}}{d^2/\alpha_{solid}} = \frac{\alpha_{solid}}{\alpha_{gas}} \approx \frac{10^{-7}\ \mathrm{m^2/s}}{10^{-5}\ \mathrm{m^2/s}} = 10^{-2}$$

as a typical value. Hence, the gas phase response is very fast and can adjust quickly to any transient changes in solid. Thus, any steady state heat transfer data for $\dot{q}''_{\rm f}$ or flame extension correlations (e.g. for upward, ceiling or wind-aided flames) can be directly applied to the transient problem of flame spread on the solid.

However, we have only considered a steady state solid model in Equation (8.4). Consider the effect if we add the transient term (Equation (3.40)), the rate of enthalpy per unit area in the CV,

$$\frac{{\rm d}H''}{{\rm d}t} \equiv \rho d c_p \frac{{\rm d}}{{\rm d}t} \int_{x=x_{\rm p}(t)}^{x=x_{\rm f}(t)} T {\rm d}x \tag{8.11}$$

where $x_{\rm f}$ is the position of the flame tip, $x_{\rm p} + \delta_{\rm f}$. Here T_∞ is a reference temperature, say 25 °C. For opposed flow spread we do not expect $\delta_{\rm f}$ to depend on $x_{\rm p}$, the region undergoing pyrolysis. Here we consider that $x = 0$ to $x = x_{\rm p}$ is pyrolyzing, although burnout may have occurred for $x = x_{\rm b} < x_{\rm p}$. We will ignore the burnout effect here. For opposed flow spread, $\delta_{\rm f}$ will depend on the fluid mechanics of the induced flow (u_∞) in combination with the slowly opposite moving flame $(v_{\rm p})$. However, for wind-aided spread it might be expected that $\delta_{\rm f}$ will depend on the extent of the pyrolysis region – in other words, a longer flame for more fuel produced over $x = 0$ to $x = x_{\rm p}$.

The transient term can be approximated as follows, where $c_{\rm p} \approx c_{\rm v}$:

$$\begin{aligned} \frac{{\rm d}H''}{{\rm d}t} &\approx \rho d c_p \frac{{\rm d}}{{\rm d}t}\left[\left(\frac{T_{\rm ig} + T_{\rm s}}{2}\right)\delta_{\rm f}\right] \\ &= \rho d c_p \left[\frac{{\rm d}T_{\rm s}}{{\rm d}t}\frac{\delta_{\rm f}}{2} + \left(\frac{T_{\rm ig} + T_{\rm s}}{2}\right)\frac{{\rm d}\delta_{\rm f}}{{\rm d}t}\right] \end{aligned} \tag{8.12}$$

If we consider cases where $T_{\rm s}$ is changing slowing with time, then

$$\frac{{\rm d}T_{\rm s}}{dt} \approx 0$$

is reasonable. For opposed flow, $\delta_{\rm f} \neq \delta_{\rm f}(x_{\rm p}(t))$, so ${\rm d}\delta_{\rm f}/{\rm d}t = 0$. However, for the example of wind-aided (upward) spread in Figure 8.8 it appears that ${\rm d}\delta_{\rm f}/{\rm d}t > 0$ as $\dot{q}''_{\rm e}$ increases, since

$$\frac{{\rm d}\delta_{\rm f}}{{\rm d}t} = v_{\rm f} - v_{\rm p}$$

Thus, for opposed flow spread, the steady state thermal flame spread model appears valid. In wind-aided flame spread, it seems appropriate to modify our governing equation for the thermally thin case as

$$\rho d c_p \left(\frac{T_{\rm ig} + T_{\rm s}}{2}\right)\frac{{\rm d}\delta_{\rm f}}{{\rm d}t} + \rho d c_v v_{\rm p}(T_{\rm ig} - T_{\rm s}) = \dot{q}''_{\rm f}\delta_{\rm f} \tag{8.13}$$

Since

$$\frac{{\rm d}\delta_{\rm f}}{{\rm d}t} = \frac{{\rm d}x_{\rm f}}{{\rm d}t} - \frac{{\rm d}x_{\rm p}}{{\rm d}t} = v_{\rm p}\left(\frac{{\rm d}x_{\rm f}}{{\rm d}x_{\rm p}} - 1\right) \tag{8.14}$$

combining gives, for wind-aided spread on a thin material,

$$v_{\mathrm{p}} \approx \frac{\dot{q}_{\mathrm{f}}'' \delta_{\mathrm{f}}}{\rho d c_p \left[(T_{\mathrm{ig}} - T_{\mathrm{s}}) + \left(\frac{\mathrm{d}x_{\mathrm{f}}}{\mathrm{d}x_{\mathrm{p}}} - 1 \right) \left(\frac{T_{\mathrm{ig}} + T_{\mathrm{s}}}{2} \right) \right]} \tag{8.15}$$

From Figure 8.8, $v_{\mathrm{f}} > v_{\mathrm{p}}$, so $\mathrm{d}x_{\mathrm{f}}/\mathrm{d}x_{\mathrm{p}} > 1$. Hence, transient effects will cause a lower speed for wind-aided spread. Again, since the gas phase response time is much faster than the solid, we can use steady gas phase results for $\dot{q}_{\mathrm{f}}''$ and δ_{f} in these formulas for flame spread on surfaces.

8.4 Surface Flame Spread for a Thermally Thick Solid

Let us now turn to the case of a thermally thick solid. Of course thickness effects can be important, but only after the thermal penetration depth due to the flame heating reaches the back face, i.e. $t = t_{\mathrm{f}}$, $\delta_{\mathrm{T}}(t_{\mathrm{f}}) = d$. As in the ignition case, if t_{f} is relatively small, say 10 to even 100 s, the thermally thick approximation could even apply to solids of $d \leq 1$ cm. Again, we represent all of the processes by a thermal approximation involving the effective properties of T_{ig}, k, ρ and c. Materials are considered homogeneous and any measurements of their properties should be done under consistent conditions of their use. Other assumptions for this derivation are listed below:

1. $\dot{q}_{\mathrm{f}}''$ is constant over region δ_{f} and zero beyond.
2. Negligible surface heat loss, i.e. $\dot{q}_f'' \gg \sigma T_{\mathrm{ig}}^4$.
3. At $y = \delta_{\mathrm{T}}$, $\dot{q}'' = 0$ and $T = T_{\infty}$.
4. Both initial and upstream temperatures are T_{∞}.
5. Steady flame spread, $v_{\mathrm{p}} = \text{constant}$.

Consider the control volume in Figure 8.9 where the thermally thick case is drawn for a wind-aided mode, but results will apply in general for the opposed case. The control

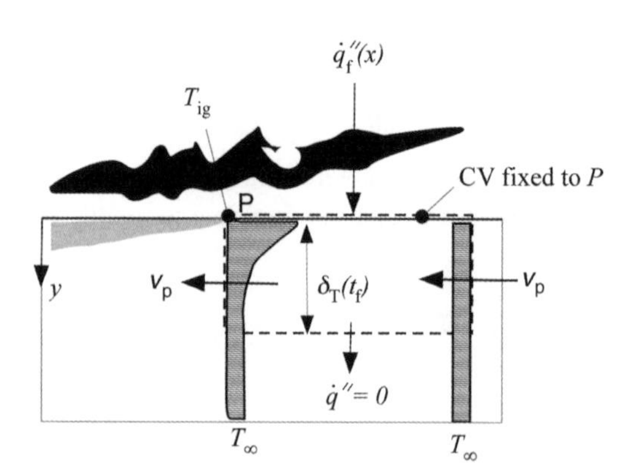

Figure 8.9 Control volume energy conservation for a thermally thick solid with flame spread steady velocity, V_{P}

volume is selected to span the flame heated length δ_f and extend to the depth δ_T. It is fixed to the point, P. The specific depth is $\delta_T = \delta_T(t_f)$, since t_f is the time for the temperature at P to have been increased from T_∞ to T_{ig}. However, t_f is also the time for the flame to traverse δ_f, i.e.

$$\delta_f = \int_0^{t_f} v_p \, dt = v_p \, t_f \tag{8.16}$$

since v_p is constant. The conservation of energy for the CV is

$$\rho c_p v_p \int_0^{\delta_T} (T - T_\infty) \, dy = \dot{q}_f'' \delta_f \tag{8.17}$$

Let us approximate the temperature profile as

$$\frac{T - T_\infty}{T_{ig} - T_\infty} = \left(1 - \frac{y}{\delta_T}\right)^2 \tag{8.18}$$

which satisfies

$$y = 0, \quad T = T_{ig}$$
$$y = \delta_T, \quad T = T_\infty \text{ and } \frac{\partial T}{\partial y} = 0$$

This approximation gives us a reasonable profile that satisfies the boundary conditions and allows us to perform the integration

$$\int_0^{\delta_T} (T - T_\infty) \, dy = (T_{ig} - T_\infty) \, \delta_T / 3 \tag{8.19}$$

We have seen that a reasonable approximation for the penetration depth is

$$\delta_T = C \sqrt{\left(\frac{k}{\rho c_p}\right) t} \tag{8.20}$$

where C can range from 1 to 4 depending on its definition of how close T is to T_∞. Let us take $C = 2.7$ as a rational estimate. Now we substitute the results of Equations (8.16), (8.19) and (8.20) into Equation (8.17) to obtain

$$\rho c_p v_p \frac{(T_{ig} - T_\infty)}{3} 2.7 \sqrt{\left(\frac{k}{\rho c_p}\right)\left(\frac{\delta_f}{v_p}\right)} \approx \dot{q}_f'' \delta_f$$

or

$$v_p \approx \frac{(\dot{q}_f'')^2 \, \delta_f}{(0.81)(k \rho c_p)(T_{ig} - T_\infty)^2} \tag{8.21}$$

Returning to the form of Equation (8.16), we can estimate

$$t_f \approx 0.81\ k\rho c_p \left(\frac{T_{ig} - T_\infty}{\dot{q}_f''}\right)^2 \tag{8.22}$$

We might just as well use the exact thermally thick high heat flux (short time) result, where 0.81 is replaced by $\pi/4 = 0.785$. We can use

$$\boxed{v_p \approx \frac{4\,(\dot{q}_f'')^2\,\delta_f}{\pi(k\rho c_p)(T_{ig} - T_s)^2}} \tag{8.23}$$

where we have generalized to T_s which could be changing under smoke and fire conditions. Thus Equation (8.23) might be used in a quasi-steady form where $\dot{q}_f''$, δ_f, and T_s might all vary with time as the flame grows and fire conditions worsen. Furthermore, Equation (8.23) follows the form of Equation (8.7) which says that flame spread is essentially the flame heating distance divided by the ignition time.

8.5 Experimental Considerations for Solid Surface Spread

We can adopt Equations (8.6) and (8.23) for practical applications. Since these results complement the corresponding thermally thin and thick ignition solutions, the thermal properties should be identical for both ignition and flame spread with respect to a given material.

8.5.1 Opposed flow

For opposed flow spread under natural convection conditions, Ito and Kashiwagi [14] have measured the $\dot{q}_f''(x)$ profile for PMMA ($d = 0.47$ cm) at several angles. Their results, shown in Figure 8.10, suggest values for

$$\dot{q}_f'' \approx 70\ \text{kW/m}^2$$

and

$$\delta_f \approx 2\ \text{mm}$$

as rough approximations for their experimental profiles, e.g. $\phi = 90°$. Furthermore, δ_f increases as spread goes from downward ($\phi = 0°$) to horizontal ($\phi = 90°$) to slightly upward ($\phi = 80°$). These angles all constitute natural convection opposed flow flame spread.

A more practical way to yield results for opposed flow spread is to recognize that the parameter

$$\Phi \equiv \frac{4}{\pi}(\dot{q}_f'')^2\,\delta_f \tag{8.24}$$

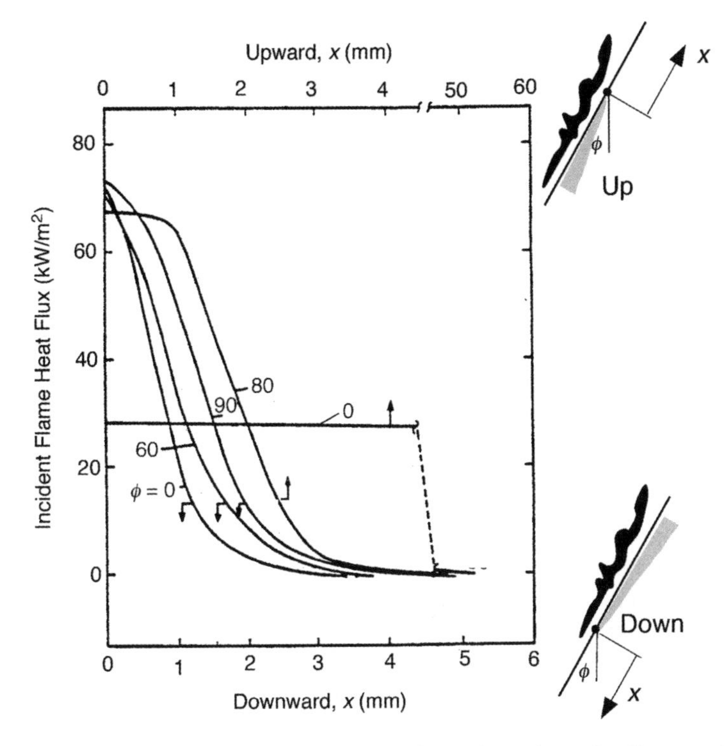

Figure 8.10 Distributions of normal heat flux at the surface along the distance ahead of the pyrolysis front for 0.47 cm thick PMMA [14]

can be regarded as a material property for a given flow condition. This can be argued from laminar premixed conditions for the leading flame with the heat flux related to flame temperature and distances related to a quenching length. Under natural convection induced flow velocities, Φ might be constant. Figure 8.11 shows steady velocity data plotted for thick PMMA as a function of incident radiant heat flux [12,15].

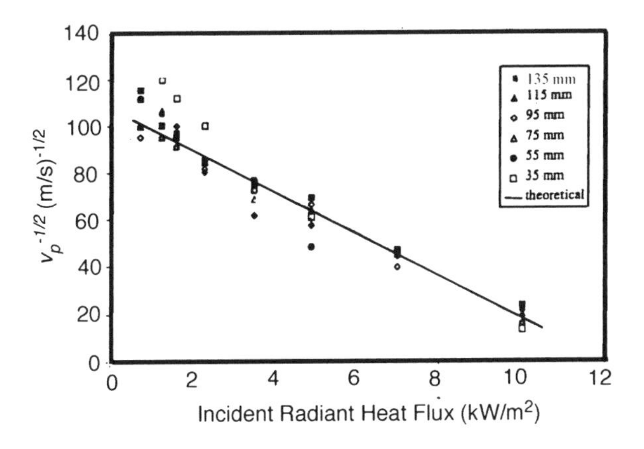

Figure 8.11 Evaluation of Φ from the LIFT [12]

Under steady conditions,

$$\boxed{v_{\mathrm{p}} = \frac{\Phi}{k\rho c_p (T_{\mathrm{ig}} - T_{\mathrm{s}})^2}, \quad T_{\mathrm{s}} > T_{\mathrm{min}}} \tag{8.25}$$

If T_{s} is caused by long-time radiant heating, we can write

$$\dot{q}''_{\mathrm{e}} = h_{\mathrm{t}}(T_{\mathrm{s}} - T_{\infty}) \tag{8.26}$$

For h_{t} approximated as a constant, we can alternatively write Equation (8.25) as

$$v_{\mathrm{p}}^{-1/2} = \left(\frac{k\rho c_p}{h_{\mathrm{t}}^2 \Phi}\right)^{1/2} (\dot{q}''_{\mathrm{ig,crit}} - \dot{q}''_{\mathrm{e}}) \tag{8.27}$$

where $\dot{q}''_{\mathrm{ig,crit}}$ is the critical heat flux for ignition given by Equation (7.28). Since $k\rho c_p$ can be evaluated from ignition data for the material, and for h_{t} evaluated at $\dot{q}''_{\mathrm{ig,crit}}$, then Φ can be determined from the slope of the data in Figure 8.11. Also the intercept in the abscissa is $\dot{q}''_{\mathrm{ig,crit}}$ and the lower flux limit gives $\dot{q}''_{\mathrm{s,crit}}$ or $T_{\mathrm{s,min}}$ from Figure 8.11. Data for a number of materials under conditions of lateral flame spread for a vertical sample in the LIFT [15,16] are given in Table 8.1 (also see Table 7.4). The value of Φ given for PMMA is 14.0 $\mathrm{kW^2/m^3}$, which can be compared to an estimate from the data of Ito and Kashiwagi [14]. We approximate using Equation (8.24):

$$\begin{aligned}\Phi &\approx \frac{4}{\pi}(70\ \mathrm{kW/m^2})^2(0.002\ \mathrm{m}) \\ &= 12.5\ \mathrm{kW^2/m^3}\end{aligned}$$

compared to a measured value of 14 $\mathrm{kW^2/m^3}$. This is probably the level of consistency to be expected for a 'property' such as Φ.

For unsteady lateral flame spread, Equation (8.25) can be applied in a quasi-steady fashion. We use the short-time approximation for the surface temperature (Equation (7.40)),

$$T_{\mathrm{s}} - T_{\infty} \approx \left(\frac{\dot{q}''_{\mathrm{e}}}{h_{\mathrm{t}}}\right) F(t) \tag{8.28}$$

Table 8.1 Opposed flow properties for lateral flame spread on a vertical surface [16]

Material	T_{ig} (°C)	$k\rho c$ ((kW/m²K)²s)	Φ((kW)²/m³)	$T_{\mathrm{s,crit}}$ (°C)
PMMA	380	1.0	14.0	~ 20
Fiberboard	330	0.46	2.3	210
Plywood	390	0.54	13.0	120
Wool carpet	435	0.25	7.3	335
Polyurethane foam	435	0.03	4.1	215
Paper on gypsum board	565	0.45	14.0	425

where

$$F(t) = \frac{2h_t}{\sqrt{\pi}} \sqrt{\frac{k\rho c_p}{t}} \leq 1$$

Then Equation (8.25) can be expressed as

$$v_p^{1/2} \approx \left(\frac{k\rho c_p}{h_t^2 \Phi}\right)^{1/2} \left[\dot{q}''_{\text{ig,crit}} - \dot{q}''_{\text{e}} F(t)\right] \tag{8.29}$$

Examine data taken under a natural convection lateral vertical surface spread for a particle board as shown in Figures 8.12(a) and (b) [17]. The correlation of these data based on Equation (8.29) suggests that the critical flux for ignition is

$$\dot{q}''_{\text{ig,crit}} \approx 16\,\text{kW/m}^2$$

and for the flame spread is

$$\dot{q}''_{\text{s,crit}} \approx 5\ \text{kW/m}^2$$

For an average $h_t = 42\ \text{W/m}^2\ \text{K}$ and $T_\infty = 25\,^\circ\text{C}$, we obtain

$$T_{\text{ig}} = 25 + \frac{16}{0.042} = 406\,^\circ\text{C}$$

and

$$T_{\text{s,min}} = 25 + \frac{5}{0.042} = 144\,^\circ\text{C}$$

From the slope of the data in Figure 8.12(b),one can estimate

$$\left(\frac{k\rho c_p}{h_t^2 \Phi}\right)^{1/2} = 240\ (\text{mm s})^{3/2}/\text{J}$$

Using $h_t = 42\ \text{W/m}^2\ \text{K}$ and $k\rho c_p = 0.93\ (\text{kW/m}^2\ \text{K})^2\ \text{s}$ (Table 7.4), Φ can be estimated:

$$\begin{aligned}\Phi &= \frac{k\rho c_p}{h_t^2\,(240)^2} \\ &= \frac{0.93\ (\text{kW/m}^2\ \ \text{K})^2\ \text{s}}{(0.042\ \text{kW/m}^2\ K)^2[240\ (\text{mm s})^{3/2}/\text{J} \times (10^{-3}\ \text{m/mm})^{3/2} \times (10^3\ \text{J/kJ})]^2} \\ &= 9.15\ \text{kW}^2/\text{m}^3\end{aligned}$$

This compares to a Φ of 5 kW^2/m^3 for another particle board in Table 7.4.

The pseudo-property Φ is a parameter that represents flow phenomena in addition to thermal and chemical properties of the material. Incidently, Φ can be utilized for

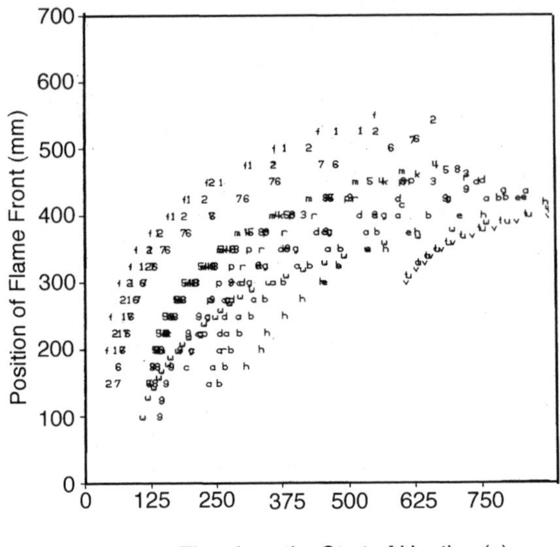

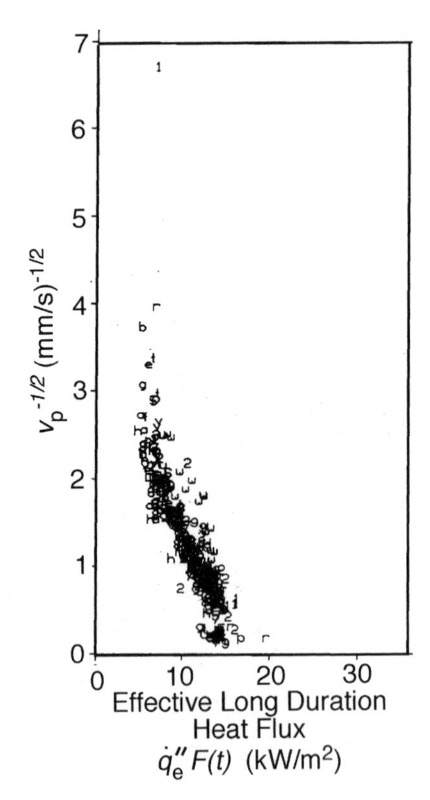

Figure 8.12 (a) Flame front movement for a wood particle board under opposed flow spread in still air under various external heating distributions. Symbols t, u, v, w are for downward spread; all the rest are for lateral spread on a vertical sample [17]. (b) Correlation based on Equation (8.29)[17]

thermally thin opposed flow spread as well as wind-aided. However, under wind-aided conditions, it is not practical to link $\dot{q}''_f$ and δ_f into a Φ function. For these flow conditions $\dot{q}''_f$ and δ_f are distinctly independent variables.

8.5.2 *Wind-aided*

For turbulent wall fires, from the work of Ahmad and Faeth [18], it can be shown that

$$\dot{q}''_f = \text{function } [x_p,\ Y_{O_2,\infty},\ Gr,\ Pr,\ \text{fuel properties, flame radiation } (X_r), \phi]$$

where

x_p = length of the pyrolyzing region
$Y_{O_2,\infty}$ = ambient oxygen mass fraction
Gr = Grashof number $= g\ \cos\phi x_p^3/\nu_\infty^2$,
Pr = Prandtl number= ν_∞/α_∞
ϕ = wall orientation angle from the vertical
X_r = flame of a radiation fraction

The radiation dependence has been simplified using only X_r, but in general it is quite complex and significant for tall wall flames (> 1 m).
The corresponding flame length can be shown to be [19]

$$x_f = x_p + \delta_f = \frac{1.02\,\ell_c}{Y_{O_{2,\infty}}(1 - X_r)^{1/3}} \tag{8.30a}$$

where a convective plume length scale is

$$\ell_c = \left(\frac{\dot{m}''_F\,\Delta h_c\,x_p}{\rho_\infty c_p T_\infty \sqrt{g\cos\phi}}\right)^{2/3} \tag{8.30b}$$

The orientation effect included as $\cos\phi$ is inferred from boundary layer theory and is not appropriate for an upward-facing surface at large ϕ where the boundary layer is likely to separate. It is more likely to apply to flame lengths on a bottom surface and not to lifted flames on a top surface. Similar results apply to laminar wall flames, but these correspond to $Gr < 0.5 \times 10^8$ or about $x_p < 100$ mm. However, irregular edges or rough spots can trip the flow to turbulent for x_p as low as 50 mm. Hence, accidental fires of any significance will quickly become turbulent. The burning rate per unit area, $\dot{m}''_F$, depends on the same variables as $\dot{q}''_f$ above.

The Ahmad and Faeth [18] data encompass alcohols saturated into an inert wall of x_p up to 150 mm and x_f up to 450 mm. Typically, $\dot{q}''_f$ is roughly constant over the visible flame extension (δ_f) with values of between 20 and 30 kW/m^2. The same behavior is seen for the radiatively enhanced burning of solid materials – again showing $\dot{q}''_f$ values of 20–30 kW/m^2 over δ_f for x_f up to 1.5 m. These data are shown in Figure 8.13. Such empirical results for the flame heat flux are useful for obtaining practical estimates for upward flame spread on a wall.

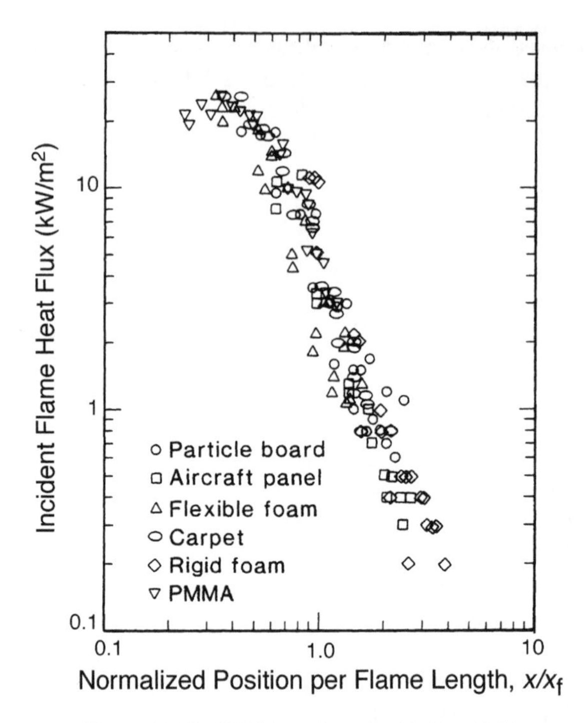

Figure 8.13 Wall heat flux distribution [20]

Example 8.1 It is found that flame length on a vertical wall can be approximated as

$$y_{\mathrm{f}} = (0.01\ \mathrm{m}^2/\mathrm{kW})(\dot{Q}'' y_{\mathrm{p}})$$

where $\dot{Q}''$ is the energy release flux in kW/m^2 of the burning wall. For thick polyurethane foam listed in Table 8.1 compute

(a) the upward spread velocity assuming that the region initially ignited is 0.1 by 0.1 m high and

(b) the lateral spread rate on the wall occurs after 220 °C.

Take $\dot{Q}'' = 425$ kW/m^2 for the foam burning in air at 25 °C. The adjacent surface immediately outside the spreading flame is heated to 225 °C.

Solution

(a) We adopt y as the vertical coordinate, as shown in Figure 8.14. Therefore Equation (8.23) becomes

$$\frac{\mathrm{d}y_{\mathrm{p}}}{\mathrm{d}t} = \frac{(\dot{q}_{\mathrm{f}}'')^2 (y_{\mathrm{f}} - y_{\mathrm{p}})}{\pi/4k\rho c_p (T_{\mathrm{ig}} - T_{\mathrm{s}})^2} = \frac{y_{\mathrm{f}} - y_{\mathrm{p}}}{t_{\mathrm{f}}}$$

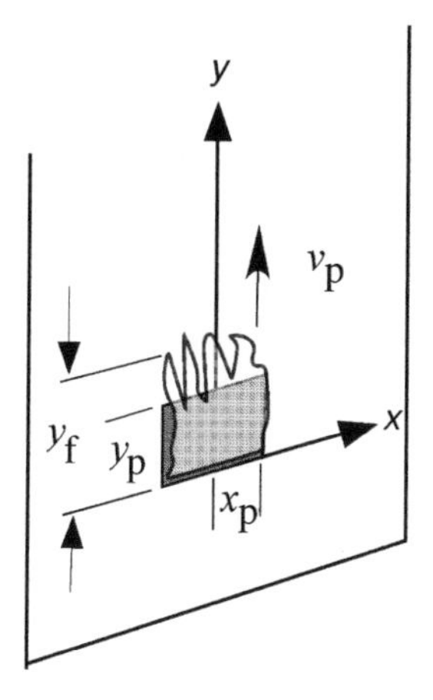

Figure 8.14 Example 8.1

We are given

$$\begin{aligned} y_f &= 0.01\dot{Q}'' y_p \\ &= (0.01)(425)y_p \\ &= 4.25y_p \end{aligned}$$

Based on Figure 8.13, we estimate

$$\dot{q}_f'' \approx 25 \text{ kW/m}^2$$

Then we can compute

$$\begin{aligned} t_f &= \frac{\pi}{4} k\rho c_p \left(\frac{T_{ig} - T_s}{\dot{q}_f''} \right)^2 \\ &= \frac{\pi}{4} (0.03 \text{ (kW/m}^2\text{ K)}^2 \text{ s}) \left(\frac{(435 - 225)\text{K}}{25 \text{ kW/m}^2} \right)^2 \\ &= 1.66\text{s} \end{aligned}$$

Substituting into the differential equation, we obtain

$$\frac{dy_p}{dt} = \frac{3.25\, y_p}{1.66} = 1.95\, y_p$$

or solving with $t = 0$, $y_p = 0.1$ m,

$$y_p(m) = 0.1\, e^{1.95\, t(s)}$$

The speed is

$$v_p = \frac{dy_p}{dt} = 0.195\, e^{1.95\, t}$$

In 2 s, $y_p = 4.9$ m, and $v_p = 9.67$ m/s. U pward spread is very fast. However, for the empirical result,

$$y_f \approx 0.01\,\dot{Q}''\,y_p$$

only $\dot{Q}'' > 100$ kW/m^2 will allow spread to occur. The energy release flux of a material is a very critical variable for upward and wind-aided flame spread in general.

(b) The lateral spread is constant and given by

$$\begin{aligned} u_p &= \frac{\Phi}{k\rho c_p (T_{ig} - T_s)^2} \\ &= \frac{4.1\ (\mathrm{kW/m})^2}{(0.03\ (\mathrm{kW/m^2\ K})^2\ \mathrm{s})(435 - 225)^2} \\ &= 0.003\ \mathrm{m/s} = 3\ \mathrm{mm/s} \end{aligned}$$

This is much, much slower than the upward flame speed and presents a significantly lower relative hazard.

8.6 Some Fundamental Results for Surface Spread

deRis [4] was the first to obtain a complete theoretical solution for opposed flame spread on a surface. His solution is too detailed to present here, but we can heuristically derive the same exact results from our formulations for the thermally thin and thick cases. Consider Figure 8.15. We estimate the heat transfer lengths as

$$\delta_g \sim \delta_f \sim \delta_{BL} \sim \frac{\alpha_g}{u_\infty} \tag{8.31}$$

where δ_{BL} is the laminar boundary layer thickness. These are order of magnitude estimates only. The effects of chemical kinetics were ignored in deRis' solution and only become a factor if the time for the chemical reaction is longer than the time for a gas particle to traverse δ_g. By our order of magnitude estimates this flow time can be represented as δ_{BL}/u_∞. The ratio of the flow to chemical time is called the Damkohler number, Da, as defined in Equations (4.23) and (4.24). Using Equation (8.31), we can express this as

$$\text{Damkohler number, } Da = \frac{t_{FLOW}}{t_{REACTION}} = \frac{\delta_{BL}/u_\infty}{\rho c_p T_\infty/[E/(RT_\infty)]\Delta h_c A e^{-E/(RT_\infty)}}$$

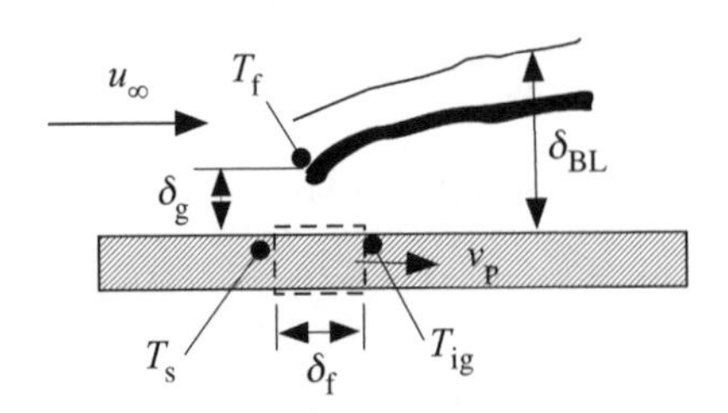

Figure 8.15 The deRis model

or

$$Da = \frac{\alpha_g A \Delta h_c}{u_\infty^2 \rho c_p T_\infty} \left(\frac{E}{RT_\infty} \right) e^{-E/(RT_\infty)} \tag{8.32}$$

where α_g, ρ, c_p are gas phase properties. For small Da we expect gas phase kinetics to be important and control the spread process. For large Da we expect heat transfer to control the process. The latter corresponds to the deRis solution and our simple thermal model. We shall examine that aspect.

From Equation (8.31), the flame heat flux is estimated by

$$\dot{q}_f'' = k_g \left(\frac{T_f - T_{ig}}{\delta_g} \right) \tag{8.33}$$

with T_f, the flame temperature. Combining and substituting into Equation (8.6) for the thermally thin case gives

$$v_p = \frac{[k_g (T_f - T_{ig})/\delta_g]\ \delta_f}{\rho c_p d(T_{ig} - T_s)}$$

If $\delta_f = \sqrt{2}\delta_g$, we obtain exactly the deRis solution for the thermally thin case,

$$\boxed{v_p = \frac{\sqrt{2} k_g (T_f - T_{ig})}{\rho c_p d(T_{ig} - T_s)}} \tag{8.34}$$

In the thermally thick case (Equation (8.21)),

$$v_p = \frac{[k_g (T_f - T_{ig})/\delta_g]^2\ \delta_f}{(k \rho c_p)(T_{ig} - T_s)^2} = \frac{(k_g^2 u_\infty / \alpha_g)(T_f - T_{ig})^2}{(k \rho c_p)(T_{ig} - T_s)^2}$$

or

$$\boxed{v_p = \frac{(k \rho c_p)_g u_\infty (T_f - T_{ig})^2}{(k \rho c_p)(T_{ig} - T_s)^2}} \tag{8.35}$$

which is exactly the theoretical result of deRis [4].

The deRis solutions were examined experimentally by Fernandez-Pello, Ray and Glassman [5,6]. Results are shown for thick PMMA in Figure 8.16 [6] as a function of u_∞ and $Y_{O_2,\infty}$. It should be recognized that T_f depends on $Y_{O_2,\infty}$, the free stream oxygen mass fraction, and A, the pre-exponential factor, depends on $Y_{F,s}$ and $Y_{O_2,\infty}$. These experimental results show that for the rise and fall of the data for large u_∞, the deRis model is not followed. This is because the chemical reaction times are becoming relatively large, $Da \to$ small. Indeed, the results can be correlated by

$$\frac{v_p (k \rho c_p)(T_{ig} - T_s)^2}{u_\infty (k \rho c_p)_g (T_f - T_{ig})^2} = f(Da) \to 1 \quad \text{as} \quad Da \to \text{large} \tag{8.36}$$

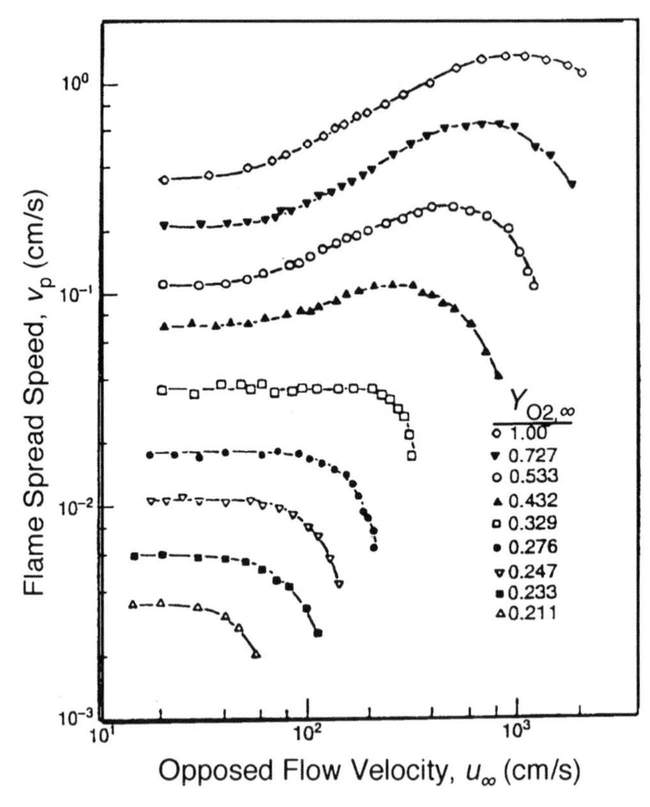

Figure 8.16 Flame spread rate over thick PMMA sheets as a function of the opposed forced flow velocity for several flow oxygen mass fractions (Fernandez-Pello, Ray and Glassman [6])

where $f(Da)$ is monotonic and asymptotic to 1 as Da becomes large [6]. The flame temperature can be approximated by

$$T_{\mathrm{f}} - T_{\mathrm{ig}} = \frac{Y_{\mathrm{O_2},\infty}\, \Delta h_{\mathrm{c}}}{c_{p,\mathrm{g}}\, r} \tag{8.37}$$

Loh and Fernandez-Pello [8] have shown that the form of Equation (8.30) appears to hold for laminar wind-aided spread over a flat plate. Data are shown plotted in Figure 8.17 and are consistent with

$$v_{\mathrm{p}} \sim u_\infty Y_{\mathrm{O_2},\infty}^2$$

following Equations (8.36) and (8.37). Note, from Figures 8.16 and 8.17 that at

$$u_\infty = 1\,\mathrm{m/s} \text{ and } Y_{\mathrm{O_2},\infty} = 1,$$
$$v_{\mathrm{p}} \approx 0.6\,\mathrm{cm/s} \quad \text{for opposed flow}$$

and

$$v_{\mathrm{p}} \approx 0.6\,\mathrm{m/s} \quad \text{for wind-aided spread}$$

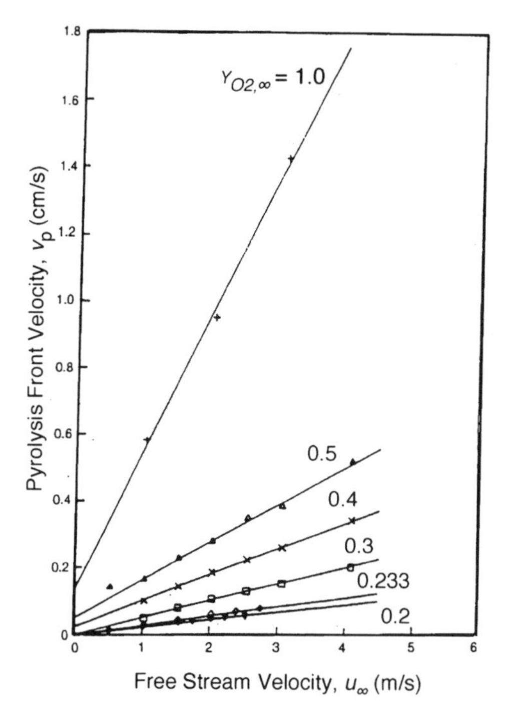

Figure 8.17 Flame spread rate over thick PMMA sheets as a function of the concurrent forced flow velocity for several flow oxygen mass fractions (Loh and Fernandez-Pello [8])

8.7 Examples of Other Flame Spread Conditions

Many fire spread problems can be formulated by the deceptively simple expression

$$\text{Flame speed}(v_\mathrm{p}) = \frac{\text{flame heated length }(\delta_\mathrm{f})}{\text{time to ignite }(t_\mathrm{f})}$$

The key to a successful derivation of a particular flame spread phenomena is to establish the mechanism controlling the spread in terms of expressions for δ_f and t_f. We will not continue to develop such approximate expressions, but will show some results that are grounded in data and empirical analyses.

8.7.1 Orientation effects

Figure 8.18 shows spread rates measured for thin films ($\sim$ 0.2 mm) of polyethylene teraphthalate (PET) and paper laid on fiberglass at various orientations spreading up and down [19]. The results can partly be explained by relating the momentum to buoyancy as

$$\rho u_\infty^2 \sim \rho(g\cos\phi)\ \delta_\mathrm{BL} \tag{8.38}$$

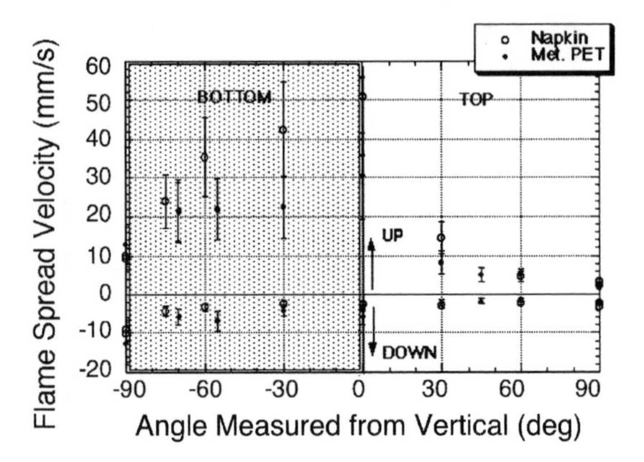

Figure 8.18 Spread as a function of orientation and angle [19]

and from Equation (8.31)

$$u_\infty \sim (\alpha_g \, g \cos \phi)^{1/3} \tag{8.39}$$

Thus, we can replace u_∞ in Equation (8.36) and apply it to both opposed and wind-aided cases. For upward or wind-aided spread the speed increases as $\cos \phi$ increases to the vertical orientation. For downward or opposed flow spread, the speed is not significantly affected by changes in ϕ until the horizontal inclination is approached for the bottom orientation $(-90 \leq \phi \leq -60)$. In that region, the downward spread becomes wind-aided as a stagnation plane flow results from the bottom. Figure 8.19 gives sketches of the

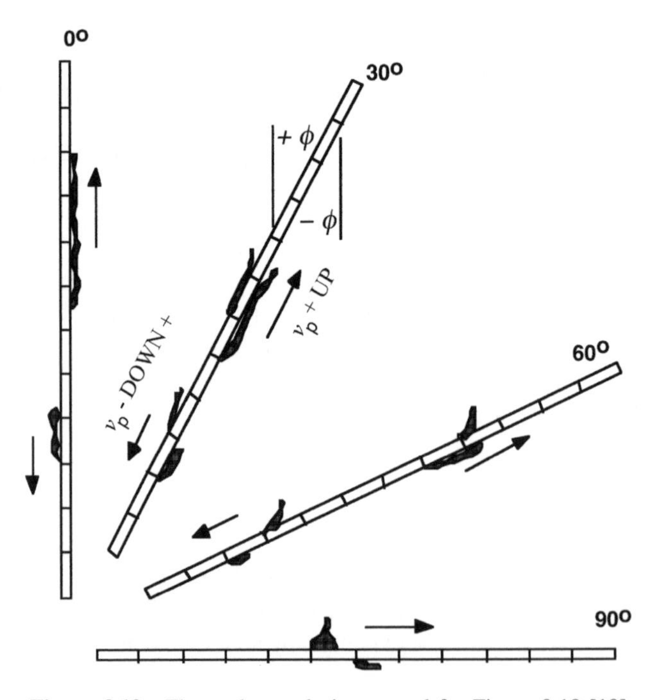

Figure 8.19 Flame shapes during spread for Figure 8.18 [19]

flame shapes for spread under some of these orientations. For spread on the bottom surface we always have a relatively large δ_f. For spread on the top, we get 'lift-off' of the flame between $\phi = 30°$ to $60°$.

8.7.2 *Porous media*

Thomas [21,22] has analyzed and correlated data for porous, woody material representative of forest debris, grass fields, cribs and even blocks of wooden houses. He finds that $v_p \sim u_\infty / \rho_b$ for the wind-aided case, and $v_p \sim \rho_b^{-1}$ for natural convection, where ρ_b is the bulk density of the fuel burned in the array. Figure 8.20 shows the wind-aided data and Figure 8.21 is the no-wind case. From Thomas [21,22], an empirical formula can be shown to approximate much of the data:

$$\boxed{\rho_b v_p = c(1 + u_\infty) \quad \text{in m, kg and s units}} \tag{8.40}$$

where c can range from about 0.05 for wood cribs ($\rho_b \sim$ 10–100 kg/m^3) up to 0.10 for forest fuels ($\rho_b \sim$ 1-5 kg/m^3). The urban house array data were represented as 10 kg/m^3 in Figure 8.20.

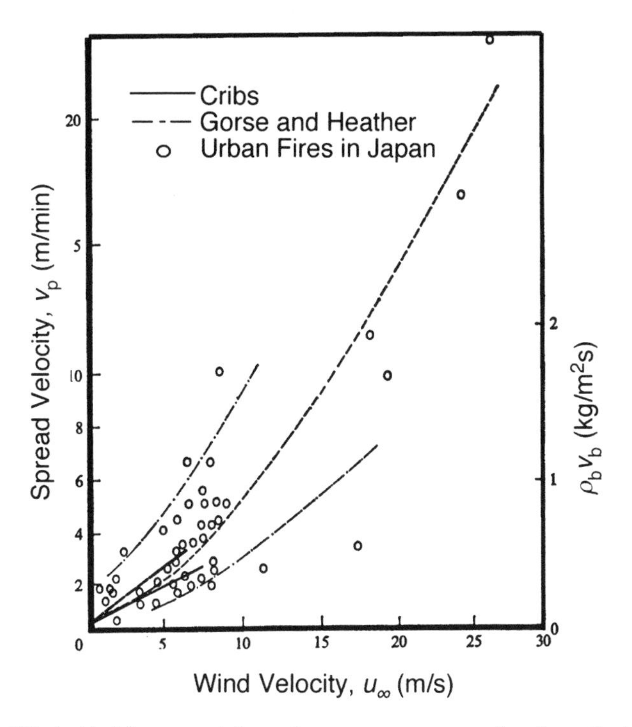

Figure 8.20 Wind-aided fire spread through porous arrays as a function of wind speed [21]

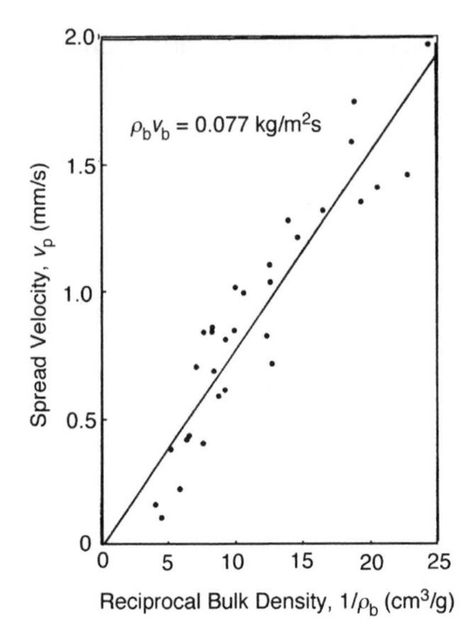

Figure 8.21 Natural convection fire spread through a porous fuel bed from Thomas [22]

8.7.3 *Liquid flame spread*

Flame spread over a deep pool of liquid fuel depends on flow effects in the liquid. These flow effects constitute: (a) a surface tension induced flow since surface tension (σ) is decreasing with temperature (T) and (b) buoyancy induced flow below the surface. For shallow pools, viscous and heat transfer effects with the floor will be factors. Figure 8.22 shows these mechanisms. To ignite the liquid, we must raise it to its flashpoint. Then the spread flame with its energy transport mechanisms must heat the liquid at temperature T to its flashpoint, T_L. Data from Akita [23] in Figure 8.23 show the effect of increasing bulk liquid temperature (T) on the spread rate. At $T = -10°C$ for the methanol liquid, surface tension and viscous effects begin to cause a combustible mixture to form ahead of the advancing flame and its ignition causes a sudden increase in the flame speed. This process is repeated as a pulsating flame until $T = 1\ °C$, where the speed again becomes steady. After $T = 11°C$, the flashpoint, the flame spread is no longer a surface

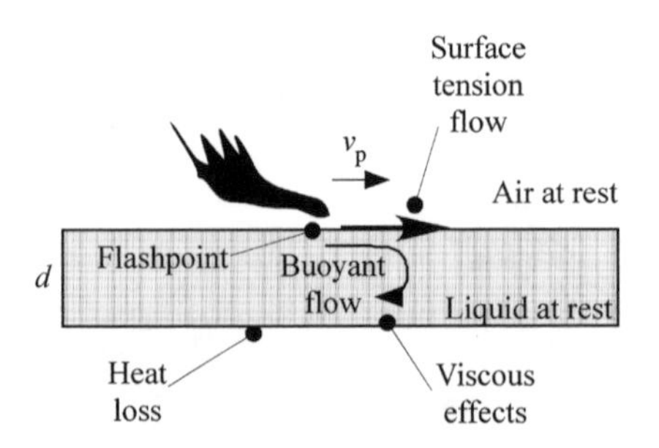

Figure 8.22 Mechanism for flame spread on a liquid

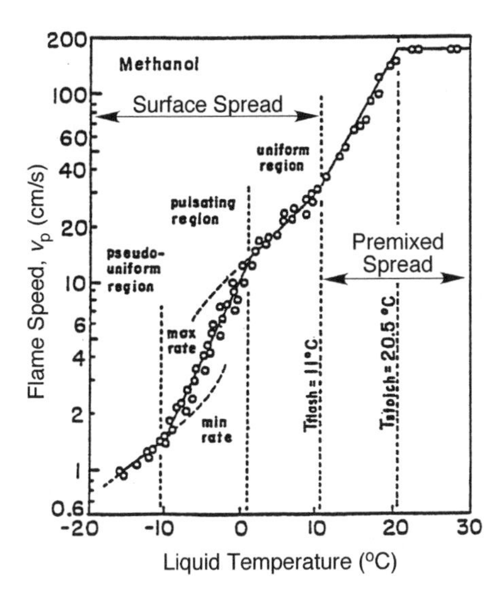

Figure 8.23 Relationship between the liquid temperature and the rate of plane flame spread of methanol in a vessel 2.6 cm wide and 1.0 cm deep [23]

phenomenon. Now it is flame propagation through the premixed fuel and air in the gas phase. Increases in speed continue with T up to the liquid temperature corresponding to stoichiometric conditions (20.5 °C), after which the mixture becomes fuel rich at a constant speed.

8.7.4 Fire spread through a dwelling

An interesting correlation of the rate of fire growth measured by flame volume was given by Labes [24] and Waterman [25]. They show that after room flashover in a room of volume V_0, the fire volume (V_f) grows as

$$\frac{dV_f}{dt} = k_f V_f \tag{8.41}$$

where k_f depends on the structure, its contents and ventilation. Two figures (Figures 8.24 and 8.25) display data taken from two dwellings. A simple explanation of Equation (8.41) can be formulated by considering fire spread illustrated for the control volume in Figure 8.26. We write a conservation of energy for the CV as

$$\rho c_p v_p A (T_{ig} - T_\infty) = \dot{q}_f'' A. \tag{8.42}$$

The flame volumetric rate of change is related to flame spread as

$$\frac{dV_f}{dt} = v_p A \tag{8.43}$$

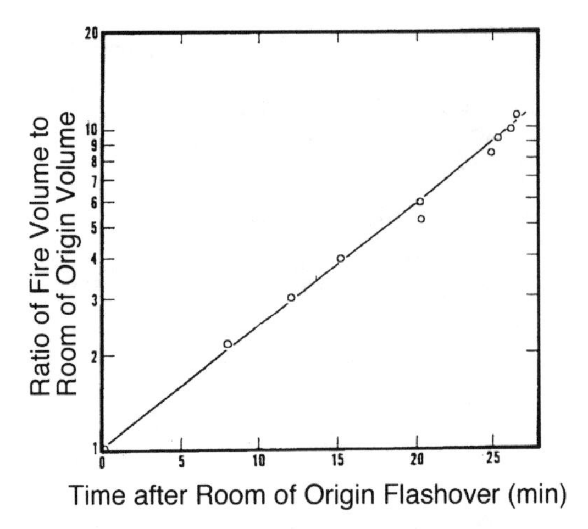

Figure 8.24 Correlation of volumetric fire growth for a dwelling with similar compartment characteristics [24]

for constant A. The heat flux for a 'thin' flame model can be represented as

$$\dot{q}''_{\mathrm{f}} = \epsilon_{\mathrm{f}} \sigma T_{\mathrm{f}}^4 \tag{8.44}$$

with

$$\epsilon_{\mathrm{f}} = 1 - \mathrm{e}^{-\kappa_{\mathrm{f}}(V_{\mathrm{f}}/A)} \approx \kappa_{\mathrm{f}} \left(\frac{V_{\mathrm{f}}}{A} \right) \tag{8.45}$$

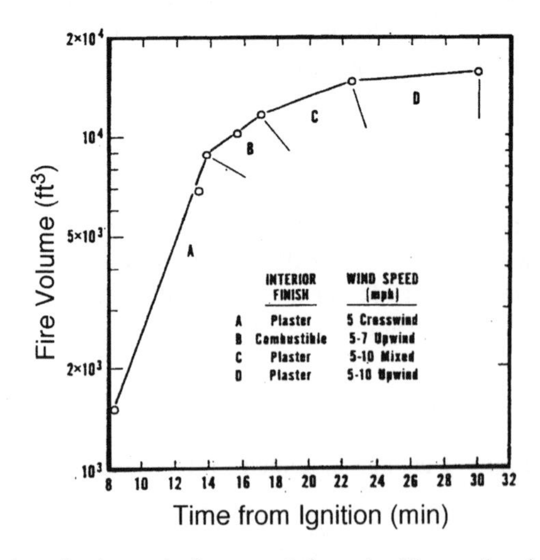

Figure 8.25 Correlation of volumetric fire growth for a dwelling under changing conditions [25]

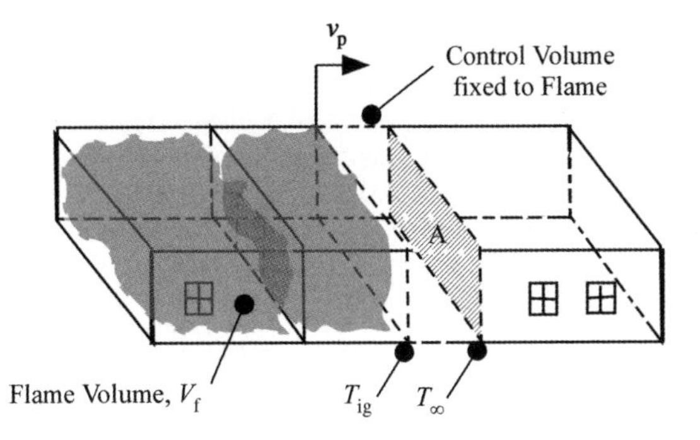

Figure 8.26 Fire spread in a dwelling after room flashover

where κ_f is a radiative absorption coefficient for the flame. Combining yields

$$\frac{dV_f}{dt} = \left[\frac{\sigma\kappa_f T_f^4}{\rho c_p(T_{ig} - T_\infty)}\right] V_f \tag{8.46}$$

or the form of Equation (8.41), the experimental correlation. It is interesting to note that the fire doubling time is constant at about 7.5 minutes for dwelling, having similar compartments in Figure 8.24. Doubling times could range from roughly 1 to 10 minutes over a typical spectrum of dwellings and wind conditions.

References

1. Williams, F.A., Mechanisms of fire spread, *Proc. Comb. Inst.*, 1976, **16**, p. 1281.
2. Fernandez-Pello, A.C., The solid phase, in *Combustion Fundamentals of Fire* (ed. G. Cox), Academic Press, London, 1995.
3. Thomas, P.H., The growth of fire – ignition to full involvement, in *Combustion Fundamentals of Fire* (ed. G. Cox), Academic Press, London, 1995.
4. deRis, J.N., Spread of a laminar diffusion flame, *Proc. Comb. Inst.*, 1969, pp. 241–52.
5. Fernandez-Pello, A.C., Ray, S.R. and Glassman, I., A study of heat transfer mechanisms in horizontal flame propagation, *J. Heat Transfer*, 1980, **102**(2), 357–63.
6. Fernandez-Pello, A.C., Ray, S.R. and Glassman, I., Flame spread in an opposed forced flow: the effect of ambient oxygen concentration, *Proc. Comb. Inst.*, 1980, **18**, pp. 579–89.
7. Altenkirch, R.A., Eichhorn, R. and Shang, P.C., Buoyancy effects on flames spreading down thermally thin fuels, *Combustion and Flame*, 1980, **37**, 71–83.
8. Loh, H.T. and Fernandez-Pello, A.C., A study of the controlling mechanisms of flow assisted flame spread, *Proc. Comb. Inst.*, 1984, **20**, pp. 1575–82.
9. Loh, H.T. and Fernandez-Pello, A.C., Flow assisted flame spread over thermally thin fuels, in Proceedings of 1st Symposium on Fire Safety Science, Hemisphere, New York, 1986, p. 65.
10. Sibulkin, M. and Kim, J., The dependence of flame propagation on surface heat transfer. II. Upward burning, *J. Combust. Sci. Technol.*, 1977, **17**, 39.
11. Rothermel, R.C., Modeling fire behavior, in 1st International Conference on *Forest Fire Research*, Coimbra, Portugal, November 1990.

12. Long Jr, R.T., An evaluation of the lateral ignition and flame spread test for material flammability assessment for micro-gravity environments, MS Thesis, Department of Fire Protection Engineering, University of Maryland, College Park, Maryland, 1998.
13. Saito, K., Williams, F.A., Wichman, I.S. and Quintiere, J.G., Upward turbulent flame spread on wood under external radiation, *Trans. ASME, J. Heat Transfer*, May 1989, **111**, 438–45.
14. Ito, A. and Kashiwagi, T., Characterization of flame spread over PMMA using holographic interferometry sampling orientation effects, *Combustion and Flame*, 1988, **71**, 189.
15. ASTM E-1321-90, *Standard Method for Determining Material Ignition and Flame Spread Properties*, American Society for Testing and Materials, Philadelphia, Pennsylvania, 1990.
16. Quintiere, J.G. and Harkleroad, M., *New Concepts for Measuring Flame Spread Properties*, NBSIR 84-2943, National Bureau of Standards, Gaithersburg, Maryland, 1984.
17. Quintiere, J.G., Harkleroad, M. and Walton, D., Measurement of material flame spread properties, *J. Combust. Sci. Technol.*, 1983, **32**, 67–89.
18. Ahmad, T. and Faeth, G.M., Turbulent wall fires, in *17th Symposium (International) on Combustion*, The Combustion Institute, Pittsburgh, Pennsylvania, *Proc. Comb. Inst.*, 1979, **17**, p. 1149–1162.
19. Quintiere, J.G., The effects of angular orientation on flame spread over thin materials, *Fire Safety J.*, 2001, **36**(3), 291–312.
20. Quintiere, J.G., Harkleroad, M. and Hasemi, Y., Wall flames and implications for upward flame spread, *Combust. Sci. Technol.*, 1986, **48**, 191–222.
21. Thomas, P.H., Rates of spread of some wind driven fires, *Forestry*, 1971, **44**(2), 155–7.
22. Thomas, P.H., Some aspects of growth and spread of fires in the open, *Forestry*, 1967, **40**(2), 139–64.
23. Akita, K., Some problems of flame spread along a liquid surface, in 14th Symposium (International) on *Combustion*, The Combustion Institute, Pittsburgh, Pennsylvania, *Proc. Comb. Inst.*, 1975, **14**, p. 1075.
24. Labes, W.G., The Ellis Parkway and Gray Dwelling burns, *Fire Technol.*, May 1966, **2**(4), 287.
25. Waterman, T.E., *Experimental Structure Fires*, IITRI Final Technical Report J6269, Chicago, Ilinois, July 1974.

Problems

8.1 A thin vertical drapery is ignited. The flame spreads up and down both sides. Note that the axis of symmetry is equivalent to an insulated boundary. The properties of the drape are given below:

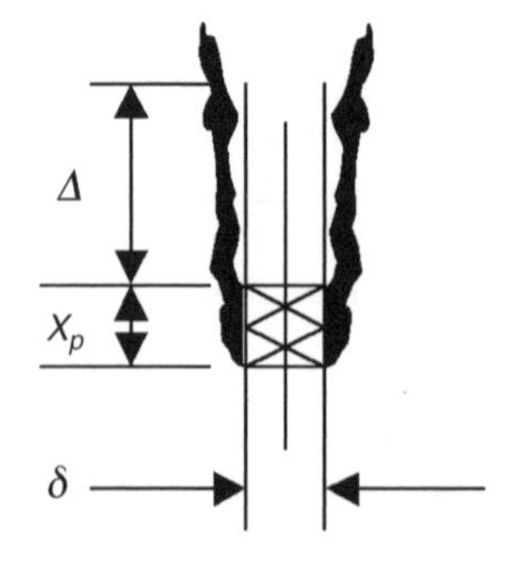

$T_\infty = 25\,°\text{C}$
$Y_{\text{ox},\infty} = 0.233$
$k = 0.08\,\text{W/m K}$
$c = 1.5\,\text{J/g K}$
$\rho = 100\,\text{kg/m}^3$
$T_{\text{ig}} = 400\,°\text{C}$
$T_{\text{o}} = 25\,°\text{C}$, initial temperature
Thickness = 2 mm

Consider the analysis in two dimensions only, i.e. per unit depth into the page.

(a) Calculate the flame temperature from thin diffusion flame theory, using $c_{p,g} = 1$ J/g K and $k_g = 0.03$ W/m K

(b) Calculate the downward spread flame speed.

(c) Calculate the initial upward spread flame speed if the flame extension $\Delta = ax_p$, where x_p is the pyrolysis length and $a = 1$. The heat flux of the flame is 4 W/cm^2 and the initial pyrolysis zone is 2 cm in length.

Note that diffusion flame theory will show the flame temperature is approximately given by $T_f \approx T_\infty + Y_{ox,\infty}(\Delta h_{ox}/c_p)$, where Δh_{ox} is the heat of combustion per unit mass of oxygen consumed.

8.2 A 1 m high slab of PMMA is ignited at the bottom and spreads upward. Ignition is applied over a length of 5 cm. If the slab is 1 mm thick and ignited on both sides, then how long will it take to be fully involved in pyrolysis? Assume the flame height is governed by $x_f \cong 0.01\dot{Q}'$, where x_f is in meters and $\dot{Q}'$ is in kW/m. The flame heat flux is 25 kW/m^2; assume $\dot{m}'' = \dot{q}''/L$, where $L = 1.6$ kJ/g. Let $T_\infty = 20\,°C$ and obtain data for PMMA from Tables 2.3, 7.5, 7.6 and 8.1.

8.3 Re-work Problem 8.2 using a PMMA thickness of 50 mm.

8.4 If the PMMA in Problem 8.2 is ignited at the top for the thick case, compute the time for full involvement. Let $\Phi = 14$ (kW)^2/m^3.

8.5 A thick wood slab spread a flame downward at 0.1 mm/s; the slab was initially at 20 °C. If the slab were preheated to 100 °C, what is the spread rate? Assume $T_{ig} = 450\,°C$.

8.6 A paper match once ignited releases 250 kW/m^2 of energy and burns for 60 seconds. The match is a thin flat configuration (0.5 mm thick), 0.5 cm wide and 4 cm long (see the figure below). The person striking the match grips it 1 cm from the top and inadvertently turns it upside down. Ignore edge effects.

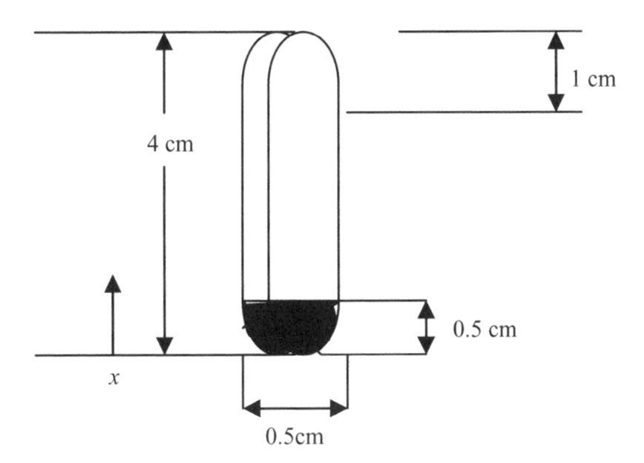

The initial region ignited is 0.5 cm and the ambient temperature is 20 °C. The flame height during flame spread is given by $x_f = (0.01\text{m}^2/\text{kW})\dot{Q}'$, where $\dot{Q}'$ is the energy release rate per unit width in kW/m.

Match properties:

Density = 100 kg/m^3
Specific heat = 1.8 J/g K
Ignition temperature = 400 °C
Flame heat flux = 25 kW/m^2

(a) If this match material is exposed to a heat flux of 25 kW/m^2 and insulated on the opposite face, how long will it take to ignite?

(b) For the finger scenario described, determine the pyrolysis position just when the flame tip reaches the finger tip.

(c) How long will it take the flame to reach the fingertip?

8.7 A thick material is tested for ignition and flame spread.

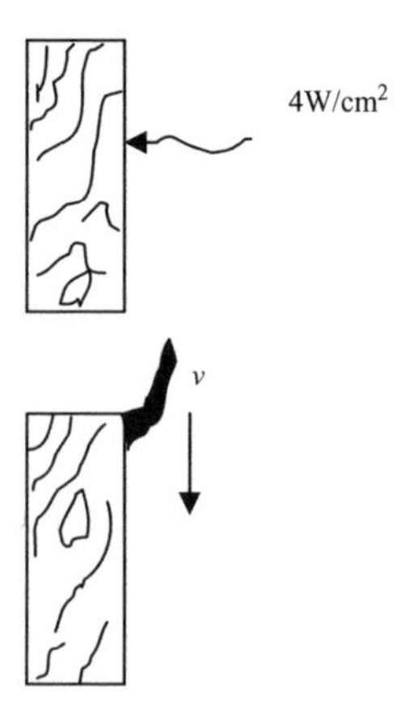

(a) An ignition test is conducted under a radiative heat flux of 4 W/cm^2. The initial temperature is 20 °C. It is found that the surface temperature rises to 100 °C in 3 seconds. The ignition (temperature?) is 420 °C.

(i) Estimate the property $k\rho c$ of the material.

(ii) When will it ignite?

(b) Two flame spread experiments are conducted to examine downward spread velocity. The material is preheated in an oven to 100 °C and to 200 °C respectively for the two experiments. Calculate the ratio of spread rates for the two experiments $(v_{100\,°\mathrm{C}}/v_{200\,°\mathrm{C}})$.

8.8 A thick polystyrene wainscotting covers the wall of a room up to 1 m from the floor. It is ignited over a 0.2 m region and begins to spread. Assume that the resulting smoke layer in the room does not descend below 1 m and no mixing occurs between the smoke layer and the lower limit. The initial temperature is 20 °C, the ambient oxygen mass fraction is 0.233 and the specific heat of air is 1 J/g K.

Polystyrene properties:

Density (ρ) = 40 kg/m^3
Specific heat (c_p) = 1.5 J/g K

Conductivity $(k) = 0.4$ W/m K
Vaporization (T_v) and ignition temperature (T_{ig}) are both equal to 400 °C
Heat of combustion $(\Delta h_c) = 39$ kJ/g
Stoichiometric O_2 to fuel ratio $(r) = 3$ g/g
Heat of gasification $(L) = 1.8$ kJ/g

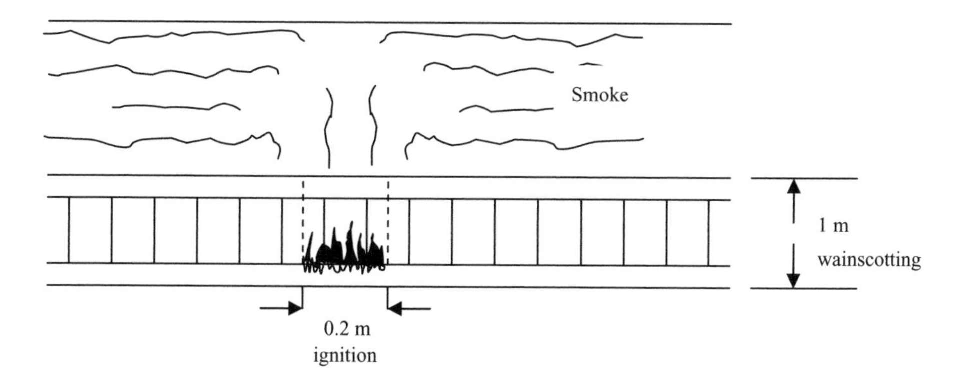

Calculate the upward spread velocity at 0.5 m from the floor. The flame height from the floor is 1.8 m and the heat flux from the flame is estimated at 3 W/cm^2.

8.9 A wall is ignited by a spill of gasoline along the wall–floor interface. The width of the spill is 0.5 m. The energy release rate of the gasoline is 300 kW/m (width). In all calculations, the dominant heat flux is from the flame to the wall. Only consider this heat flux; it is 30 kW/m^2 for all flames. The initial temperature is 20 °C. The relationship between the wall flame height and the energy release rate per unit width $(\dot{Q}')$ is

$$y_f(m) = 0.01\dot{Q}'(\text{kW/m})$$

The properties of the wall material are:

$T_{ig} = 350$ °C
$T_{s,min} = 20$ °C
$\Delta h_c = 25$ kJ/g
$\Phi = 15$ kW^2/m^3
$L = 1.8$ kJ/g
$k\rho c = 0.45$ (kW/m^2 K)^2 s

(a) When does the wall ignite after the gasoline fire starts, and what is the region ignited?

(b) What is the maximum width of the wall fire 60 s after ignition of the wall?

(c) What is the flame height 60 s after the wall ignition?

(d) What is the maximum height of the pyrolysis front after 60 s from wall ignition?

(e) What is the energy release rate of the wall 60 s after its ignition?

(f) Sixty seconds after ignition of the wall, the gasoline flame ceases. The wall continues to burn and spread. What is the maximum pyrolysis height 30 s from the time the gasoline flame ceases?

8.10 Flame spread is initiated on a paneled wall in the corner of a family room. Later, a large picture window, opposite the corner, breaks. A plush thick carpet covers the floor and a large sofa is in front of the picture window. The ceiling is noncombustible plaster. Explain the most probable scenario for causing the window to break and shatter.

8.11 A vat of flammable liquid ignites and produces a flame that impinges on a 1 m radius of the paper-facing insulation on the ceiling. The flame temperature is 500 °C and convectively heats the ceiling with a heat transfer coefficient of 50 W/m^2 K. The initial temperature is 35 °C. The paper-face on the ceiling is 1 mm thick with properties as shown:

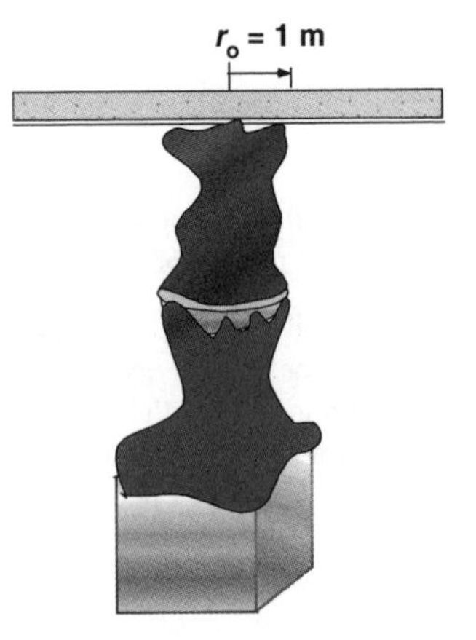

Ignition temperature, $T_{ig} = 350\,°\text{C}$
Density, $\rho = 250$ kg/m^3
Specific heat, $c = 2100$ J/kg K

(a) Compute the time to ignite the paper.

(b) Compute the initial flame spread rate under the ceiling. The flame stretches out from the ignited region by 0.5 m and applies a heat flux of 30 kW/m^2 to the surface.

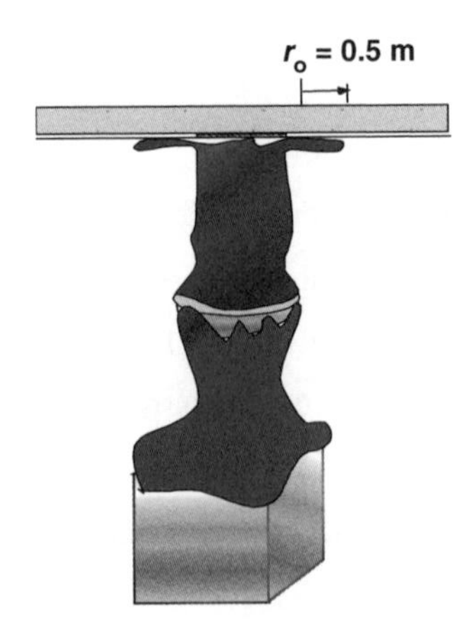

8.12 Flame spread *into the wind* can depend on (check the best answer):

	Yes	*Not necessarily*	*No*
(a) Wind speed	—	—	—
(b) Flame temperature	—	—	—
(c) Flame length in the wind direction	—	—	—
(d) Pyrolysis region heat flux	—	—	—

8.13 The temperature at the edge of a thick wood element exposed to a uniform heat flux can be expressed, in terms of time, by the approximate equation given below for small time:

$$T_s - T_\infty \approx \frac{4.2}{\sqrt{\pi}} \dot{q}''_e \sqrt{t/k\rho c}$$

The notation is consistent with that of the text. The properties of the wood are:

Thermal conductivity	0.12 W/m K
Density	510 kg/m^3
Specific heat	1380 J/kg K
Piloted ignition temperature	385 °C
Ambient temperature	25 °C

After ignition of the edge, the external heat flux is removed and it is no longer felt by the wood, in any way. A researcher measures the heat flux of the flame, ahead of the burning wood, as it spreads steadily and horizontally along the edge. That measurement is depicted in the plot below. Compute the flame speed on the wood, based on this measurement.

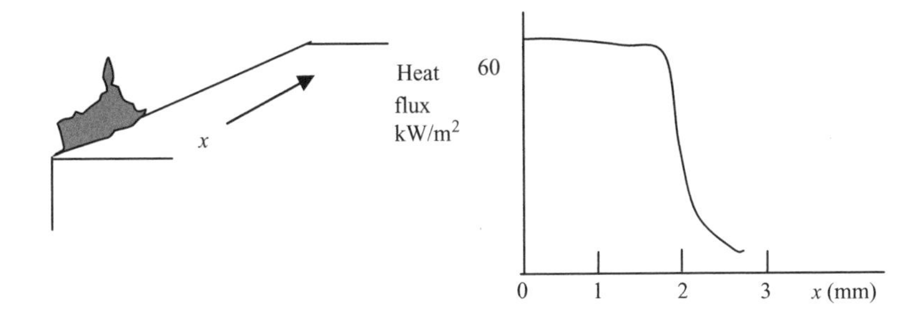

8.14 The ASTM Steiner Tunnel Test ignites a material over 1.5 m at its burner with a combined flame length of 3.2 m from the start of the ignition region. As the flame moves away from the burner influence, its burning rate becomes steady over a 0.3 m length and flame extends only 0.2 m beyond this burning length. The flame heat flux remains constant at 25 kW/m^2. Ignore any radiant feedback and increases in the tunnel air temperature. What is the ratio of steady speed to its initial speed induced by the burner? The material is considered to be thick.

9

Burning Rate

9.1 Introduction

Burning rate is strictly defined as the mass rate of fuel consumed by the chemical reaction, usually but not exclusively in the gas-phase. For the flaming combustion of solids and liquids, the burning rate is loosely used to mean the mass loss rate of the condensed phase fuel. However, these two quantities – mass loss rate and burning rate – are not necessarily equal. In general,

$$\begin{aligned}[\text{Fuel mass loss rate}] = {} & [\text{fuel burning rate}] \\ & + [\text{rate of inert gases released with the fuel}] \\ & + [\text{rate of fuel gases (and soot) not burned in the flame}]\end{aligned}$$

or

$$\dot{m}_{\mathrm{F}} = \dot{m}_{\mathrm{F,R}} + \dot{m}_{\mathrm{F,I}} + \dot{m}_{\mathrm{F,U}} \tag{9.1}$$

In this chapter we shall consider only

$$\dot{m}_{\mathrm{F}} = \dot{m}_{\mathrm{F,R}} \tag{9.2}$$

assuming all of the fuel is burned and there are no inerts in the evolved fuel. The latter assumption could be relaxed since $Y_{\mathrm{F,o}}$ (the mass concentration of fuel in the condensed phase) will be a parameter, and in general $Y_{\mathrm{F,o}} \leq 1$. In reality, all fuels will have unburned products of combustion to some degree. This will implicitly be taken into account since chemical properties for fuels under natural fire conditions are based on measurements of the fuel mass loss rate. Hence, in this chapter, and hereafter for fire

Fundamentals of Fire Phenomena James G. Quintiere

conditions, we regard the heat of combustion as the chemical rate of energy released (firepower) per unit fuel mass loss,

$$\Delta h_c \equiv \frac{\dot{Q}_{\text{chem}}}{\dot{m}_F} \tag{9.3}$$

Similarly, yields of species, y_i, are given as

$$y_i \equiv \frac{\dot{m}_{i,\text{chem}}}{\dot{m}_F} \tag{9.4}$$

where $\dot{m}_{i,\text{chem}}$ is the mass rate of production of species i from the fire.

Factors that influence the burning rate of 'fuel packages' include (1) the net heat flux received at the surface, $\dot{q}''$, and (2) the evaporative or decomposition relationships for the fuel. We can represent each as

(1) Net heat flux:

$$\begin{aligned}\dot{q}'' = &\text{ flame convective heat flux,} \quad \dot{q}''_{\text{f,c}} \\ &+ \text{flame radiative heat flux,} \quad \dot{q}''_{\text{f,r}} \\ &+ \text{external environmental radiative heat flux,} \quad \dot{q}''_{\text{e}} \\ &- \text{surface radiative heat loss,} \quad \epsilon\sigma T_s^4\end{aligned} \tag{9.5}$$

(2) Fuel volatilization mass flux:

$$\begin{aligned}\dot{m}''_F = &\text{ function (chemical decomposition kinetics,} \\ &\text{heat conduction, charring and possibly other} \\ &\text{physical, chemical and thermodynamic properties)}\end{aligned}$$

Some of the thermal and fluid factors affecting the burning rate are illustrated in Figure 9.1. Convective heat flux depends on the flow conditions and we see that both

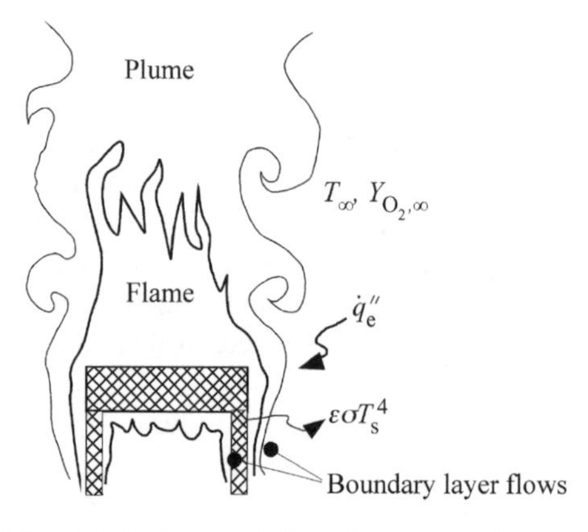

Figure 9.1 Factors influencing the burning rate

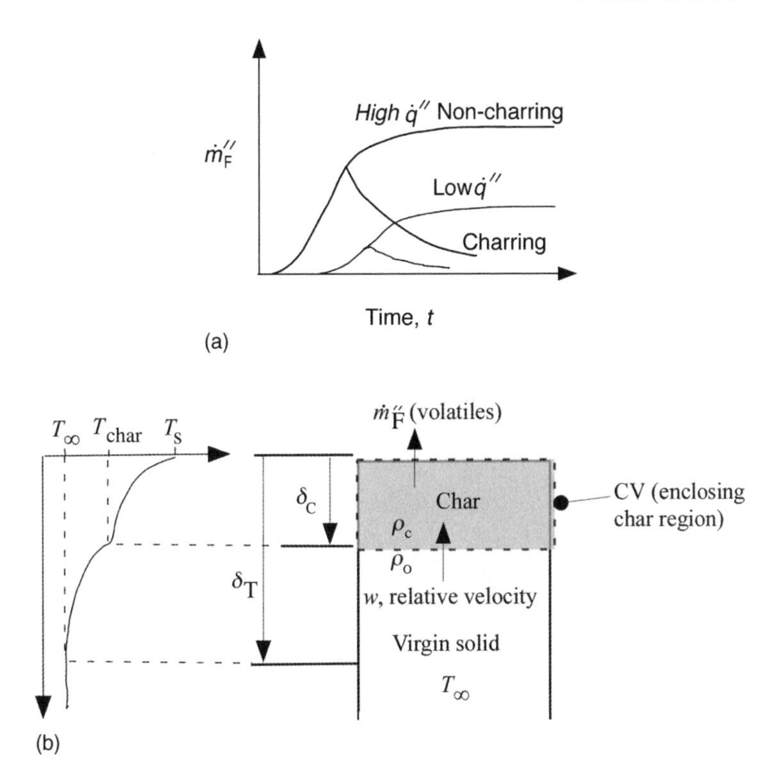

Figure 9.2 Idealized burning rate behavior of thick charring and noncharring materials: (a) mass loss behaviour and (b) temperature behavior

boundary layer and plume flows are possible for a burning fuel package. For flames of significance thickness (e.g. ≥ 5 cm, then $\epsilon_f \geq 0.05$), flame radiation will be important and can dominate the flame heating. Since the particular flow field controls the flame shape, flame radiation will depend on flow as well as fire size ($\dot{m}_F$). The external radiative heating could come from a heated room subject to a fire; then it will vary with time, or it could come from a fixed heat source of a test, where it might vary with distance. The radiative loss term depends on surface temperature, and this may be approximated by a constant for liquids or noncharring solids. For liquids, this surface temperature is nearly the boiling point, and will be taken as such. For charring materials the surface temperature will increase as the char layer grows and insulates the material. It can range from 500 to 800 °C depending on heat flux $(\approx \dot{q}''_r/\sigma)^{1/4}$ or char oxidation.

The condensed phase properties that affect fuel volatilization are complex and the most difficult to characterize for materials and products, in general. The mass loss rate depends on time as well as the imposed surface heat flux. Figure 9.2 illustrates idealized experimental data for the mass loss rate of charring and noncharring materials. Figures 9.3 and 9.4 give actual data for these material classes along with surface temperature information. Initially, both types of materials act in the same way since a thin char layer has little effect. However, at some char thickness, the mass loss rate abruptly begins to decrease for the charring material as the effect of char layer thickness becomes important. The decreasing behavior in burning rate roughly follows $t^{-1/2}$. This behavior

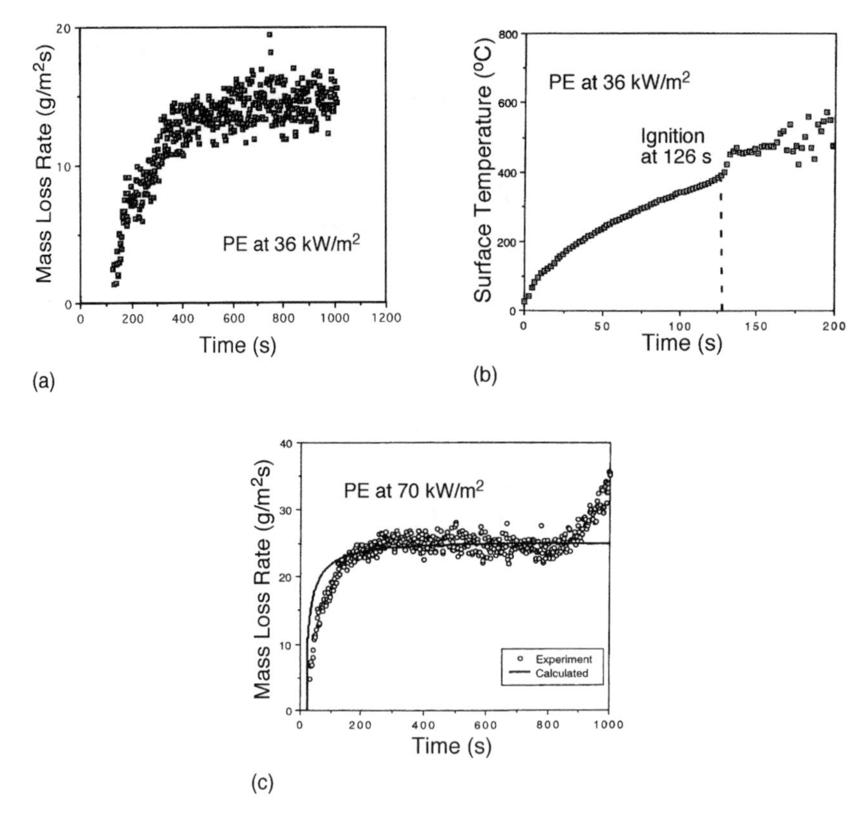

Figure 9.3 (a) Transient mass loss rate for polyethylene exposed to $36\,\text{kW/m}^2$. (b) surface temperature results for polyethylene with $36\,\text{kw/m}^2$ irradiance. (c) Transient mass loss rate exposed to $70\,\text{kw/m}^2$ irradiance [1]

can be derived from considering a thermal penetration depth that is defined by a fixed temperature, indicative of the onset of charring, T_{char}. For a pure conduction model, we know $\delta_T \sim \sqrt{\alpha t}$ so we would expect $\delta_\text{c} \sim \sqrt{\alpha_\text{c} t}$ where α_c is the char diffusivity. By a mass balance for the control volume (Figure 9.2b), we approximate

$$\frac{\text{d}}{\text{d}t}(\rho_\text{c}\delta_\text{c}) + \dot{m}_\text{F}'' - \rho_\text{o} w = 0$$

where $w = \text{d}\delta_\text{c}/\text{d}t$ and $\rho_\text{o} = \rho_\text{c} + \rho_{\text{volatiles}}$. Therefore, $\dot{m}_\text{F}'' = (\rho_\text{o} - \rho_\text{c})(\text{d}\delta_\text{c}/\text{d}t)$ or $\dot{m}_\text{F}'' \sim [(\rho_\text{o} - \rho_\text{c})/2]\sqrt{\alpha_\text{c}/t}$, where t is the time after reaching the peak.

In contrast, the noncharring 'liquid evaporative-type' material reaches a steady peak burning rate. For both types of materials, finite thickness effects and their backface boundary condition will affect the shape of these curves late in time. This is seen by the increased burning rate of the 2.5 cm thick polypropylene at $70\,\text{W/m}^2$ in Figure 9.3(c), but not for the 5 cm thick red oak material at $75\,\text{kW/m}^2$ in Figure 9.4(a).

A crude, but effective way to deal with the complexities of transient burning and fuel package properties is to consider liquid-like, quasi-steady burning. By Equation (6.34),

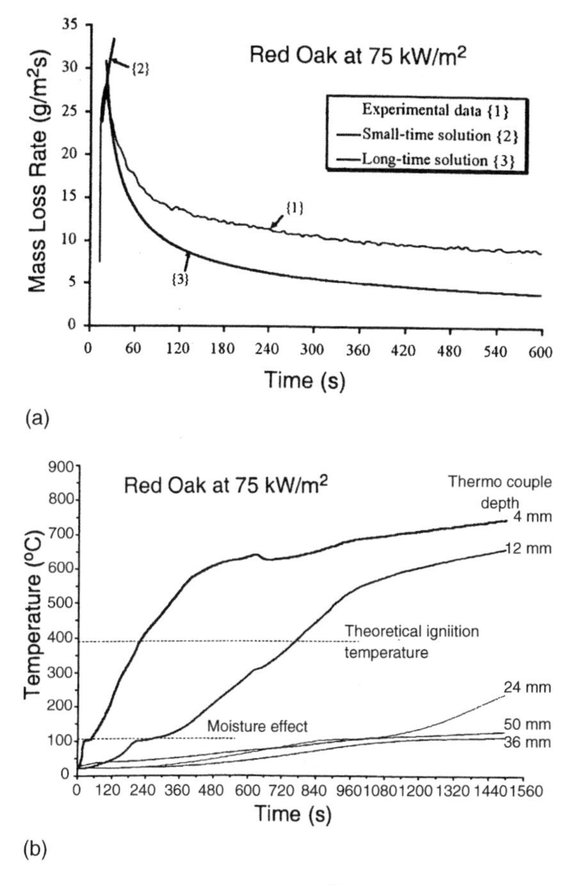

Figure 9.4 (a) Burning rate of red oak at 75 kW/m^2 (b) Temperatures measured for red oak at 75 kW/m^2 [2]

we derived for the steady burning of a liquid that

$$\dot{m}''_{\mathrm{F}} = \frac{\dot{q}''}{L} \tag{9.6}$$

where here we call L an effective heat of gasification. This overall property can be determined from the slope $(1/L)$ of peak $\dot{m}''_{\mathrm{F}}$ versus $\dot{q}''$ data. Such data are shown in Figure 9.5 for polypropylene burning a 10 cm square horizontal sample under increasing radiant heat flux [1]. Thus, $L = 3.07\,\mathrm{kJ/g}$ for this polypropylene specimen. For peak data on Douglas fir in Figure 9.6, the best fit lines give L of 6.8 kJ/g for heating along the grain and 12.5 kJ/g across the grain. It is easier to volatilize in the heated along-the-grain orientation, since the wood fibers and nutrient transport paths are in this heated direction (see Figure 7.10). While the L representation for noncharring materials may be very valid under steady burning, the L for a charring material disguises a lot of phenomena. It is interesting to point out that a noncharring material might be regarded as having all fuel converted to vapor, whereas for a charring material only a fraction is

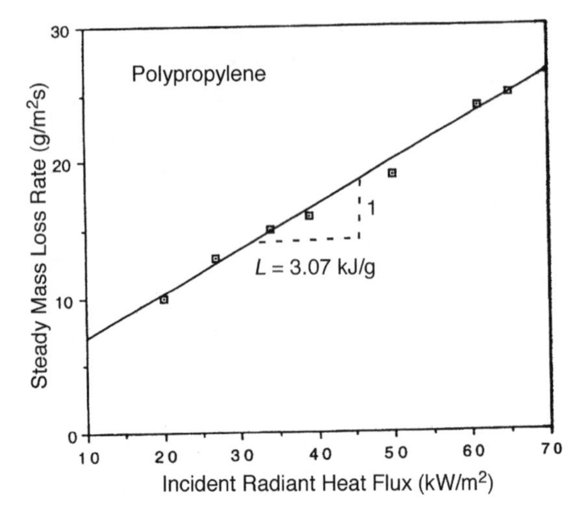

Figure 9.5 Peak mass loss rate of polypropylene [1]

converted. Hence an L-value, based on the original fuel mass. (L_o), is related to the common value based on mass loss as

$$L_o(\text{kJ/g original}) = L(\text{kJ/g volatiles})(1 - X_c) \tag{9.7}$$

where X_c is the char fraction, the mass of char per unit mass of original material. The char fraction was measured for the Douglas fir specimen as 0.25 ± 0.07 along and 0.45 ± 0.15 across the grain orientations. Thus,

$$L_o = \begin{cases} (6.8)(1.0 - 0.25) & = 5.1\,\text{kJ/g} \quad \text{or} \\ (12.5)(1.0 - 0.45) & = 6.9\,\text{kJ/g} \end{cases}$$

In this form, the energy needed to break the original polymer bonds to cause 'unzipping' or volatilization with char is closer to values representative of noncharring solid polymers. Table 9.1 gives some representative values found for the heats of gasification.

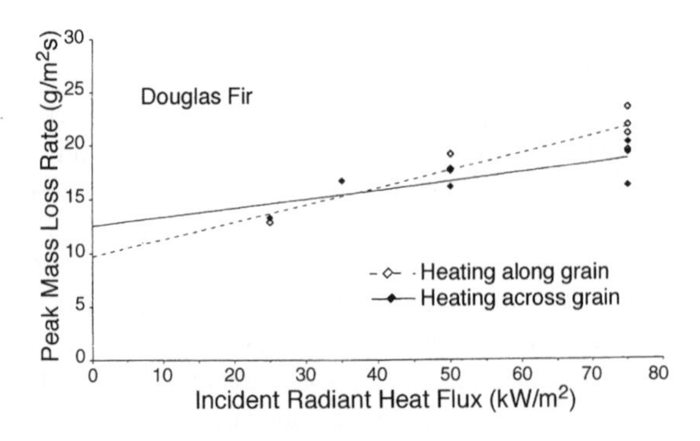

Figure 9.6 Peak burning rate as a function of the incident heat flux for Douglas fir [2]

Table 9.1 Effective heats of gasification

Material	Source	kJ/g mass lost
Polyethylene	[1]	3.6
Polyethylene	[3]	1.8, 2.3
Polypropylene	[1]	3.1
Polypropylene	[3]	2.0
Nylon	[1]	3.8
Nylon	[3]	2.4
Polymethylmethacrylate	[1]	2.8
Polymethylmethacrylate	[3]	1.6
Polystyrene	[3]	1.3–1.9
Polystyrene	[4]	4.0, 7.3
Polyurethane foam, flexible	[3]	1.2–2.7
Polyurethane foam, rigid	[3]	1.2
Polyurethane foam, rigid	[4]	5.6
Douglas fir	[2]	12.5,(6.8)[a]
Redwood	[2]	9.4,(4.6)
Red oak	[2]	9.4,(7.9)
Maple	[2]	4.7,(6.3)

[a] Heat transfer parallel to the wood grain.

It should be realized that the use of L values from Table 9.1 must be taken as representative for a given material. The values can vary, not just due to the issues associated with the interpretation of experimental data, but also because the materials listed are commercial products and are subject to manufacturing and environmental factors such as purety, moisture, grain orientation, aging, etc.

We shall formalize our use of L by considering analyses for the steady burning of liquid fuels where the heat of gasification is a true fuel property.

9.2 Diffusive Burning of Liquid Fuels

9.2.1 *Stagnant layer*

In the study of steady evaporation of liquid, we derived in Equation (6.34), and given as Equation (9.6),

$$\dot{m}''_{\mathrm{F}} = \frac{\dot{q}''}{L}$$

where $L = h_{\mathrm{fg}} + \int_{T_{\mathrm{i}}}^{T_{\mathrm{s}}} c_{p,\mathrm{l}}\mathrm{d}T$ with

h_{fg} = energy required to vaporize a unit mass of liquid at temperature T_{s}

$\int_{T_{\mathrm{i}}}^{T_{\mathrm{s}}} c_{p,\mathrm{l}}\mathrm{d}T$ = energy needed to raise a unit mass of liquid from its original temperature T_{i} to the vaporization temperature on the surface, T_{s}. This latter energy is sometimes referred to as the 'sensible energy', since it is realized by a temperature change

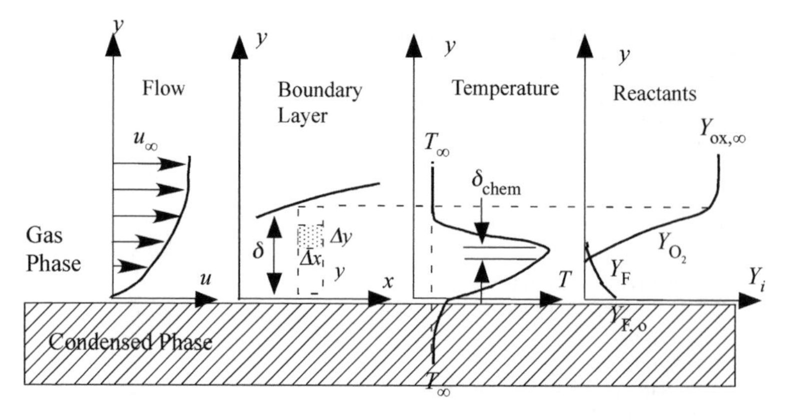

Figure 9.7 Stagnant layer model – pure convective burning

We shall consider at first only steady burning. Moreover, we shall restrict our heating to only that of convection received from the gas phase. This is flame heating. Therefore, purely convective heating is given as

$$\dot{q}'' = -\left(-k\frac{\mathrm{d}T}{\mathrm{d}y}\right)_{y=0} = \dot{m}_{\mathrm{F}}'' L \tag{9.8}$$

where k is the gas phase thermal conductivity; the coordinate system is shown in Figure 9.7. A special approximation is made for the gas- phase equations which is known as the stagnant layer model. The region of interest is the control volume depicted. It is a planar region of width Δx and thickness δ. Here δ can be imagined to represent the fluid-dynamical boundary layer where all of the diffusional changes of heat, mass and momentum are occurring. In Figure 9.7, we illustrate such changes by the portrayal of the temperature, fuel and oxygen concentration profiles. Combustion is occurring within this boundary layer. The rate of combustion can be governed by an Arrhenius law

$$\dot{m}_{\mathrm{F}}''' = A(Y_{\mathrm{F}}, Y_{\mathrm{O_2}})\mathrm{e}^{-E/(RT)} \tag{9.9}$$

for each point in the boundary layer with the pre-exponential parameter, A, a function of reactant concentrations and temperature. However, as depicted for this diffusion flame, the fuel and oxygen concentration may not penetrate much past each other. Indeed, if the chemical reaction time is very fast relative to the diffusion time (Equation (8.32))

$$\frac{t_{\mathrm{chem}}}{t_{\mathrm{Diff}}} = \frac{1}{Da} = \frac{u_\infty^2 \rho c_p T}{\alpha A \Delta h_{\mathrm{c}}[E/(RT)]\mathrm{e}^{-E/(RT)}}$$

or

$$\begin{aligned}\frac{t_{\mathrm{chem}}}{t_{\mathrm{Diff}}} &= \left(\frac{u_\infty x}{\nu}\right)^2 \frac{\nu^2 \rho c_p T}{\alpha x^2 A \Delta h_{\mathrm{c}}[E/(RT)]\mathrm{e}^{-E/(RT)}} \\ &= Re_x^2\, Pr^2 \frac{kT}{A\Delta h_{\mathrm{c}} x^2 [E/(RT)]\mathrm{e}^{-E/(RT)}}\end{aligned} \tag{9.10}$$

where Re_x is the Reynolds number, $u_\infty x/\nu$, and Pr is the Prandtl number, ν/α. Alternatively, a chemical length can be deduced by inspection as

$$\delta_{\text{chem}} = \sqrt{\frac{kT}{A\Delta h_c}} \tag{9.11}$$

Accordingly,

$$\frac{t_{\text{chem}}}{t_{\text{Diff}}} = \frac{Re_{\text{chem}}^2 \, Pr^2}{[E/(RT)\text{e}^{-E/(RT)}} \tag{9.12}$$

where $Re_{\text{chem}} \equiv u_\infty \delta_{\text{chem}}/\nu$, a chemical Reynolds number.

The physical interpretation of δ_{chem} is a region much thinner than δ, the boundary layer thickness, in which most of the chemical reaction is experienced. Outside of this region, A is nearly zero, while within the region A is relatively large. From our experience with the functions A and $E/(RT)$, we expect and will require that $t_{\text{chem}} \ll t_{\text{Diff}}$. In the flame region ($\delta_{\text{chem}}$), Re_{chem} is small and we might expect that diffusion and viscous effects dominate the characteristics of this reaction region. Indeed, if the chemical time is approximated as zero, then δ_{chem} would be an infinitesimal sheet – a flame sheet approximation. The concentration profiles of Y_{O_2} and Y_F would no longer overlap, but would be zero at the sheet. We will use this result.

Since diffusional effects are most important, we wish to emphasize these processes in the gas phase. For the control volume selected in Figure 9.7, the bold assumption is made that transport processes across the lateral faces in the x direction do not change – or change very slowly. Thus we only consider changes in the y direction. This approximation is known as the stagnant layer model since the direct effect of the main flow velocity (u) is not expressed. A differential control volume $\Delta y \times \Delta x \times$ unity is selected.

By a process of applying the conservation of mass, species and energy to the control volume ($\Delta y \times \Delta x \times$ unit length), and expressing all variables at $y + \Delta y$ as

$$f(y + \Delta y) \approx f(y) + \frac{\text{d}f}{\text{d}y}\Delta y + O(\Delta y)^2$$

using Taylor's expansion theorem, we arrive at the following:

Conservation of mass

$$\frac{\text{d}}{\text{d}y}(\rho v) = 0 \tag{9.13}$$

Conservation of species

$$\frac{\text{d}}{\text{d}y}(\rho v Y_i) = \frac{\text{d}}{\text{d}y}\left(\rho D \frac{\text{d}Y_i}{\text{d}y}\right) + \dot{m}_i''' \tag{9.14}$$

where Fick's law of diffusion has been used; $\dot{m}_i'''$ is the production rate per unit volume of species i.

Diffusion mass flux

$$\dot{m}''_{i,\mathrm{Diff}} = -\rho D \frac{\mathrm{d}Y_i}{\mathrm{d}y} \tag{9.15}$$

Conservation of energy

$$c_p \frac{\mathrm{d}}{\mathrm{d}y}(\rho v T) = \frac{\mathrm{d}}{\mathrm{d}y}\left(k \frac{\mathrm{d}T}{\mathrm{d}y}\right) + \dot{m}_{\mathrm{F}}''' \Delta h_{\mathrm{c}} \tag{9.16}$$

with only heat transfer by conduction considered from Fourier's law:

$$\dot{q}'' = -k \frac{\mathrm{d}T}{\mathrm{d}y} \tag{9.17}$$

where $\dot{m}_{\mathrm{F}}'''$ is the gas phase fuel consumption rate per unit volume and Δh_{c} is the heat of combustion.

We have made the further assumptions:

1. Radiant heat transfer is ignored.
2. Specific heat is constant and equal for all species.
3. Only laminar transport processes are considered in the transverse y direction.

Furthermore, we will take all other properties as constant and independent of temperature. Due to the high temperatures expected, these assumptions will not lead to accurate quantitative results unless we ultimately make some adjustments later. However, the solution to this stagnant layer with only pure conduction diffusion will display the correct features of a diffusion flame. Aspects of the solution can be taken as a guide and to give insight into the dynamics and interaction of fluid transport and combustion, even in complex turbulent unsteady flows. Incidentally, the conservation of momentum is implicitly used in the stagnant layer model since:

Conservation of momentum

$$\frac{\mathrm{d}p}{\mathrm{d}y} = 0 \tag{9.18}$$

is an expression for the conservation of momentum provided δ is small. Recall for boundary layer flows that

$$\frac{\delta}{x} \sim \frac{1}{Re_x^n} \tag{9.19}$$

where $n = \frac{1}{2}$ for laminar flows and about $\frac{4}{5}$ for turbulent flow. Under natural convection we can replace Re_x by $G_{r_x}{}^{1/2}$ with $Gr_x \equiv gx^3/\nu^2$. Thus, we see for boundary layer flows that δ is small if Re_x is large enough, i.e. $Re_x > 10^3$.

9.2.2 *Stagnant layer solution*

The simple but representative chemical reaction is considered with one product, P, as

$$1\,\mathrm{g\,F(fuel)} + r\,\mathrm{g\,O_2(oxygen)} \rightarrow (r+1)\mathrm{g\,P(product)} \tag{9.20}$$

The inclusion of more products only makes more details for us with no change in the basic content of the results.

Let us examine the integration of the governing differential equations. From Equation (9.13),

$$\int_0^y \frac{\mathrm{d}}{\mathrm{d}y}(\rho v) = 0$$

gives

$$\dot{m}_\mathrm{F}'' = \rho v = \text{constant} \tag{9.21}$$

Since at $y = 0$, the mass flux ρv is the mass loss rate of burning rate evolved from the condensed phase. From Equation (9.18) we also realize that we have a constant pressure process

$$p(y) = p_\infty \tag{9.22}$$

The remaining energy and species equations can be summarized as

$$c_p \dot{m}_\mathrm{F}'' \frac{\mathrm{d}T}{\mathrm{d}y} - k\frac{\mathrm{d}^2T}{\mathrm{d}y^2} = \dot{m}_\mathrm{F}'''\Delta h_\mathrm{c} \tag{9.23}$$

$$\dot{m}_\mathrm{F}'' \frac{\mathrm{d}Y_i}{\mathrm{d}y} - \rho D\frac{\mathrm{d}^2Y_i}{\mathrm{d}y^2} = \begin{cases} -\dot{m}_\mathrm{F}''', & i = \mathrm{F} \quad (9.24\mathrm{a}) \\ -r\dot{m}_\mathrm{F}''', & i = \mathrm{O_2} \quad (9.24\mathrm{b}) \\ (r+1)\dot{m}_\mathrm{F}''', & i = P \quad (9.24\mathrm{c}) \end{cases}$$

We have four equations but *five* unknowns. Although a constant, in this steady state case, $\dot{m}_\mathrm{F}''$ is not known. We need to specify two boundary conditions for each variable. This is done by the conditions at the wall ($y = 0$) and in the free stream of the enviornment outside of the boundary layer ($y = \delta$). Usually the environment conditions are known. At $y = \delta$,

$$\begin{aligned} T &= T_\infty \\ Y_{\mathrm{O_2}} &= Y_{\mathrm{O_2},\infty}(0.233 \text{ for air}) \\ Y_\mathrm{P} &= Y_{\mathrm{P},\infty}(\approx 0 \text{ for air}) \\ Y_\mathrm{F} &= 0 \text{ (if no fuel is left)} \end{aligned} \tag{9.25}$$

At $y = 0$,

$$\begin{aligned} T &= T_\mathrm{v}, && \text{the vaporization temperature} \\ Y_\mathrm{F} &= Y_{\mathrm{F,o}} && \text{within the condensed phase, i.e. } y = 0^- \end{aligned} \tag{9.26a}$$

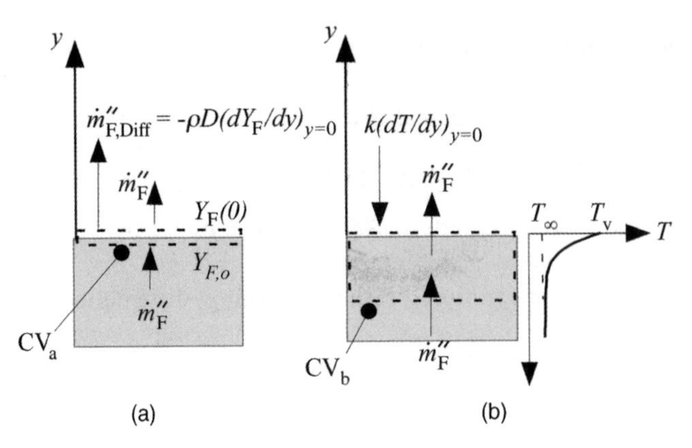

Figure 9.8 Surface boundary conditions (a) and (b)

However, we wish to express Y_F for the gas phase $y = 0^+$. This requires a more careful analysis for a control volume bounding the surface (Figure 9.8). From Figure 9.8(a), the mass flow rate of the liquid-condensed phase fuel must balance out the gas phase bulk transport at the flow rate $\dot{m}''_F$ and the diffusion flow rate due to a concentration gradient in the gas phase. Note that there is no concentration gradient in the liquid since it is of a uniform concentration. If we have a pure condensed phase fuel, $Y_{F,o} = 1$; in general, the surface concentration of fuel gives

$$\text{Across} \quad y = 0: \qquad \dot{m}''_F Y_{F,o} = \dot{m}''_F Y_F(0) - \left(\rho D \frac{dY_F}{dy}\right)_{y=0} \tag{9.26b}$$

Similarly, we can write for no P and O_2 absorbed in the condensed phase,

$$y = 0, \quad 0 = \dot{m}''_F Y_{O_2} - \rho D \frac{dY_{O_2}}{dy} \tag{9.26c}$$

$$y = 0, \quad 0 = \dot{m}''_F Y_P - \rho D \frac{dY_P}{dy} \tag{9.26d}$$

The corresponding boundary condition for energy is essentially Equation (9.8a), which follows from Figure 9.8(b), i.e.

$$y = 0, \quad 0 = \dot{m}''_F L - k \frac{dT}{dy} \tag{9.26e}$$

A certain similarity marks these equations and boundary conditions for the T and Y_i values. This recognition compels us to seek some variable transformations that will bring simplicity to the equations, and ultimately their solution. We seek to normalize and to simplify.

First we recognize that for air and for diffusion in air

$$D \approx \nu \approx \alpha \approx 10^{-5}\ \text{m}^2/\text{s}$$

or

$$Pr = \frac{\nu}{\alpha} \approx 1$$

and Schmidt number, $Sc \equiv \nu/D \approx 1$, and therefore

$$\text{Lewis number,} \quad Le \equiv \frac{Sc}{Pr} = \frac{\alpha}{D} \approx 1 \tag{9.27}$$

This is called the $Le = 1$ assumption, and is a reasonable approximation for combustion in air. Consequently, we define $\gamma \equiv \rho D = k/c_p$.

Our transformation will take several steps:

1. Let

$$\theta_{\mathrm{T}} \equiv T - T_{\infty}, \qquad \theta_i \equiv Y_i - Y_{i,\infty} \tag{9.28}$$

2. Subtract one equation from another among Equations (9.23) and (9.24). For example, for T and $i = F$,

$$\dot{m}_{\mathrm{F}}'' \frac{\mathrm{d}\theta_{\mathrm{T}}}{\mathrm{d}y} - \gamma \frac{\mathrm{d}^2\theta_{\mathrm{T}}}{\mathrm{d}y^2} = \dot{m}_{\mathrm{F}}''' \frac{\Delta h_{\mathrm{c}}}{c_p} \tag{9.29a}$$

Operating by

$$\left(\frac{\Delta h_{\mathrm{c}}}{c_p}\right) \times \left[\left(\dot{m}_{\mathrm{F}}'' \frac{\mathrm{d}\theta_{\mathrm{F}}}{\mathrm{d}y} - \gamma \frac{\mathrm{d}^2\theta_{\mathrm{F}}}{\mathrm{d}y^2}\right) = -\dot{m}_{\mathrm{F}}'''\right] \tag{9.29b}$$

gives

$$\dot{m}_{\mathrm{F}} \frac{\mathrm{d}\beta_{\mathrm{FT}}}{\mathrm{d}y} - \gamma \frac{\mathrm{d}^2\beta_{\mathrm{FT}}}{\mathrm{d}y^2} = 0 \tag{9.30}$$

where

$$\beta_{\mathrm{FT}} \equiv \theta_{\mathrm{T}} + \frac{\Delta h_{\mathrm{c}}}{c_p}\theta_{\mathrm{F}}$$

or

$$\beta_{\mathrm{FT}} = (T - T_{\infty}) + \frac{\Delta h_{\mathrm{c}}}{c_p} Y_{\mathrm{F}} \tag{9.31}$$

(recall that $Y_{\mathrm{F},\infty} = 0$).

The β-equation is now homogeneous (RHS = 0) with boundary conditions,

$$\begin{aligned} y &= \delta, \quad \theta = \theta_{\mathrm{F}} = 0 \quad \text{or} \quad \beta_{\mathrm{FT}} = 0 \\ y &= 0, \quad 0 = \frac{\dot{m}_{\mathrm{F}}'' L}{c_p} - \gamma \frac{\theta_{\mathrm{T}}}{\mathrm{d}y} \end{aligned} \tag{9.32a}$$

and

$$\dot{m}_{\mathrm{F}}''Y_{\mathrm{F,o}} = \dot{m}_{\mathrm{F}}''Y_{\mathrm{F}}(0) - \gamma\frac{\mathrm{d}\theta_{\mathrm{F}}}{\mathrm{d}y}$$

In the β format at $y = 0$,

$$0 = \dot{m}_{\mathrm{F}}''\left[\frac{L}{c_p} + \frac{\Delta h_{\mathrm{c}}}{c_p}(Y_{\mathrm{F}}(0) - Y_{\mathrm{F,o}})\right] - \gamma\frac{\mathrm{d}\beta_{\mathrm{FT}}}{\mathrm{d}y} \tag{9.32b}$$

Since the term in brackets on the RHS is constant, we can conveniently normalize β_{FT} as

$$b_{\mathrm{FT}} \equiv \frac{\beta_{\mathrm{FT}}}{L/c_p + (\Delta h_{\mathrm{c}}/c_p)(Y_{\mathrm{F}}(0) - Y_{\mathrm{F,o}})}$$

or

$$\boxed{b_{\mathrm{FT}} = \frac{c_p(T - T_\infty) + \Delta h_{\mathrm{c}}Y_{\mathrm{F}}}{L - \Delta h_{\mathrm{c}}(Y_{\mathrm{F,o}} - Y_{\mathrm{F}}(0))}} \tag{9.33}$$

Recall $Y_{\mathrm{F,o}} = 1$ for a pure fuel condensed phase, and $Y_{\mathrm{F}}(0)$ is not known. For a liquid fuel, $Y_{\mathrm{F}}(0)$ is found from the Clausius–Clapeyron equation provided we know $T(0)$.

In the b format, all of the equations – three from appropriate algebraic additions – are of the form

$$\dot{m}_{\mathrm{F}}''\frac{\mathrm{d}b}{\mathrm{d}y} - \gamma\frac{\mathrm{d}^2b}{\mathrm{d}y^2} = 0 \tag{9.34}$$

with

$$y = \delta, \quad b = 0$$

and

$$y = 0, \quad 0 = \dot{m}_{\mathrm{F}}'' - \gamma\frac{\mathrm{d}b}{\mathrm{d}y}$$

It can be shown that the other b values are

$$\boxed{b_{\mathrm{OT}} = \frac{c_p(T - T_\infty) + (Y_{\mathrm{O_2}} - Y_{\mathrm{O_2,\infty}})(\Delta h_{\mathrm{c}}/r)}{L}} \tag{9.35}$$

$$\boxed{b_{\mathrm{OF}} = \frac{Y_{\mathrm{F}} - (Y_{\mathrm{O_2}} - Y_{\mathrm{O_2,\infty}})/r}{Y_{\mathrm{F}}(0) - Y_{\mathrm{F,o}}}} \tag{9.36}$$

which follows the same conditions as Equation (9.34). Note that if the more complete boundary layer equations or the full Navier–Stokes equations were considered, Equation (9.34) would be a partial differential equation containing the additional independent variables x, z and time t. However, it could still be put into this more compact form. This variable transformation, which has essentially eliminated the nonlinear chemical source terms involving $\dot{m}_F'''$, is known as the Shvab–Zeldovich variable transformation (S–Z).

Let us integrate Equation (9.34) using the boundary conditions. Integrating once with $a = \dot{m}_F''/\gamma$,

$$ab - \frac{db}{dy} = ac_1$$

Letting $\eta = b - c_1$, so that

$$\int \frac{d\eta}{\eta} = \int_0^y a dy$$

gives

$$\eta = \eta(0)e^{ay}$$

from

$$y = \delta, \quad 0 - c_1 = \eta(0)e^{a\delta}$$

and

$$y = 0, \quad 0 = a - \eta(0)\ a\ e^0$$

Therefore, $\eta(0) = 1$ and $c_1 = -e^{a\delta}$ or

$$\boxed{b = e^{ay} - e^{a\delta}} \tag{9.37}$$

is our solution. However, we are far from a meaningful answer. Our objective is to find the 'burning rate', $\dot{m}_F''$, and its dependencies.

9.2.3 Burning rate – an eigenvalue

There is only one value of $\dot{m}_F''$ – an eigenvalue – that will satisfy a given set of environmental and fuel property conditions. How can this be found? We must be creative. We have already pushed the chemical kinetic terms out of the problem, but surely they must not have disappeared in reality.

Let us recall from the discussion of liquid evaporation that thermodynamically we have a property relationship between fuel vapor concentration and surface temperature,

$$Y_F(0) = f(T(0)) \qquad \text{or} \qquad b_{OT}(0) = b_{OF}(0)$$

From Equation (9.37), we evaluate b at $y = 0$ and express

$$\dot{m}''_{\mathrm{F}} = \left(\frac{k}{c_p \delta}\right) \ln[1 - b(0)] \tag{9.38}$$

If we know $b(0)$, we have our answer, but we have several alternatives for b: $b_{\mathrm{FT}}, b_{\mathrm{OT}}$ or b_{OF}. Any of these is fine. We chose b_{OT} because it gives us the most expedient approximate answer:

$$b_{\mathrm{OT}}(0) = \frac{c_p(T(0) - T_\infty) + (Y_{\mathrm{O}_2}(0) - Y_{\mathrm{O}_2,\infty})(\Delta h_{\mathrm{c}}/r)}{L} \tag{9.39}$$

We do expect $Y_{\mathrm{O}_2}(0)$ to be zero at the surface for combustion within the boundary layer since the flame reaction is fast and no oxygen is left. This must be clearly true even if the chemistry is not so fast. Moreover, since we are heating the surface with a nearby flame that approaches an adiabatic flame temperature, we would expect a high surface temperature. For a liquid fuel, we must have

$$T(0) \leq T_{\mathrm{boil}} \quad \text{at} \quad p_\infty$$

and we use the equality to the boiling point as an approximation. For a solid, we expect

$$T(0) \geq T_{\mathrm{ig}}$$

and again might use the equality or some slightly higher temperature. In general,

$$T(0) = T_{\mathrm{v}} \tag{9.40}$$

a specified vaporization temperature. This choice is empirical, unless we have steady burning of pure liquid, where the boiling point is valid, although we have not proven it. To prove it, we could equate $b_{\mathrm{OT}}(0)$ and $b_{\mathrm{OF}}(0)$ using the thermodynamic equilibrium to obtain $Y_{\mathrm{F}}(T)$, the saturation condition.

Proceeding, we obtain from Equation (9.38),

$$\boxed{\dot{m}''_{\mathrm{F}} = \left(\frac{k}{c_p \delta}\right) \ln(1 + B)} \tag{9.41}$$

The dimensionless group in the log term is called the Spalding B number, named after Professor Brian D. Spalding who demonstrated its early use [5]. From Equation (9.39), the surface conditions give

$$\boxed{B \equiv \frac{Y_{\mathrm{O}_2,\infty}(\Delta h_{\mathrm{c}}/r) - c_p(T_{\mathrm{v}} - T_\infty)}{L}} \tag{9.42}$$

This group expresses the ratio of chemical energy released to energy required to vaporize the fuel (per unit mass of fuel). It is clear it must be greater than 1 to achieve sustained burning. B contains both environmental and fuel property factors. Note a subtle point that c_p in the numerator is for the gas phase while in L, where

$$L \equiv h_{\mathrm{fg}} + c_{p,\mathrm{o}}(T_{\mathrm{v}} - T_{\mathrm{i}})$$

$c_{p,\mathrm{o}}$ being the specific heat of the condensed phase.

9.3 Diffusion Flame Variables

9.3.1 Concentrations and mixture fractions

Equation (9.41) constitutes a fundamental solution for purely convective mass burning flux in a stagnant layer. Sorting through the S–Z transformation will allow us to obtain specific stagnant layer solutions for T and Y_i. However, the introduction of a new variable – the mixture fraction – will allow us to express these profiles in mixture fraction space where they are universal. They only require a spatial and temporal determination of the mixture fraction f. The mixture fraction is defined as the mass fraction of original fuel atoms. It is as if the fuel atoms are all painted red in their evolved state, and as they are transported and chemically recombined, we track their mass relative to the gas phase mixture mass. Since these fuel atoms cannot be destroyed, the governing equation for their mass conservation must be

$$\dot{m}''_{\mathrm{F}} \frac{\mathrm{d}f}{\mathrm{d}y} - \rho D \frac{\mathrm{d}^2 f}{\mathrm{d}y^2} = 0 \tag{9.43}$$

since these atoms have bulk transport and diffusion of the species containing the fuel atoms. Although the mixture fraction has physical meaning and can be chemically measured, it can also be thought of as a fictitious property that satisfies a conservation of mass equation as an inert species composed of the original fuel atoms. The boundary conditions must be

$$\text{At} \quad y = \delta: \qquad f = 0 \text{ (no fuel in the free stream)} \tag{9.44a}$$

$$\text{At} \quad y = 0: \qquad f = f(0) \tag{9.44b}$$

Also from Equation (9.26b), the transport of fuel at $y = 0$ is

$$\dot{m}''_{\mathrm{F}} Y_{\mathrm{F,o}} = \dot{m}''_{\mathrm{F}} f(0) - \rho D \frac{\mathrm{d}f}{\mathrm{d}y} \tag{9.44c}$$

If we define

$$\boxed{b_{\mathrm{f}} \equiv \frac{f}{f(0) - Y_{\mathrm{F,o}}}} \tag{9.45}$$

then it can be verified that b_f perfectly satisfies Equation (9.34) with the identical boundary conditions as do all of the b values. Thus,

$$b_{FT} = b_{OT} = b_{OF} = b_{PF} = \cdots = b_f = e^{c_p \dot{m}_F'' y/k} - e^{c_p \dot{m}_F'' \delta/k} \tag{9.46}$$

Before bringing this analysis to a close we must take care of some unfinished business. The chemical kinetic terms, i.e. $\dot{m}_F'''$, have been transformed away. How can they be dealt with?

Recall that we are assuming $t_{chem} \ll t_{Diff}$ (or t_{mix}, if turbulent flow). Anyone who has carefully observed a laminar diffusion flame – preferably one with little soot, e.g. burning a small amount of alcohol, say, in a whiskey glass of Sambucca – can perceive of a thin flame (sheet) of blue incandescence from CH radicals or some yellow from heated soot in the reaction zone. As in the premixed flame (laminar deflagration), this flame is of the order of 1 mm in thickness. A quenched candle flame produced by the insertion of a metal screen would also reveal this thin yellow (soot) luminous cup-shaped sheet of flame. Although wind or turbulence would distort and convolute this flame sheet, locally its structure would be preserved provided that $t_{chem} \ll t_{mix}$. As a consequence of the fast chemical kinetics time, we can idealize the flame sheet as an infinitessimal sheet. The reaction then occurs at $y = y_f$ in our one dimensional model.

The location of y_f is determined by the natural conditions requiring

$$\text{At} \quad y = y_f: \qquad Y_F = Y_{O_2} = 0 \tag{9.47}$$

This condition defines the thin flame approximation, and implies that the flame settles on a location to accommodate the supply of both fuel and oxygen. This condition also provides a mathematical closure for the problem since all other solutions now follow. We will illustrate these.

Let us consider the solutions for Y_F and Y_{O_2} in the mixture fraction space. From Equations (9.45) and (9.46), for any b,

$$b = \frac{f}{f(0) - Y_{F,o}}$$

and

$$b(0) = \frac{f(0)}{f(0) - Y_{F,o}}$$

or

$$f(0) = \frac{b(0)Y_{F,o}}{b(0) - 1}$$

and finally

$$b = \frac{f}{b(0)Y_{F,o}/[b(0) - 1] - Y_{F,o}} = \frac{(b(0) - 1)f}{Y_{F,o}}$$

or

$$\boxed{f = \frac{Y_{F,o} b}{b(0) - 1}} \tag{9.48}$$

Now select $b = b_{OF}$. Therefore, from Equation (9.36),

$$f = \frac{Y_{F,o}\left[\frac{Y_F - (Y_{O_2} - Y_{O_2,\infty})/r}{Y_F(0) - Y_{F,o}}\right]}{\frac{Y_F(0) - (Y_{O_2}(0) - Y_{O_2,\infty})/r}{Y_F(0) - Y_{F,o}} - \frac{Y_F(0) - Y_{F,o}}{Y_F(0) - Y_{F,o}}}$$

and rearranging, taking $Y_{O_2}(0) = 0$,

$$f = \frac{Y_F - (Y_{O_2} - Y_{O_2,\infty})/r}{Y_{O_2,\infty}/(rY_{F,o}) + 1} \tag{9.49}$$

Since at the flame sheet $Y_{O_2} = Y_F = 0$, this is the definition of a stoichiometric system so that

$$\boxed{f(y_f) \equiv f_{st} = \frac{Y_{O_2,\infty}/r}{Y_{O_2,\infty}/(rY_{F,o}) + 1}} \tag{9.50}$$

By substituting Equation (9.50) into Equation (9.49), it can be shown that

$$f = \left(Y_F - \frac{Y_{O_2}}{r}\right)\left(\frac{Y_{F,o} - f_{st}}{Y_{F,o}}\right) + f_{st} \tag{9.51}$$

We can now choose to represent Y_{O_2} and Y_F in y or in the $f(y)$ variable as given by Equation (9.46). On the fuel side of the flame, from Eq. (9.51),

$$\begin{gathered} 0 \le y \le y_f, \quad Y_{O_2} = 0 \\ Y_F = \frac{f - f_{st}}{1 - f_{st}/Y_{F,o}} \end{gathered} \tag{9.52}$$

Similarly, on the oxygen side, $y_f \le y \le \delta$,

$$Y_F = 0$$

and

$$Y_{O_2} = \frac{-r(f - f_{st})}{1 - f_{st}/Y_{F,o}} = \frac{Y_{O_2,\infty}(f_{st} - f)}{f_{st}} \tag{9.53}$$

Note at $y = 0$ that

$$Y_F(0) = \frac{f(0) - f_{st}}{1 - f_{st}/Y_{F,o}}$$

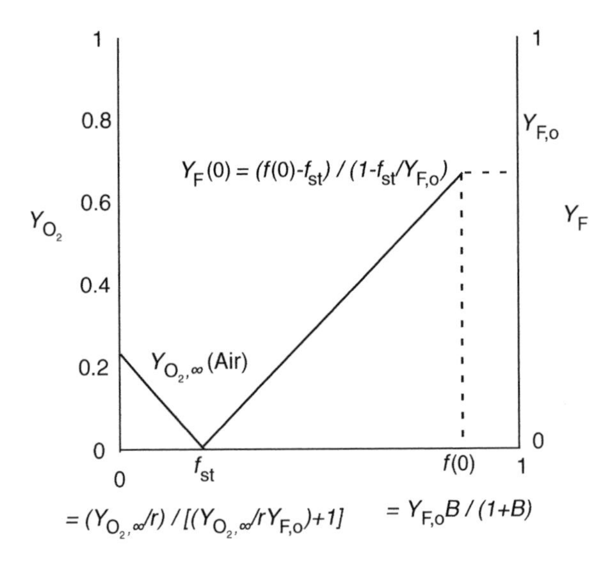

Figure 9.9 Oxygen–fuel mass concentrations in mixture fraction space

by Equation (9.45b), and at $y = \delta$,

$$Y_{O_2} = \frac{+rf_{st}}{1 - f_{st}/Y_{F,o}}$$

$$= \frac{Y_{O_2}/[Y_{O_2}/rY_{F,o} + 1]}{1 - \dfrac{[Y_{O_2,\infty}/(rY_{F,o})]}{[Y_{O_2,\infty}/(rY_{F,o}) + 1]}}$$

$$= \frac{Y_{O_2,\infty}}{Y_{O_2,\infty}/(rY_{F,o}) + 1 - Y_{O_2,\infty}/(rY_{F,o})} = Y_{O_2,\infty}$$

The results are plotted in Figure 9.9. Typical values of f_{st} are, for air $(Y_{O_2,\infty} = 0.233)$ and $r \sim 1$ to 3, roughly 0.05–0.2. The results are linear in f but nonlinear in y. This is true for all of the variables Y_{P_i} and T. The relationships for Y_i and T in f are universal, independent of geometry, flow conditions, etc., provided the thin flame – laminar – assumption holds. Even for unsteady, turbulent flows this conclusion still holds under a local thin flame approximation.

Let us recapitulate. We have achieved a solution to boundary-layer-like burning of a steady liquid-like fuel. A thin flame or fast chemistry relative to the mixing of fuel and oxygen is assumed. All effects of radiation have been ignored – a serious omission for flames of any considerable thickness. This radiation issue cannot easily be resolved exactly, but we will return to a way to include its effects approximately.

9.3.2 *Flame temperature and location*

The position of the flame sheet, y_f, can be determined in a straightforward manner. From Equations (9.45), (9.46) and (9.50),

$$e^{\mu} - e^{\mu(y_f/\delta)} = \frac{Y_{O_2,\infty}/r}{[Y_{O_2,\infty}/(rY_{F,o}) + 1](Y_{F,o} - f(0))} \tag{9.54}$$

where $\mu \equiv c_p \dot{m}''_{\mathrm{F}} \delta / k$. From Equations (9.39), (9.42) and (9.46), $b_{\mathrm{f}}(0) = B$, the Spalding number. Therefore, it can be shown from Equation (9.45) that

$$Y_{\mathrm{F,o}} - f(0) = \frac{Y_{\mathrm{F,o}}}{1 + B}$$

Realizing that $\mathrm{e}^{\mu} = 1 + B$ from Equation (9.41), it follows that

$$\mathrm{e}^{\mu y_{\mathrm{f}}/\delta} = \frac{1 + B}{Y_{\mathrm{O_2},\infty}/(rY_{\mathrm{F,o}}) + 1}$$

or

$$\boxed{\frac{y_{\mathrm{f}}}{\delta} = \frac{\ln[(1 + B)/(Y_{\mathrm{O_2},\infty}/(rY_{\mathrm{F,o}}) + 1)]}{\ln(1 + B)}} \tag{9.55}$$

The flame temperature is found from using Equations (9.35), (9.48) and (9.50), evaluated at $y = y_{\mathrm{f}}$ or $f = f_{\mathrm{st}}$,

$$\frac{Y_{\mathrm{O_2},\infty}/r}{\frac{Y_{\mathrm{O_2},\infty}}{rY_{\mathrm{F,o}}} + 1} = \frac{\dfrac{Y_{\mathrm{F,o}}[c_p(T_{\mathrm{F}} - T_\infty) - Y_{\mathrm{O_2},\infty}\Delta h_{\mathrm{c}}/r]}{L}}{\dfrac{c_p(T_{\mathrm{v}} - T_\infty) - Y_{\mathrm{O_2},\infty}(\Delta h_{\mathrm{c}}/r)}{L} - 1}$$

After rearranging, we find the flame enthalpy rise as

$$\boxed{c_p(T_{\mathrm{f}} - T_\infty) = \frac{Y_{\mathrm{F,o}}\Delta h_{\mathrm{c}} - L + c_p(T_{\mathrm{v}} - T_\infty)}{1 + rY_{\mathrm{F,o}}/Y_{\mathrm{O_2},\infty}}} \tag{9.56a}$$

Alternatively, this can be expressed as

$$\frac{c_p(T_{\mathrm{f}} - T_\infty)}{L} = \frac{B - \tau_{\mathrm{o}}(r_{\mathrm{o}} - 1) - r_{\mathrm{o}}}{1 + r_{\mathrm{o}}}$$

$$r_{\mathrm{o}} = \frac{Y_{\mathrm{O_2},\infty}}{rY_{\mathrm{F,o}}}$$

$$\tau_{\mathrm{o}} = \frac{c_p(T_{\mathrm{v}} - T_\infty)}{L} \tag{9.56b}$$

The quantity $rY_{\mathrm{F,o}}/Y_{\mathrm{O_2},\infty}$ is the ratio of stoichiometric oxygen to fuel mass ratio divided by the available or supplied oxygen to fuel mass fractions. In contrast, for the premixed adiabatic flame temperature,

$$c_p(T_{\mathrm{f,ad}} - T_\infty) = Y_{\mathrm{F,u}}\Delta h_{\mathrm{c}} \tag{9.57}$$

the analog of Equation (4.54). As in estimating the lower limit concentration that permits a premixed flame to exist by requiring $T_{\mathrm{f,ad}} = 1300\,^{\circ}\mathrm{C}$, a similar requirement could be imposed to define conditions for maintaining a diffusion flame. We will use this to estimate extinction conditions.

9.4 Convective Burning for Specific Flow Conditions

The stagnant layer analysis offers a pedagogical framework for presenting the essence of diffusive burning. For the most part the one-dimensional stagnant layer approximated a two-dimensional boundary layer in which $\delta = \delta(x)$, with x the flow direction. For a convective boundary layer, the heat transfer coefficient, h_c, is defined as

$$h_c(T_R - T_s) = -k\left(\frac{dT}{dy}\right)_{y=0} \tag{9.58}$$

where T_R is a reference temperature and T_s is the wall temperature. For a flame this might be approximated with $T_R = T_f$, as the flame temperature,

$$h_c(T_f - T_s) \approx \frac{k(T_f - T_s)}{\delta}$$

for the stagnant layer, or

$$h_c \approx \frac{k}{\delta} \tag{9.59}$$

This crude approximation allows us to extend the stagnant layer solution to a host of convective heat transfer counterpart burning problems. Recall that for Equations (9.41) and (9.42), we can write

$$\boxed{\dot{m}_F'' \approx \frac{h_c}{c_p}\ln(1+B)} \tag{9.60}$$

We wish to write this as

$$\dot{m}_F'' = \frac{h_c}{c_p}\left[\frac{\ln(1+B)}{B}\right]B \tag{9.61}$$

where the significance of the blocking factor, $\ln(1+B)/B$ will become apparent. Substituting for B and recalling the heat flux from Equation (9.6), Equation (9.61) becomes

$$\dot{q}'' = h_c\left[\frac{\ln(1+B)}{B}\right]\left[\frac{Y_{O_2,\infty}\Delta h_c}{c_p r} - (T_v - T_\infty)\right] \tag{9.62}$$

Thus we see that the reference temperature should have been $Y_{O_2,\infty}\Delta h_c/(c_p r) + T_\infty$, which is slightly different from Equation (9.56a). Nevertheless, we see that

$$\dot{q}'' = h_c\left[\frac{\ln(1+B)}{B}\right](T_R - T_v)$$

The blocking factor can alternatively be written from Equation (9.60) as

$$\frac{\ln(1+B)}{B} = \frac{c_p \dot{m}''_{\mathrm{F}}/h_{\mathrm{c}}}{\exp(c_p \dot{m}''_{\mathrm{F}}/h_{\mathrm{c}}) - 1} \tag{9.63}$$

which approaches 1 as $\dot{m}''_{\mathrm{F}} \to 0$. Hence the heat transfer coefficient in the presence of mass transfer is smaller than h_{c}, the heat transfer coefficient without mass transfer. The blocking factor acts to enlarge the boundary later due to blowing caused by the vaporized fuel. This transverse flow makes it more difficult to transfer heat across the boundary layer. Moreover, in Equation (9.61), it is seen that for mass transfer in combustion, the mass transfer coefficient is

$$h_{\mathrm{m}} \equiv \frac{h_{\mathrm{c}}}{c_p} \frac{\ln(1+B)}{B} \tag{9.64}$$

and the driving force for this transfer is B. B acts like ΔT for heat, ΔY for mass or Δ voltage for electrical current flow.

To obtain approximate solutions for convection heating problems, we only need to identify a heat transfer problem that has a given theoretical or empirical correlation for h_{c}. This is usually given in the form of the Nusselt number (*Nu*),

$$\frac{h_{\mathrm{c}} x}{k} \equiv Nu_x = f(Re_{x_1}, Gr_{x_2} Pr) \tag{9.65}$$

The corresponding dimensionless burning rate (called the Sherwood number (Sh) in pure mass transfer) should then give a functional relationship of the form

$$\frac{\dot{m}''_{\mathrm{F}} c_p x}{k} = f(Nu_x, B) \tag{9.66}$$

or more generally may contain other factors that we have seen, e.g. $\tau_{\mathrm{o}} = c_p(T_{\mathrm{v}} - T_\infty)/L$, $r_{\mathrm{o}} = Y_{\mathrm{O_2},\infty}/(rY_{\mathrm{F,o}})$, etc. Figure 9.10 gives some pure convective heat transfer formulas and their associated geometries. Since the heat transfer formula will contain $\Delta T/T_\infty$ in the Grashof number, i.e.

$$Gr_x = \frac{g \Delta T x^3}{T_\infty \nu^2}$$

it is suggested to take $\Delta T/T_\infty \approx 3$ since it could be argued that values of 1 to 7 should be used for this type of approximation. Remember, this is only an approximate procedure. The effect of turbulence is not just in h_{c}, but is also in r. In turbulent flow, due to unmixedness of the fuel and oxidizer, r can be replaced by nr, where n is the excess oxidizer fraction.

There have been a number of direct solutions in the form of Equation (9.65) for convective combustion. These have been theoretical – exact or integral approximations to the boundary layer equations, or empirical – based on correlations to experimental data. Some examples are listed below:

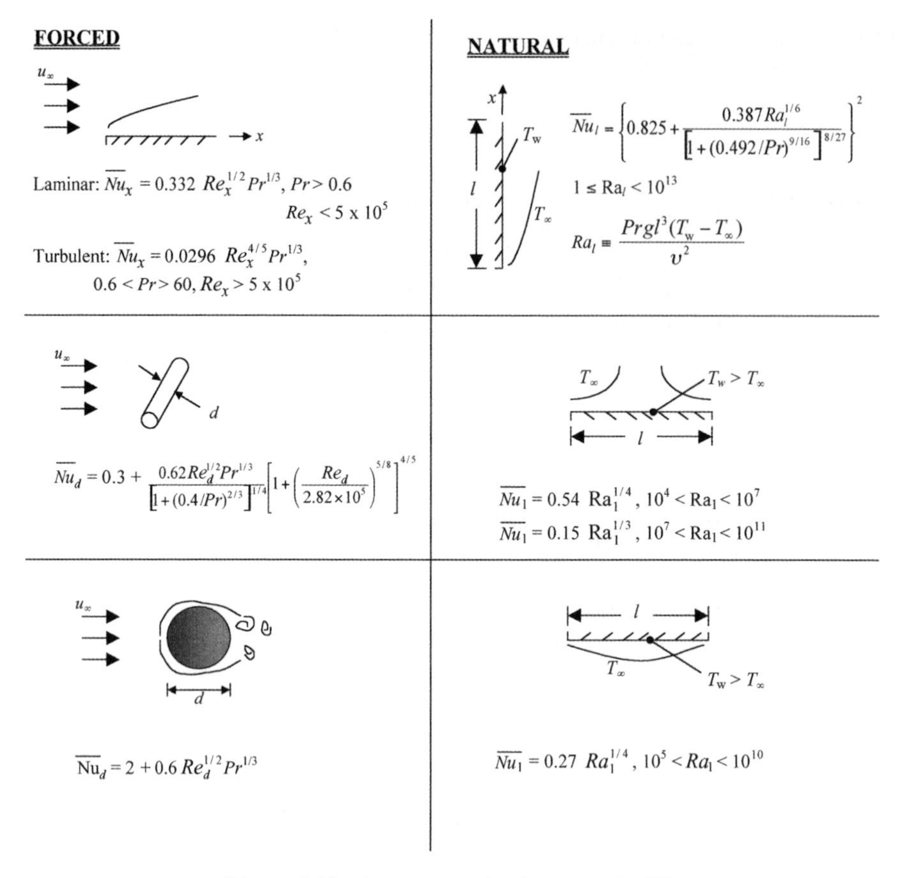

Figure 9.10 Pure convective heat transfer [6]

1. *Forced convection*, burning of a flat plate. This classical exact analysis following the well-known Blasius solution for incompressible flow was done by Emmons in 1956 [7]. It includes both variable density and viscosity. Glassman [8] presents a functional fit to the Emmons solution as

$$\frac{\dot{m}''_F c_p x}{k} = 0.385 \left(\frac{u_\infty x}{\nu}\right)^{1/2} Pr \frac{\ln(1 + B)}{B^{0.15}} \tag{9.67}$$

2. The corresponding *laminar natural convection* burning rate on a vertical surface was done by Kosdon, Williams and Buman [9] and by Kim, deRis and Kroesser [10]. The latter gives

$$\frac{\dot{m}''_F c_p x}{k_w} = 3\, Pr_w\, Gr_w^{*1/4}\, F(B, \tau_o, r_o) \tag{9.68}$$

where the transport properties are evaluated at T_w and F is given in Figure 9.11(a).

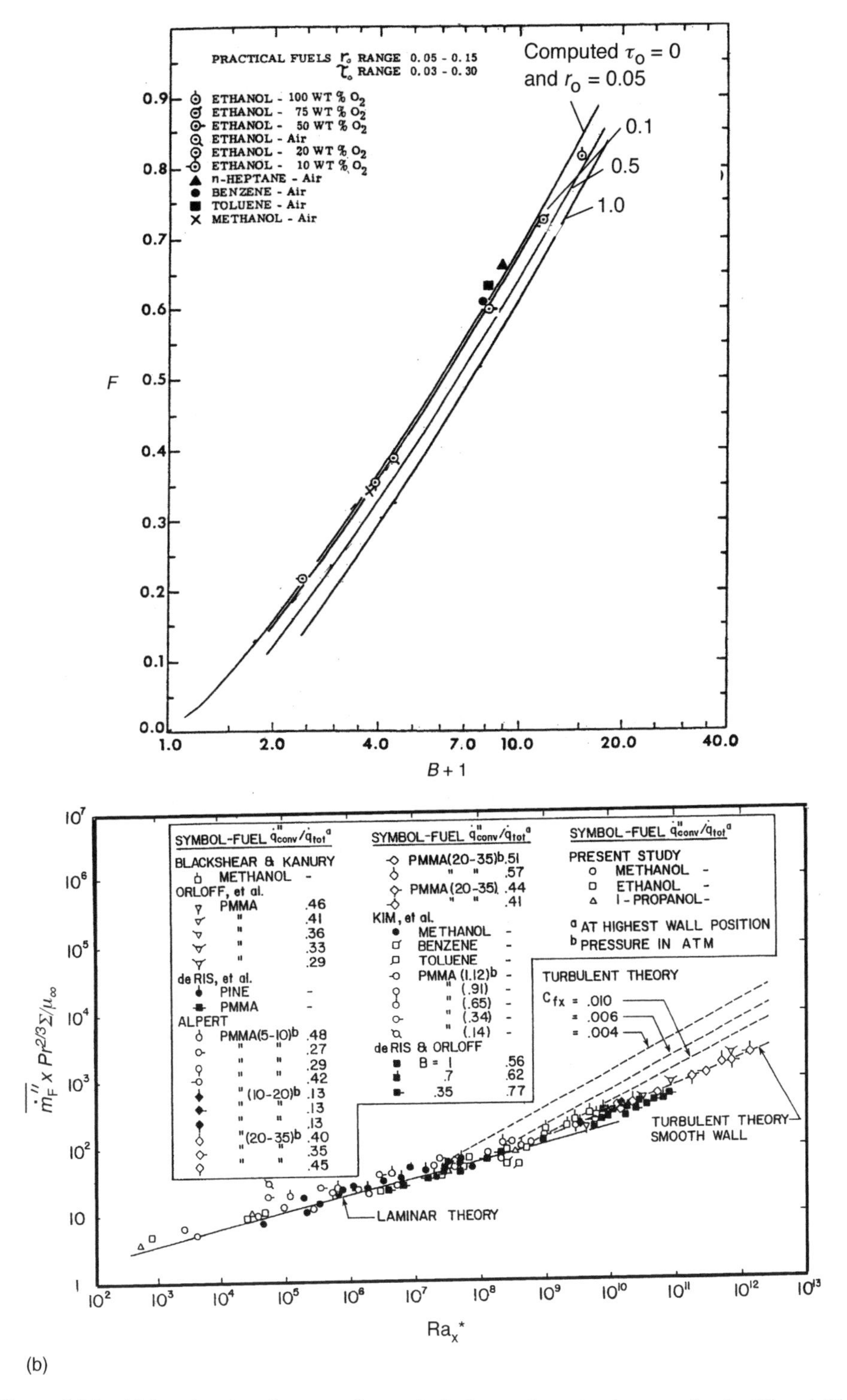

(b)

Figure 9.11 (a) Laminar burning rate of a vertical plate under natural convection by Kim, deRis and Kroesser [10]. (b) Turbulent burning rate of a vertical plate under natural connection by Ahmad and Faeth [11]

3. *Turbulent convective burning of vertical plate*. An approximate solution matched to data was given by Ahmad and Faeth [11]. The average burning rate $\overline{\dot{m}''_{\mathrm{F}}}$ for a distance x measured from the start of the plate is given by the formula below:

$$\frac{\overline{\dot{m}''_{\mathrm{F}}}xc_p}{k} = \frac{0.0285 Gr_x^{*0.4}\,Pr^{0.73}}{\Sigma} \tag{9.69}$$

where

$$Gr^* = \left(\frac{L}{4c_pT_\infty}\right)\left(\frac{g\cos\phi x^3}{\nu^2}\right) \quad (\phi \text{ measured from the vertical})$$

$$\Sigma = \left[\frac{1+B}{B\ln(1+B)}\right]^{1/2}\left[\frac{1+0.5\,Pr/(1+B)}{3(B+\tau_{\mathrm{o}})\eta_{\mathrm{f}}+\tau_{\mathrm{o}}}\right]^{1/4}$$

$$\tau_{\mathrm{o}} = \frac{c_p(T_{\mathrm{v}}-T_\infty)}{L}$$

$$\eta_{\mathrm{f}} = 1-\theta_{\mathrm{FO_f}}^{1/3}$$

$$\theta_{\mathrm{FO_f}} = \frac{r_{\mathrm{o}}(B+1)}{B(r_{\mathrm{o}}+1)}$$

$$r_{\mathrm{o}} = \frac{Y_{\mathrm{O_2},\infty}}{rY_{\mathrm{F,o}}}$$

$$\overline{\dot{m}''_{\mathrm{F}}} = \frac{1}{x}\int_0^x \dot{m}''_{\mathrm{F}}\,\mathrm{d}x$$

This solution is a bit tedious, but complete and consistent with data on liquid fuels saturated into inert plates, as displayed in Figure 9.11(b).

Table 9.2 gives some typical estimated or exact properties in order to compute the B numbers. It is seen in some cases, particularly for charring materials, that the B numbers for burning in air are less than 1. This implies that purely convective burning is not possible. Indeed, we have ignored surface radiation loss, which can be relatively considerable (at 350 °C, 8.5 kW/m^2). The only hope of achieving 'steady' burning is to provide additional heat flux if burning is not possible on its own. Therefore, not only is there a minimum heat flux to achieve ignition, but there is also a minimum external heat flux needed to maintain burning.

Example 9.1 Methanol is saturated into a thick board of porous ceramic fibers. The board is maintained wet with a supply temperature at ambient of 20 °C. A steady air flow is directed across the horizontal board at 3 m/s. The board is 10 cm in the flow directions, and is placed flush with the floor. Use the following property data:

For air:

$$\nu = 15\times 10^{-6}\,\mathrm{m^2/s}$$

$$k = 25\times 10^{-3}\,\mathrm{W/m\,K}$$

$$Pr = 0.7$$

$$\rho = 1.1\,\mathrm{kg/m^3}$$

$$c_p = 1.05\,\mathrm{kJ/kg\,K}$$

Table 9.2 Estimated fuel properties for B-number estimates (from various sources)

Material	Δh_c^a (kJ/g)	L (kJ/g)	T_v^b (°C)	B-estimate[c]
		Liquids		
n-Hexane	42	0.45	69	6.7
n-Heptane	41	0.48	98	6.2
n-Octane	41	0.52	125	5.7
Benzene	28	0.48	80	6.2
Toulene	28	0.50	110	5.9
Naphthalene	30	0.55	218	5.2
Methanol	19	1.2	64	2.5
Ethanol	26	0.97	78	3.1
n-Butanol	35	0.82	117	3.6
Acetone	28	0.58	56	5.2
		Solids		
Polyethylene	38	3.6	360	0.75
Polypropylene	38	3.1	330	0.89
Nylon	27	3.8	500	0.68
Polymethylmethacrylate	24	2.0	300	1.4
Polystyrene	27	3.0	350	0.91
		Solids, charring		
Polyurethane foam, rigid	17	5.0	300	0.56
Douglas fir	13	12.5	380	0.22
Redwood	12	9.4	380	0.29
Red oak	12	9.4	300	0.30
Maple	13	4.7	350	0.58

[a]Based on flaming conditions and the actual heat of combustion.
[b]Boiling temperature for liquids; estimated ignition temperature for solids.
[c]$Y_{F,o} = 1,\ Y_{O_2,\infty} = 0.233,\ T_\infty = T_i = 25\,°C,\ c_p = 1$ kJ/kg K.

Methanol boiling point = 337 K = 64 °C
Heat of vaporization, $h_{fg} = 1.1$ kJ/g
Liquid specific heat = 2.5 kJ/kg K
Assume steady burning and ignore radiation effects.

(a) Assess that the flow is laminar.

(b) Calculate the B number.

(c) What is the steady burning rate g/s if the plate is 20 cm wide? Use the Emmons result given by Equation (9.67).

(d) Calculate the average net heat flux to the surface.

(e) What is the surface radiation heat flux emitted assuming a blackbody?

(f) What is the role of the ceramic board in ignition and steady burning?

Solution

(a) $\mathrm{Re}_x = \dfrac{u_\infty x}{\nu} = \dfrac{(3\,\mathrm{m/s})(0.1\,\mathrm{m})}{15 \times 10^{-6}\,\mathrm{m^2/s}} = 2 \times 10^4$

Flow is laminar since Re $< 5 \times 10^5$, transition.

(b) $B = \dfrac{Y_{O_2,\infty}\Delta h_c/r - c_p(T_v - T_\infty)}{L}$

From Table 2.3,

$$\Delta h_c = 20.0\,\text{kJ/g}$$

From the stoichiometric equation for complete combustion,

$$\begin{aligned}
CH_4O + 1.5\,O_2 &\rightarrow CO_2 + 2\,H_2O \\
r &= \frac{(1.5)(32)}{32} = 1.5\,\text{g}\,O_2/\text{g fuel} \\
Y_{O_2,\infty}\Delta h_c/r &= (0.233)(20.0\,\text{kJ/g})/1.5 = 3.11\,\text{kJ/g} \\
c_p(T_v - T_\infty) &= (1.05\,\text{kJ/kg K} \times 10^{-3}\,\text{kg/g})(64\text{–}20) = 0.046\,\text{kJ/g} \\
L &= h_{fg} + c_l(T_v - T_\infty) \\
&= 1.10\,\text{kJ/g} + (2.5 \times 10^{-3}\,\text{kJ/g K})(64\text{–}20) = 1.21\,\text{kJ/g} \\
B &= (3.11 - 0.046)/1.21 = 2.3
\end{aligned}$$

(c) From Equation (9.67), using air properties for the gas,

$$\begin{aligned}
\dot{m}''_F(x) &= \left(\frac{25 \times 10^{-3}\ \text{W/m K}}{1.05\,\text{J/g K}}\right)(0.385)\left(\frac{3\,\text{m/s}}{15 \times 10^{-6}\,\text{m}^2/\text{s}}\right)^{1/2}(0.7) \\
&\quad \times \frac{\ln(1 + 2.53)}{(2.53)^{0.15}} x^{-1/2}, \quad x \quad \text{in m} \\
&= 1.097/\sqrt{x(m)} \quad \text{in} \quad \text{g/m}^2\,\text{s}
\end{aligned}$$

The burning rate over $x = 0.10$ m of width 0.20 m,

$$\begin{aligned}
\dot{m} &= (0.20\ \text{m})\int_0^{0.10\,\text{m}} \frac{1.097}{x^{1/2}}\,dx \\
\dot{m} &= 0.139\ \text{g/s}
\end{aligned}$$

(d) From Equation (9.6),

$$\begin{aligned}
\overline{\dot{q}''} &= \overline{\dot{m}''}\,L \\
&= \left(\frac{0.139\,\text{g/s}}{0.20 \times 0.10\,\text{m}^2}\right)(1.21\ \text{kJ/g}) \\
&= 8.4\,\text{kW/m}^2
\end{aligned}$$

(e) $\dot{q}''_r = \sigma(T_v^4 - T_\infty^4)$, assuming ambient surroundings

$= 5.67 \times 10^{-11}\,\text{kW/m}^2\,\text{K}^4[(337)^4 - (293)^4]$

$= 0.313\,\text{kW/m}^2$

Hence, re-radiation is negligible.

(f) In steady burning, the ceramic board functions as a wick and achieves the temperature of the methanol liquid in depth. For the ignition process, the ceramic board functions as a heat sink along with the absorbed methanol. The surface temperature must achieve the flash (or fire) point of the methanol to ignite with a pilot. The ignition process cannot be ignored in the unsteady burning problem since it forms the initial condition.

9.5 Radiation Effects on Burning

The inclusion of radiative heat transfer effects can be accommodated by the stagnant layer model. However, this can only be done if *a priori* we can prescribe or calculate these effects. The complications of radiative heat transfer in flames is illustrated in Figure 9.12. This illustration is only schematic and does not represent the spectral and continuum effects fully. A more complete overview on radiative heat transfer in flame can be found in Tien, Lee and Stretton [12]. In Figure 9.12, the heat fluxes are presented as incident (to a sensor at T_∞) and absorbed (at T_v) at the surface. Any attempt to discriminate further for the radiant heating would prove tedious and pedantic. It should be clear from heat transfer principles that we have effects of surface and gas phase radiative emittance, reflectance, absorptance and transmittance. These are complicated by the spectral character of the radiation, the soot and combustion product temperature and concentration distributions, and the decomposition of the surface. Reasonable approximations that serve to simplify are:

1. Surfaces are opaque and approach emittance and absorptance of unity.
2. External radiant flux is not affected by the flame, e.g. thin flame.
3. Flame incident radiant flux ($\dot{q}''_{f,r}$) and total radiant energy ($X_r\dot{Q}$) might be estimated from measurements.
4. External radiant heat flux ($\dot{q}''_e$) is expressed as that measured by a heat flux meter maintained at T_∞. Correspondingly, the surface loss is given with respect to T_∞ as $\sigma(T_v^4 - T_\infty^4)$.

With the heat flux specifications in Figure 9.12, the energy boundary condition at $y = 0$ becomes, for steady burning (see Equations (9.6) and (9.8)),

$$\dot{m}''_F L = k\left(\frac{dT}{dy}\right)_{y=0} + \dot{q}''_{f,r} + \dot{q}''_e - \sigma(T_v^4 - T_\infty^4) \tag{9.70}$$

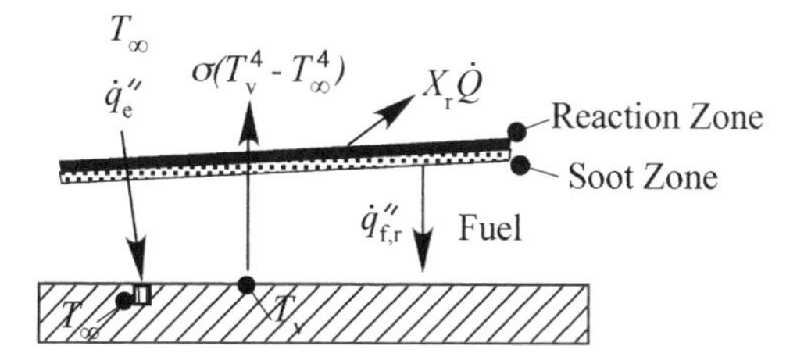

Figure 9.12 Radiative heat transfer in burning

If we can define the radiative terms *a priori*, we are simply adding a constant to the RHS. An expedient transformation reduces the boundary condition to its original convective form:

$$\dot{m}''_{\mathrm{F}} L_{\mathrm{m}} = k \frac{\mathrm{d}T}{\mathrm{d}y} \quad \text{at} \quad y = 0 \tag{9.71}$$

where

$$L_{\mathrm{m}} \equiv L - \frac{\dot{q}''_{\mathrm{f,r}} + \dot{q}''_{\mathrm{e}} - \sigma(T_{\mathrm{v}}^4 - T_{\infty}^4)}{\dot{m}''_{\mathrm{F}}} \tag{9.72}$$

is defined as an equivalent, modified 'L'. The net addition of radiant heat makes the burning rate increase as one would expect. It might also make it burn where a material will not normally. This is an important point because many common materials and products will not burn without additional heat flux. While this addition may seem artificial, it is the key to disastrous fire growth since special configurations favor heat feedback ($\dot{q}''_{\mathrm{e}}$) that can promote fire growth. Limited testing of a material by applying a small flame might cause ignition, but not necessarily sustained burning or spread; this can be misleading to the evaluation of a material's general performance in fire.

The introduction of Equation (9.71) for Equation (9.26e) makes this a new problem identical to what was done for the pure diffusion/convective modeling of the burning rate. Hence L is simply replaced by L_{m} to obtain the solution with radiative effects. Some rearranging of the stagnant layer case can be very illustrative. From Equations (9.61) and (9.42) we can write

$$\dot{m}''_{\mathrm{F}} = \frac{h_{\mathrm{c}}}{c_p} \frac{\ln(1+B)}{B} \frac{Y_{\mathrm{O_2},\infty} \Delta h_{\mathrm{c}}/r - c_p(T_{\mathrm{v}} - T_{\infty})}{L - [\dot{q}''_{\mathrm{f,r}} + \dot{q}''_{\mathrm{e}} - \sigma(T_{\mathrm{v}}^4 - T_{\infty}^4)]/\dot{m}''_{\mathrm{F}}}$$

or

$$\begin{aligned} \dot{m}''_{\mathrm{F}} L = {} & \frac{h_{\mathrm{c}}}{c_p} \left(\frac{\lambda}{\mathrm{e}^{\lambda} - 1} \right) \left[\frac{Y_{\mathrm{O_2},\infty} \Delta h_{\mathrm{c}}}{r} (1 - X_{\mathrm{r}}) - c_p(T_{\mathrm{v}} - T_{\infty}) \right] \\ & + \dot{q}''_{\mathrm{f,r}} + \dot{q}''_{\mathrm{e}} - \sigma(T_{\mathrm{v}}^4 - T_{\infty}^4) \end{aligned} \tag{9.73}$$

with $\lambda = c_p \dot{m}''_{\mathrm{F}}/h_{\mathrm{c}}$ and where we have added $(1 - X_{\mathrm{r}})$ to account for radiated energy ($X_{\mathrm{r}}\dot{Q}$) eliminated from the flame. Although this requires an iterative solution for $\dot{m}''_{\mathrm{F}}$, its form clearly displays the heat flux contributions to vaporization. The LHS is the energy required to vaporize under steady burning and the RHS is the net absorbed surface heat flux, the first term being the convective heat flux. Typical convection ranges from 5 to 10 kW/m^2 for turbulent buoyant flows but as high as 50–70 kW/m^2 for laminar natural convection. For a forced convection laminar flame, such as a small jet flame, the convective heat flux can be 100–300 kW/m^2. Therefore, convective effects cannot generally be dismissed although they are a relatively small component in turbulent natural fires.

Figure 9.13 displays the separation of average flame convective and radiative components for burning pools of PMMA at diameters ranging from 0.15 to 1.2 m.

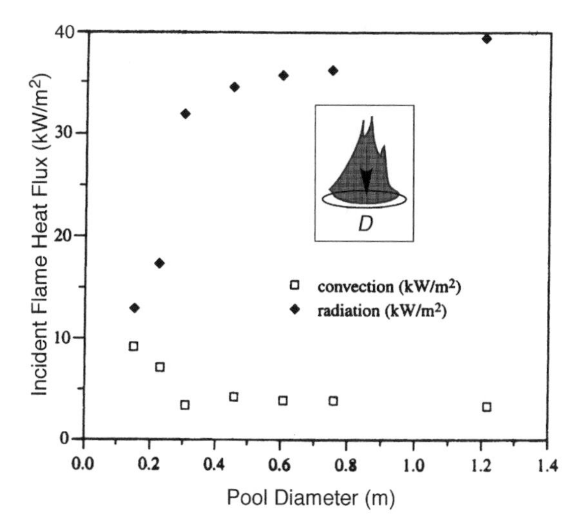

Figure 9.13 Average flame heat flux in PMMA pool fires (from Modak and Croce [13] and Iqbal and Quintiere [14])

Radiation increases with scale while convection drops as the diameter increases. Local resolution of these flame heat fluxes were estimated by Orloff, Modak and Alpert [15] for a burning 4 m vertical wall of PMMA. It is clearly seen that flame radiative effects are significant relative to convection.

In Figure 9.14 we see a more classical demonstration of the range of burning rate behavior of a pool fire. Below a diameter of 25 cm the burning rate is laminar with $h_c \propto D^{-1/4}$; afterwards it is turbulent as $h_c \propto D^0$ or $D^{1/5}$ at most. However, with

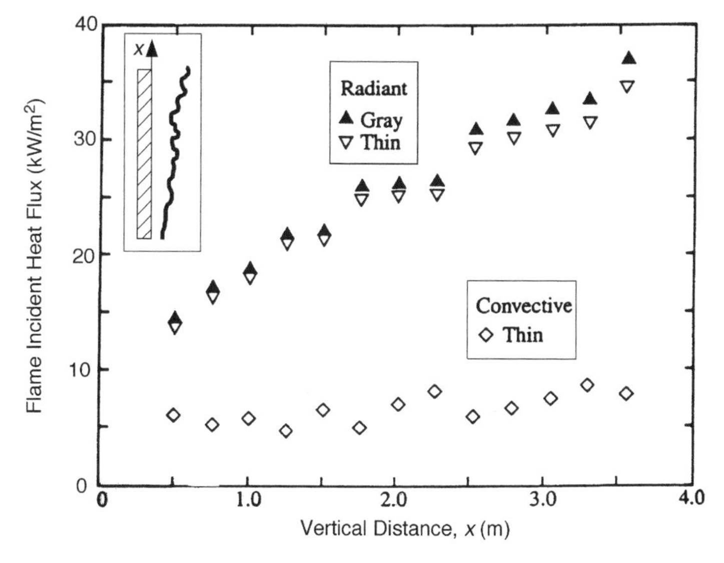

Figure 9.14 Local flame heat flux to a burning PMMA vertical surface (from Orloff, Modak and Alpert [15])

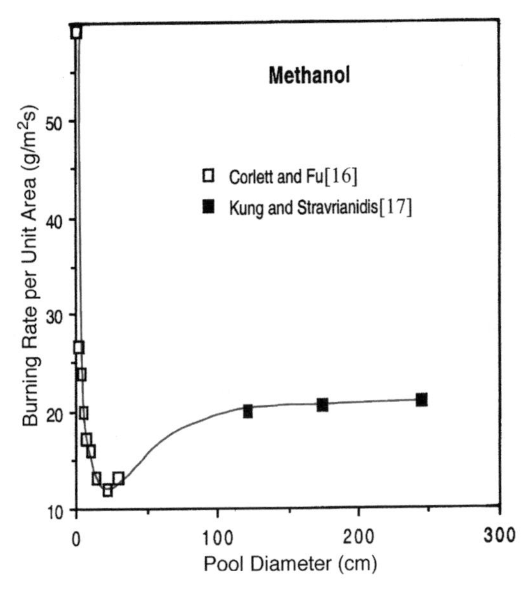

Figure 9.15 Regimes of the steady burning rate for methanol: $D < 25\,\text{cm}$ laminar, $D > 25\,\text{cm}$ turbulent, $D > 100\,\text{cm}$ radiation saturated [16, 17]

increasing diameter, the primary cause of the increase in the burning rate is due to flame radiation. This follows as the flame emittance behaves as

$$\epsilon_{\mathrm{f}} \approx 1 - \mathrm{e}^{-\kappa D}$$

which approaches 1 at about $D \geq 100\,\text{cm}$. Since $L = 1.21\,\text{kJ/g}$ for methanol, we see from Figure 9.15 that (1) the maximum net surface heat flux at about $D = 5\,\text{cm}$ is

$$\begin{aligned} \dot{q}'' &= (27\,\text{g/m}^2\text{s})(1.21\,\text{kJ/g}) \\ &= 32.7\,\text{kW/m}^2 \end{aligned}$$

(2) a minimum at the onset of turbulent flow of

$$\begin{aligned} \dot{q}'' &= (12\,\text{g/m}^2\text{s})(1.21\,\text{kJ/g}) \\ &= 14.5\,\text{kW/m}^2 \end{aligned}$$

and (3) a maximum under 'saturated' flame radiation conditions of

$$\begin{aligned} \dot{q}'' &= (21\,\text{g/m}^2\text{s})(1.21\,\text{kJ/g}) \\ &= 25.4\,\text{kW/m}^2 \end{aligned}$$

Note that the re-radiation loss from the methanol surface at $T_{\mathrm{b}} = 64\,^{\circ}\text{C}$ is only about $0.3\,\text{kW/m}^2$.

Table 9.3 Asymptotic burning rates (from various sources)

	g/m^2 s
Polyvinyl chloride (granular)	16
Methanol	21
Flexible polyurethane (foams)	21–27
Polymethymethacrylate	28
Polystyrene (granular)	38
Acetone	40
Gasolene	48–62
JP-4	52–70
Heptane	66
Hexane	70–80
Butane	80
Benzene	98
Liquid natural gas	80–100
Liquid propane	100–130

The asymptotic burning rate behavior under saturated flame radiation conditions is a useful fact. It provides an upper limit, at pool-like fuel configurations of typically 1–2 m diameter, for the burning flux. Experimental values exist in the burning literature for liquids as well as solids. They should be thoughtfully used for design and analysis purposes. Some maximum values are listed in Table 9.3.

Equation (9.73) can be used to explain the burning behavior with respect to the roles of flame radiation and convection. It can also explain the effects of oxygen and the addition of external radiant heat flux. These effects are vividly shown by the correlation offered from data of Tewarson and Pion [18] of irradiated horizontal small square sheets of burning PMMA in flows of varying oxygen mole fractions. The set of results for steady burning are described by a linear correlation in $\dot{q}''_e$ and X_{O_2} for $L = 1.62\,\text{kJ/g}$ in Figure 9.16. This follows from Equation (9.73):

$$\dot{m}''_F - \dot{m}''_{F,\infty} = \frac{h_c\left(\frac{\lambda}{e^{\lambda}-1}\right)\left(\frac{\Delta h_c}{c_p r}\right)(1 - X_r)(Y_{O_2} - Y_{O_2,\infty}) + \dot{q}''_e}{L}$$

where $\dot{m}''_{F,\infty}$ is the burning rate under ambient air conditions with $Y_{O_2,\infty} = 0.233$ and $\dot{q}''_e = 0$. This result assumes that the mass transfer coefficient maintains a fairly constant value over this range of $\dot{m}''_F$, and that $\dot{q}''_f$ is constant for all conditions of oxygen and radiation. A valid argument can be made for the constancy of $\dot{q}''_f$ as long as the flame is tall ($L_f > 2D$ at least) and does not change color (the soot character is consistent). A tall gray cylindrical flame will achieve a constant emittance for $L_f > 2D$ with respect to its base.

9.6 Property Values for Burning Rate Calculations

The polymer PMMA has been used to describe many aspects of steady burning rate theory. This is because it behaves so ideally. PMMA decomposes to its base monomer; although it melts, its transition to the vaporized monomer is smooth; its decomposition

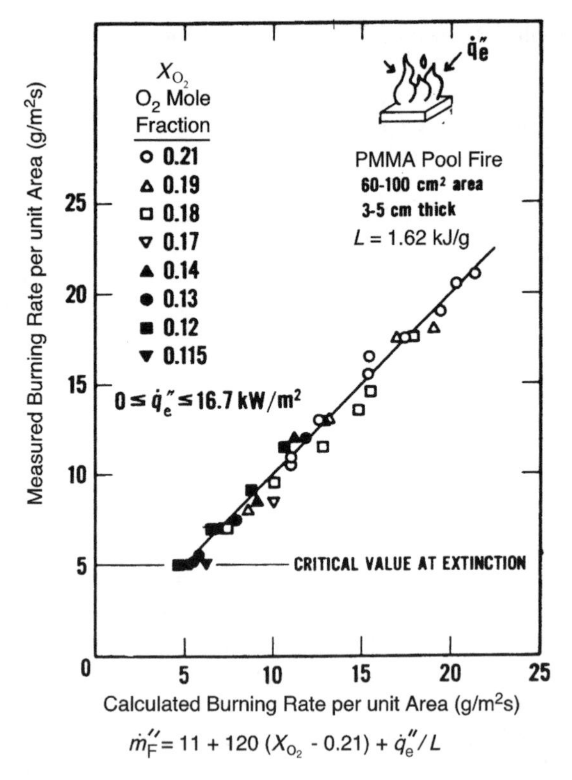

Figure 9.16 Pool fire burning rates for PMMA as a function of oxygen and external radiation [18]

kinetics give a reasonably constant surface temperature during burning; and steady burning is easy to achieve. However, all brands of PMMA are not identical and values of L may range from 1.6 to 2.8 kJ/g. As materials become more complex in their decomposition, the idealizations made in our theory cannot be fulfilled. Specifically, for charring materials, steady burning is not physically possible. Yet, provided we define consistent 'properties' for the steady burning theory, it can yield rational time-average valid burning rates for even charring materials. The approximate nature of this steady modeling approach must always be kept in mind and used appropriately.

We have already mentioned that practical data under fire conditions are commonly presented in terms of mass loss of the fuel package (i.e. Equations (9.3), (9.4) and (9.6)). Effectively, this means for fuels in natural fire scenarios, such as furnishings, composites, plastic commodities, etc., we must interpret the steady burning theory in the following manner:

1. Δh_c and L are both based on mass loss, i.e. kJ/g mass loss.
2. $Y_{F,o}$ is taken as unity.
3. r is evaluated by assuming that $\Delta h_c/r \approx 13\,\text{kJ/g}\,O_2$ used.
4. X_r, the radiative fraction, depends on flame size and may increase and then decrease with fire diameter. It can range from 0.10 to 0.60.

5. Fluid properties can be taken as that of air with or without attention to temperature effects. Because of the large temperature range found in combustion, true property effects with temperature are rarely perfectly taken into account. However, in keeping with standard heat transfer practices of adopting properties at a film temperature, the following are recommended:

Air at 1000 °C:

$$\begin{aligned} \rho &= 0.26\,\text{kg/m}^3 \\ c_p &= 1.2\,\text{kJ/kg K} \\ k &= 0.081\,\text{W/m K} \\ \nu &= 1.8 \times 10^{-4}\,\text{m}^2/\text{s} \\ Pr &= 0.70 \end{aligned}$$

9.7 Suppression and Extinction of Burning

Equation (9.73) offers a basis for addressing the effects of variables on a reduction in the burning rate. We have already seen in Figure 9.16 that the reduction of oxygen can reduce the burning rate and lead to an extinction value, namely about 5 g/m^2 s for PMMA. Moreover, the correlation shows that if the critical mass flux is indeed constant then there is a relationship between oxygen concentration (X_{O_2}) and external heat flux $(\dot{q}''_e)$ at extinction. The higher $\dot{q}''_e$, the lower X_{O_2} can be before extinction occurs. For $\dot{q}''_e = 0$, the extinction value for X_{O_2} by the correlation in Figure 9.16 is 0.16 or 16 %. Note that the ambient temperature T_∞ was about 25 °C in these experiments.

9.7.1 Chemical and physical factors

We will try to generalize these effects for suppression and will adopt a temperature criterion for the extinction of a diffusion flame. Clearly at extinction, chemical kinetic effects become important and the reaction 'quenches'. The heat losses for the specific chemical dynamics of the reaction become too great. This can be qualitatively explained in terms of Equation (9.12):

$$\frac{t_{\text{chem}}}{t_{\text{Diff}}} \approx \frac{(Re_{\text{chem}}^2 Pr^2)}{R_{\text{T}}} \tag{9.74}$$

where the dimensionless parameters are as follows:

1. Chemical Reynold's number,

$$Re_{\text{chem}} = \frac{u_\infty \delta_{\text{chem}}}{\nu}$$

$$\delta_{\text{chem}} = \sqrt{\frac{kT}{A\Delta h_c}}$$

where δ_{chem} is a reaction length scale, not unlike the reaction width of a premixed flame.

2. Dimensionless reaction rate parameter

$$R_T = \frac{E}{RT} e^{-E/(RT)}$$

3. Prandtl number, $Pr = \nu/\alpha$.

Extinction becomes likely as t_{chem}/t_{Diff} becomes large, or R_T is small or Re_{chem} is large.

Provided u_∞ is not as high and A (the pre-exponential coefficient, dependent on $Y_{F,o}$ and $Y_{O_2,\infty}$) is in a normal range for nonretarded hydrocarbon-based fuels, the principal factor controlling the relative times is R_T. The empirical rule for the limiting conditions to achieve the propagation of premixed flames is for $T \approx 1300\,°C$ at least. At lower temperatures, t_{chem} becomes too great with respect to t_{Diff} (or t_{mix}) and the reaction rate is too slow. This results because local heat loss dominates for the slow reaction and slows it further as T drops, causing a cascading downward spiral toward extinction.

On the other hand, a high value of u_∞ will achieve the same result. This can be referred to as a 'flame stretch' effect, usually expressed as a velocity gradient, e.g. u_∞/δ for a boundary layer. Blowing out a flame is representative of this effect – it also serves to separate the fuel from the oxygen by this velocity gradient across the flame sheet. For natural fire conditions flame stretch is relatively small compared with commercial combustors. However, flame stretch can still be a factor in turbulent fires, serving to locally extinguish the convoluted flame sheet in high shear regions. However, the fluctuating character of species and temperature fields in turbulent flames play a more complex role than described by flame stretch alone.

In addition, the effects of gas phase retardants can change both A and E. If E is increased, our critical temperature criterion for extinction must accordingly be increased to maintain effectively a critical constancy for E/T. These chemical effects are complex and specific, and we will not be able to adequately quantify them. It is sufficient to remember that both velocity (flame stretch) and chemistry (retardant kinetics) can affect extinction. We will only examine the temperature extinction criterion.

9.7.2 *Suppression by water and diluents*

Since water is used as a common extinguishing agent, it will be included explicitly in our analysis of suppression. Other physical (nonchemically acting) agent effects could also be included to any degree one wishes. We shall only address water. We assume water acts in two ways:

1. Droplets are evaporated in the flame.
2. Droplets are evaporated on the surface.

The addition of water vapor is not addressed. We would have to specifically consider the species conservation equations for at least the products CO_2 and H_2O. Only water that is evaporated is considered, realizing that much water can be put on a fire but only a fraction finds its mark. To consider the flame evaporation effect (1), the energy equation (Equation (9.23)) must contain the energy loss from the gas phase due to water droplet evaporation. The net energy release rate in the gas phase is

$$\dot{Q}'''_{\text{net}} = \dot{m}'''_{\text{F}} \Delta h_{\text{c}} - X_{\text{r}} \dot{m}'''_{\text{F}} \Delta h_{\text{c}} - \dot{m}'''_{\text{w}} L_{\text{w}} \tag{9.75}$$

where we have accounted for radiation loss by the flame radiation fraction X_{r} and the water energy loss by $\dot{m}'''_{\text{w}}$, the rate of water evaporated per unit volume in the flame, and L_{w}, the heat of gasification for water. We take $L_{\text{w}} = h_{\text{fg}} + c_p(T_{\text{b}} - T_\infty) = 2.6\,\text{kJ/g}$ for water. It is clear that the determination of $\dot{m}'''_{\text{w}}$ is not easy, but well-controlled experiments or CFD (computer fluid dynamic) models have the potential to develop this information. For now we simply represent the water heat loss in the flame as a fraction $(X_{\text{w,f}})$ of the flame energy,

$$\dot{Q}'''_{\text{net}} = (1 - X_{\text{r}} - X_{\text{w,f}}) \dot{m}'''_{\text{F}} \Delta h_{\text{c}} \tag{9.76}$$

From the surface evaporation effect (2), we reformulate the surface energy balance (Equation (9.26e)) as the heat transferred (net) must be used to evaporate the fuel and the water at the surface,

$$y = 0, \qquad k\frac{\text{d}T}{\text{d}y} = \dot{m}''_{\text{F}} L + \dot{m}''_{\text{w}} L_{\text{w}} \tag{9.77}$$

In the generalization leading to Equation (9.73) we can directly write

$$\boxed{\dot{m}''_{\text{F}} = \left(\frac{h_{\text{c}}}{c_p}\right) \ln(1 + B)}$$

where

$$\boxed{B = \frac{Y_{\text{O}_2,\infty}(1 - X_{\text{r}} - X_{\text{w,f}})\Delta h_{\text{c}}/r - c_p(T_{\text{v}} - T_\infty)}{L_{\text{m}}}} \tag{9.78}$$

and

$$\boxed{L_{\text{m}} = L - \frac{\dot{q}''_{\text{f,r}} + \dot{q}''_{\text{e}} - \sigma(T_{\text{v}}^4 - T_\infty^4) - \dot{m}''_{\text{w}} L_{\text{w}}}{\dot{m}''_{\text{F}}}}$$

In other words, we replace Δh_{c} by $(1 - X_{\text{r}} - X_{\text{w,f}})\Delta h_{\text{c}}$ and L by L_{m} everywhere in our stagnant layer solutions. The solution for flame temperature becomes, from Equation (9.56),

$$\boxed{c_p(T_{\text{f}} - T_\infty) = \frac{Y_{\text{F,o}}(1 - X_{\text{r}} - X_{\text{w,f}})\Delta h_{\text{c}} - L_{\text{m}} + c_p(T_{\text{v}} - T_\infty)}{1 + rY_{\text{F,o}}/Y_{\text{O}_2,\infty}}} \tag{9.79}$$

These equations give us the means to estimate physical effects by the reduction of oxygen or addition of water on suppression. The equation for flame temperature with a critical value selected as 1300 °C gives a basis to establish extinction conditions.

Example 9.2 Estimate the critical ambient oxygen mass fraction needed for the extinction of PMMA. No external radiation or water is added and $T_\infty = 25\,°\mathrm{C}$. Compute the mass burning flux just at extinction assuming $h_c = 8\,\mathrm{W/m^2\,K}$. For PMMA use a value of $L = 1.6\,\mathrm{kJ/g}$, $\Delta h_c = 25\,\mathrm{kJ/g}$ and $T_v = 360\,°\mathrm{C}$.

Solution At extinction we expect the flame to be small so we will assume no flame radiative heat flux and $X_r = 0$. This is not a bad approximation since near extinction soot is reduced and a blue flame is common. For PMMA, $Y_{F,o} = 1$. Substituting values into Equation (9.79), we obtain (approximating $L_m = L$)

$$(1.2 \times 10^{-3}\,\mathrm{kJ/g\,K})(1300 - 25)\mathrm{K} = \frac{(1)(1)25\,\mathrm{kJ/g} - 1.6\,\mathrm{kJ/g} + (1.2 \times 10^{-3}\,\mathrm{kJ/g\,K})(360 - 25)\mathrm{K}}{1 + (25/13)(1)/Y_{O_2,\infty}}$$

$$1.53 = \frac{23.8}{1 + 1.92/Y_{O_2,\infty}}$$

$$1.53 + \frac{2.94}{Y_{O_2,\infty}} = 23.8$$

$$Y_{O_2,\infty} = 0.132$$

or

$$X_{O_2,\infty} = (0.132)(29/32) = 0.12\,\mathrm{molar}$$

This molar value compares to 0.16 for correlation of the data in Figure 9.16. This follows from the correlation at zero external flux and for the critical mass flux at extinction, 5 g/m^2 s.

The burning rate is found from our basic general equation (Equation (9.78)), where

$$B = \frac{Y_{O_2,\infty}\Delta h_c/r - c_p(T_v - T_\infty)}{L} = \frac{(0.132)(13) - (1.2 \times 10^{-3})(360 - 25)}{1.6} = 0.821$$

Then

$$\dot{m}''_F = \frac{h_c}{c_p}\ln(1 + B) = \frac{(8\,\mathrm{W/m^2\,K})}{(1.2\,\mathrm{J/g\,K})}\ln(1.82) = 4.0\,\mathrm{g/m^2\,s} \quad \text{at extinction}$$

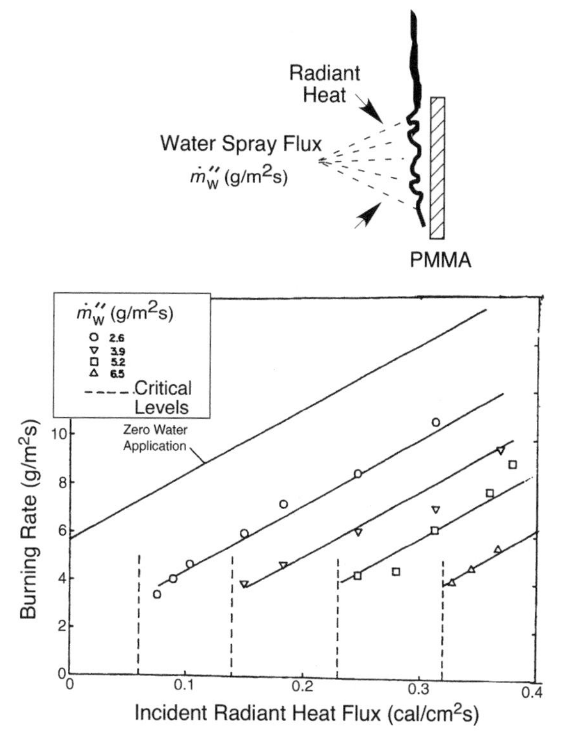

Figure 9.17 Burning rate of vertical PMMA slabs versus external radiant flux for various water application rates [19]

This compares to 5 g/m^2 s of Figure 9.16.

The effect of a water spray on the burning of a vertical PMMA slab was investigated by Magee and Reitz [19]. Their results show a linear behavior of the burning rate with external radiant heating and the applied water rate per unit area of PMMA surface as displayed in Figure 9.17. Assuming steady burning, which was sought in the experiments, we can apply Equations (9.78) and (9.79). We assume the following:

$$X_r = 0$$
$$X_{w,f} = 0$$

for simplicity, since we have no information on these quantities. However, for a wall flame, we expect them to be small. By Equation (9.78) we can write

$$\dot{m}''_F L = \frac{h_c}{c_p}\left(\frac{\dot{m}''_F c_p/h_c}{e^{\dot{m}''_F c_p/h_c} - 1}\right)\left[\frac{Y_{O_2,\infty}\Delta h_c}{r - c_p(T_v - T_\infty)}\right] + \dot{q}''_{f,r} + \dot{q}''_e - \sigma(T_v^4 - T_\infty^4) - \dot{m}''_w L_w \tag{9.80}$$

and from Equation (9.79) by substituting for L_m,

$$\dot{m}''_F\left[c_p(T_f - T_\infty)\left(1 + \frac{rY_{F,o}}{Y_{O_2,\infty}}\right) - c_p(T_v - T_\infty)\right]$$
$$= \dot{m}''_F Y_{F,o}\Delta h_c - \dot{m}''_F L + \dot{q}''_{f,r} + \dot{q}''_e - \sigma(T_v^4 - T_\infty^4) - \dot{m}''_w L_w$$

or

$$\dot{m}''_F\left[c_p(T_f - T_\infty)\left(1 + \frac{rY_{F,o}}{Y_{O_2,\infty}}\right) - c_p(T_v - T_\infty) + L - Y_{F,o}\Delta h_c\right]$$
$$= \dot{q}''_{f,r} + \dot{q}''_e - \sigma(T_v^4 - T_\infty^4) - \dot{m}''_w L_w \tag{9.81a}$$

With $T_f = 1300\,°C$, $T_\infty = 25\,°C$ and $Y_{F,o} = 1$, we have two equations in two unknowns: $\dot{m}''_w$ and $\dot{m}''_F$ for the extinction conditions. However, under only suppression of the burning rate, Equation (9.80) applies, giving a nearly linear result in terms of $\dot{q}''_e$ and $\dot{m}''_w$ as well as $Y_{O_2,\infty}$. The nonlinearity of the blocking effect can be ignored as a first approximation, and the experimental results can be matched to this theory. By subtracting Equation (9.81a) from Equation (9.80), we express the critical mass loss flux as

$$\dot{m}''_{F,crit}\left[Y_{F,o}\Delta h_c + c_p(T_v - T_\infty) - c_p(T_f - T_\infty)\left(1 + \frac{rY_{F,o}}{Y_{O_2,\infty}}\right)\right]$$
$$= \frac{h_c}{c_p}\left(\frac{\dot{m}''_F c_p/h_c}{e^{\dot{m}''_F c_p/h_c} - 1}\right)\left[\frac{Y_{O_2,\infty}\Delta h_c}{r - c_p(T_v - T_\infty)}\right] \tag{9.81b}$$

Ignoring the blocking factor this gives, for $h_c = 8\,\mathrm{W/m^2\,K}$,

$$\dot{m}''_{F,crit}\Big[(1)(25\,\mathrm{kJ/g}) + (1.2 \times 10^{-3}\,\mathrm{kJ/g\,K})(360 - 25)\,\mathrm{K}$$
$$- (1.2 \times 10^{-3})(1300 - 25)\left(1 + \frac{25/13}{0.233}\right)\Big]$$
$$= \left(\frac{8\,\mathrm{W/m^2\,K}}{1.2\,\mathrm{J/g\,K}}\right)(1)[0.233(13) - (1.2 \times 10^{-3})(260 - 25)]$$
$$11.24\,\dot{m}''_{F,crit} = 17.51$$
$$\dot{m}''_{F,crit} = 1.56\,\mathrm{g/m^2\,s}$$

To have this result match the data of Magee and Reitz [19] of about 4 $\mathrm{g/m^2s}$ at extinction requires a value for h_c of about 16–20 $\mathrm{W/m^2\,K}$ – a reasonable selection over the original estimate of 8.

The critical water flow rate required for extinguishment can be found as well. To match the data more appropriately, select 5.7 $\mathrm{g/m^2\,s}$ for $\dot{m}''_F$ corresponding to $\dot{m}''_w$ and $\dot{q}''_e$ equal to 0 from Figure 9.17. This requires the total flame net surface heat flux to be, for $L = 1.6\,\mathrm{kJ/g}$,

$$\dot{q}''_{f,net} = (5.7\,\mathrm{g/m^2\,s})(1.6\,\mathrm{kJ/g}) = 9.12\,\mathrm{kW/m^2}$$

Therefore, if this flame heat flux does not appreciably change, reasonable for a vertical surface, from Equation (9.80)

$$\dot{m}''_{\rm w} L_{\rm w} = \dot{q}''_{\rm f,net} + \dot{q}''_{\rm e} - \dot{m}''_{\rm F} L$$

or

$$\dot{m}''_{\rm w\,crit} = \frac{9.12 + \dot{q}''_{\rm e} - (1.56\,{\rm g/m^2\,s})(1.6\,{\rm kJ/g})}{2.6\,{\rm kJ/g}}$$
$$= 0.38\,\dot{q}''_{\rm e} + 2.55$$

Accordingly, at a water application of 5.2 g/m^2 s (0.0076 gpm/ft^2), the critical (minimum) external flux to allow the flame to survive is

$$\dot{q}''_{\rm e} = \frac{5.2 - 2.55}{0.38} = 7.0\,{\rm kW/m^2}$$

This is roughly 0.23 cal/cm^2 s or 9.6 kW/m^2, suggested from the data of Magee and Reitz. It demonstrates that the theory does well at representing extinction. It should also be noted that the range of water flux for extinguishment is roughly 10–100 times smaller than required by common sprinkler design specifications. The Magee application rates are about 0.005 gpm/ft^2 while common prescribed sprinkler densities range from 0.05 gpm/ft^2 (light) to 0.5 gpm/ft^2 (extremely hazardous).

9.8 The Burning Rate of Complex Materials

A burning rate of common materials and products in fire can only be specifically and accurately established through measurement. Estimates from various sources are listed in Table 9.4. Also, oxygen consumption calorimeters are used to measure the energy release rate of complex fuel packages directly. In most cases, the results are a combination of effects: ignition, spread and burning rate.

Table 9.4 Estimates of burning rates for various fuel systems [20]

		g/s
Waste containers	(office size)	2–6
Waste containers	(tall industrial)	4–12
Upholstered chairs (single)		10–50
Upholstered sofas		30–80
Beds/mattresses		40–120
Closet, residential		~ 40
Bedroom, residential		~ 130
Kitchen, residential		~ 180
House, residential		$\sim 4 \times 10^4$

The burning of wood and charring materials are not steady, as we have portrayed in our theory up to now. Figure 9.4 shows such burning behavior in which

$$\dot{m}_F'' \sim \frac{1}{\sqrt{t}} \quad (t \text{ is time})$$

as the charring effects become important. However, the average or peak burning under flaming conditions has been described for wood simply by

$$\overline{\dot{m}_{\text{wood}}''} = C_w D^{-n} \tag{9.82}$$

where n has been found empirically to be about $\frac{1}{2}$ and D is the diameter or equivalent thickness dimension of a stick. This result is empirical with C_w as $0.88\,\text{mg/cm}^{1.5}$ s for Sugar pine and $1.03\,\text{mg/cm}^{1.5}$ s for Ponderosa pine [21]. Even a C_w of 1.3 was found to hold for high density ($0.34\,\text{g/cm}^3$) rigid polyurethane foam [22].

Although Equation (9.82) appears to hold for a single stick, it is not necessarily the case. These charring materials with an effective B of roughly $(0.233)(12)/6 \approx 0.46$ cannot easily sustain flaming without an external heat flux. In wood fuel systems, this is accomplished by the arrangement of logs in a fireplace, the thermal feedback in room fires or in furniture/pallet style arrays, known as 'cribs'. A wood crib is an ordered array of equal sized wood sticks in equally spaced layers. The sticks in the wood burn either according to Equation (9.82) and the exposed stick surface area (A) or by the heat flux created within the crib by the incomplete combustion of pyrolyzed fuel in a low porosity crib. The average gas temperature in the crib is related to the air mass flow rate for this ventilation-limited latter case. In that case, the crib average-peak flaming mass loss rate has been experimentally found to be proportional to

$$\dot{m}_{\text{air}} \sim \rho_{\text{air}} A_o \sqrt{H}$$

where

H = height of the crib
A_o = cross-sectional area of the vertical crib shafts

Thus, it is found that the time-average burning rate can be represented by

$$\frac{\overline{\dot{m}}_{\text{crib}}}{C_w D^{-1/2} A} = f\left(\frac{\rho_{\text{air}} A_o \sqrt{H}}{C_w D^{-1/2} A}\right) = f\left(\frac{A_o \sqrt{sD}}{A}\right) \tag{9.83}$$

where s is the spacing between the sticks. This is shown in Figure 9.18 by the correlation established by Heskestad [23] and compared to laboratory data taken by students in a class at the University of Maryland [24]. Note that for a tall crib its height could be a factor according to the theory, but it is not included in the correlation.

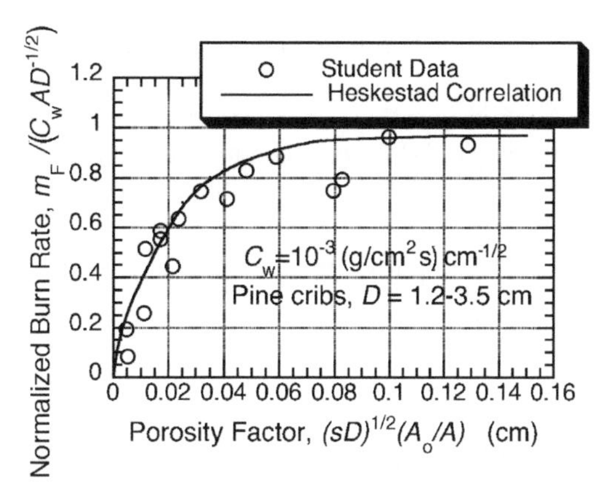

Figure 9.18 Burning rate correlation for wood cribs according to Heskestad [23] with student data [24]

9.9 Control Volume Alternative to the Theory of Diffusive Burning

This chapter began by discussing the steady burning of liquids and then extended that theory to more complex conditions. As an alternative approach to the stagnant layer model, we can consider the more complex case from the start. The physical and chemical phenomena are delineated in macroscopic terms, and represented in detailed, but relatively simple, mathematics – mathematics that can yield algebraic solutions for the more general problem.

The approach is to formulate the entire burning problem using conservation laws for a control volume. The condensed phase will use control volumes that move with the vaporization front. This front is the surface of a regressing liquid or solid without char, or it is the char front as it extends into the virgin material. The original thickness, l, does not change. While the condensed phase is unsteady, the gas phase, because of its lower density, is steady or quasi-steady in that its steady solution adjusts to the instantaneous input of the condensed phase.

The condensed phase (solid or liquid) is considered as a general phase that will vaporize to a gaseous fuel with a mass fraction $Y_{F,o}$, and can possibly form char with properties designated by $()_c$. The virgin fuel properties are designated without any subscript.

Water is considered and is sprayed into the system, some reaching the surface $(1\text{-}X_w)$ and some being evaporated in the flame (X_w = fraction of water evaporated in the flame). The water spray considered is that which is fully evaporated either in the flame or on the surface. Of course in reality some water applied never reaches the fuel, and this must ultimately be considered as an efficiency factor. Therefore the water considered is that which completely contributes to suppression. The water is applied per unit area of the condensed phase surface and is assumed not to affect the flow rate of gaseous fuel from the surface. Any blockage effect of the water, or any other extinguishing agent that might be considered, is an effect ignored here, but might be included in general. This effect must be considered for agents that act by blocking the surface as well as absorbing energy.

The control volume relationships that are considered are based on a moving control volume surface with velocity, $\boldsymbol{w}$, while the medium velocity is $\boldsymbol{v}$. The conservation of mass is given as

$$\frac{\mathrm{d}}{\mathrm{d}t}\iiint_{\mathrm{CV}} \rho \mathrm{d}V + \oiint_{\mathrm{CS}} \rho(\boldsymbol{v}-\boldsymbol{w})\cdot\boldsymbol{n}\,\mathrm{d}S = 0 \tag{9.84}$$

The conservation of species is

$$\frac{\mathrm{d}}{\mathrm{d}t}\iiint_{\mathrm{CV}} \rho Y_i\,\mathrm{d}V + \oiint_{\mathrm{CS}} \rho Y_i(\boldsymbol{v}-\boldsymbol{w})\cdot\boldsymbol{n}\,\mathrm{d}S = \iiint_{\mathrm{CV}} (-\dot{m}_{\mathrm{F}}''')s_i\,\mathrm{d}V \tag{9.85}$$

where Y_i is the mass fraction of species i, $\dot{m}_{\mathrm{F}}'''$ is the consumption rate of fuel per unit volume and s_i is the stoichiometric coefficient for the ith species per unit mass of fuel reacted. The conservation of energy is based on the sensible enthalpy per unit mass of the medium, $h(= c_p T$, based on Equation (3.40)):

$$\frac{\mathrm{d}}{\mathrm{d}t}\iiint_{\mathrm{CV}} \rho h \mathrm{d}V + \oiint_{\mathrm{CS}} \rho h(\boldsymbol{v}-\boldsymbol{w})\cdot\boldsymbol{n}\mathrm{d}S - V\frac{\mathrm{d}p}{\mathrm{d}t} = \iiint_{\mathrm{CV}} \dot{m}_{\mathrm{F}}'''\Delta h_{\mathrm{c}}\,\mathrm{d}V - \oiint_{\mathrm{CS}} \dot{\boldsymbol{q}}''\cdot\boldsymbol{n}\,\mathrm{d}S \tag{9.86}$$

where the pressure term is neglected in the following since the pressure is constant, except for minor hydrostatic variations, in both the condensed and the gas phases. The unit normal, $\boldsymbol{n}$, is taken as positive in the outward direction from the control volume surface, S. Hence the last term in Equation (9.86) is the net heat added to the control volume.

The specific geometry of the problem is illustrated in Figure 9.19. The freestream is designatated by $()_{\infty}$ and the backface by $()_{\mathrm{o}}$. The char front, or surface regression if no char, velocity is given by $\mathrm{d}\delta_{\mathrm{c}}/\mathrm{d}t$. First we will consider the condensed phase and three control volumes: the char, the vaporizing interfacial surface and the virgin material. Two variations of the virgin material control volume will be used: one where the process is steady within the material and there is no influence of the backface (thermally thick) and the other where the backface has an influence (thermally thin). A separate control volume will be considered for the water in which liquid water arrives at the control volume and is converted into vapor at the boiling point of water, T_{b}. The water is assumed to evaporate

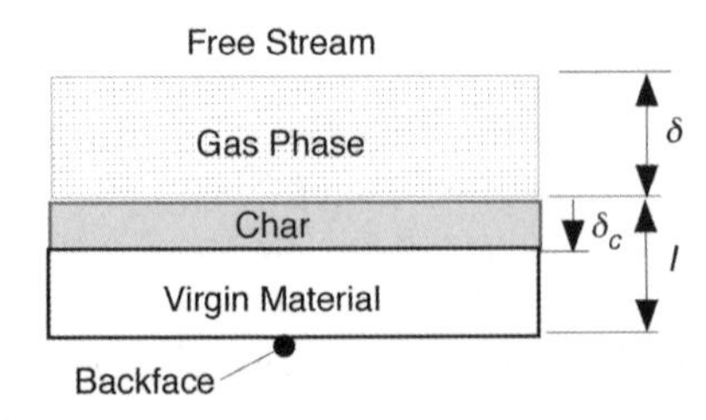

Figure 9.19 General gas phase and solid phase configuration

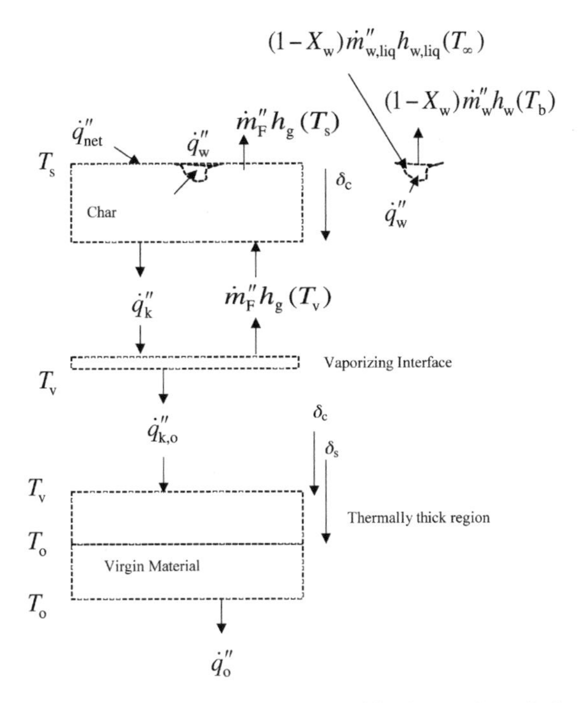

Figure 9.20 Control volumes selected in the condensed phase

in a steady process. Finally, a control volume will be considered for the gas phase whose thickness is representative of a constant boundary layer or flame zone.

9.9.1 Condensed phase

The control volumes for the condensed phase are shown in Figure 9.20. The control volumes are selected to include the water evaporated on the surface that is depicted as a region, but is assumed uniformly distributed over the surface. The surface area is really a differential, small area. The char control volume only contains the char at a constant density, ρ_c. Similarly, the virgin material has a constant density, ρ. The interfacial control volume has zero volume, and it is assumed that the transition from the condensed phase to the gaseous fuel (and char) occurs quickly in this small region. It is generally found that for high heat flux conditions in fire, this is a good assumption, but it is recognized that this process is distributed over a region. In the virgin solid, there are two alternative control volumes: (a) the entire region, which includes the effect of the backface, and can represent a thermally thin solid, or (b) the region bounded by δ_c and δ_s, which both move at the same speed $d\delta_c/dt$. The material within this volume is fixed and unchanged with time as long as δ_s does not reach the backface. As long as this condition is met, the solid can be considered thermally thick. The time t over which this condition is met is fulfilled when $l - \delta_c \geq \sqrt{\alpha t}$, where α is the thermal diffusivity. It should be noted that the surfaces coinciding with δ_c are moving control surfaces with $\boldsymbol{w} = d\delta_c/dt\,\boldsymbol{j}$, where $\boldsymbol{j}$ is positive downward in the y direction. The details follow. The symbol notation is that used throughout the text. The analysis follows:

Conservation of energy for the char

$$\frac{\mathrm{d}}{\mathrm{d}t}\int_0^{\delta_c} \rho_c h_c \,\mathrm{d}y + \sum_{\mathrm{CS}} \rho h (\boldsymbol{v} - \boldsymbol{w}) \cdot \boldsymbol{n} = \sum_{\mathrm{CS}} \dot{q}''_{\mathrm{added}}$$

$$\int_0^{\delta_c} \frac{\partial}{\partial t}(\rho_c h_c)\,\mathrm{d}y + \rho_c h_c(T_v)\frac{\mathrm{d}\delta_c}{\mathrm{d}t} + \rho_c h_c(T_v)\left(0 - \frac{\mathrm{d}\delta_c}{\mathrm{d}t}\right)$$
$$+ \dot{m}''_F(h_g(T_s) - h_g(T_v)) = \dot{q}''_{\mathrm{net}} - \dot{q}''_k - \dot{q}''_w \tag{9.87}$$

Conservation of energy for the water

$$0 + (1 - X_w)\dot{m}''_w h_{w,g}(T_b) - (1 - X_w)\dot{m}''_w h_w(T_\infty) = \dot{q}''_w \tag{9.88}$$

The enthalpies can be expressed in terms of the heat of vaporization of the water:

$$h_{w,g}(T_b) - h_w(T_\infty) = c_{w,\mathrm{lig}}(T_b - T_\infty) + h_{w,\mathrm{fg}} \equiv L_w \tag{9.89}$$

where $h_{w,\mathrm{fg}}$ is the heat of vaporization.

Conservation of energy for the vaporization interface

$$0 + \dot{m}''_F h_g(T_v) + \rho_c\left(0 - \frac{\mathrm{d}\delta_c}{\mathrm{d}t}\right)(-1)h_c(T_v) + \rho\left(0 - \frac{\mathrm{d}\delta_c}{\mathrm{d}t}\right)(+1)h(T_v) = \dot{q}''_k - \dot{q}''_{k,o} \tag{9.90}$$

Conservation of energy for the virgin solid including the backface

$$\frac{\mathrm{d}}{\mathrm{d}t}\int_{\delta_c}^{l} \rho h \,\mathrm{d}y + \rho\left(0 - \frac{\mathrm{d}\delta_c}{\mathrm{d}t}\right)(-1)h(T_v) + 0 = \dot{q}''_{k,o} - \dot{q}''_o \tag{9.91}$$

Differentiating the integral realizing δ_c depends on time,

$$\int_{\delta_c}^{l} \frac{\partial}{\partial t}(\rho h)\,\mathrm{d}y = \dot{q}''_{k,o} - \dot{q}''_o$$

Conservation of mass for the vaporizing interface

$$0 + \dot{m}''_F + \rho_c\left(0 - \frac{\mathrm{d}\delta_c}{\mathrm{d}t}\right)(-1) + \rho\left(0 - \frac{\mathrm{d}\delta_c}{\mathrm{d}t}\right)(+1) = 0$$

or

$$\dot{m}''_F = (\rho - \rho_c)\frac{\mathrm{d}\delta_c}{\mathrm{d}t} \tag{9.92}$$

Conservation of energy in the virgin solid for steady, thermally thick conditions

$$l - \delta_c \geq \sqrt{\alpha t}$$

In this case there is no heat transfer at $y = \delta_s$:

$$\frac{d}{dt}\int_{\delta_c}^{\delta_s} \rho h \, dy + \rho\left(0 - \frac{d\delta_c}{dt}\right)(-1)h(T_v) + \rho\left(0 - \frac{d\delta_c}{dt}\right)(+1)h(T_o) = \dot{q}''_{k,o} \tag{9.93}$$

Combining with Equation (9.92) and defining the char fraction, $X_c = \rho_c/\rho$ gives

$$\frac{\dot{m}''_F}{1 - X_c} c(T_v - T_o) = \dot{q}''_{k,o}$$

where c is the specific heat of the solid.

The above equations are combined to form the condition for the burning rate at the surface, $y = 0$. This becomes a boundary condition for the gas phase where combustion is occurring. This is done best by adding Equations (9.87), (9.88), (9.90) and (9.91). In doing so, a grouping of the enthalpies at the vaporization interface arises which represents the enthalpy of vaporization defined in terms of energy per unit mass of the virgin solid (by convention):

$$\Delta h_{py} \equiv (1 - X_c)h_g(T_v) + X_c h_c(T_v) - h(T_v) \tag{9.94}$$

The final result for the burning rate can be expressed in terms of the heat of gasification of the virgin solid, L_o:

$$L_o \equiv \Delta h_{py} + c(T_v - T_o) \tag{9.95}$$

The net surface heat flux is given explicitly as

$$\dot{q}''_{net} = \dot{q}''_{f,c} + \dot{q}''_{f,r} + \dot{q}''_{e,r} - \sigma(T_s^4 - T_\infty^4) \tag{9.96}$$

where

$$\dot{q}''_{f,c} = \text{flame convective heat flux}$$
$$\dot{q}''_{f,r} = \text{flame radiative heat flux}$$
$$\dot{q}''_{e,r} = \text{the external raditive heat flux}$$
$$\sigma(T_s^4 - T_\infty^4) = \text{re-radiative loss}$$

$$\frac{\dot{m}''_F L_o}{1 - X_c} = \dot{q}''_{f,c} + \dot{q}''_{f,r} + \dot{q}''_{e,r} - \sigma(T_s^4 - T_\infty^4) - \int_0^{\delta_c} \frac{\partial}{\partial t}(\rho_c h_c)dy - (1 - X_w)\dot{m}''_w L_w$$
$$+ \left[\frac{\dot{m}''_F c(T_v - T_o)}{1 - X_c} - \int_{\delta_c}^{l} \frac{\partial}{\partial t}(\rho h)dy - \dot{q}''_o\right] \tag{9.97}$$

where the quantity in the [ ] is zero for the steady, thermally thick solid case. If the virgin solid is thick enough, it can equilibrate to the steady state, but the increase in the surface temperature as the char layer thickens is a continuing unsteady effect. The unsteady term involving the char enthalpy will not be very large since the char density is small. The decrease in the burning rate for the charring case is apparent, especially considering the surface temperature increase over T_v to T_s and the reduction due to the char fraction. One can also see why the experimentally measured time-average, effective heat of gasification of charring materials (L) from mass loss or calorimetry apparatuses is higher than that of noncharring solids, i.e. $L = L_o/(1 - X_c)$, Equation (9.7).

9.9.2 Gas phase

The control volume in the gas phase (Figure 9.19) is considered as stationary, steady and responds fast to any changes transmitted from the condensed phase. A portion of the water used in suppression is evaporated in the flame. The flame is a thickness of δ_R within the boundary layer. Kinetic effects are important in this region where essentially all of the combustion occurs. The conservation relationships follow:

Conservation of mass

$$\dot{m}''_P = \dot{m}''_F + \dot{m}''_\infty + X_w \dot{m}''_w + (1 - X_w)\dot{m}''_w \tag{9.98}$$

emphasizing the separate contributions of water vapor due evaporation from the flame and the surface.

Conservation of species
The chemical equation is represented in terms of the stoichiometric mass ratio of oxygen to fuel reacted, r:

$$1\,\mathrm{g\,F} + r\,\mathrm{g\,O_2} \rightarrow (r+1)\,\mathrm{g\,P} \tag{9.99}$$

(If this reaction were in a turbulent flow, excess oxidizer will be present due to the 'unmixedness' of the fuel and oxygen in the turbulent stream. Hence, r will effectively be larger for a turbulent reaction than the molecular chemical constraint.)

Assuming a uniform reaction rate in the flame, we obtain:

$$\text{Fuel}: \quad -Y_{F,o}\dot{m}''_F = -\dot{m}'''_F \delta_R \tag{9.100}$$

$$\text{Oxygen}: \quad -Y_{ox,\infty}\dot{m}''_\infty = -r\dot{m}'''_F \delta_R \tag{9.101}$$

Kinetics
Arrhenius kinetics apply and these are represented as

$$\dot{m}'''_F = A\,\mathrm{e}^{-E/(RT)} \tag{9.102}$$

The thickness of the flame zone can be estimated in a manner similar to that used for the premixed flame. A control volume is selected between the condensed phase surface and the point just before the reaction zone. This is the 'preheat zone', which is heated to T_i.

Heat transfer is received in this zone from the flame by conduction. The result is

$$\dot{m}_{\mathrm{F}}'' c_p (T_{\mathrm{i}} - T_{\mathrm{s}}) = k(T_{\mathrm{f}} - T_{\mathrm{i}})\delta_{\mathrm{R}} \tag{9.103}$$

Equation (9.103) can give an estimate for the flame thickness that is similar to that of the premixed flame by using Equation (9.100) and an estimate of 3 for the temperature ratio:

$$\delta_{\mathrm{R}} \approx \left(\frac{3kY_{\mathrm{F,o}}}{c_p \dot{m}_{\mathrm{F}}'''}\right)^{1/2} \tag{9.104}$$

We will avoid the kinetics subsequently in solving for the burning rate and the flame temperature, but it is important in understanding extinction. The solutions for the burning rate and flame temperature are quasi-steady solutions for the gas phase. Where steady conditions do not exist, we have extinction ensuing. Extinction will be addressed later in Section 9.10.

Conservation of energy

The products leave the flame at the flame temperature, T_{f}, and water vapor flows in from the water evaporated on the surface and water evaporated in the flame. The control volume excludes the water droplets in the flame which receive heat, $\dot{q}_{\mathrm{f,w}}''$, from the flame control volume. As computed in Equation (9.88) for the fraction of water evaporated now in the flame,

$$\dot{q}_{\mathrm{f,w}}'' = X_{\mathrm{w}} \dot{m}_{\mathrm{w}}'' h_{\mathrm{w,g}}(T_{\mathrm{b}}) + X_{\mathrm{w}} \dot{m}_{\mathrm{w}}'' h_{\mathrm{w}}(T_{\infty}) = X_{\mathrm{w}} \dot{m}_{\mathrm{w}}'' L_{\mathrm{w}} \tag{9.105}$$

The energy balance for the flame control volume is given as

$$\begin{aligned} &\dot{m}_{\mathrm{P}}'' c_p T_{\mathrm{f}} - \dot{m}_{\infty}'' c_p T_{\infty} - X_{\mathrm{w}} \dot{m}_{\mathrm{w}}'' c_p T_{\mathrm{b}} - \dot{m}_{\mathrm{F}}'' c_p T_{\mathrm{f}} - (1 - X_{\mathrm{w}}) \dot{m}_{\mathrm{w}}'' c_p T_{\mathrm{b}} \\ &= -\dot{q}_{\mathrm{f,c}}'' - \dot{q}_{\mathrm{f,w}}'' - X_{\mathrm{r}} \dot{m}_{\mathrm{F}}''' \delta_{\mathrm{R}} \Delta h_{\mathrm{c}} + \dot{m}_{\mathrm{F}}''' \delta_{\mathrm{R}} \Delta h_{\mathrm{c}} \end{aligned}$$

or

$$\dot{m}_{\mathrm{P}}'' c_p T_{\mathrm{f}} - \dot{m}_{\infty}'' c_p T_{\infty} - \dot{m}_{\mathrm{w}}'' c_p T_{\mathrm{b}} - \dot{m}_{\mathrm{F}}'' c_p T_{\mathrm{f}} = -\dot{q}_{\mathrm{f,c}}'' - X_{\mathrm{w}} \dot{m}_{\mathrm{w}}'' L_{\mathrm{w}} + (1 - X_{\mathrm{r}}) \dot{m}_{\mathrm{F}}''' \delta_{\mathrm{R}} \Delta h_{\mathrm{c}} \tag{9.106}$$

In this case, all of the radiation loss from the flame is accounted for by X_{r}, including that received by the surface of the condensed phase. Therefore flame radiation heat flux does not show up here.

Equation (9.106) can be modified by eliminating all of the mass terms in favor of the burning rate (flux) term. This gives

$$\begin{aligned} &\dot{m}_{\mathrm{F}}'' \left\{ c_p (T_{\mathrm{f}} - T_{\infty}) - c_p (T_{\mathrm{s}} - T_{\infty}) + \left(\frac{rY_{\mathrm{F,o}}}{Y_{\mathrm{ox},\infty}}\right) c_p (T_{\mathrm{f}} - T_{\infty}) \right. \\ &\left. + \left(\frac{\dot{m}_{\mathrm{w}}''}{\dot{m}_{\mathrm{F}}''}\right) [c_p (T_{\mathrm{f}} - T_{\infty}) - c_p (T_{\mathrm{b}} - T_{\infty}) + X_{\mathrm{w}} L_{\mathrm{w}}] - (1 - X_{\mathrm{r}}) Y_{\mathrm{F,o}} \Delta h_{\mathrm{c}} \right\} \\ &= -\dot{q}_{\mathrm{f,c}}'' \end{aligned} \tag{9.107}$$

This equation along with Equation (9.97) allows us to eliminate the convective flame heat flux to develop an equation for the flame temperature. This equation will still contain the burning rate in terms of the effective heat of gasification, L_m. From Equation (9.97), we define L_m as the 'modified' heat of gasification by the following:

$$\begin{aligned}\dot{q}''_{f,c} &= \dot{m}''_F L_m \\ &\equiv \dot{m}''_F\left(\frac{L_o}{1-X_c} - \frac{1}{\dot{m}''_F}\left\{\dot{q}''_{f,r} + \dot{q}''_{e,r} - \sigma(T_s^4 - T_\infty^4)\right.\right. \\ &\quad - \int_0^{\delta_c}\frac{\partial}{\partial t}(\rho_c h_c)\,dy - (1-X_w)\dot{m}''_w L_w \\ &\quad \left.\left. + \left[\frac{\dot{m}''_F c(T_v - T_o)}{1-X_c} - \int_{\delta_c}^{l}\frac{\partial}{\partial t}(\rho h)\,dy - \dot{q}''_o\right]\right\}\right)\end{aligned} \tag{9.108}$$

This shows that this modified heat of gasification includes all effects that augment or reduce the mass loss rate. Recall that the term in the [ ] becomes zero if the solid is thermally thick and the virgin solid equilibrates to the steady state. Equating Equations (9.107) and (9.108) gives an equation for the flame temperature:

$$\begin{aligned}c_p(T_f - T_\infty) &= \frac{1}{1 + (rY_{F,o}/Y_{ox,\infty}) + (\dot{m}''_w/\dot{m}''_F)} \\ &\times\left\{(1-X_r)Y_{F,o}\Delta h_c - \frac{L_o}{1-X_c} + c_p(T_s - T_\infty) - \left(\frac{\dot{m}''_w}{\dot{m}''_F}\right)[L_w + c_p(T_b - T_\infty)]\right. \\ &\quad [\dot{q}''_{f,r} + \dot{q}''_{e,r} - \sigma(T_s^4 - T_\infty^4) - \int_0^{\delta_c}(\partial/\partial t)(\rho_c h_c)\,dy] \\ &\quad \left. + \frac{-[\dot{m}''_F c(T_v - T_o)/(1-X_c) - \int_{\delta_c}^{l}(\partial/\partial t)(\rho h)\,dy - \dot{q}''_o]}{\dot{m}''_F}\right\}\end{aligned} \tag{9.109}$$

This exactly reduces to the stagnant layer one-dimensional pure diffusion problem (Equation (9.56)).

In the stagnant layer solution, an accounting for the water evaporated in the flame and the radiation loss would have modified the energy equation as (Equation (9.23)).

$$\begin{aligned}c_p\dot{m}''_F\frac{dT}{dy} - k\frac{d^2T}{dy^2} &= \dot{m}'''_F(1-X_r)\Delta h_c - X_w\dot{m}'''_w L_w \\ &= \dot{m}'''_F\left[(1-X_r)\Delta h_c - \left(\frac{\dot{m}''_w}{\dot{m}''_F}\right)X_w L_w\right]\end{aligned} \tag{9.110}$$

The term in the bracket can be regarded as an equivalent heat of combustion for the more complete problem. If this effect is followed through in the stagnant layer solution of the ordinary differential equations with the more complete boundary condition given by

Equation (9.108), then the B number would become

$$B = \frac{Y_{\text{ox},\infty}[(1 - X_\text{r})\Delta h_\text{c} - (\dot{m}''_\text{w}/\dot{m}''_\text{F})X_\text{w}L_\text{w}]/r - c_p(T_\text{s} - T_\infty)}{L_\text{m}} \tag{9.111}$$

The burning rate is then given as

$$\dot{m}''_\text{F} = \frac{h_\text{c}}{c_p}\ln(1 + B) \tag{9.112a}$$

or

$$\frac{\dot{m}''_\text{F}c_p}{h_\text{c}} \equiv \gamma = \frac{\gamma}{\text{e}^\gamma - 1}B \tag{9.112b}$$

and

$$\frac{\dot{m}''_\text{F}c_p x}{k} = \frac{h_\text{c}x}{k}\frac{\gamma}{\text{e}^\gamma - 1}B \tag{9.112c}$$

Alternatively, in Equation (9.108), the convective heat flux could be approximated as

$$\dot{q}''_\text{c} = h_\text{c}(T_\text{f} - T_\text{s}) \tag{9.113}$$

and we obtain an approximate result for the burning rate using Equation (9.109). In summary, Equations (9.111) and (9.112) represent a more complete solution for the burning rate, and Equation (9.109) gives the corresponding flame temperature.

9.10 General Considerations for Extinction Based on Kinetics

Let us reconsider the critical flame temperature criterion for extinction. Williams [25], in a review of flame extinction, reports the theoretical adiabatic flame temperatures for different fuels in counter-flow diffusion flame experiments. These temperatures decreased with the strain rate (u_∞/x), and ranged from 1700 to 2300 K. However, experimental measured temperatures in the literature tended to be much lower (e.g. Williams [25] reports 1650 K for methane, 1880 K for iso-octane and 1500 K for methylmethracrylate and heptane). He concludes that 1500 ± 50 K can represent an approximate extinction temperature for 'many carbon–hydrogen–oxygen fuels burning in oxygen–nitrogen mixtures without chemical inhibitors'.

Macek [27] examined the flammability limits for premixed fuel–air systems and small diffusion flames under natural convection conditions, and computed the equilibrium flame temperature for these flame systems. Data were considered for the alkanes and alcohols at their measured premixed lower flammability limits, and at their measured

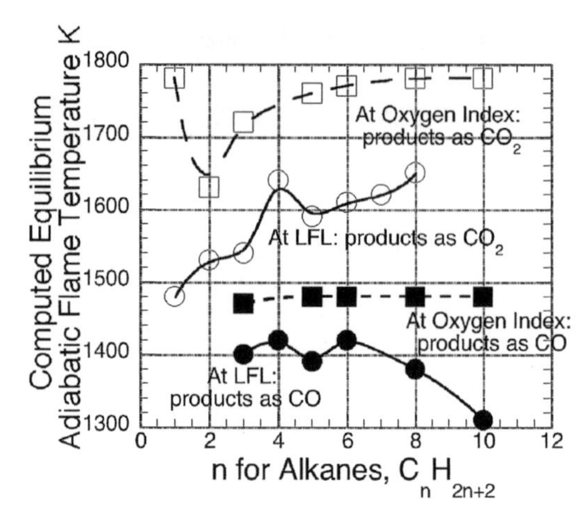

Figure 9.21 Adiabatic flame temperatures at extinction for premixed and diffusion flames (from Macek [26])

oxygen concentration for extinction in the standard limiting-oxygen index (LOI) method. The LOI method is a small-scale test method for diffusion flames in an inverted hemispherical burner arrangement within a chamber of controlled oxygen. Figure 9.21 shows the computed adiabatic flame temperatures at these limit conditions for complete combustion and for the alternative extreme condition of burning only to CO and water. Macek argues that extinction of the diffusion flames should occur somewhere between the two limits of stoichiometry, and therefore the implication is that the critical flame temperature is about 1600 K, as generally suggested for the premixed flames in the literature. This criterion is often applied for premixed flames in which the adiabatic flame temperature for constant specific heat follows from Equation (4.55) as

$$c_p(T_{\mathrm{f,ad}} - T_\infty) = Y_{\mathrm{F,o}} \Delta h_{\mathrm{c}} \tag{9.114}$$

where the fuel mass fraction supplied is $Y_{\mathrm{F,o}}$. The corresponding expression for a diffusion flame is more complicated, as shown by Equation (9.109). The LOI conditions correspond to, at least, near-stoichiometric fuel concentrations as characteristic of a diffusion flame, and hence the temperatures for extinction are higher than those for the LFL premixed flames in Figure 9.21. The value of 1600 K appears to be consistent with the results in Figure 9.21, and suggests that premixed flames may burn more completely than diffusion flames at extinction. Since the heat transfer processes are not influenced by whether the system is premixed or not, the same kinetics should apply to both flame configurations. Flames in natural fire conditions are affected by gravity-induced flows that are relatively low speed, with their inherent turbulent structure. Therefore, it might be assumed that the flow field has a small effect on natural convection flames, and the critical extinction temperature is more consistent with the 1200–1300 °C suggested by Williams [25]; we will use 1300 °C ($\sim$ 1600 K) in the following for diffusion flames. Indeed, Roberts and Quince [27] concluded 1600 K is an approximate exinction limit as they successfully predicted the liquid temperature needed for sustained burning (firepoint). They equated b_{OF} and b_{OT} and used the Clausius-Clapeyron equation along with this critical flame temperature.

9.10.1 A demonstration of the similarity of extinction in premixed and diffusion flames

The control volume analysis of the premixed flame of Section 4.5.4 can be used together with the analysis here in Section 9.9 for the diffusion flame to relate the two processes. We assume the kinetics is the same for each and given as in Equation (9.102). Since we are interested in extinction, it is reasonable to assume the heat loss from the flame to be by radiation from an optically thin flame of absorption coefficient, κ:

$$\dot{q}''_{\text{loss}} = 4\kappa\sigma(T_{\text{f}}^4 - T_\infty^4)\delta_{\text{R}} \tag{9.115}$$

Then for the premixed case, following Equation (4.42),

$$\left(\frac{\dot{m}'''_{\text{F}}}{Y_{\text{F,o}}}\right)[Y_{\text{F,o}}\Delta h_{\text{c}} - c_p(T_{\text{f}} - T_{\text{o}})] = 4\kappa\sigma(T_{\text{f}}^4 - T_\infty^4) \tag{9.116}$$

The analogous equation can be derived for the diffusion flame. Here, we will consider, for simplicity, no application of water and no charring. The optically thin flame heat loss allows an explicit expression for

$$X_{\text{r}}\dot{m}''_{\text{F}}\Delta h_{\text{c}} = 4\kappa\sigma(T_{\text{f}}^4 - T_\infty^4)\delta_{\text{R}} \tag{9.117}$$

and half of this, due to symmetry, is the incident surface flame radiation, $\dot{q}''_{\text{f,r}}$. Hence, from Equations (9.100) and (9.107), it follows that

$$\left(\frac{\dot{m}'''_{\text{F}}\delta_{\text{R}}}{Y_{\text{F,o}}}\right)\left[Y_{\text{F,o}}\Delta h_{\text{c}} - \left(1 + \frac{rY_{\text{F,o}}}{Y_{\text{ox},\infty}}\right)c_p(T_{\text{f}} - T_\infty) + c_p(T_{\text{s}} - T_\infty)\right] = \dot{q}''_{\text{f,c}} + 4\kappa\sigma(T_{\text{f}}^4 - T_\infty^4)\delta_{\text{R}} \tag{9.118}$$

By substituting for the flame convective heat flux from Equation (9.108) for steady conditions and no external heat flux (and no char and water), the analogue to Equation (9.116) follows for the diffusion flame:

$$\begin{aligned}&\left(\frac{\dot{m}'''_{\text{F}}}{Y_{\text{F,o}}}\right)\left[Y_{\text{F,o}}\Delta h_{\text{c}} - L - \left(1 + \frac{rY_{\text{F,o}}}{Y_{\text{ox},\infty}}\right)c_p(T_{\text{f}} - T_\infty) + c_p(T_{\text{s}} - T_\infty)\right]\\ &= 2\kappa\sigma(T_{\text{f}}^4 - T_\infty^4) + \frac{\sigma(T_{\text{v}}^4 - T_\infty^4)}{\delta_{\text{R}}}\end{aligned} \tag{9.119}$$

The right-hand side (RHS) of Equations (9.116) and (9.119) represent the net heat loss and the left-hand side (LHS) represents the energy gain. The gain and the loss terms can be plotted as a function of the flame temperature for both the diffusion and premixed flames as 'Semenov' combustion diagrams. Intersection of the gain and loss curves indicates a steady solution, while a tangency indicates extinction.

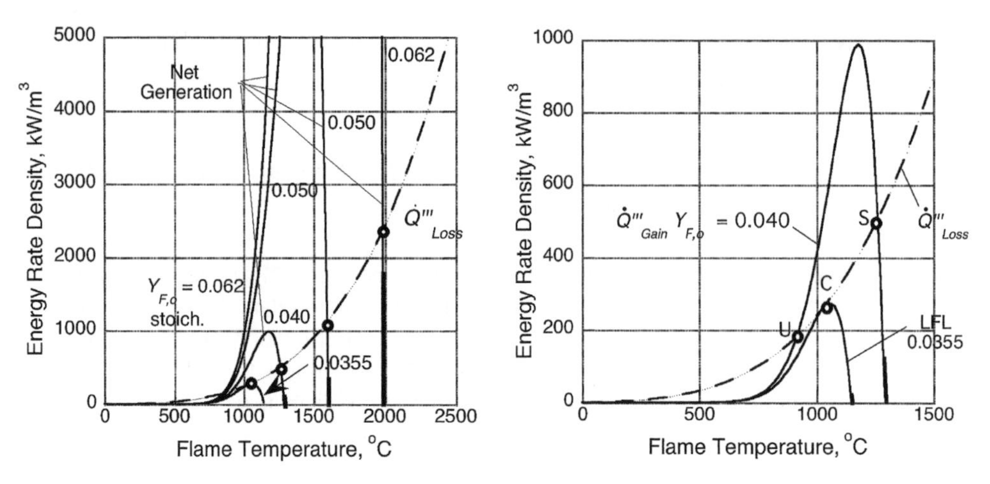

Figure 9.22 (a) Computed steady state solutions for premixed *n*-heptane propagation. (b) Computed critical condition (LFL) for heptane propagation

To illustrate these Semenov diagrams in reasonable quantitative terms, *n*-heptane was selected with the following properties:

$$\begin{aligned}
\Delta h_c &= 41.2\,\text{kJ/g (Table2.3)}\\
r &= 3.15\\
T_v &= 98\,^\circ\text{C (Table 9.2)}\\
k &= 0.117\,\text{W/m K}\\
L &= 0.477\,\text{kJ/g}\\
c_p &= 1.3\,\text{J/g K}\\
\rho_o &= 1.18\,\text{kg/m}^3\\
T_o &= T_\infty = 25\,^\circ\text{C}\\
E/R &= 19\,133\,\text{K} \quad \text{(from reference [25])}\\
A &= 1.41 \times 10^8\,\text{g/m}^3\,\text{s}
\end{aligned}$$

The A value was selected to match the lower flammability limit of $Y_F = 0.0355$ using Equation (4.40). The computed results are shown in Figures 9.22 and 9.23. The figures show the steady state solutions for the resulting flame temperatures. For premixed propagation, the computed temperature at stoichiometric fuel conditions is nearly the adiabatic flame temperature due to the steep nature of the gain curve. As the fuel concentration drops, so does the flame temperature. In Figure 9.22(b), it is clearly shown that at a fuel concentration of 0.035, there is a tangency of the loss and gain curves, which indicates a critical condition below which there is no longer a steady solution. This is the computed LFL and agrees well with the experimental value. The figure also clearly shows that, in general, two steady solutions are possible (S and U), but that the upper one is stable (S) and the lower one is unstable (U). The critical temperature for no sustained

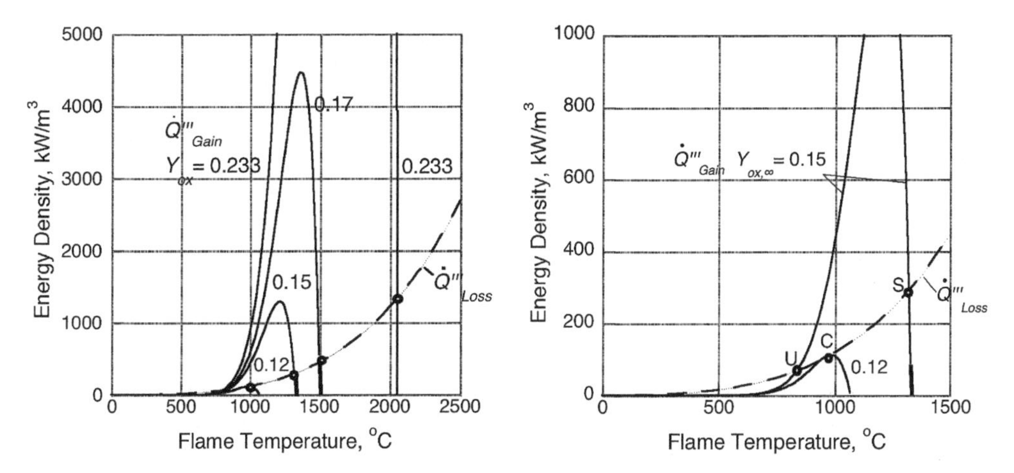

Figure 9.23 (a) Computed steady burning for *n*-heptane as a function of the ambient oxygen mass fraction. (b) Critical conditions at extinction

propagation is about 1050 °C, while the corresponding adiabatic flame on the abscissa is about 1150 °C. This is in contrast to the value of 1300 °C commonly assumed and indicated earlier. However, it should be emphasized that the computation is approximate, and is especially subject to the constant value for specific heat used in the calculation.

Similar results are computed for the diffusion flame under steady burning of *n*-heptane liquid. The resulting flame temperatures are given in Figures 9.23(a) and (b). The corresponding burning rates could be determined from Equation (9.112). Again, at ambient air conditions (0.233), the flame temperature is similar to the adiabatic flame temperature. As the ambient oxygen concentration drops, the flame temperature is reduced until a critical value, *c*, or about 950 °C results. Its corresponding adiabatic flame temperature is about 1100 °C. Again these results are approximate. However, they do show that the critical temperatures for extinction of premixed and diffusion flames are similar here for the heptane. Hence, the empirical critical value of about 1300 °C is credible.

9.11 Applications to Extinction for Diffusive Burning

Let us examine the 1300 °C criterion as a condition of flame extinction in diffusive burning to see how these critical conditions depend on oxygen and external radiation. After all, these two parameters – oxygen and radiation – are the two significant variables that make most solids burn. We will look to some data for anchoring this application.

Tewarson and Pion [18] conducted tests on the burning of materials in the FMRC flammability apparatus. Horizontal samples of 60–100 cm^2 were tested in order to measure their steady burning rate under radiant heating and varying oxygen concentrations as fed by O_2/N_2 steady flows of 50–70 L/min or speeds of about 0.5 cm/s. These speeds constitute basically natural convection burning conditions, and a heat transfer coefficient of about 9 W/m^2 K was estimated in applying the theoretical results of Section 9.10. The coefficient is not precisely known for the apparatus, while contributes to some uncertainty in the results. They determined the oxygen concentration to cause extinction,

Table 9.4 Material properties

Material	Δh_c (kJ/g)	L (kJ/g)	T_v (°C)	r (g O_2/g fuel)
Heptane	41.2	0.477	98	3.17
Styrene	27.8	0.64	146	2.14
PMMA	24.2	1.6	397	1.86
Methyl alcohol	19.1	1.195	64.5	1.51
Polyoxymethylene (POX)	14.4	2.43	425	1.107
Wood (Douglas fir)	12.4	~6.8	382	0.95

and did this for several external radiant heating conditions in some cases. The latter was done for PMMA in some detail. Computations based on Section 9.10 were performed for some of the materials considered by Tewarson and Pion and will be described here [28]. The following assumptions are made in applying the extinction theory:

1. The burning rate is relatively small, so the B number is small and $\ln(1+B) = B$.
2. The flame radiation is taken as zero since the flame size diminishes.
3. The fuel is pure so $Y_{F,o} = 1$.
4. The flame temperature at extinction is 1300 °C.
5. The heat transfer coefficient is taken at 9 W/m^2 K.
6. The ambient temperature is 25 °C unless otherwise specified.
7. The properties needed are taken from Reference [28] for consistency and are given in Table 9.4. The heat of gasification of wood is approximated since it is unsteady and charring.

In order to apply this extinction criterion, it will further be assumed that the steady state form of Equation (9.112) applies with no char, no water and no backface loss. As a consequence the burning rate can be described as

$$\dot{m}''_F = \frac{1}{L}\left[\frac{h_c}{c_p}\left[\frac{Y_{ox,\infty}\Delta h_c}{r - c_p(T_v - T_\infty)}\right] + \dot{q}''_{e,r} - \sigma(T_v^4 - T_\infty^4)\right] \tag{9.120}$$

Apply Equation (9.109) for these pure materials under conditions as steady, no char and no water:

$$c_p(T_f - T_\infty) = \frac{\Delta h_c - L + c_p(T_v - T_\infty) + [\dot{q}''_{e,r} - \sigma(T_v^4 - T_\infty^4)]/\dot{m}''_F}{1 + (r/Y_{ox,\infty})} \tag{9.121}$$

Eliminate the burning rate from Equations (9.120) and (9.121) in order to obtain the following relationship between external radiation, oxygen and the flame temperature:

$$\frac{(h_c/c_p)\left[(Y_{ox,\infty}\Delta h_c/r) - c_p(T_v - T_\infty)\right](Y_{ox,\infty}/r)}{\begin{array}{c}\langle\{[(Y_{ox,\infty}\Delta h_c/r - c_p(T_v - T_\infty)] - [1 + (r/Y_{ox,\infty})]c_p(T_f - T_\infty)\}\\ \times\{(h_c/c_p)\left[(Y_{ox,\infty}\Delta h_c/r) - c_p(T_v - T_\infty)\right] + \dot{q}''_{e,r} - \sigma(T_v^4 - T_\infty^4)\}\rangle\end{array}} = \frac{1}{L} \tag{9.122}$$

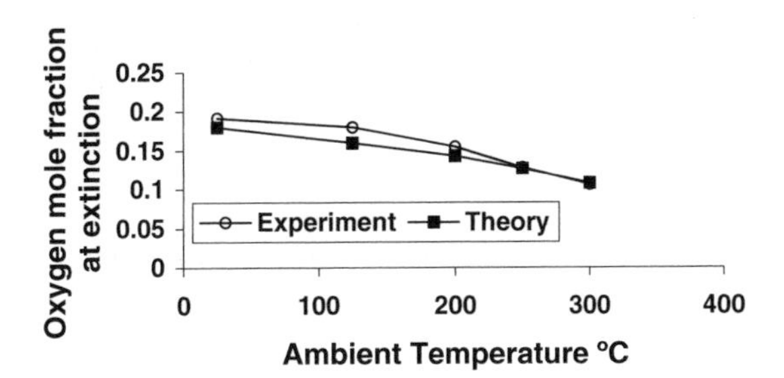

Figure 9.24 Oxygen index for polystyrene as a function of ambient temperature [18, 27]

In the denominator, the first term in the braces is the net energy above the minimum value to sustain the flame and the second term in the braces is the net heat flux to the fuel. Both of these terms must be positive for a valid physical solution.

For no external radiant heating, and for all of the following terms as small: $\sigma(T_v^4 - T_\infty^4)$, $c_p(T_v - T_\infty)$, $Y_{ox,\infty}/r$, then Equation (9.122) can be solved to give an approximate solution for the critical oxygen concentration:

$$Y_{ox,\infty} \approx \frac{c_p(T_f - T_\infty)}{(\Delta h_c - L)/r} = \frac{1.53}{13.1 - L/r} \tag{9.123}$$

This gives the approximate oxygen mass fraction at extinction. For liquid fuels, where typically $L < 1$ kJ/g, the lowest the critical oxygen mass fraction can be is 0.12, or 0.11 as a mole fraction. Also, this approximate result explains why the critical oxygen concentration increases with L and decreases with the ambient temperature. Figure 9.24 gives experimental results for polystyrene, compared to the theoretical value based on the criterion using Equation (9.122) [28].

Equation (9.122) gives a general relationship between the critical extinction relationship and the external heat flux and oxygen concentration at 25 °C. Experimental results are compared to the theory of Equation (9.122) for polymethyl methracrylate (PMMA) in Figure 9.25. The heat transfer coefficient has a small influence on the theoretical results, but a more interesting effect is a singularity at a mass fraction of about 0.12. This occurs when the energy needed to sustain the flame is just equal to the energy from combustion, i.e. the first brace in the denominator of Equation (9.122) is zero. This singularity is approximately given as

$$Y_{ox,\infty} \approx \frac{c_p(T_f - T_\infty)}{\Delta h_c/r} = 0.12 \tag{9.124}$$

This singularity is independent of the fuel. Figure 9.26 shows results over a complete range of oxygen concentrations for the fuels listed in Table 9.4. It shows that Douglas fir requires about 18 kW/m^2 to burn in air, while most of the other fuels burn easily in air with no external heating. Heptane will continue to burn with no external heating to the asymptotic oxygen fraction of 0.12. Below an oxygen fraction of 0.18, external heating is needed to sustain PMMA.

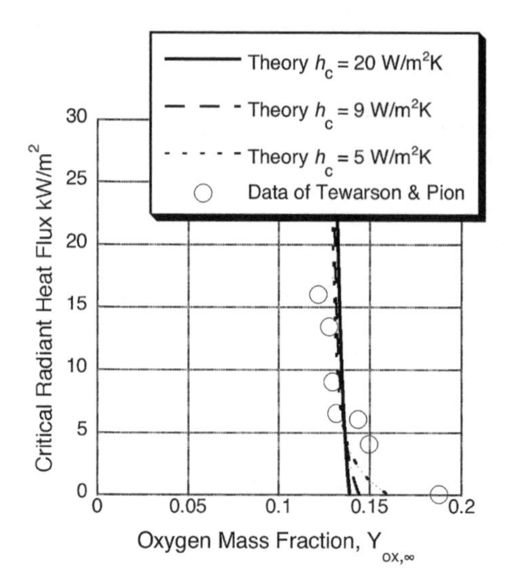

Figure 9.25 Critical heat flux and oxygen concentration at extinction for PMMA [18, 28]

The critical burning mass flux at extinction is between 2 and 4 g/m^2 s when burning in air, as seen in the theoretical plot of Figure 9.27. A curious minimum mass flux at extinction appears to occur for the oxygen associated with air for nearly all of the fuels shown. The critical mass flux at exinction is difficult to measure accurately, but experimental literature values confirm this theoretical order-of-magnitude for air. The asymptote at 0.12 has not been verified. These theoretical renditions should be taken as qualitative for now.

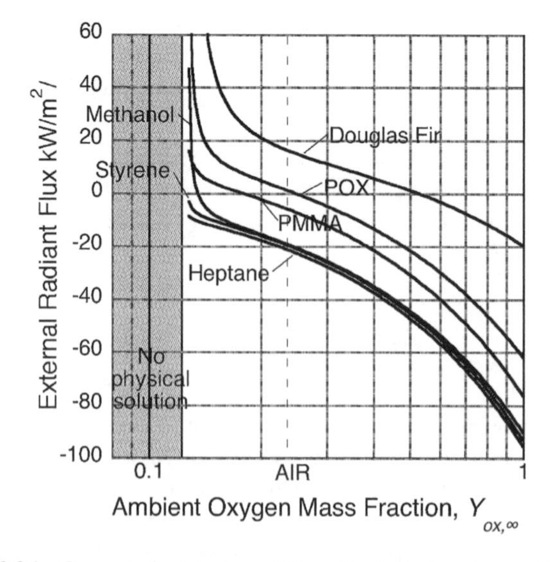

Figure 9.26 Computed critical conditions for fuels burning at 25 °C [28]

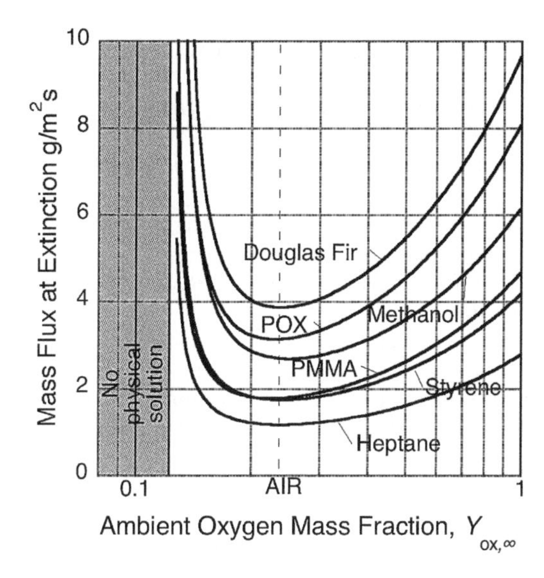

Figure 9.27 Critical mass flux at extinction [28]

References

1. Hopkins, D. and Quintiere, J.G., Material fire properties and predictions for thermoplastics, *Fire Safety J.*, 1996, **26** 241–268.
2. Spearpoint, M.J. and Quintiere, J.G., Predicting the ignition and burning rate of wood in the cone calorimeter using an integral model, Masters Thesis, Department of Fire Protection Engineering, University of Maryland, 1999.
3. Tewarson, A., Generation of heat and chemical compounds in fires, in *The SFPE Handbook of Fire and Protection Engineering*, 2nd edn (eds P. J. DiNenno), Section 3, National Fire Protection Association Quincy, Massachusetts, 1995, p. **3**–68.
4. Dillon, S.E., Quintiere, J.G. and Kim, W.H., Discussion of a model and correlation for the ISO 9705 room-corner test, in *Fire Safety Science – Proceedings of the 6th International Symposium*, IAFSS, 2000, p. 1015–1026.
5. Spalding, D.B., The combustion of liquid fuels, *Proc. Comb. Inst.*, 1953, **4**, pp. 847–864.
6. Incoperra, F.P. and deWitt, D.P., *Fundamentals of Heat and Mass Transfer*, 4th edn, John Wiley & Sons, New York, 1996.
7. Emmons, H.W., The film combustion of liquid fuel, *Z. Angew Math. Mech.*, 1956, **36**(1), 60–71.
8. Glassman, I., *Combustion*, Academic Press, New York, 1977, p. 185.
9. Kosdon, F.J., Williams, F.A. and Buman, C., Combustion of vertical cellulosic cylinders in air, *Proc. Comb. Inst.*, 1968, **12**, pp. 253–64.
10. Kim, J.S., deRis, J. and Kroesser, F.W., Laminar free-convective burning fo fuel surfaces, *Proc. Comb. Inst.*, 1969, **13**, pp. 949–61.
11. Ahmad, T. and Faeth, G.M., Turbulent wall fires, *Proc. Comb. Inst.*, 1979, **17**, pp. 1149–60.
12. Tien, C.L., Lee, K.Y. and Stretton, A.J., Radiation heat transfer, in *The SFPE Handbook of Fire Protection Engineering*, 2nd edn (eds P.J. DiNenno *et al.*), Section 1, National Fire Protection Aassociation, Quincy, Massechusetts, 1995, pp. 1–65 to 1–79.
13. Modak, A. and Croce, P., Plastic pool fires, *Combustion and Flame*, 1977 **30**, 251–65.

14. Iqbal, N. and Quintiere, J.G., Flame heat fluxes in PMMA pool fires, *J. Fire Protection Engng*, 1994, **6**(4), 153–62.
15. Orloff, L., Modak, A. and Alpert, R.L., Burning of large-scale vertical surfaces, *Proc. Comb. Inst.*, 1977, **16**, pp. 1345–54.
16. Corlett, R.C. and Fu, T.M., Some recent experiments with pool fires, *Pyrodynamics*, 1966, **4**, 253–69.
17. Kung, H.C. and Stavrianidis, P., Buoyant plumes of large - combustion scale pool fires, *Proc. Comb. Inst.*, 1982, **19**, pp. 905–12.
18. Tewarson, A. and Pion, R.F., Flammability of plastics. I. Burning intensity, *Combustion and Flame*, 1978, **26**, 85–103.
19. Magee, R.S. and Reitz, R.D., Extinguishment of radiation-augmented plastic fires by water sprays, *Proc. Comb. Inst.*, 1975, **15**, pp. 337–47.
20. Quintiere, J.G., The growth of fire in building compartments, in *Fire Standards and Safety* (ed. A. F. Robertson), ASTM STP614, American Society for Testing and Materials, Philadelphia, Pennsylvania, 1977, pp. 131–67.
21. Block, J., A theoretical and experimental study of non-propagating free-burning fires, *Proc. Comb. Inst.*, 1971, **13**, pp. 971–78.
22. Quintiere, J.G. and McCaffrey, B.J., *The Burning of Wood and Plastic Cribs in an Enclosure*, Vol. I., NBSIR 80-2054, National Bureau of Standards, US Department of Commerce, Gaithersburg, Maryland, November 1980.
23. Heskestad, G., Modeling of enclosure fires, *Proc. Comb. Inst.*, 1973, **14**, p. 1021–1030.
24. Students, ENFP 620, Fire Dynamics Laboratory, University of Maryland, College Park, Maryland, 1998.
25. Williams, F.A., A review of flame extinction, *Fire Safety J.*, 1981, **3**, 163–75.
26. Macek, A., *Flammability Limits: Thermodynamics and Kinetics*, NBSIR 76-1076, National Bureau of Standards, US Department of Commerce, Gaithersburg, Maryland, May 1976.
27. Roberts, A.F. and Quince, B.W. A limiting condition for the burning of flammable liquids, *Combustion and Flame*, 1973, **20**, 245–251.
28. Quintiere, J.G. and Rangwala, A.S., A theory for flame extinction based on flame temperature, *Fire and Materials*, 2004, **28**, 387–402.

Problems

9.1 For pure convection burning of PMMA in air at 20 °C and steady state, determine the mass loss rate per unit area for a vertical slab of height $l_1 = 2\,\text{cm}$ and $l_2 = 20\,\text{cm}$. Use the gas properties for air. PMMA properties from Table 2.3 are: $\Delta h_c = 25\,\text{kJ/g}$, $L = 1.6\,\text{kJ/g}$, $T_v = 380\,°\text{C}$ and $\Delta h_{ox} = 13\,\text{kJ/g}\,O_2$.

9.2 Polyoxymethylene burns in a mixture of 12 % oxygen by mass and nitrogen is present. Assume it burns purely convectively with a heat transfer coefficient of 8 W/m^2 K. Find $\dot{m}''$.

Polyoxymethylene properties:

$$T_\infty = 20\,°\text{C} \qquad L = 2.4\,\text{kJ/g}$$
$$T_v = 300\,°\text{C} \qquad \Delta h_{ox} = 14.5\,\text{kJ/g}\,O_2$$

9.3 n-Octane spills on hot pavement during a summer day. The pavement is 40 °C and heats the octane to this temperature. Wind temperature is 33 °C and pressure is 1 atm.

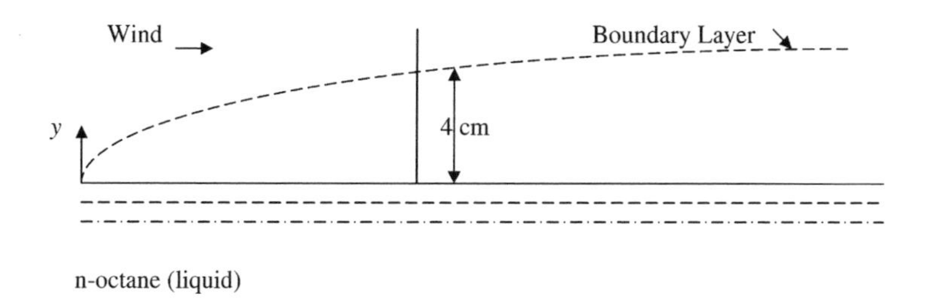

n-octane (liquid)

n-Octane (liquid)

Data:

Ambient conditions are at a temperature of 33 °C and pressure of 760 mmHg. Density of air is 1.2 kg/m^3 and has a specific heat of 1.04 J/g K.

Octane properties:
Boiling temperature is 125.6 °C.
Heat of vaporization is 0.305 kJ/g.
Specific heat (liquid) is 2.20 J/g K.
Specific heat (gas) is 1.67 J/g K.
Density (liquid) is 705 kg/m^3.
Lower flammability limit in air is 0.95 %.
Upper flammability limit in air is 3.20 %.

Use data from Table 6.1.

The octane spill is burning in wind that causes a convective heat transfer coefficient of 20 W/m^2 K. The same data as given above still apply.

(a) Compute the heat of gasification of the octane.
(b) Compute the B-number for purely convective burning but ignore the radiation effects.
(c) Compute the steady burning rate per unit area ignoring all effects from radiation.
(d) Compute the convective heat flux from the flame.
(e) Extinction occurs, at a burning rate of 2 g/m^2 s, due to a heat flux loss from the spill surface as the spill thickness decreases. Calculate the heat flux loss to cause the burning rate to reach the extinction condition. Radiation effects are still unimportant.

9.4 A candle burns at a steady rate (see the drawing below). The melted wax along the wick has a diameter $D = 0.5$ mm and pyrolysis occurs over a length of $l_p = 1$ mm. Treat the wick as a flat plate of width πD and a height of l_p which has a convective heat transfer coefficient $h = 3\ \mathrm{W/m^2\ K}$. Ignore all radiative effects.

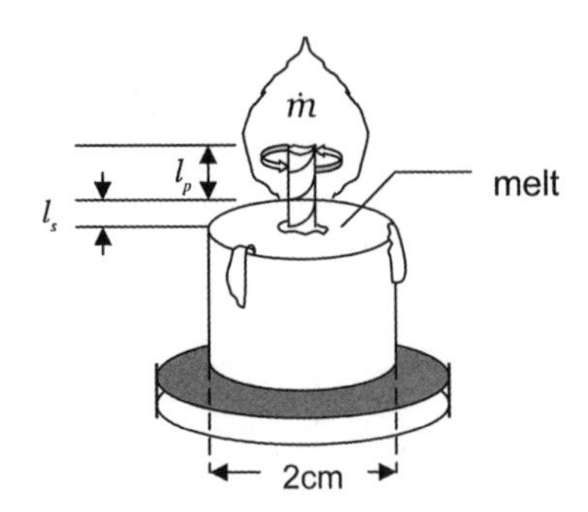

Properties of the wax:

Heat of combustion of liquid wax = 48.5 kJ/g
Stoichiometric air to fuel mass ratio = 15
Density of the solid and liquid phases = 900 kg/m^3
Specific heat of the solid and liquid phases = 2.9 J/g K
Heat of vaporization = 318 J/g
Heat of fusion (melting) = 146 J/g
Temperature at melting = 52 °C
Temperature at vaporization = 98 °C
Diameter of the candle, as shown is 2 cm

Also:

Initial and ambient temperature in air = 25 °C

$$c_{p,g} \quad \text{(for air)} = 1\,\text{J/g K}$$
$$k_g = 0.03\,\text{W/m K}$$

(a) Determine the mass burning rate in g/s. Assume the initial temperature in the wick is the melt temperature.

(b)Assuming pure conduction heat transfer from the candle flame to the melted wax pool, and a thin constant melt layer on the wax, estimate the distance between the flame and the melt layer (i.e. l_s). Also, assume an effective heat transfer area based on the candlewick diameter, i.e. 0.5 mm.

9.5 A thick polystyrene wainscotting covers the wall of a room up to 1 m from the floor. It is ignited over a 0.2 m and begins to spread. Assume that the resulting smoke layer in the room does not descend below 1 m and no mixing occurs between the smoke layer and the lower air.

Polystyrene properties:
Density, $\rho = 40\,\text{kg/m}^3$
Specific heat, $c_p = 1.5\,\text{J/g K}$
Conductivity, $k = 0.4\,\text{W/m K}$
Vaporization and ignition temperature, $T_v = 400\,°\text{C}$
Heat of combustion, $\Delta h_c = 39\,\text{kJ/g}$
Stoichiometric oxygen to fuel ratio, $r = 3\,\text{g/g}$
Heat of gasification, $L = 1.8\,\text{kJ/g}$

Other information:
Initial temperature, $T_\infty = 20\,°C$
Ambient oxygen mass fraction, $Y_{ox,\infty} = 0.233$
Specific heat of air, $c_p = 1\,J/g\,K$

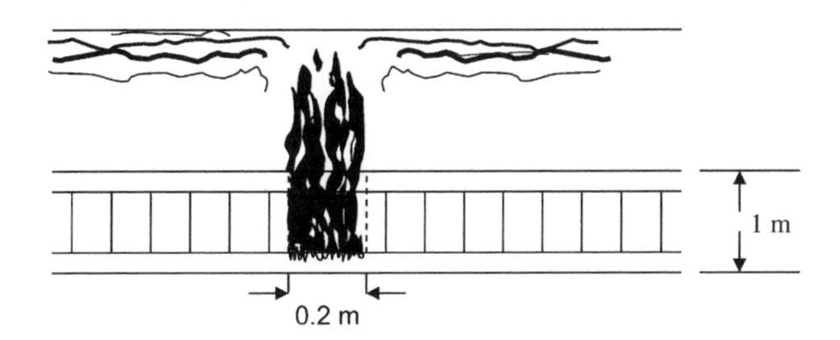

(a) The fire spreads to the top of the wainscotting, but does not spread laterally. What is the burning (pyrolysis) rate of the polystyrene over this region assuming a steady state. The vertical natural convection heat transfer coefficient, $h = 15\,W/m^2\,K$.

(b) Later the fire spreads laterally to 2 m and the surface experiences a heat flux due to radiation from the heated room of $1.5\,W/cm^2$. Using the small B-number approximation, account for all radiation effects and calculate the mass loss rate per unit area of polystyrene.

9.6 PMMA burns in air at an average $\dot{m}'' = 15\,g/m^2\,s$. Air temperature is 20 °C and the heat transfer coefficient is $15\,W/m^2\,K$. Determine the net radiative heat flux to the fuel surface. What amount comes from the flame?

9.7 Calculate $\dot{m}''$ for pools of heptane burning in air having diameters $D = 1$ cm, 10 cm and 1 m. Take $\Delta h_c/r = 13\,kJ/g$, $T_v = 98\,°C$, properties for air are at 20 °C and use Tables 9.2 and 10.4. Use the small B-number approximation. Assume that $\dot{q}''_{r,f} = \sigma(1 - e^{-\kappa D})T_f^4$, where

$T_f = 1600\,K$, an effective radiation temperature

$\kappa = 12\,m^{-1}$, a flame absorption coefficient

9.8 A vertical slab of PMMA burns in air at an average mass loss rate of $15\,g/m^2\,s$. The air temperature is 20 °C. Assume that the heat transfer coefficient of the slab is $15\,W/m^2\,K$. Determine the net radiative heat flux to the surface. What amount comes from the flame?

PMMA properties are:

$$\Delta h_c = 25\,kJ/g$$
$$L = 1.6\,kJ/g$$
$$T_{ig} = 380\,°C$$

9.9 Properties needed:
Liquid n-decane:
Boiling point = 447 K
Heat of vaporization = 0.36 kJ/g
Flashpoint = 317 K
Lower flammability limit (by volume) = 0.6 %
Specific heat = 2.1 kJ/kg K

Density = 730 kg/m^3
Heat of combustion = 44.7 kJ/g
Stoichiometric oxygen to fluid mass ratio = 3.4
Thermal conductivity = 0.5 W/m K
Air:
Kinematic viscosity = 0.5 W/m K
Density = 1.2 kg/m^3
Specific heat = 1.0 kJ/kg K
Gravitational acceleration, $g = 9.81\ \mathrm{m/s^2}$
Stefan–Boltzman constant, $\sigma = 5.67 \times 10^{-11}\ \mathrm{kW/m^2\,K^4}$
n-Decane is exposed to an atmospheric temperature of 20 °C and a pressure of 1 atm

Answer the following using the above data:

(a) The decane is at 20 °C. Determine the fuel vapor concentration at the liquid surface. Is the decane flammable under these conditions?

(b) If the decane is at the lower flammability limit for an air atmosphere, compute the necessary surface temperature. Will the liquid autoignite at this temperature?

(c) Compute the heat of gasification for the decane initially at 20 °C and at the boiling point.

(d) For convective heat transfer from air to 80 °C and with a convective heat transfer coefficient (h) of 15 W/m^2 K, compute the mass rate of evaporation per unit area of the decane at the temperature where it just becomes flammable; this is before combustion occurs. Assume the bulk temperature of the liquid is still at 20 °C.

(e) If the decane were exposed to an external radiant heat flux of 30 kW/m^2, how long would it take to ignite by piloted ignition? Assume $h = 15\ \mathrm{W/m^2\,K}$, $T_\infty = 20\,°\mathrm{C}$; also, the liquid is very deep, remains stagnant and acts like a solid.

(f) For purely convective-controlled burning with $h = 15\ \mathrm{W/m^2}$ compute the burning rate per unit area under steady conditions for a deep pool of decane in air at 20 °C and an atmosphere of 50% oxygen by mass at 20 °C.

(g) A 1.5 m diameter deep pool of decane burns steadily in air ($T_\infty = 20\,°\mathrm{C}$). Its radiative fraction is 0.25 of the total chemical energy release and 5 % of this radiative loss is uniformly incident over the surface of the pool. Accounting for all the radiative effects, what is the total energy release of the pool fire? The heat transfer coefficient is 15 W/m^2 K. (Use the small B approximation.)

(h) Decane is ignited over region $x_o = 0.5$ m in a deep tray. A wind blows air at 20 °C over the tray at $v_\infty = 10\ \mathrm{m/s}$. The convective heat transfer coefficient is found to be 50 W/m^2 K.

(i)

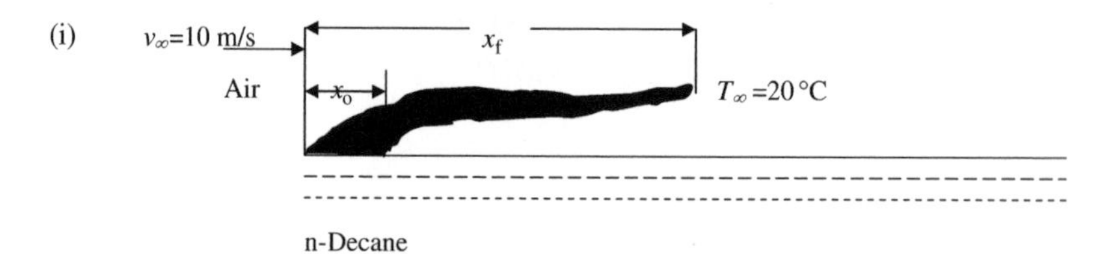

Assume the decane flame spreads like that over a solid surface; i.e. ignore circulation effects in the liquid. Assume pure convective burning. Compute the convective heat flux from the flame to the surface and also compute the flame spread velocity just following ignition over x_o. It is known that the flame length, x_f, can be found by

$$\frac{x_f}{x_o} = 11.6\left(\frac{\dot{m}''}{\rho_\infty \sqrt{g x_o}}\right)^{2.7} \left(\frac{v_\infty}{\sqrt{g x_o}}\right)^{-0.25} \left(v_\infty \frac{x_o}{\nu}\right)^{0.89}$$

where

$\dot{m}''$ = mass burning rate per unit area
ρ_∞ = air density
g = gravitational acceleration
ν = kinematic viscosity

9.10 D. B. Spalding published a seminal paper, 'The combustion of liquid fuels' in the 4th International Symposium on *Combustion* (Reference [5]). In it he addressed the burning of droplets by using a porous sphere under both natural and forced convective flows. The figure below shows the burning of kerosene on a 1.5 inch diameter sphere under natural convection conditions. Spalding empirically found from his data that the burning rate per unit area could expressed as

$$\frac{\dot{m}_F'' D c_p}{k} = 0.45\, B^{3/4} \left(\frac{g D^3}{\alpha^2}\right)^{1/4}$$

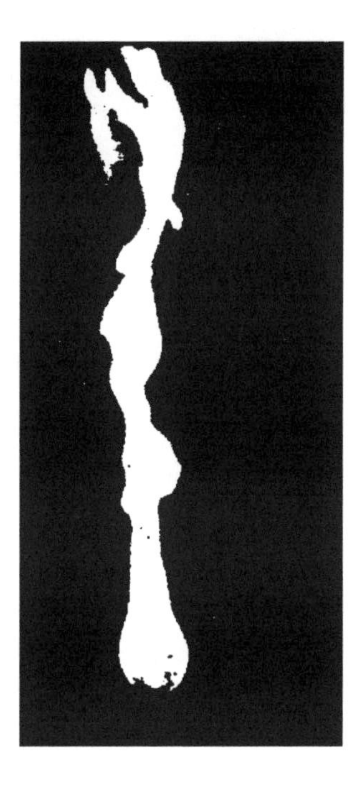

The symbols are consistent with Section 9.2 of the text. We are to use the stagnant layer theory of burning with suitable approximations to analyze the burning of a droplet in natural convection. The droplet, suspended and burning in air, is assumed to remain spherical with diameter, D.

(a) From the stagnant layer theory in pure convective burning, develop an alternative, but similar, formula for $\dot{m}''_{\mathrm{F}} D c_p / k$. It is known from heat transfer results that

$$Nu_D = 2 + \frac{0.589(Gr\,Pr)^{1/4}}{[1 + (0.469\,Pr)^{9/16}]^{4/9}}$$

(b) A droplet of n-decane, $C_{10}H_{22}$, beginning at a diameter of 10 mm, burns in air which is at 25 °C. The properties of n-decane are given below:

Boiling point	174 °C
Liquid density	730 kg/m^3
Liquid specific heat	1.2 J/g K
Heat of vaporization	0.360 kJ/g
Heat of combustion	44.2 kJ/g
Stoichiometric ox/fuel mass ratio	3.5 g/g

From your formula, compute the initial burning rate in g/s for pure convective burning.

(c) The following additional information is available. The flame height in the wake region above the burning droplet can be estimated by

$$z_{\mathrm{f}} = 0.23\dot{Q}^{2/5} - 1.02\,D$$

where z_{f} is measured from the center of the droplet in m, $\dot{Q}$ is the firepower in kW and D is the diameter in m. The flame absorption coefficient, κ, can be estimated as 0.45 m^{-1}. The mean beam length of the flame relative to its base can be expressed as

$$\frac{L_{\mathrm{e}}}{D} = 0.65(1 - \mathrm{e}^{-2.2\,z_{\mathrm{f}}/D})$$

The overall radiation fraction, X_{r}, is 0.15. Compute the initial burning rate, now including radiation effects. Show all your work and clearly indicate any assumptions.

(d) The droplet burns until no liquid fuel remains. How long does this take? Use the most complete burning rate description. Show how you include the variation of D over time. An integration process is need here.

9.11 A pool of PMMA with a diameter of 1 m burns in air. Radiation effects are to be included. In addition to using the properties from Problem 9.1, let $X_{\mathrm{r}} = 0.25$, $T_{\mathrm{f}} = 1500$ K and $\kappa = 0.5$ m^{-1}. If it is known that the critical mass loss rate for PMMA is 4 g/m^2 s, calculate the flux of water ($\dot{m}''_{\mathrm{w}}$) needed to extinguish this fire. If instead the oxygen is reduced, at what mass fraction will extinction occur?

9.12 Polyoxymethylene is burning in the upper smoke layer of a room fire. The critical mass loss rate per unit area at extinction is known to be 2 g/m^2 s for polyoxymethylene. The smoke layer temperature is 700 °C.

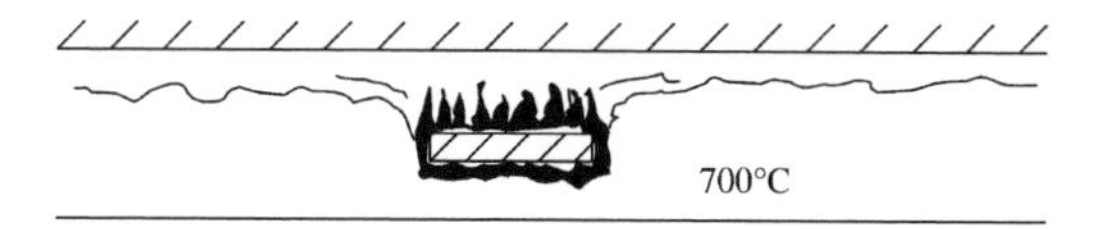

Properties:

Heat of gasification = 2.43 kJ/g
Heat of combustion = 15.5 kJ/g
Convective heat transfer coefficient for the configuration = 15 W/m^2 K
Specific heat of gas layer = 1 kJ/kg K
Heat of combustion per unit mass of oxygen consumed = 14.5 kJ/g O_2
Vaporization temperature = 250 °C

Assume purely convective burning and negligible radiation.

(a) Calculate the oxygen mass fraction to cause extinction.

(b) For 1 m^2 of surface burning, calculate the energy release rate if the layer smoke oxygen mass fraction is 0.20.

9.13 For a standard business card, in the vertical length dimension, determine the steady burning rate (g/s) for one side of the card saturated with ethanol. Only the ethanol burns. Show your analysis and all assumptions. This is a calculation, not an experimental determination, though experiments can be conducted. State all data and sources used. You will have to make approximations and estimates for quantities in your analysis.

9.14 A polypropylene square slab, 0.3 m on a side, burns in a steady manner. Polypropylene does not char, and it can be considered that its radiation characteristics do not vary, with the flame radiative fraction at 0.38 and the flame incident heat flux to the fuel surface at 25 kW/m^2. It burns in a wind as shown in various conditions as specified. Its convective heat transfer coefficient can be taken as 50 W/m^2 K. The properties of polypropylene are listed as follows.

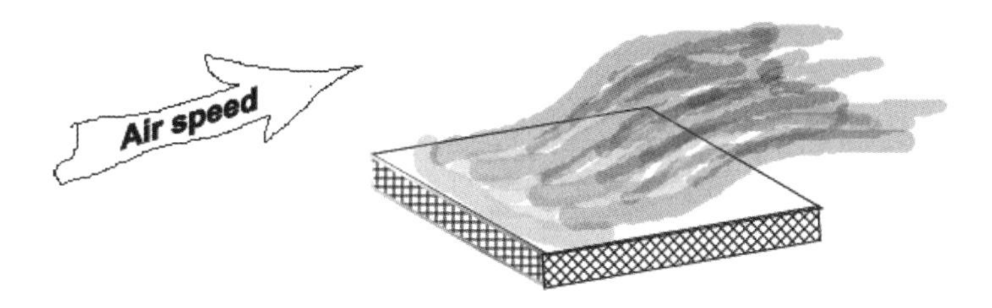

Properties:

Heat of combustion, $\Delta h_c = 38\,\text{kJ/g}$

Heat of gasification, $L = 3.1\,\text{kJ/g}$

Vaporization temperature, $T_v = 330\,°\text{C}$

Any other data or properties should be cited from the text. Assume that the B-number or $\dot{m}''_F c_p / h_c$ is small for your calculations in all of the following.

(a) If it burns in air at an ambient temperature of −10 °C, compute the burning flux in g/m^2 s.

(b) It is placed in a flow-through oven with the same wind conditions. The air temperature is 60 °C. The oven radiates to the burning slab with 45 kW/m^2. Compute the firepower in kW.

(c) How much water spray per unit area is needed to reduce the firepower by 50 % in problem (b)?

(d) Now pure oxygen is used in the oven of problem (b) with the same ambient and flow and heating conditions. Compute the burning rate in g/s.

(e) What is the convective heat flux in problem (d)?

9.15 PMMA is burned, in normal air at 25 °C, in a horizontal configuration (30 × 30 cm) under an external irradiation of 20 kW/m^2. You are to quantitatively evaluate the ability of dry ice (CO_2) and ice (H_2O) both in solid flakes at their respective freezing point (at 1 atm) to extinguish the fire.

Assume steady burning with the sample originally at 25 °C with a perfectly insulated bottom. At extinction you can ignore the flame radiation. Assume that all of the flakes hit the surface and ignore the gas phase effects of the extinguishment agents. Use thermodynamic properties of the CO_2 and H_2O, and the property data of PMMA from Table 9.2.

Compute:

(a) The ratio of flow rates for dry ice and water ice to the burning rate needed for extinction.

(b) The critical burning rate per unit area at extinction for both ice and dry ice.

(c) The flow rate of dry and ice needed to extinguish the fire.

9.16 Use the control volume of the diffusion flame shown to derive the flame temperature of the outgoing products.

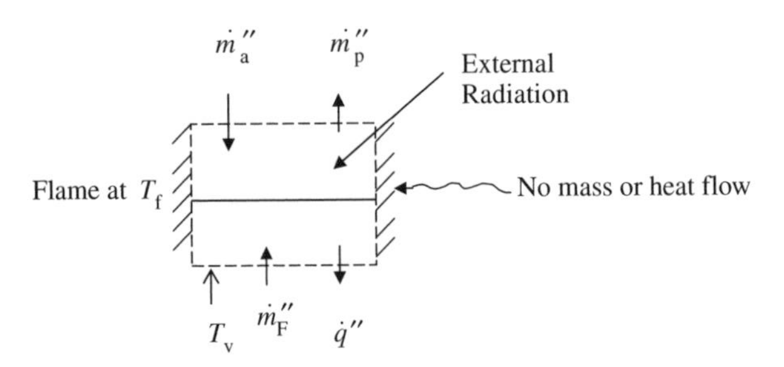

where

$\dot{m}''_a$ = mass flux of air diffuses toward the flame

$\dot{m}''_p$ = mass flux of products diffusing away from the flame

$\dot{q}'' = \dot{m}'' L$, heat loss to the fuel

$\dot{m}''_F$ = evaporated fuel diffuses toward the flame

$$c_g(T_f - T_v) = \frac{Y_{ox,\infty}[(\Delta h_c - L)/r] + c_g(T_\infty - T_v)}{1 + Y_{ox,\infty}/r}$$

9.17 Consider the following plastic materials and their properties. These materials do not char; they melt and vaporize.

Material	L^a (kJ/g)	L^b (kJ/g)	L^c (kJ/g)	Δh_c^b (kJ/g)	T_v^b (°C)	T_v^c (°C)	X_r^b	Rate without external radiation $\dot{m}''_{F,o}$ (g/m^2 s)
Polymethyl-methracrylate (PMMA)	1.61	1.63	2.0	24.2	391	300	0.31	5.8 v, 7.5 h[a]
Polyoxy-methylene (POM)	2.93	2.43	—	14.4	419	—	0.22	6.2 v, 6.0 h
Polystyrene (PS)	2.09	1.70	3.0	27	419	350	0.59	10.8 h
Polyethylene (PE)	2.31	3.0	3.6	38.4	444	360	0.43	3.4 h

[a] R. S. Magee and R. D. Reitz [19].
[b] A. Tewarson [3].
[c] J. G. Quintiere [1].
[d] Orientation: v, vertical sample; h, horizontal sample.

You are to use these data in the following calculations. Use the Magee data for L and the Tewarson data for T_v, since these values have some uncertainty, as indicated by their variation from the different sources. These differences can be due to variations in the measurements or in the materials. Plastics can have additives in varying amounts, yet still be listed as one polymer.

For each material and burning orientation compute the following. Assume steady burning in all your calculations.

(a) Based on the burning rates measured for the horizontal and vertical samples, without any additional radiant heating, find the flame convective and radiative heat flux components. The vertical sample was 7 inches wide and 14 inches high; the horizontal was 7 inches square.

(b) Assuming the flame heating is constant, write expressions for the burning rates in terms of the external radiative heat flux, $\dot{q}''_{e,r}$, and the water application flux, $\dot{m}''_w$. In these experiments, all of the water reached the surface and none was absorbed in the flame.

(c) Determine, and plot if you wish, the relationship between the external radiative heat flux and the water application flux at the point of extinction. At extinction, you can consider the mass flux of the burning fuel to be small.

(d) Compute the burning flux just at extinction, and show how it depends on the radiative heat flux and water flux.

9.18 The following data are given for the WTC towers:

Building:

110 stories, 11 ft per story
208 ft × 208 ft floor space
7 lb/ft^2 of floor area of wood-based furnishings

Jet fuel, JP4:
Liquid density = 760 kg/m^3
Heat of combustion = 43.5 kJ/g
Heat of gasification = 0.80 kJ/g
Effective vaporization temperature = 230 °C
Radiative loss fraction = 0.20
Wood:
Heat of combustion = 12.0 kJ/g
Heat of gasification = 6.0 kJ/g
Effective vaporization temperature = 380 °C
Radiative loss fraction = 0.35
Conversion constants:
1 ft = 0.308 m
1 gal = 0.003 75 m^3
1 lb/ft^2 = 4.88 kg/m^2
Atmosphere:
No wind
25 °C

Dry air, density = 1.1 kg/m^3

(a) 20,000 gal of JP 4 fuel spills entirely over the 87th floor of the building. The fuel experiences a convective heat transfer coefficient of 15 W/m^2 K, and a total incident radiative heat flux of 150 kW/m^2 due to the flame and compartment effects. Assume a blocking factor of 1.

  (i) Estimate the burning rate.

  (ii) Calculate the energy release rate.

  (iii) Calculate the duration of the JP4 fire.

(b) Following the burning off of the jet fuel, the wood-based furnishings burn. The wood fuel loading is distributed such its exposed area to the flames is 0.65 of the floor area.

  (i) Compute the burning rate of the wood fuel on a floor under identical flame heating conditions as the JP4 in 1.

  (ii) How long will this fuel burn?

10

Fire Plumes

10.1 Introduction

A fire plume is loosely described as a vertically rising column of gases resulting from a flame. The term plume is generally used to describe the noncombustion region, which might dominate the flow away from the combustion source, especially if the source is small. A plume is principally buoyancy driven. This means flow has been induced into the plume due to an increase in temperature or consequent reduction in density. A plume is also likely to be turbulent rather than laminar. Anyone who has observed a rising streak of white smoke from a cigarette left in an ash tray should recall seeing the wavy laminar emitted streak of smoke break-up into a turbulent wider plume in less than 1 foot of height.* Thus, any significant accidental fire will have an associated turbulent plume – even a smoldering fire. Near the fire we will have to deal with a combusting plume, but 'far' from the fire it might suffice to ignore the details of combustion entirely. The far-field fire plume acts as if it had been heated by a source of strength $\dot{Q}$(kW). Since radiated energy from the flame will be lost to the plume, the plume effects will only respond to $(1 - X_r)\dot{Q}$, where X_r is the fraction of radiant energy lost with respect to the actual chemical energy release of the first. We shall see that X_r is an empirical function of fuel and fire geometry. Indeed, much about fire plumes is empirical, meaning that its results are based on experimental correlations. Often the correlations are incomplete with some secondary variables not included since they were not addressed in the experimental study. In other cases, the correlations can be somewhat ambiguous since they represent time-averaged properties, or length scales (e.g. plume width, flame height), that were not precisely or consistently defined. However, because fire plume results are grounded in measurement – not purely theory – they are practical and invaluable for design and hazard assessment.

*$Gr = \frac{\Delta T}{T_\infty}\frac{gz^3}{\nu^2} \approx (5)\frac{(9.81\,\text{m/s}^2)(0.3\,\text{m})^3}{(20\times10^{-6}\,\text{m}^2/\text{s})^2} = 3.3 \times 10^9$. Expect turbulent plumes for $Gr > 10^9$, especially with atmospheric disturbances.

Fundamentals of Fire Phenomena James G. Quintiere

Some of the foundational studies for turbulent fire plumes include the work of the following investigators:

1. The measurements of Rouse, Yih and Humphries (1952) [1] helped to generalize the temperature and velocity relationships for turbulent plumes from small sources, and established the Gaussian profile approximation as adequate descriptions for normalized vertical velocity (w) and temperature (T), e.g.

$$\frac{w}{w_m} = \exp\left[-\left(\frac{r}{b(z)}\right)^2\right] \tag{10.1a}$$

and

$$\frac{T - T_\infty}{T_m - T_\infty} = \exp\left[-\beta\left(\frac{r}{b(z)}\right)^2\right] \tag{10.1b}$$

where $()_m$ implies maximum or centerline line values and r is the radius. Here, for an axisymmetric plume, b is an effective plume radius that depends on the vertical coordinate z. Such a profile in a scaled variable, $\eta = r/b(z)$, is known as a similarity solution; i.e. the profiles have the same shape at any position z. Near the source where the streamlines will not necessarily follow the vertical direction, a similarity solution will not be possible. The constant β in Equation (10.1b) is nearly unity.
2. The theoretical analysis by Morton, Taylor and Turner (1956) [2] established approximate similarity solutions for an idealized point source in a uniform and stably stratified atmosphere.
3. Yokoi (1960) [3], singly produced a 'small book' as a report, which carefully investigated point, line and finite heat sources with eventual applications to the hazard from house fire and window flame plumes.
4. Steward (1964) [4] and (1970) [5] presented a rigorous but implicit approximate analysis for both axisymmetric and two-dimensional strip fire plumes respectively, including combustion and variable density effects.

All of the solutions are approximate since only a complete solution to the Navier–Stokes unsteady equations with combustion effects would yield fundamental results. This is not currently possible, but will be in the future. These fire plumes have all the scales of turbulent frequencies and specific features due to buoyancy: fine scales – high-frequency turbulence is responsible for the fuel–oxygen mixing and combustion that occurs locally; large turbulent scales – relatively low frequencies – are responsible for the global 'engulfment' of air swept into the plume by the large eddy or vortex structures that 'roll-up' along the outside of a buoyant plume. This is shown schematically in Figure 10.1, and is also shown in a high-speed sequence of photographs by McCaffrey [6] for the luminous flame structure (Figure 10.2). The size of the eddies depend on the diameter of the fire (D) and the energy release rate ($\dot{Q}$). The eddy frequency, f, has been shown to

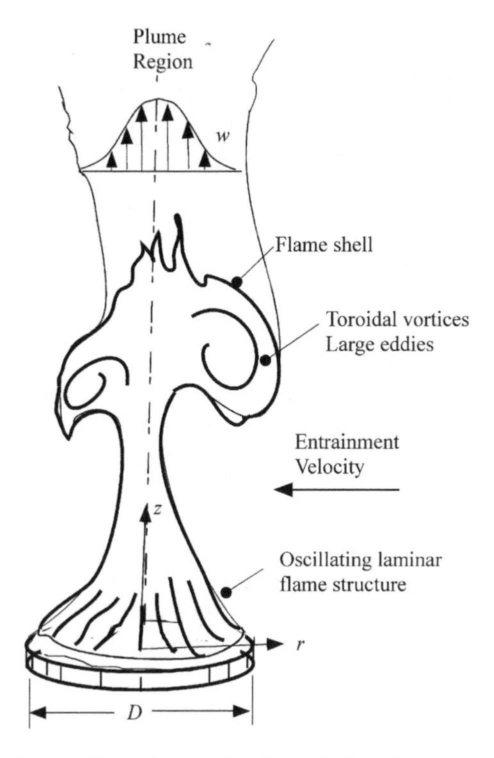

Figure 10.1 Schematic of a turbulent fire plume

correlate with the inverse of a characteristic time,

$$f = 0.48\sqrt{\frac{g}{D}}\,(\text{Hz}) \tag{10.2}$$

by Pagni [7] in Figure 10.3.

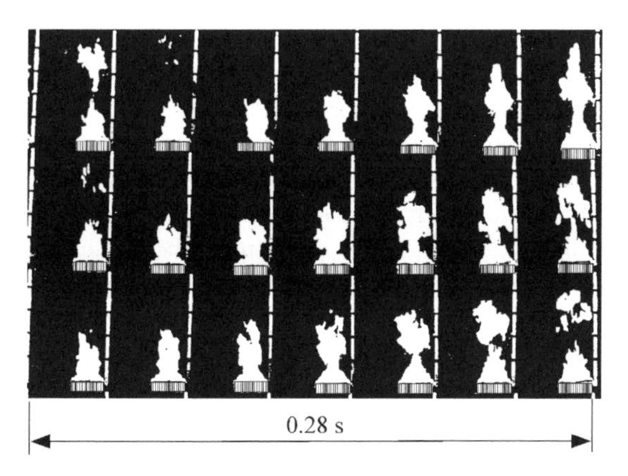

Figure 10.2 Intermittency of a buoyant diffusion flame burning on a 0.3 m porous burner. The sequence represents 1.3 s of cine film, showing 3 Hz oscillation (from McCaffrey [6])

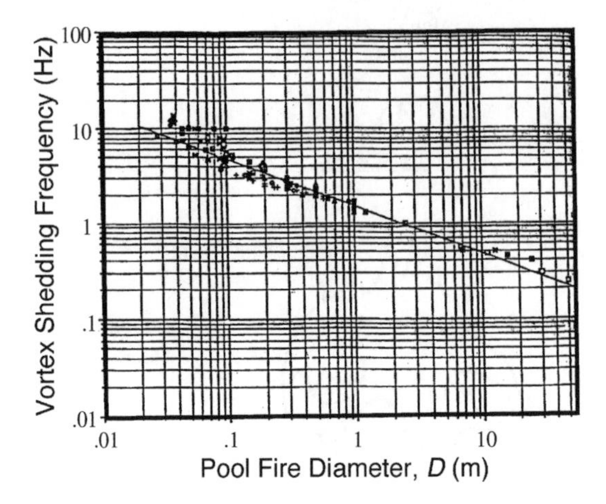

Figure 10.3 Eddy shedding frequency of buoyant axisymmetric plumes from Pagni [7]

The mechanism of turbulent mixing which brings air into the buoyant plume is called entrainment. It has been described empirically by relating the momentum of the induced air proportionally to the vertical momentum (mean or centerline),

$$-\lim_{r\to\infty}(\rho_\infty rv^2) \propto \frac{\mathrm{d}}{\mathrm{d}z}(\rho_\mathrm{m} w_\mathrm{m}^2 b^2)$$

or dimensionally, the entrainment velocity is traditionally given as

$$-\lim_{r\to\infty}(rv) = \alpha_\mathrm{o}\sqrt{\frac{\rho_\mathrm{m}}{\rho_\infty}}w_\mathrm{m}b \qquad (10.3)$$

where α_o is a constant, the entrainment constant for a Boussinesq or 'constant density' plume (i.e. a plume whose density constant is everywhere except in the buoyancy term $\Delta\rho gz$). Many analyses and correlations have ignored the density ratio term in Equation (10.3), treating it as unity. The entrainment constant is not universal, but depends on the mathematical model used. Nominally its value is $O(10^{-1})$. It must be determined by a modeling strategy together with consistent data.

Since all of the theoretical results for plumes rely on experimental data, we must realize that the accuracy of the correlation depends on the accuracy of the data. Measurements of temperature with thermocouples are subject to fouling by soot and radiation error. At flame temperatures, the error can be considerable ($\sim$ 100 °C). Moreover, the fluctuating nature of the properties can lead to fluctuations at a point of $\pm$500 °C) or more. We will see that the maximum time-averaged temperature in turbulent flames can be 800–1200 °C, or possibly higher, depending on their size, yet laminar flames can only survive for roughly a flame temperature of >1300 °C. Hence, a turbulent flame must be looked on as a collection of highly convoluted, disjointed laminar flame sheets flapping rapidly within a turbulent flow field of large and small eddies. Extinction occurs intermittently due to thermal quenching, flame stretch (shear) and depletion of fuel and oxygen locally. Indeed, the term 'turbulent diffusion flame' has lost its favor having been superceded by the term 'turbulent nonpremixed flames'.

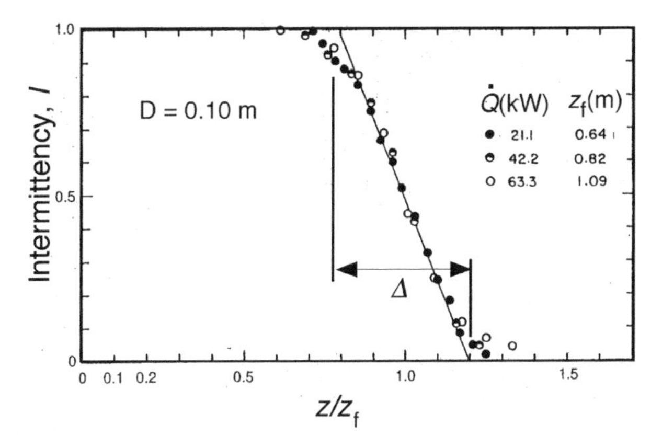

Figure 10.4 Flame length intermittency from Zukoski [8]

A buoyancy-dominated turbulent flame has a very complex structure, raising from an oscillating laminar base to a core of luminosity and ending in an intermittent flaming region of broken flame remnants (Figure 10.2). Note that the frequency of flame structure oscillation measured by McCaffrey for a $D = 0.3$ m burner at 3 Hz is consistent with the vortex shedding frequencies of the Pagni correlation (Equation (10.2)). The intermittency (I = duration of visible flame/flame cycle time) presented by Zukoski [8] in Figure 10.4 shows the flame fluctuations about the mean at $I = 0.5$. We see about this mean that the flame length, z_f, varies by about $\pm 20\,\%$ for $D = 0.1$ m. However, the intermittent region expands with D: $z_f \pm \Delta_f/2$, $\Delta_f = 0.40\,z_f$, $0.60\,z_f$ and $0.80\,z_f$ for $D = 0.10$, 0.19 and 0.50 m respectively. Furthermore, the 'eye-averaged' z_f tends to be about 20 % higher than z_f at $I = 0.50$, according to Zukoski [8].

Measurements of velocity in a plume have more difficulties. Pitot tube measurements rely on temperature and can have ambiguous directional effects. Hot-wire anemometry can be fouled and require buoyancy corrections. Nonintrinsic measurements also offer challenges but have not been sufficiently used in fire measurements. The low-speed flows, especially at the edges of the plume, offer a big measurement challenge. Where a pressure measurement is needed for velocity, appropriate time-averaging of pressure and temperature signals need to be processed. Dynamic pressures can range from 10 to 10^3 pascals (Pa) in plumes. Realizing that 1 atm is approximately 10^5 Pa, it is readily approximated that plumes are constant pressure processes. Only vertical static pressure differences are important since they are the source of buoyancy. For large-diameter plumes, especially near their base, it is likely that pressure differences due to the curvature of streamlines can also be responsible for induced flow from the ambient entrainment, but such studies of the base entrainment effect are unknown to this author. Of course, away from the base where streamlines are more parallel to the vertical, pressure effects are nil, and entrainment for the turbulent plume is an engulfment process based on buoyancy and turbulence.

Other measurements such as gas species and soot all have importance in fire plumes but will not be discussed here. As we have seen for simple diffusion flames, the mixture fraction plays a role in generalizing these spatial distributions. Thus, if the mixture fraction is determined for the flow field, the prospect of establishing the primary species concentration profiles is possible.

As stated earlier, much has been written about fire plumes. Excellent reviews exist by Zukoski [8], Heskestad [9], Delichatsios [10] and McCaffrey [11]. The student is encouraged to read the literature since many styles of presentation exist, and one style might be more useful than another. We cannot address all here, but will try to provide some theoretical framework for understanding the basis for the many alternative correlations that exist.

10.2 Buoyant Plumes

We will derive the governing equations for a buoyant plume with heat added just at its source, approximated here as a point. For a two-dimensional planar plume, this is a line. Either could ideally represent a cigarette tip, electrical resistor, a small fire or the plume far from a big fire where the details of the source are no longer important. We list the following assumptions:

1. There is an idealized point (or line) source with firepower, $\dot{Q}$, losing radiant fraction, X_r.
2. The plume is Boussinesq, or of constant density ρ_∞, except in the body force term.
3. The pressure at any level in the plume is due to an ambient at rest, or 'hydrostatic'.
4. The ambient is at a uniform temperature, T_∞.
5. Properties are assumed uniform across the plume at any elevation, z. This is called a top-hat profile as compared to the more empirically correct Gaussian profile given in Equation (10.1).
6. All properties are constant, and $c_{p_i} = c_p$.

10.2.1 Governing equations

Later we shall include combustion and flame radiation effects, but we will still maintain all of assumptions 2 to 5 above. The top-hat profile and Boussinesq assumptions serve only to simplify our mathematics, while retaining the basic physics of the problem. However, since the theory can only be taken so far before experimental data must be relied on for its missing pieces, the degree of these simplifications should not reduce the generality of the results. We shall use the following conservation equations in control volume form for a fixed CV and for steady state conditions:

Conservation of mass (Equation (3.15))

$$\sum_{\text{net out}}^{j} \dot{m}_j = 0 \tag{10.4}$$

Conservation of vertical (z) *Momentum* (Equation (3.23))

$$+z \uparrow \sum F_z = \sum_{\text{net out}}^{j} \dot{m}_j w_j \tag{10.5}$$

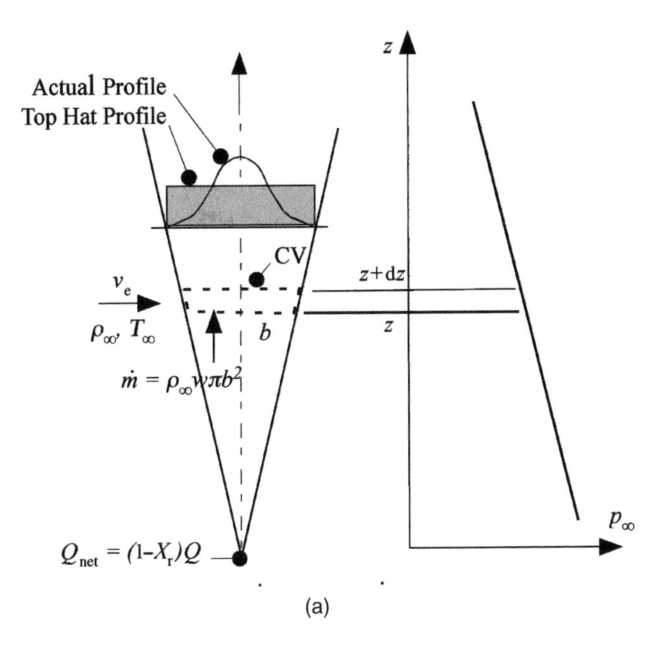

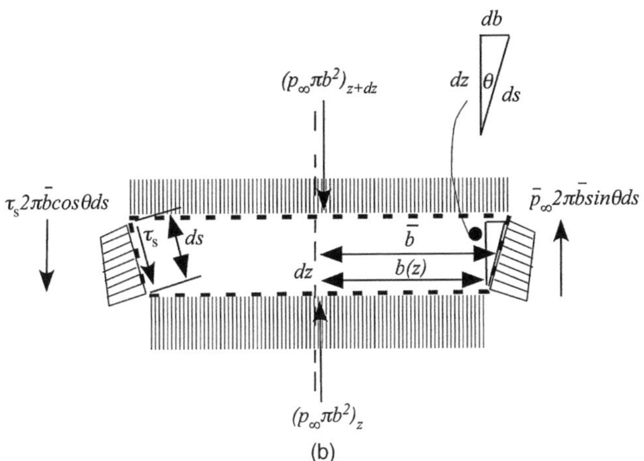

Figure 10.5 (a) Schematic of a point source plume and (b) surface focus

Conservation of energy (Equation (3.48))

$$c_p \sum_{\text{net out}}^{j} \dot{m}_j T_j = \dot{Q} - \dot{q}_{loss} \tag{10.6}$$

We will apply these to a small control volume of radius b and height dz, as shown in Figure 10.5, and we will employ the Taylor expansion theorem to represent properties at $z + \mathrm{d}z$ to those at z:

$$F(z + \mathrm{d}z) = F(z) + \frac{\mathrm{d}F}{\mathrm{d}z}\mathrm{d}z + O(\mathrm{d}z)^2 \tag{10.7}$$

Using the entrainment relationship (Equation (10.3)), in terms of the top-hat profile,

$$v_e = \alpha_o w \tag{10.8}$$

The conservation of mass can then be written as

$$\dot{m}(z + dz) - \dot{m}(z) - \rho_\infty v_e 2\pi b = 0 \tag{10.9}$$

The mass flow rate in terms of the Boussinesq top-hat assumption is given by

$$\boxed{\dot{m} = \rho_\infty w \pi b^2} \tag{10.10}$$

Therefore, the conservation of mass using Equations (10.7) to (10.10) with the operation

$$\lim_{dz \to 0} \frac{1}{dz}$$

gives

$$\frac{d}{dz}(\rho_\infty w \pi b^2) = \rho_\infty \alpha_o w \pi b$$

or

$$\frac{d}{dz}(wb^2) = 2\alpha_o wb \tag{10.11}$$

The conservation of momentum in the z direction with positive forces in the $+z$ directions can be written as

$$\bar{p}_\infty 2\pi\bar{b}\, db + (p_\infty \pi b^2)_z - (p_\infty \pi b^2)_{z+dz} - \rho g \pi \bar{b}^2 dz - \tau_s 2\pi\bar{b} dz = (w\dot{m})_{z+dz} - (w\dot{m})_z \tag{10.12}$$

The pressure forces are derived in Figure 10.5(b), taking into account the pressure at the perimeter region $2\pi\bar{b}$ ds, where $\bar{b}$ is a mean between z and $z + dz$. Likewise, so is $\bar{p}_\infty$. The vertical component of the shear force also follows from similar arguments. However, at the edge of the plume we have a static fluid (approximately) or $dw/dr = 0$ for the actual profile. Since

$$\tau_s \propto \frac{dw}{dr}$$

realizing that we are considering turbulent time-averaged quantities, then $\tau_s = 0$ at $r = b$. Therefore, in the limit as $dz \to 0$,

$$p_\infty 2\pi b \frac{db}{dz} - \frac{d}{dz}(p_\infty \pi b^2) - \rho g \pi b^2 = \frac{d}{dz}(\rho_\infty w^2 \pi b^2) \tag{10.13}$$

However, in the static ambient,

$$\frac{\mathrm{d}p_\infty}{\mathrm{d}z} = -\rho_\infty g \tag{10.14}$$

Combining gives

$$\frac{\mathrm{d}}{\mathrm{d}z}(\rho_\infty w^2 \pi b^2) = (\rho_\infty - \rho) g \pi b^2 \tag{10.15}$$

Employing the perfect gas equation of state,

$$\rho T = \rho_\infty T_\infty \tag{10.16}$$

To reduce non-linear effects, we make another assumption:

7. The change in density or temperature is small. (This completes the full Boussinesq model.)

We shall retain this assumption even for the high-temperature combustion case, realizing that it is quite severe. However, its neglect will undoubtedly be compensated through ultimate experimental correlations. With assumption 7 we approximate

$$\frac{\rho_\infty - \rho}{\rho_\infty} = 1 - \frac{T_\infty}{T} = \frac{T - T_\infty}{T} = \left(\frac{T - T_\infty}{T_\infty}\right)\left(\frac{T_\infty}{T}\right)$$
$$\frac{\rho_\infty - \rho}{\rho_\infty} \approx \frac{T - T_\infty}{T_\infty} \quad \text{for} \quad T/T_\infty \to 1 \tag{10.17}$$

Therefore, Equation (10.15) becomes

$$\boxed{\frac{\mathrm{d}}{\mathrm{d}z}(w^2 b^2) = \left(\frac{T - T_\infty}{T_\infty}\right) g b^2} \tag{10.18}$$

The conservation of energy for the differential control volume in Figure 10.5 becomes, following Equation (10.6),

$$c_p\left[(\dot{m}T)_{z+\mathrm{d}z} - (\dot{m}T)_z\right] - \rho_\infty v_\mathrm{e} c_p T_\infty \; 2\pi b \; \mathrm{d}z \tag{10.19}$$

The heat added is zero because we have no combustion in the CV, and all of the heat is ideally added at the origin, $z = 0$. Operating as before, we obtain

$$c_p \frac{\mathrm{d}}{\mathrm{d}z}\left[(\rho_\infty w \pi b^2) T\right] - \rho_\infty v_\mathrm{e} c_p T_\infty \; 2\pi b = 0 \tag{10.20}$$

From Equation (10.11), it can be shown that

$$c_p \frac{\mathrm{d}}{\mathrm{d}z}(\rho_\infty w \pi b^2 T) - c_p \frac{\mathrm{d}}{\mathrm{d}z}(\rho_\infty w \pi b^2 T_\infty) = 0$$

or

$$\rho_\infty c_p \frac{\mathrm{d}}{\mathrm{d}z}[w\pi b^2(T - T_\infty)] = 0 \tag{10.21}$$

This can be integrated, and using a control volume enclosing $z = 0$ yields

$$\rho_\infty c_p w\pi b^2(T - T_\infty) = \dot{Q}(1 - X_r)$$

where $\dot{Q}$ is the origin term representing the power of the source. Hence,

$$\boxed{wb^2 \frac{T - T_\infty}{T_\infty} = \frac{\dot{Q}(1 - X_r)}{\rho_\infty c_p T_\infty \pi}} \tag{10.22}$$

We now have three equations, (10.11), (10.18) and (10.22), to solve for w, b, and $\phi = (T - T_\infty)/T_\infty$. The initial conditions for this ideal plume are selected as (at $z = 0$)

$$b = 0 \quad \text{(a point source)}$$

and

$$w^2 b = 0 \quad \text{(no momentum)}$$

If we had significant momentum or mass flow rate at the origin, the plume would resort to a jet and the initial momentum would control the entrainment as described by Equation (10.3). This jet behavior will have consequences for the behavior of flame height compared to the flame height from natural fires with negligible initial momentum.

10.2.2 *Plume characteristic scales*

It is very useful to minimize significant variables by expressing the equations in dimensionless terms. To do this we need a characteristic length to normalize z. If we had a finite source diameter, D, it would be a natural selection; however, it does not exist in the point source problem. We can determine this natural length scale, z_c, by exploring the equations dimensionally. We equate dimensions. From Equation (10.22),

$$w_c z_c^2 \stackrel{D}{=} \frac{\dot{Q}}{\rho_\infty c_p T_\infty}$$

or

$$w_c \stackrel{D}{=} \frac{\dot{Q}}{\rho_\infty c_p T_\infty z_c^2}$$

Substituting into Equation (10.18) gives

$$w_c^2 z_c \stackrel{D}{=} g z_c^2$$

or

$$\left(\frac{\dot{Q}}{\rho_\infty c_p T_\infty z_c^2}\right)^2 z_c^{-1} \stackrel{D}{=} g$$

giving

$$\boxed{z_c = \left(\frac{\dot{Q}}{\rho_\infty c_p T_\infty \sqrt{g}}\right)^{2/5}} \tag{10.23}$$

as the natural length scale. This length scale is an important variable controlling the size of the large eddy structures and the height of the flame, in addition to the effect by D for a finite fire. We might even speculate that the dimensionless flame height should be expressed as

$$\frac{z_f}{D} = f\left(\frac{z_c}{D}\right)$$

Incidentally, if we were examining a line source in which the source strength would need to be expressed as $\dot{Q}'$, i.e. energy release rate per unit length of the linear source, Equation (10.23) could be reexamined dimensionally as

$$z_c \stackrel{D}{=} \left(\frac{\dot{Q}/L}{\rho_\infty c_p T_\infty \sqrt{g}/z_c}\right)^{2/5}$$

or

$$\boxed{z_{c,\text{line}} = \left(\frac{\dot{Q}'}{\rho_\infty c_p T_\infty \sqrt{g}}\right)^{2/3}} \tag{10.24}$$

This is the characteristic length scale for a line plume.

To complete the point source normalization, we determine, in general, a characteristic velocity as

$$w_c = \sqrt{g z_c} \tag{10.25a}$$

or specifically for a point source,

$$w_c = g^{1/2}\left(\frac{\dot{Q}}{\rho_\infty c_p T_\infty \sqrt{g}}\right)^{1/5} \tag{10.25b}$$

where $\dot{Q}$ is the firepower in this scaling.

10.2.3 Solutions

Solutions for a Gaussian profile, with Equation (10.1) for an axisymmetric point source, and for a line source

$$\frac{w}{w_{\mathrm{m}}} = \exp\left[-\left(\frac{y}{b}\right)^2\right] \tag{10.26a}$$

and

$$\frac{T - T_\infty}{T_{\mathrm{m}} - T_\infty} = \exp\left[-\beta\left(\frac{y}{b}\right)^2\right] \tag{10.26b}$$

with dimensions as shown in Figure 10.6, are tabulated in Table 10.1 [12]. The dimensionless variables are

$$\begin{aligned} \Phi &= (T_{\mathrm{m}} - T_\infty)/T_\infty \\ W &= w_{\mathrm{m}}/w_{\mathrm{c}} \\ B &= b/z_{\mathrm{c}} \\ \zeta &= z/z_{\mathrm{c}} \end{aligned} \tag{10.27}$$

The solutions in Table 10.1 were fitted to data by Yokoi [3] from small fire sources with an estimation of X_{r}, the radiation fraction, for his fuel sources. Fits by Zukoski [8], without accounting for radiation losses ($X_{\mathrm{r}} = 1$), give values of $\alpha_{\mathrm{o}} = 0.11$ and $\beta = 0.91$ for the axisymmetric source. Yuan and Cox [13] have shown that α_{o} and β can range from 0.12 to 0.16 and from 0.64 to 1.5 respectively, based on a review of data from many studies. Figures 10.7 and 10.8 show the similarity results and centerline characteristics respectively by Yokoi [3] for the point and line sources.

It is interesting to note some features of these ideal cases since they match the far-field data of large fire plumes. The velocity and temperature at the centerline of a point plume decreases with height and both are singular at the origin. Only the temperature has this behavior in the line plume, with the velocity staying constant along the centerline. A

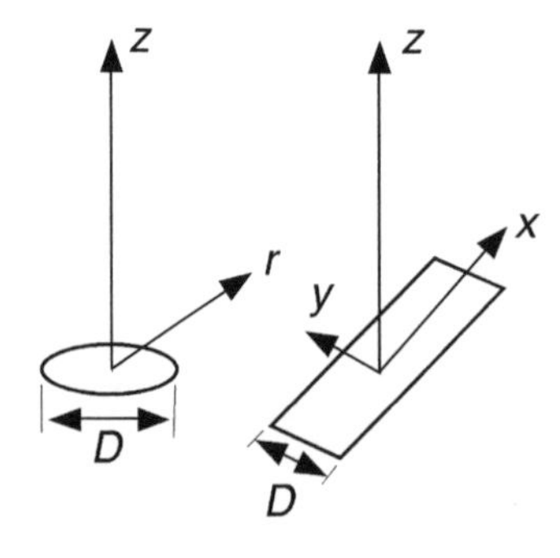

Figure 10.6 Finite sources

Table 10.1 Far-field Gaussian plume correlations [12]

Dimensionless variable	Axisymmetric ($X_r = 0.20$ for data)	Infinite line ($X_r = 0.30$ for data)
B	$C_1\zeta$	$C_1\zeta$
	$C_1 = \frac{6}{5}\alpha$, 0.118	$C_1 = \frac{2}{\sqrt{\pi}}\alpha$, 0.103
W	$C_v\zeta^{-1/3}$	$C_v\zeta^{0}$
	$C_v = \left[\left(\frac{25}{24\pi}\right)\left(\frac{\beta+1}{\beta}\right)\alpha^{-2}(1-X_r)\right]^{1/3}$	$C_v = \left[\left(\frac{\beta+1}{2\beta}\right)^{1/6}\alpha^{-1/3}(1-X_r)^{1/3}\right]$
	$C_v = 4.17(1-X_r)^{1/3}$	$C_v = 2.25(1-X_r)^{1/3}$
Φ	$C_T\zeta^{-5/3}$	$C_T\zeta^{-1}$
	$C_T = \left[\frac{2}{3}\left(\frac{25}{24\pi}\right)^{2/3}\frac{(\beta+1)^{2/3}}{\beta^{-1/3}\alpha^{4/3}}(1-X_r)^{2/3}\right]$	$C_T = \left[\frac{(\beta+1)^{1/3}}{2^{5/6}\beta^{-1/6}}\alpha^{-2/3}(1-X_r)^{2/3}\right]$
	$C_T = 10.58(1-X_r)^{1/3}$	$C_T = 3.30(1-X_r)^{2/3}$
$\frac{W^2}{\Phi\zeta}$	$\frac{3/2}{\beta}$, 1.64	$\sqrt{\frac{2}{\beta}}$, 1.54
α	0.098	0.091
β	0.913	0.845

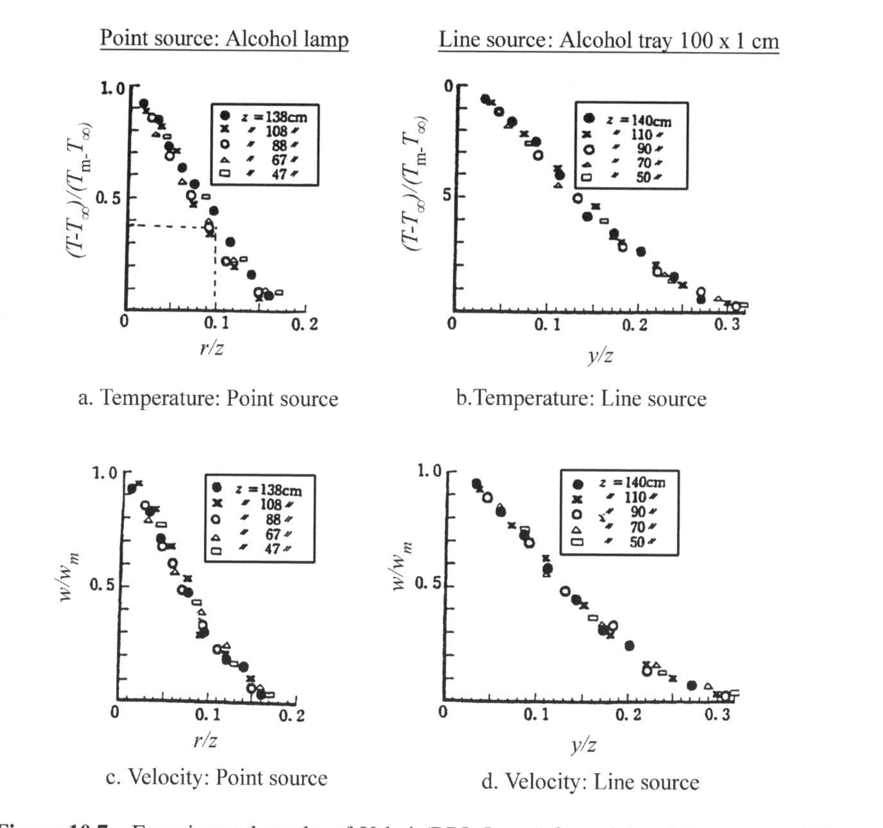

a. Temperature: Point source

b. Temperature: Line source

c. Velocity: Point source

d. Velocity: Line source

Figure 10.7 Experimental results of Yokoi (BRI, Japan) for point and line sources [3]

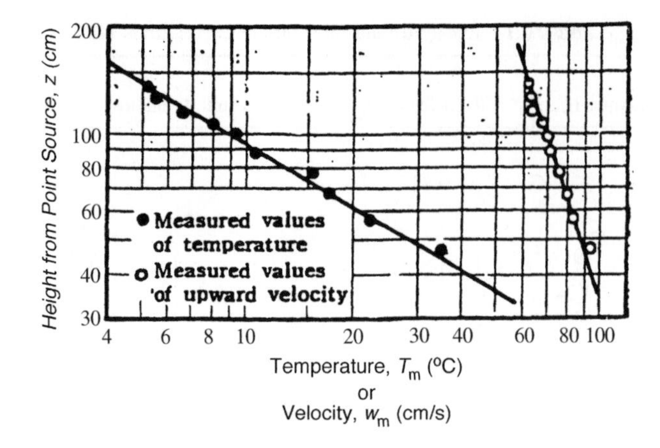

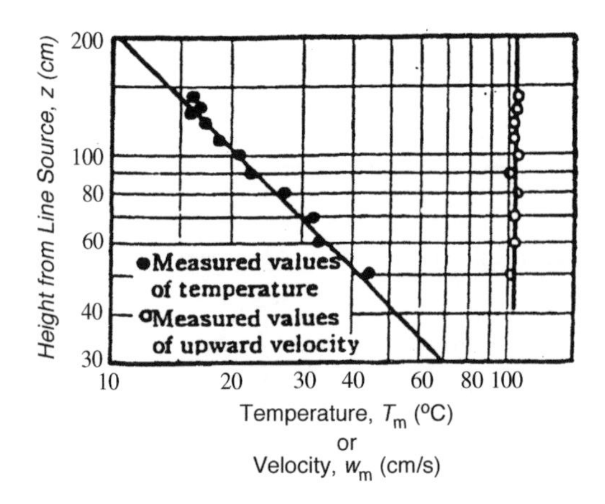

Figure 10.8 Yokoi data for time-averaged axial values of velocity and temperature rise [3]

local Froude number *(Fr)* is invariant and nearly the same for both plumes. This local *Fr* number is represented as

$$Fr \equiv \frac{\text{kinetic energy}}{\text{potential energy}}$$
$$= \frac{w_m^2/g z_c}{[(T_m - T_\infty)/T_\infty](z/z_c)} = \frac{w_m^2}{g[(T_m - T_\infty)/T_\infty]z} = \frac{W^2}{\Phi\zeta} \approx 1.6 \pm 0.05 \tag{10.28}$$

as seen from Table 10.1. This shows that velocity is directly related to buoyancy. Note that no effect of viscosity has come into the analysis for the free plume. Indeed, an application of Bernoulli's equation to the center streamline would give a value of $Fr = 2.0$. For wall plumes, or ceiling–plume interactions, we might expect to see effects of $Gr = (gz^3/\nu^2)[(T_m - T_\infty)/T_\infty]$. For laminar plumes, viscous effects must be present since viscosity is the only mechanism allowing ambient air to be 'entrained' into the plume flow.

10.3 Combusting Plumes

We will now examine the combustion aspect of plumes and attempt to demonstrate more realistic results. The preceding approach provides a guide for how we might proceed. Since the complexities of combustion are too great to consider in any fundamental way, we simply adopt the same 'point source' approach with the additional assumption:

8. Combustion occurs uniformly throughout the plume so long as there is fuel present.

The equations become the same as before, except with a net effective heat addition per unit volume as $(1 - X_r)\dot{Q}'''$ within the differential element of Figure 10.5. From Equations (10.11), (10.18) and (10.21), the conservation of mass, momentum and energy become

$$\frac{d}{dz}(wb^2) = 2\alpha_o wb \tag{10.29a}$$

$$\frac{d}{dz}(w^2b^2) = \left(\frac{T - T_\infty}{T_\infty}\right)gb^2 \tag{10.29b}$$

and

$$\rho_\infty c_p \frac{d}{dz}[w\pi b^2(T - T_\infty)] = (1 - X_r)\dot{Q}'''\pi b^2 \tag{10.29c}$$

for the top-hat axisymmetric case.

The maximum rate of fuel that can burn in the control volume dz in Figure 10.5(a) is that which reacts completely with the entrained oxygen or with a known stoichiometric ratio of oxygen to fuel, r. Thus, we can write

$$\dot{Q}''' = \frac{\rho_\infty(\alpha_o w)(2\pi b\, dz)(\Delta h_c\, Y_{O_2,\infty})}{(\pi b^2\, dz)(r)} \tag{10.30}$$

Substituting in Equation (10.29c) gives

$$\rho_\infty c_p \frac{d}{dz}[wb^2(T - T_\infty)] = \frac{(1 - X_r)2\alpha_o\rho_\infty wb\Delta h_c\, Y_{O_2,\infty}}{r} \tag{10.31}$$

Using Equation (10.29a), this becomes

$$\frac{d}{dz}[wb^2(T - T_\infty)] = (1 - X_r)\frac{\Delta h_c\, Y_{O_2,\infty}}{c_p r}\frac{d}{dz}(wb^2) \tag{10.32}$$

which integrates to give, with $b = 0$ at $z = 0$,

$$T - T_\infty = (1 - X_r)Y_{O_2,\infty}\left(\frac{\Delta h_c}{rc_p}\right) \tag{10.33}$$

However, it is not likely that all of the air entrained will react with the fuel. The turbulent structure of the flame does not allow the fuel and oxygen to instantaneously mix. There is a delay, sometimes described as 'unmixedness'. Hence, some amount of oxygen (or air) more than stoichiometric will be entrained but not burned. Call this excess entrainment fraction

$$n \equiv \frac{\text{mass of air entrained}}{\text{mass of air reacted}} \tag{10.34}$$

Equation (10.33) should be modified to give

$$T - T_\infty = \frac{(1 - X_r) Y_{O_2,\infty} \Delta h_c}{n c_p r} \tag{10.35}$$

a lower temperature in the flame due to dilution by the excess air. This factor n has been found to be about 10, and may be a somewhat universal constant for a class of fire plumes. The conditions for this constancy have not been established. The main point of this result is that the time and spatial average flame temperature is a constant that depends principally on X_r alone since $Y_{O_2,\infty} \Delta h_c / r \approx 2.9\,\text{kJ/g}$ air for a wide range of fuels. Indeed, the flame temperature is independent of the fuel except for its radiative properties, and those flames that radiate very little have a higher temperature. This result applies to all fire plumes configurations, except that n will depend on the specific flow conditions. In dimensionless terms, for buoyancy-dominated turbulent fires, we would expect

$$\frac{T_m - T_\infty}{T_\infty} = f(\Psi / n) \tag{10.36}$$

where

$$\Psi \equiv \frac{(1 - X_r) Y_{O_2,\infty} \Delta h_c}{c_p T_\infty r}$$

is a dimensionless parameter representing absorbed chemical to sensible enthalpy.

Table 10.2 gives correlation results based on Gaussian profiles with β selected as 1 for the axisymmetric and line-fire plumes [12]. It is indeed remarkable that the 'local'

Table 10.2 Combusting region flame correlations [12]

Dimensionless variable	Axisymmetric ($X_r = 0.20$ for data)	Infinite line ($X_r = 0.30$ for data)
B	$C_l \zeta$	$C_l \zeta$
	$C_l = \frac{4}{5}\alpha,\ 0.179$	$C_l = \frac{4\alpha}{3\sqrt{\pi}},\ 0.444$
W	$C_v \zeta^{1/2}$	$C_v \zeta^{1/2}$
	$C_v = 2.02$ or $0.720\sqrt{\Psi}$	$C_v = 2.3$ or $0.877\sqrt{\Psi}$
Φ	$C_T \zeta^0$	$C_T \zeta^0$
	$C_T = 2.73$ or 0.347Ψ	$C_T = 3.1$ or 0.450Ψ
$\frac{W^2}{\Phi\zeta}$	1.49	1.70
α	0.22	0.590
β	1	1

[a] from McCaffrey data [6], but $0.50\sqrt{\Psi}$ is recommended.

Froude number $W^2/(\Phi\zeta)$ is nearly the same for the near and far field plumes, 1.60 ± 0.1. The entrainment coefficients are much larger, which probably includes pressure effects near the base for the axisymmetric fires and 'tornado-flame filament' effects for the line fire, which are actually three dimensional. It should be realized that the data corresponding to these correlations contain results for finite fires: $D > 0$, not idealized sources. The correlations in Tables 10.1 and 10.2 are one set of formulas; others exist with equal validity.

10.4 Finite Real Fire Effects

Let us examine some alternative results and consider the effects of the base fire dimension D. Measurements are given by Yokoi [3] for fire sources of Figure 10.9 and 20 cm radius ($r_o = D/2$). The temperature profiles

$$\frac{T - T_\infty}{T_m - T_\infty} \quad \text{versus} \quad \frac{z}{r_o} \quad \text{and} \quad \frac{r}{r_o}$$

are plotted in Figure 10.9. Similarity is preserved, although the Gaussian profile is not maintained in the fire zone. Also the plume radius grows more slowly in the combustion zone. This slower growth is inconsistent with theory, as presented in Table 10.2, where C_ℓ values are constant since the simple entrainment theory is not adequate in the combustion zone.

10.4.1 Turbulent axial flame temperatures

McCaffrey [6] gives results for axisymmetric fires with a square burner of side 0.3 m using methane. The results in Figure 10.10 clearly delineates three regions: (1) the

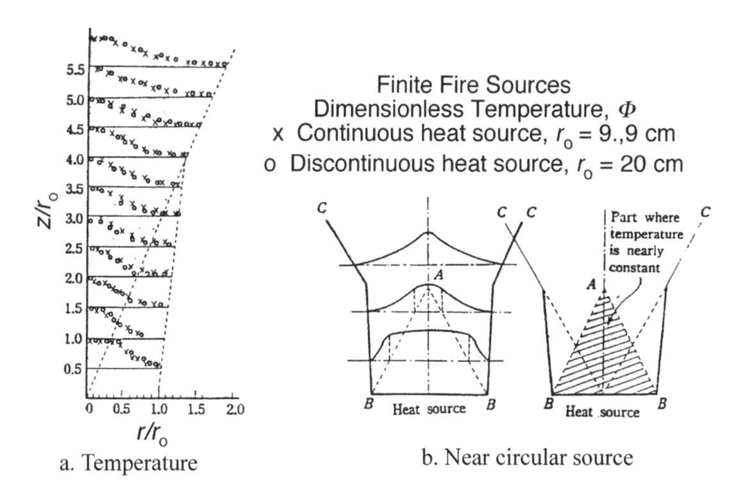

Figure 10.9 Dimensionless temperature profiles, $(T - T_\infty)/(T_m - T_\infty)$, for finite fire sources (from Yokoi [3])

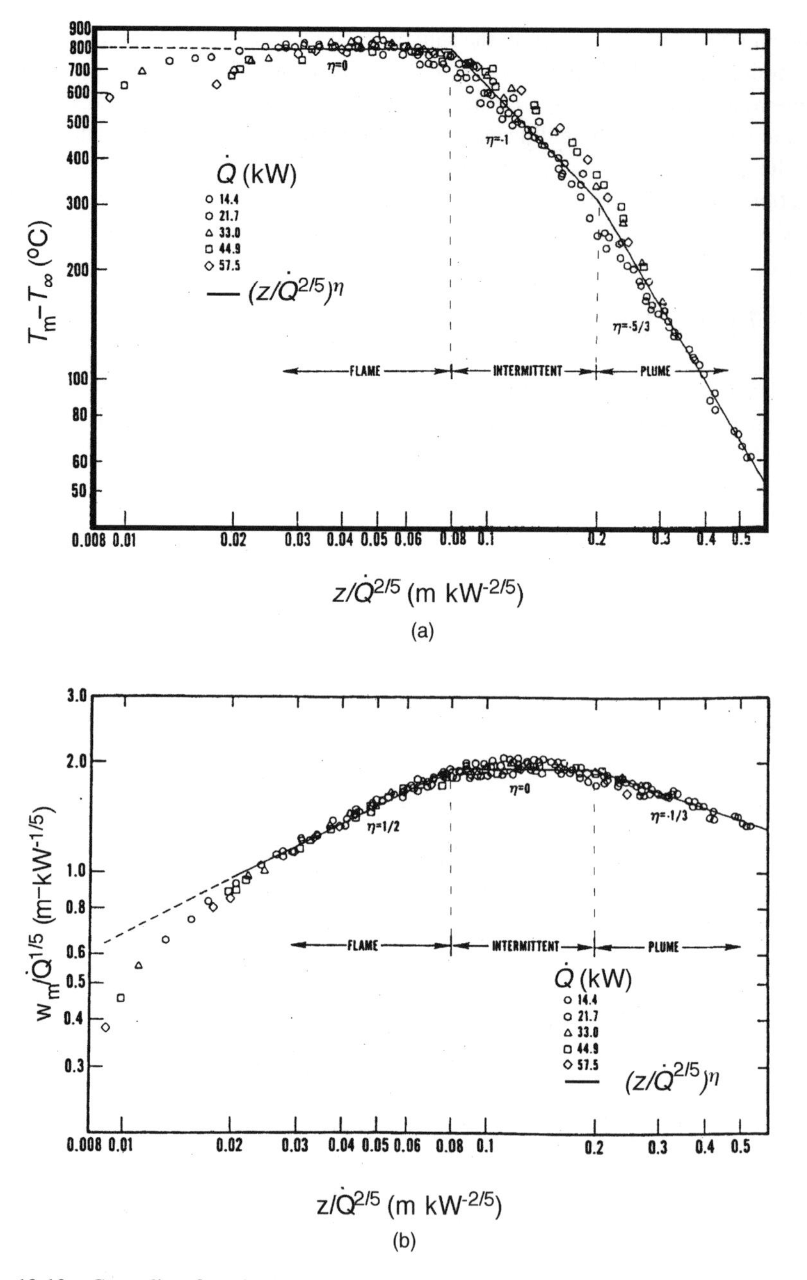

Figure 10.10 Centerline fire plume (a) temperature rise and (b) velocity (from McCaffrey [6])

continuous luminous flame, (2) the oscillating flame zone and (3) the noncombusting plume region. Similar results are displayed by Quintiere and Grove [12] in Figure 10.11 for infinite line fires of widths D ranging from about 0.1 mm to 20 cm and $\dot{Q}'$ from about 10^{-2} to 350 kW/m.

Note that the maximum flame temperature appears to be 800–900 °C for these scale fires. For larger scale fires, this flame temperature can increase with diameter; e.g. for $D \sim 2$ m,

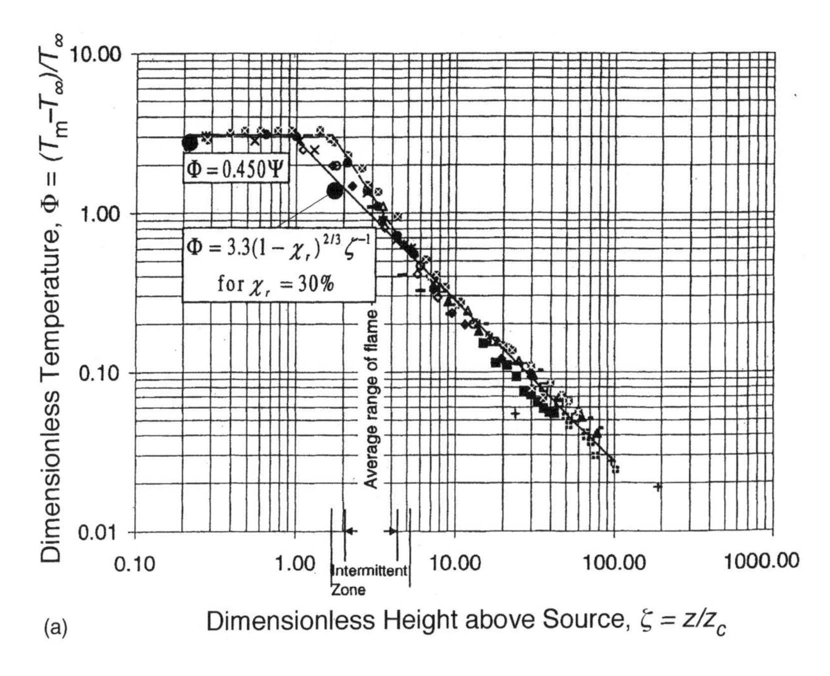

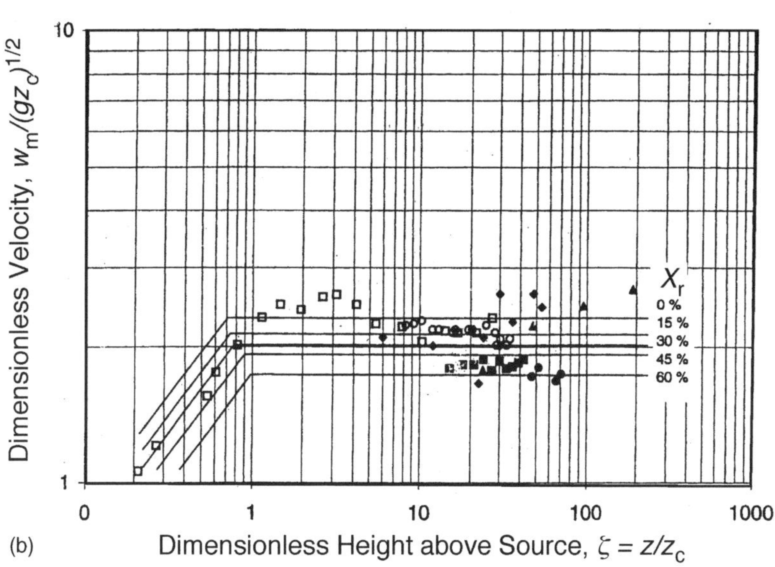

Figure 10.11 Dimensionless centerline fire plume (a) temperature rise and (b) velocity for infinite line fires of width D [12]

900 °C; 6 m, 1000 °C; 15 m, 1100 °C; and 30 m, 1200 °C [14]. The explanation is provided by Koseki [15] (Figure 10.12), showing how X_r decreases for large-diameter fires as eddies of black soot can obscure the flame. The eddy size or soot path length increases as the fire diameter increases, causing the transmittance of the external eddies to decrease and block radiation from leaving the flame. From Table 10.2,

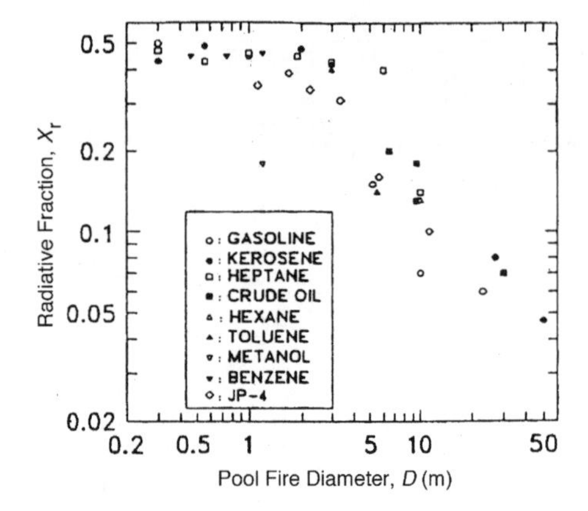

Figure 10.12 Relationship between the radiative fraction and pan diameter for various fuel fires (from Koseki [15])

the flame temperature can be found from

$$\boxed{\frac{T_\mathrm{f} - T_\infty}{T_\infty} = C_{\mathrm{T,f}}(1 - X_\mathrm{r})\left(\frac{Y_{\mathrm{O_2},\infty}\Delta h_\mathrm{c}}{rc_pT_\infty}\right)} \tag{10.37}$$

Table 10.3 gives values of $C_{\mathrm{T,f}}$ for carefully measured temperatures (50 μm Pt–Rh [16] and 1 mm [17] thermocouples); $C_{\mathrm{T,f}} = 0.50$ is recommended for $c_\mathrm{p} = 1\,\mathrm{kJ/kg\,K}$. This holds in the flame core region where combustion always exists, $z < z_\mathrm{f}$. From Koseki's Figure 10.12,

$$\frac{(T_\mathrm{f} - T_\infty)_{D=30\,\mathrm{m}}}{(T_\mathrm{f} - T_\infty)_{D=0.3\,\mathrm{m}}} = \frac{1 - X_\mathrm{r}(30\,\mathrm{m})}{1 - X_\mathrm{r}(0.3\,\mathrm{m})} \approx \frac{1 - 0.06}{1 - 0.45} = 1.7$$

This is consistent with the temperature measurements reported by Baum and McCaffrey [14],

$$\frac{(T_\mathrm{f} - T_\infty)_{D=30\,\mathrm{m}}}{(T_\mathrm{f} - T_\infty)_{D=0.3\,\mathrm{m}}} \approx \frac{1200\,^\circ\mathrm{C}}{800\,^\circ\mathrm{C}} = 1.5$$

Application of Equation (10.37) gives a maximum value of the flame temperature rise corresponding to $X_\mathrm{r} = 0$ as

$$T_\mathrm{f} - T_\infty \approx 1450\,^\circ\mathrm{C}$$

Table 10.3 Maximum measured centerline flame temperature

Maximum time averaged centerline temperature (°C)	Fuel	$\Delta h_c/s$ (kJ/g air)	D (m)	$\dot{Q}$ (kW)	X_r	Source	$C_{\mathrm{T,f}}$
1260	Natural gas	2.91	0.3	17.9	0.15	Smith and Cox [17]	0.50
1100	Natural gas	2.91	0.3	70.1	0.25	Smith and Cox [17]	0.48
930	Heptane	2.96	1.74	7713	0.43	Kung and Stavrianidis [18]	0.52
990	Heptane	3.07	1.74	973	0.18	Kung and Stavrianidis [18]	0.59

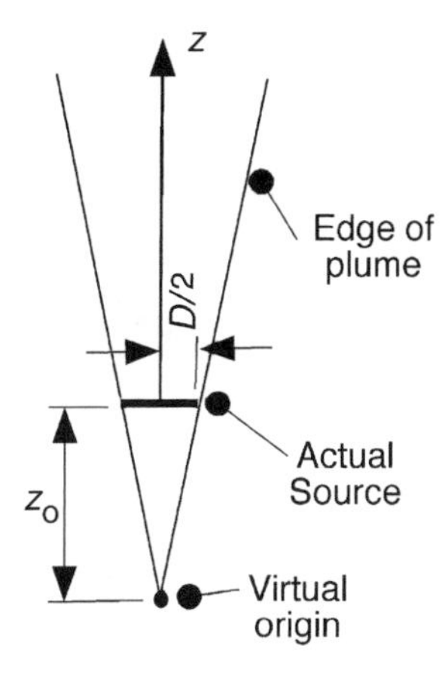

Figure 10.13 Virtual origin

Although such a value has not been reported, sufficient care has not generally been taken to obtain accurate measurements in large fires.

10.4.2 Plume temperatures

It is obvious that idealized analyses of point or line source models for fire plumes have limitations. Only at large enough distances from the fire source should they apply. One approach to bring these idealized models into conformance with finite diameter fire effects has been to modify the analysis with a virtual origin, z_o. For a finite source, a simple geometric adaption might be to locate an effective point source of energy release at a distance z_o below the actual source of diameter, D. This is illustrated in Figure 10.13 where the location of z_o is selected by making the plume width to coincide with D at $z = 0$. Various criteria have been used to determine z_o from fire plume data. From the far-field solution for temperature modified for the virtual (dimensionless) origin below the actual source, we can write from Table 10.1 in general

$$\frac{T - T_\infty}{T_\infty} = C_T(1 - X_r)^{2/3}(\zeta + \zeta_o)^{-5/3} \tag{10.38}$$

where $\zeta_o = z_o/[\dot{Q}/(\rho_\infty c_p T_\infty \sqrt{g})]^{2/5}$. This equation can be rearranged so that a new temperature variable can be plotted as a linear function of z:

$$z = -z_o + (10.58)^{3/5} \frac{\left[\frac{\dot{Q}(1-X_r)}{\rho_\infty c_p T_\infty \sqrt{g}}\right]^{2/5}}{\left(\frac{T-T_\infty}{T_\infty}\right)^{3/5}} \tag{10.39}$$

Such a plot of data is illustrated by Cox and Chitty [18] in Figure 10.14. The regression line favors the data in the intermittent flame region and in the plume immediately above the flame. For these data the virtual origin is 0.33 m below the burner. In the continuous

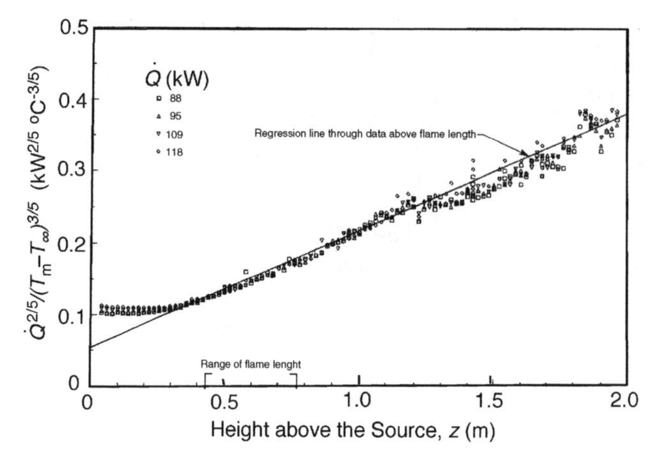

Figure 10.14 Axial temperature data plotted to reveal the virtual origin (from Cox and Chitty [18])

flame zone below $z = 0.4$ m, the flame temperature is constant, as expected. For data above $z = 1.2$ m, a linear fit for those data could yield a z_o close to zero – more consistent for far-field data. The level of uncertainty of such deductions for the correction factor z_o should be apparent.

A result for z_o must be used in the context of the specific data correlation from which it was obtained. For example, Heskestad [19] derives a correlation for the virtual origin as

$$\boxed{\frac{z_o}{D} = 1.02 - \frac{0.083\,\dot{Q}^{2/5}}{D}} \tag{10.40a}$$

where z_o and D are in m and $\dot{Q}$ is in kW. In dimensionless form this is

$$\boxed{\frac{z_o}{D} = 1.02 - 1.38\,Q_D^{*\,2/5}} \tag{10.40b}$$

where

$$Q_D^* \equiv \frac{\dot{Q}}{\rho_\infty c_p T_\infty \sqrt{g} D^{5/2}} \tag{10.41}$$

Heskestad [19] determined this result for use with the centerline far-field plume temperature correlation for $z \geq z_f$, given as

$$\boxed{\frac{T - T_\infty}{T_\infty} = 9.1(1 - X_r)^{2/3}(\zeta + \zeta_o)^{-5/3}} \tag{10.42}$$

which represents the best version of Equation (10.38). Heskestad limits the use of this equation to $T - T_\infty \leq 500\,°\text{C}$, or above the mean flame height. The data examined in

Table 10.4 Axisymmetric fire plume data for varying fuels and diameter taken from References [16], [17] and [19]

Fuel	Q_D^*	D (m)	Q (kW)	X_r	z_o/D	z_f/D
Silicone fluid	0.038	2.44	406	0.19	−0.30	0.14
Silicone fluid	0.049	1.74	211	0.19	−0.50	0.11
Natural gas, sq.[a]	0.19	0.34	14.4	—	−0.29	—
Methanol	0.19	2.44	1963	0.24	0.14	0.98
Methanol	0.22	1.74	973	0.18	−0.16	0.98
Methanol	0.26	1.22	467	0.12	−0.36	0.74
Natural gas, sq.[a]	0.29	0.34	21.7	—	−0.18	—
Natural gas, sq.[a]	0.44	0.34	33.0	—	−0.06	—
Hydrocarbon fluid	0.49	1.74	2151	0.28	0.43	2.1
Natural gas, sq.[a]	0.60	0.34	44.9	—	0.09	—
Hydrocarbon fluid	0.61	1.22	1101	0.28	0.10	1.8
Natural Gas, sq.[a]	0.77	0.34	57.5	—	0.09	—
Heptane	1.75	1.74	7713	0.43	0.71	3.2
Heptane	1.93	1.22	3569	0.41	0.43	3.2
Heptane	1.93	1.22	3517	0.36	0.43	3.2

[a] sq. implies a square rather than a circular burner was used.

deriving Equations (10.40) and (10.42) are displayed in Table 10.4 and represent a wide range of conditions.

The parameter Q_D^* is a dimensionless energy release rate of the fire. It represents the chemical energy release rate divided by an effective convective energy transport flow rate due to buoyant flow associated with the length scale D. (It was popularized by Zukoski [8] in his research and might be aptly designated the Zukoski number, $Zu \equiv Q^*$.) For natural fires, the energy release rate for pool fires of nominal diameter, D, depends on the heat transfer from the flames in which the energy release rate for a fixed diameter (D) can be widely varied by controlling the fuel supply rate. Therefore burner flames are somewhat artificial natural fires. Hasemi and Tokunaga [20] plots a range of natural fires in terms of fuel type, nominal diameter D and Q_D^*, as shown in Figure 10.15. The results show that the range for Q_D^* for natural fires is roughly 0.1 to 10.

10.4.3 Entrainment rate

Equation (10.10) gives the mass flow rate in the plume. This is exactly equal to the entrainment rate if we neglect the mass flow rate of the fuel. The latter is small, especially as z increases. For an idealized point source, the entrainment rate consistent with the far-field results of Table 10.1 is

$$\frac{\dot{m}_e}{\rho_\infty \sqrt{g}\, z^{5/2}} = C_e \zeta^{-5/6} \tag{10.43a}$$

or in terms of Q_D^*, far-field entrainment is found from

$$\boxed{\frac{\dot{m}_e}{\rho_\infty \sqrt{g}\, z^{5/2}} = C_e Q_D^{*1/3} \left(\frac{z}{D}\right)^{-5/6}} \tag{10.43b}$$

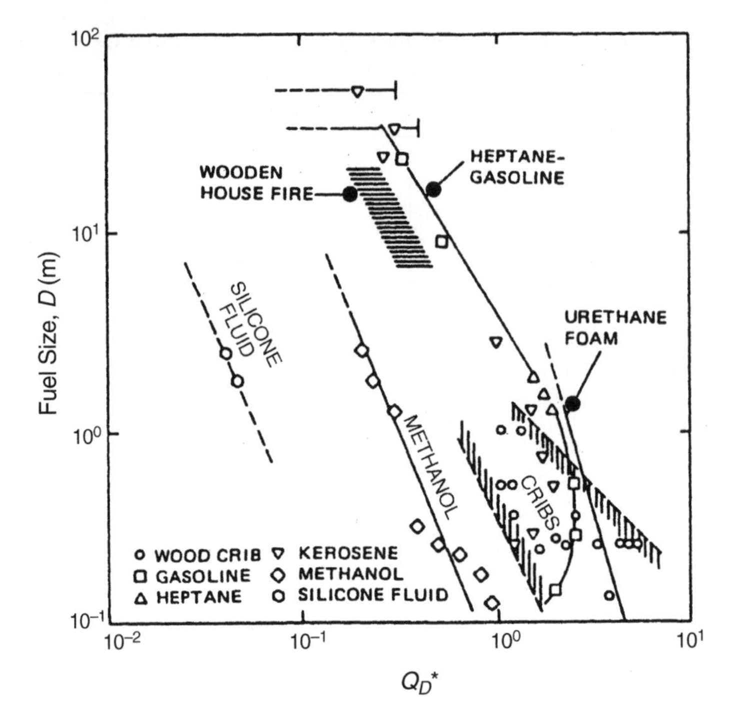

Figure 10.15 Natural fires in terms of Q_D^* by Hasemi and Tokunaga [20]

The coefficient C_e depends on X_r and is 0.17 for $X_r = 0.2$ for the empirical constants of Table 10.1; Zukoski [8] reports 0.21 and Ricou and Spalding [21] report 0.18. For the far-field, valid above the flame height, Heskestad [22] developed the empirical equation:

$$\dot{m}_e(\text{kg/s}) = 0.071[(1-X_r)\dot{Q}]^{1/3}(z+z_o)^{5/3} \times \left\{1 + 0.026[(1-X_r)\dot{Q}]^{2/3}(z+z_o)^{-5/3}\right\} \quad (10.44)$$

where $\dot{Q}$ is in kW and z and z_o are in m; z_o is given by Equation (10.40). While Eq. (10.43) is independent of D, Equation (10.44) attempts to correct for the finite diameter of the fire through the virtual origin term.

The development of an entrainment correlation for the flame region is much more complicated. Data are limited, and no single correlation is generally accepted. Delichatsios [10] shows that the flame entrainment data for burner fires of 0.1–0.5 m in diameter can be correlated over three regions by power laws in z/D depending on whether the flame is short or tall. We shall develop an equivalent result continuous over the three regions. We know the plume width is a linear function of z by Table 10.2, as well as suggested by the data of Yokoi in Figure 10.9. Although the visible flame appears to 'neck' inward and must close at its tip, the edge of the turbulent flame plume generally expands from its base. Therefore, for a fire of finite diameter, it is reasonable to represent the plume radius, b, as

$$b = \frac{D}{2} + C_\ell z \tag{10.45}$$

where C_ℓ is an empirical constant. Due to the constancy of $W^2/\Phi\zeta$, w is proportional to $z^{1/2}$ in the flame zone (see Table 10.2 and Figure 10.10(b)). Hence, substituting these forms for b and w into Equation (10.10) gives

$$\dot{m}_e = \pi \rho_\infty C_v z^{1/2} \left(\frac{D}{2} + C_\ell z \right)^2 \tag{10.46a}$$

or in dimensionless form, entrainment in the flame is found from

$$\boxed{\frac{\dot{m}_e}{\rho_\infty \sqrt{g}\, D^{5/2}} = C_e \left(\frac{z}{D} \right)^{1/2} \left[1 + 2C_\ell \left(\frac{z}{D} \right) \right]^2} \tag{10.46b}$$

where $C_\ell = 0.179$ and $C_e = 0.057\Psi^{1/2}$, approximately. These coefficients have been found to fit the data for burners of 0.19 and 0.5 m as given by Zukoski [8]. The results are plotted in Figure 10.16. Since velocity depends on plume temperature and temperature depends on radiation loss, it should be expected that C_e depends on X_r. However, it is not entirely obvious, nor is it generally accepted, that this flame entrainment rate is independent of $\dot{Q}$. This fact was demonstrated some time ago by Thomas, Webster and Raftery [23] and is due to the constancy of the flame temperature regardless of $\dot{Q}$. Essentially, the perimeter flame area is significant for entrainment.

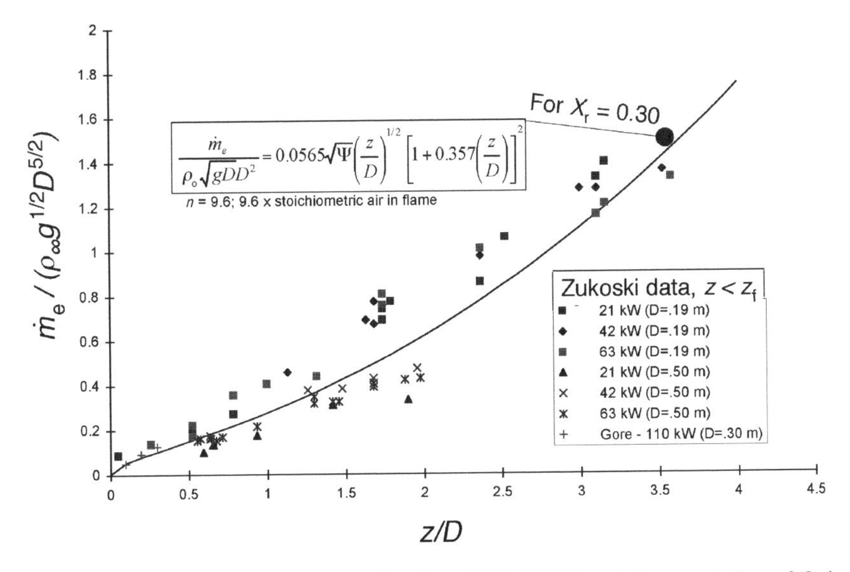

Figure 10.16 Flame entrainment rate by data from Zukoski [8] and the correlation of Quintiere and Grove [12]

10.4.4 Flame height

The flame height is intimately related to the entrainment rate. Indeed, one is dependent on the other. For a turbulent flame that can entrain n times the air needed for combustion (Equation (10.34)), and r, the mass stoichiometric oxygen to fuel ratio, the mass rate of fuel reacted over the flame length, z_f, is

$$\dot{m}_F = \frac{\dot{m}_e Y_{O_2,\infty}}{nr} \tag{10.47}$$

where $\dot{m}_e$ is taken from Equation (10.46) for $z = z_f$ and $Y_{O_2,\infty}$ is the oxygen mass fraction of the ambient. By multiplying by Δh_c, the heat of combustion (Equation (10.47)) can be put into dimensionless form as

$$Q_D^* = \frac{\left(\dfrac{\dot{m}_e}{\rho_\infty \sqrt{g}\, D^{5/2}}\right)\left(\dfrac{Y_{O_2,\infty}\Delta h_c}{r c_p T_\infty}\right)}{n} \tag{10.48}$$

Substitution of Equation (10.46b) with $z = z_f$ gives a correlation for flame height. The value n is found by a best fit of data in air $(Y_{O_2,\infty} = 0.233)$ as 9.6. The implicit equation for z_f in air is

$$\boxed{Q_D^* = 0.005\,90\left(\frac{\Psi^{3/2}}{1 - X_r}\right)\left(\frac{z_f}{D}\right)^{1/2}\left[1 + 0.357\left(\frac{z_f}{D}\right)\right]^2} \tag{10.49a}$$

In air for $z_f/D > 0.1$, an approximation for the function in z_f/D gives

$$\frac{z_f}{D} \approx 16.8 \frac{Q_D^{*2/5}}{[(\Delta h_c/s)/(c_p T_\infty)]^{3/5}(1 - X_r)^{1/5}} - 1.67 \tag{10.49b}$$

and for $z_f/D < 0.1$, emphasizing the half-power,

$$\frac{z_f}{D} \approx 2.87 \times 10^4 \frac{Q_D^{*2}}{[(\Delta h_c/s)/(c_p T_\infty)]^3(1 - X_r)} \tag{10.49c}$$

Here s is the stoichiometric air to fuel ratio.

An alternative, explicit equation for z_f given by Heskestad [24], based on the virtual origin correction (see Equation (10.40)), is given as

$$\boxed{\frac{z_f}{D} = 15.6\, Q_D^{*2/5}\left(\frac{s c_p T_\infty}{\Delta h_c}\right)^{3/5} - 1.02} \tag{10.50a}$$

or

$$z_f(\text{m}) = 0.23\,\dot{Q}^{2/5} - 1.02\, D(\text{m}), \quad \dot{Q} \text{ in kW} \tag{10.50b}$$

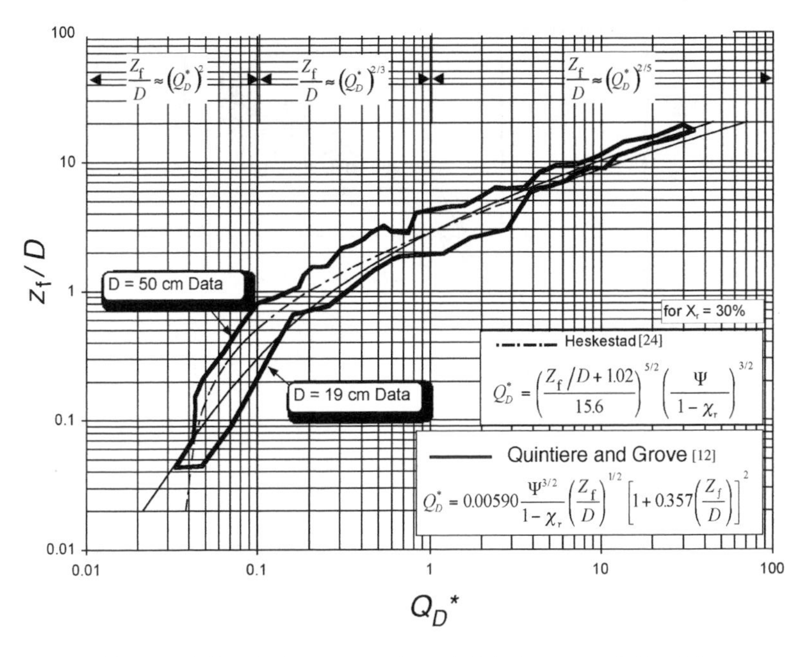

Figure 10.17 Axisymmetric flame height in terms of Q_D^* [12]

The equations are compared in Figure 10.17 along with a region of data taken from Zukoski [8]. Additional data presented by Zukoski [25] for the small Q_D^* range are presented in Figure 10.18. The straight lines show that the dependencies, consistent with Equation (10.49), are for small Q_D^*, $z_f/D \sim Q_D^{*2}$, and for large Q_D^*, $z_f/D \sim Q_D^{*2/5}$. However, this large Q_D^* behavior only applies as long as initial fuel momentum effects are negligible compared to the buoyancy velocity, i.e. w_F (fuel velocity) $\ll \sqrt{gz}$. Once the fuel injection velocity is large, the process becomes a jet flame.

10.4.5 *Jet flames*

Hawthorne, Weddell and Hottel [26] showed that the height of a turbulent jet flame is approximately constant for a given fuel and nozzle diameter. If it is truly momentum dominated, the flame length will not change with increasing fuel flow rate. This can approximately be seen by using similar reasoning as in Section 10.4.4, but realizing that in the flame region the fuel nozzle velocity, w_F, dominates. From Equations (10.10) and (10.45), with $w = w_F$, we write for the entrainment of the jet flame:

$$\dot{m}_e \approx \pi\rho_\infty w_F\left(\frac{D}{2} + C_{\ell,j}z\right)^2 \tag{10.51}$$

Turbulent jet flames have $z_f \gg D$, so we can neglect D in Equation (10.51). The fuel mass flow rate is given by

$$\dot{m}_F = \rho_F\frac{\pi}{4}D^2 w_F \tag{10.52}$$

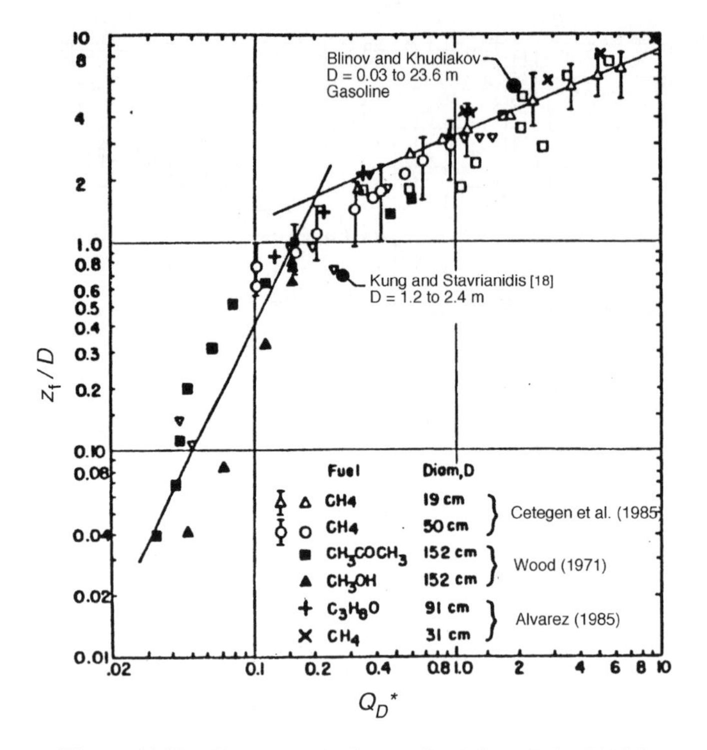

Figure 10.18 Flame lengths for small Q^* from Zukoski [25]

where ρ_F is the fuel density. Substituting Equations (10.51) and (10.52) into Equation (10.47) gives

$$\dot{m}_F \approx \left(\frac{\rho_\infty}{\rho_F}\right) \dot{m}_F Y_{O_2,\infty} \frac{C_{\ell,j}^2 (z_f/D)^2}{n_j r}$$

or

$$\boxed{\frac{z_f}{D} \approx \frac{\sqrt{n_j}}{2C_{\ell,j}} \left(\frac{\rho_F}{\rho_\infty} \frac{r}{Y_{O_2,\infty}}\right)^{1/2}} \tag{10.53}$$

where n_j and $C_{\ell j}$ are constants. Hence, turbulent jet flame height depends on nozzle diameter, fuel density and stoichiometry. McCaffrey [6] shows that this behavior occurs between Q_D^* of 10^5 to 10^6. Note

$$Q_D^* = \frac{\pi}{4} \left(\frac{\rho_F}{\rho_\infty}\right) \left(\frac{w_F}{\sqrt{gD}}\right) \left(\frac{\Delta h_c}{c_p T_\infty}\right) \tag{10.54}$$

showing Q_D^* is related to a fuel Froude number $w_F^2/(gD)$.

10.4.6 Flame heights for other geometries

It is important to be able to determine the turbulent flame heights for various burning configurations and geometrical constraints. Since flame length is inversely related to air or oxygen entrainment, constraints that limit entrainment, such as a fire against a wall or in a corner, will result in taller flames for the same $\dot{Q}$. For example, Hasemi and Tokunaga [20] empirically find, for a square fire of side D placed in a right-angle corner,

$$z_f/D = C_f\, Q_D^{*2/3}, \quad \dot{Q} \text{ in kW} \tag{10.55}$$

where $C_f = 4.3$ for the flame tip and $C_f = 3.0$ for the continuous flame height.

The interaction of axisymmetric flames with a ceiling suggests that the radial flame extension is approximately

$$R_f \approx C_f(z_{f,o} - H) \tag{10.56}$$

where C_f can range from 0.5 to 1, $z_{f,o}$ is the height of the free flame with no ceiling and H is the height of the fuel source to the ceiling [27,28].

For narrow line fires of $D/L \leq 0.1$ (see Figure 10.6), it can be shown that

$$\frac{z_f}{D} = \text{function}\left(\frac{z_{c,\text{line}}}{D}\right)$$

following Equation (10.24). An empirical fit to the collection of data shown in Figure 10.19 gives [12]

$$\frac{\dot{Q}'}{\rho_\infty c_p T_\infty \sqrt{g}\, D^{3/2}} = 0.005\,90 \frac{\Psi^{3/2}}{1 - X_r}\left(\frac{z_f}{D}\right)^{1/2}\left[1 + 0.888\left(\frac{z_f}{D}\right)\right] \tag{10.57}$$

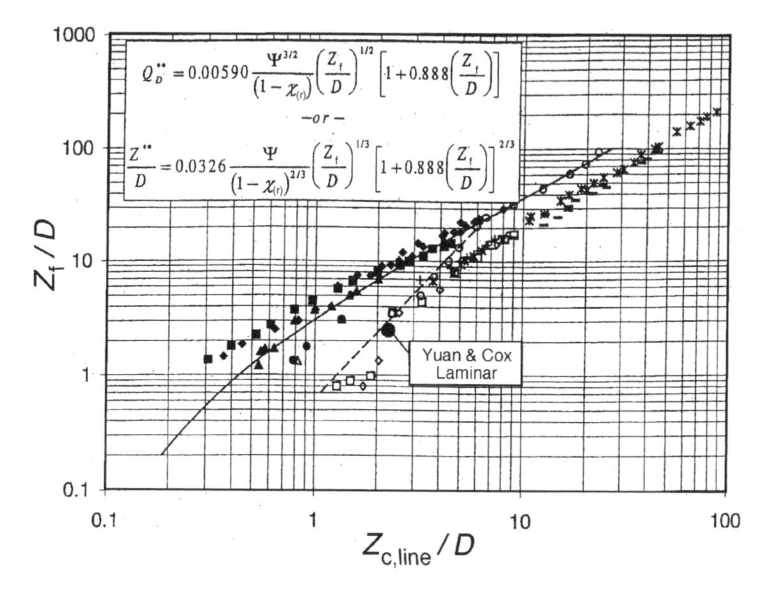

Figure 10.19 Flame height for a line fire of width D [12]

Some data for small D (1.5 cm) and $z_f/D < 10$ have been observed to be laminar; other data for z_f/D fall below the correlation, suggesting inconsistent definitions for flame height. Equation (10.57) is the counterpart to Equation (10.49) for the axisymmetric case, and was developed in a similar manner.

Wall fires are an extension of the free plume line and axisymmetric fires. Based on the data of Ahmad and Faeth [29], for turbulent wall fires, it has been shown that

$$z_f = \frac{1.02}{Y_{O_2,\infty}(1 - X_r)^{1/3}} \left(\frac{\dot{Q}'}{\rho_\infty c_p T_\infty \sqrt{g}} \right)^{2/3} \tag{10.58}$$

For a square fire of side D against a wall Hasemi and Tokunaga [20] find that

$$\frac{z_f}{D} = C_f Q_D^{*2/3} \tag{10.59}$$

where $C_f = 3.5$ for the flame tip and $C_f = 2.2$ for the continuous flame.

10.5 Transient Aspects of Fire Plumes

For simplicity let us consider an idealized point (or line) transient plume. Then by an extension of Equation (10.27), it can be shown that the plume dimensionless variables would depend on the axial position and time as

$$\{\Phi, W, B\} \quad \text{as functions of} \quad \left\{ \frac{z}{z_c}, \frac{t}{\sqrt{z_c/g}} \right\}$$

One approach is to approximate the unsteady solution by a quasi-steady solution where $\dot{Q}(t)$ is a function of time and time is adjusted. This can be approximated by determining the transport time (t_o) from the origin to position z by

$$t_o = \int_o^z \frac{dz}{w} \tag{10.60}$$

where w is taken as the steady state solution with $\dot{Q} = \dot{Q}(t)$. Then the unsteady solution can be approximated by adjusting t to $t + t_o$ at z, e.g.

$$\Phi\left(\frac{z}{z_c}, \frac{(t + t_o)}{\sqrt{z_c/g}} \right) \approx \phi_{\text{steady}}\left(\frac{z}{z_c}, \dot{Q}(t) \right) \tag{10.61}$$

More specific results are beyond the scope of our limited presentation for plumes. However, we will examine some gross features of transient plumes; namely (a) the rise of a starting plume and (b) the dynamics of a 'fire ball' due to the sudden release of a finite burst of gaseous fuel. Again, our philosophy here is not to develop exact solutions, but to represent the relevant physics through approximate analyses. In this way, experimental correlations for the phenomena can be better appreciated.

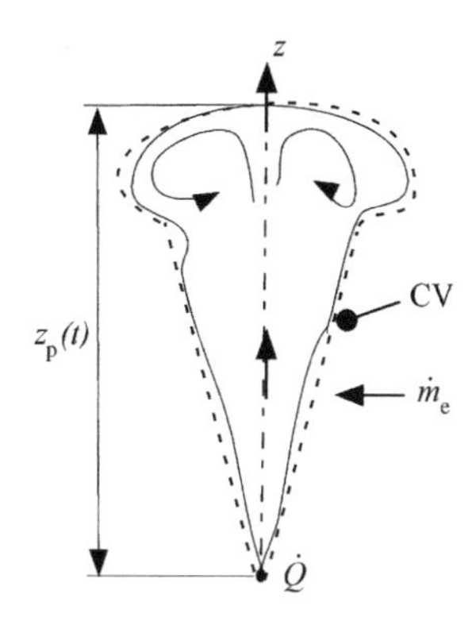

Figure 10.20 Starting plume

10.5.1 Starting plume

We consider a point source axisymmetric example as illustrated in Figure 10.20. We shall examine the rise of the plume, z_p, as a function of time. As the hot gases rise due to the source $\dot{Q}$, initiated at time $t = 0$, the gases at the 'front' or 'cap' encounter cooler ambient air. The hot gases in the cap, impeded by the air, form a recirculating zone, as illustrated in Figure 10.20. Entrainment of air occurs over the vertical plume column and the cap. The warmed, entrained air forms the gases in the plume.

The height of the rising plume can be determined from a conservation of mass for a control volume, enclosing the plume as it rises. From Equation (3.15), it follows that

$$\frac{\mathrm{d}m_\mathrm{p}}{\mathrm{d}t} = \dot{m}_\mathrm{e} + \dot{m}_\mathrm{F} \tag{10.62}$$

where

m_p = mass of the plume
$\dot{m}_e$ = rate of air entrained
$\dot{m}_F$ = rate of fuel mass supplied

We make the following assumptions in order to estimate each term:

1. The rate of fuel supply is small compared to the rate of air entrained, $\dot{m}_F \approx 0$.
2. The plume properties can be treated as quasi-steady. This is equivalent to applying Equation (10.60) with t_o small or negligible compared to t. Then at an instant of time the steady state solutions apply.
3. In keeping with assumption 2, the plume cap is ignored, and the top of the rising plume is treated as a steady state plume truncated at z_p.
4. The rising plume has a uniform density, ρ_∞, equal to the ambient.
5. The rising plume can be treated as an inverted cone of radius, b, as given in Table 10.1, $b = \frac{6}{5}\alpha_o z$.

As a consequence, the mass of the plume can be expressed as

$$m_{\mathrm{p}} = \rho_\infty \frac{\pi}{3} \left(\frac{6}{5} \alpha_{\mathrm{o}} z_{\mathrm{p}} \right)^2 z_{\mathrm{p}} \tag{10.63}$$

Substituting into Equation (10.62), along with Equation (10.43) for $\dot{m}_{\mathrm{e}}$, gives

$$\rho_\infty \frac{13\pi}{25} \alpha_{\mathrm{o}}^2 \frac{\mathrm{d}}{\mathrm{d}t}(z_{\mathrm{p}}^3) = \rho_\infty \sqrt{g} C_{\mathrm{e}} z_{\mathrm{c}}^{5/6} z_{\mathrm{p}}^{5/3} \tag{10.64}$$

where z_{c} is the characteristic plume length scale, given in Equation (10.23). Rearranging, and realizing that α_{o} and C_{e} are particular constants associated with the entrainment characteristics of the rising plume, we write

$$\frac{\mathrm{d}z_{\mathrm{p}}^3}{\mathrm{d}t} = C'_{\mathrm{p}} \sqrt{g} z_{\mathrm{c}}^{5/6} z_{\mathrm{p}}^{5/3} \tag{10.65}$$

where C'_{p} is an empirical constant. Integrating from $t = 0$, $z_{\mathrm{p}} = 0$, we obtain

$$\int_0^{z_{\mathrm{p}}} z_{\mathrm{p}}^{1/3} \mathrm{d}z_{\mathrm{p}} = C'_{\mathrm{p}} \sqrt{g} z_{\mathrm{c}}^{5/6}\, t$$

or

$$\frac{3}{4} z_{\mathrm{p}}^{4/3} = C'_{\mathrm{p}} \sqrt{g} z_{\mathrm{c}}^{5/6}\, t \tag{10.66}$$

In dimensionless form, with a new empirical constant, C_{p},

$$\boxed{\left(\frac{z_{\mathrm{p}}}{z_{\mathrm{c}}} \right)^{4/3} = C_{\mathrm{p}} \left(\frac{\sqrt{g}\, t}{\sqrt{z_{\mathrm{c}}}} \right)} \tag{10.67}$$

Zukoski [8] claims that C_{p} can range from 3 to 6, and suggests a value from experiments by Turner [30] of 3.5. Recent results by Tanaka, Fujita and Yamaguchi [31] from rising fire plumes, shown in Figure 10.21, find a value of approximately 2.

These results apply to plume rise in a tall open space of air at a uniform temperature. The results can be important for issues of fire detection and sprinkler response. Plume rise in a thermally stratified stable ($\mathrm{d}T_\infty/\mathrm{d}z > 0$) atmosphere will not continue indefinitely. Instead, it will slow and eventually stop and form a horizontal layer. It will stop where its momentum becomes zero, roughly when the plume temperature is equal to the local ambient temperature.

10.5.2 Fireball or thermal

The cloud of rising hot gas due to the sudden release of a finite amount of energy is called a thermal. If this occurs due to the sudden release of a finite volume of gaseous fuel that is ignited, the thermal then initially rises as a combusting region. This is depicted by the

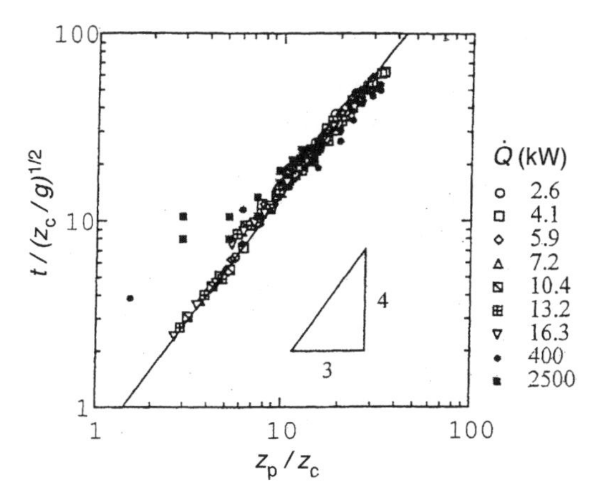

Figure 10.21 Rise time of starting fire plumes (from Tanaka, Fujita and Yamaguchi [31])

photographs in Figure 10.22 presented by Fay and Lewis [32]. Before burnout, the combusting region approximates a spherical shape, or fireball. We shall follow the analysis by Fay and Lewis [32] to describe the final diameter of the fireball (D_b), its ultimate height at burnout (z_b) and the time for burntout (t_b). Consider the rising fireball

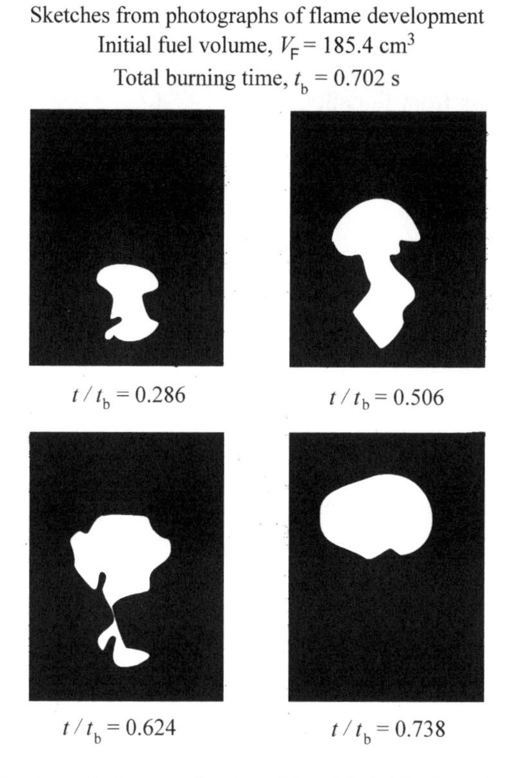

Figure 10.22 Sketches of photographs of a rising 'fireball' (from Fay and Lewis [32])

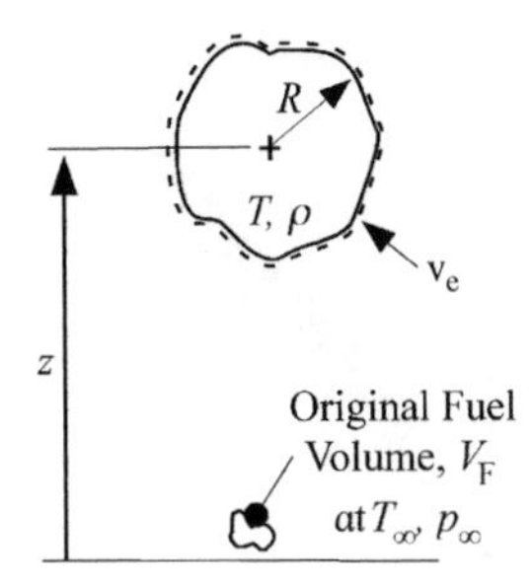

Figure 10.23 Dynamics of a spherical fireball

in a control volume moving with the cloud, as shown in Figure 10.23. The release fuel had a volume V_F at temperature T_∞ and atmospheric pressure p_∞. The following approximations are made:

1. The fireball has uniform properties.
2. The relative, normal entrainment velocity is uniform over the fireball and proportional to the rising velocity, dz/dt, i.e. $v_e = \alpha_o dz/dt$, with α an entrainment constant.
3. The pressure on the fireball is constant except for pressure effects due to buoyancy.
4. The actual air entrained is n times the stoichiometric air needed for combustion.

By conservation of mass from Equation (3.15),

$$\frac{d}{dt}\left(\rho \frac{4}{3}\pi R_b^3\right) = \rho_\infty \alpha_o \frac{dz}{dt} 4\pi R_b^2 \tag{10.68}$$

Conservation of energy gives, from Equation (3.48), which is similar to the development of Equation (10.35),

$$\rho c_p \frac{4}{3}\pi R_b^3 \frac{dT}{dt} + \dot{m}_e c_p (T - T_\infty) = (1 - X_r)\dot{m}_e Y_{O_2,\infty} \frac{\Delta h_c}{nr} \tag{10.69}$$

Since the mass of the fireball is small and changes in temperature are likely to be slow, we will ignore the transient term in Equation (10.69). These lead to a constant fireball temperature identical to the fire temperature of Equation (10.35):

$$T - T_\infty = \frac{(1 - X_r) Y_{O_2,\infty} \Delta h_c}{nr\, c_p} \tag{10.70}$$

By the ideal gas law, assuming no change in molecular weight, $p = \rho RT/M$, or density must also be constant. Consequently, Equation (10.68) becomes

$$\rho \frac{dR_b}{dt} = \rho_\infty \alpha_o \frac{dz}{dt} \tag{10.71}$$

or integrating with $R_b = \left(\frac{3}{4\pi} V_b\right)^{1/3} \approx 0$ at $t = 0$:

$$z \approx \left(\frac{\rho}{\alpha_o \rho_\infty}\right) R_b = \left(\frac{T_\infty}{\alpha_o T}\right) \left(\frac{3}{4\pi} V_b\right)^{1/3} \tag{10.72}$$

We will now apply conservation of vertical momentum. Following Example 3.3, the buoyant force counteracting the weight of the fireball is equal to the weight of ambient atmosphere displaced. For the fireball of volume, $V_b = \frac{4}{3}\pi R_b^3$,

$$\frac{d}{dt}\left(\rho V_b \frac{dz}{dt}\right) = (\rho_\infty - \rho) g V_b \tag{10.73}$$

where drag effects have been ignored. By substituting for z from Equation (10.72) and integrating, it can be shown that

$$R_b = \left(\frac{\alpha_o}{14}\right) \left(\frac{T}{\rho_\infty}\right) \left(\frac{T}{T_\infty} - 1\right) t^2 \tag{10.74}$$

The maximum height and diameter for the fire occur when all of the fuel burns out at $t = t_b$. The mass of the fireball at that time includes all of the fuel (m_F) and all of the air entrained (m_e) up to that height. The mass of air entrained can be related to the excess air factor, n, the stoichiometric oxygen to fuel mass ratio, r, and the ambient oxygen mass fraction, $Y_{O_2,\infty}$. Thus the fireball mass at burnout is

$$\begin{aligned} m_b &= m_F + m_e \\ &= m_F \left(1 + \frac{nr}{Y_{O_2,\infty}}\right) \end{aligned} \tag{10.75}$$

In terms of the original fuel volume, V_F, at T_∞ and p_∞, by the perfect gas relationship at constant pressure and approximately constant molecular weights,

$$\begin{aligned} V_b &= \frac{m_b}{m_F} \frac{T}{T_\infty} V_F \\ &= \left(1 + \frac{nr}{Y_{O_2,\infty}}\right) \left(\frac{T}{T_\infty}\right) V_F \end{aligned} \tag{10.76}$$

For a given fuel, Equations (10.70), (10.72), (10.74) and (10.76) suggest that the maximum fireball height and diameter behaves as

$$\boxed{z_b \quad \text{and} \quad D_b \sim V_F^{1/3}} \tag{10.77}$$

and the burnout time follows:

$$\boxed{t_b \sim V_F^{1/6}} \tag{10.78}$$

This is supported by the data of Fay and Lewis in Figures 10.24(a) to (c) [32].

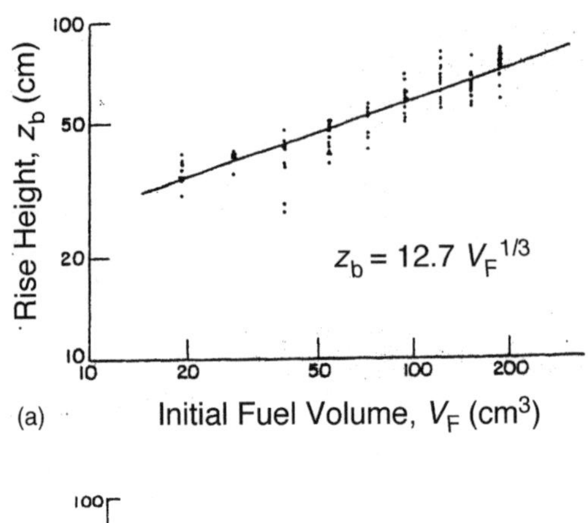

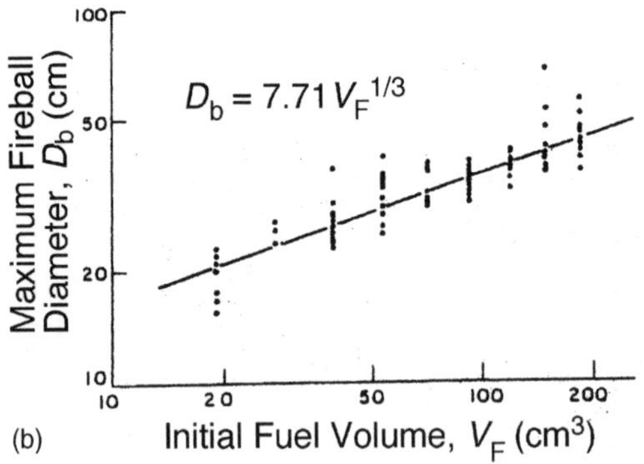

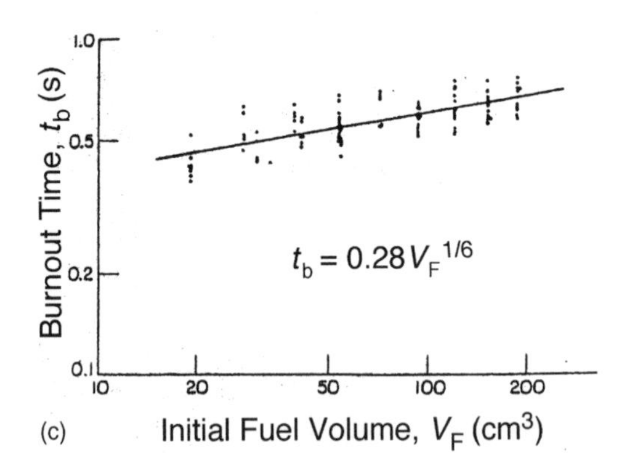

Figure 10.24 Fireball dynamics for propane (from Fay and Lewis [32]): (a) flame height, z_b, (b) burn-out time, t_b, (c) maximum fireball diameter, D_b

References

1. Rouse, H., Yih, C.S. and Humphries, H.W., Gravitational convection from a boundary source, *Tellus*, 1952, **4**, 202–10.
2. Morton, B.R., Taylor, G.I. and Turner, J.S. Turbulent gravitational convection from maintained and instanteous sources, *Proc. Royal Soc. London, Ser. A*, 1956, **234**, 1–23.

3. Yokoi, S., Study on the prevention of fire spread caused by hot upward current, Building Research Report 34, Japanese Ministry of Construction, 1960.
4. Steward, F.R., Linear flame heights for various fuels, *Combustion and Flame*, 1964, **8**, 171–78.
5. Steward, F.R., Prediction of the height of turbulent buoyant diffusion flames, *Combust. Sci. Technol.*, 1970, **2**, 203–12.
6. McCaffrey, B.J., *Purely Buoyant Diffusion Flames: Some Experimental Results*, NBSIR 79–1910, National Bureau of Standards, Gaithersburg, Maryland, 1979.
7. Pagni, P.J., Pool fire vortex shedding frequencies, *Applied Mechanics Review*, 1990, **43**, 153–70.
8. Zukoski, E.E., Properties of fire plumes, in *Combustion Fundamentals of Fire* (ed. G. Cox), Academic Press, London, 1995.
9. Heskestad, G., Fire plumes, in *The SFPE Handbook of Fire Protection Engineering*, 2nd edn, (eds P.J. DiNenno *et al.*) Section 2, National Fire Protection Association, Quincy, Massachusetts, 1995, pp. **2**-9 to **2**-19.
10. Delichatsios, M.A., Air entrainment into buoyant jet flames and pool fires, in *The SFPE Handbook of Fire Protection Engineering*, 2nd edn (eds P.J. DiNenno *et al.*), Section 2, National Fire Protection Association, Quincy, Massachusetts, 1995, pp. **2**-20 to **2**-31.
11. McCaffrey, B., Flame heights, in *The SFPE Handbook of Fire Protection Engineering*, 2nd edn (eds P.J. DiNenno *et al.*), Section 2, National Fire Protection Association, Quincy, Massachusetts, 1995, pp. **2**-1 to **2**-8.
12. Quintiere, J.G. and Grove, B.S., A united analysis for fire plumes, *Proc. Comb. Inst.*, 1998, **27**, pp. 2757–86.
13. Yuan, L. and Cox, G., An experimental study of some fire lines, *Fire Safety J.*, 1996, **27**, 123–39.
14. Baum, H.R. and McCaffrey, B.J., Fire induced flow field: theory and experiment, in *2nd Symposium (International) on Fire Safety Science*, Hemisphere, New York, 1988, pp. 129–48.
15. Koseki, H., Combustion characteristics of hazardous materials, in The Symposium for the NRIFD 56th Anniversary, Mitaka, Tokyo, National Research Institute of Fire and Disaster, 1 June 1998.
16. Smith, D.A. and Cox, G., Major chemical species in buoyant turbulent diffusion flames, *Combustion and Flame, 1992,* **91**, 226–38.
17. Kung, H.C. and Stavrianidis, P., Buoyant plumes of large scale pool fires, *Proc. Comb. Inst.*, 1983, **19**, p. 905
18. Cox, G. and Chitty, R., Some source-dependent effects of unbounded fires, *Combustion and Flame*, 1985, **60**, 219–32.
19. Heskestad, G., Virtual origins of fire plumes, *Fire Safety J.*, 1983, **5**, 109–14.
20. Hasemi, Y. and Tokunaga, T., Flame geometry effects on the buoyant plumes from turbulent diffusion flames, *Combust. Sci. Technol.*, 1984, **40**, 1–17.
21. Ricou, F.P. and Spalding, B.P., Measurements of entrainment by axial symmetric turbulent jets, *J. Fluid Mechanics*, 1961, **11**, 21–32.
22. Heskestad, G., Fire plume air entrainment according to two competing assumptions, *Proc. Comb. Inst.*, 1986, **21**, pp. 111–20.
23. Thomas, P.H., Webster, C.T. and Raftery, M.M., Experiments on buoyant diffusion flames, *Combustion and Flame*, 1961, **5**, 359–67.
24. Heskestad, G., A reduced-scale mass fire experiment, *Combustion and Flame*, 1991, **83**, 293–301.
25. Zukoski, E.E., Fluid dynamics of room fires, in *Fire Safety Science, Proceedings of the 1st International Symposium*, Hemisphere, Washington, DC, 1986.
26. Hawthorne, W.R., Weddell, D.S. and Hottel, H.C., Mixing and combustion in turbulent gas jets, *Proc. Comb. Inst.*, 1949, **3**, pp. 266–88.

27. You, H.Z. and Faeth, G.M., Ceiling heat transfer during fire plume and fire impingment, *Fire and Materials*, 1979, **3**, 140–7.
28. Gross, D., Measurement of flame length under ceilings, *Fire Safety J.*, 1989, **15**, 31–44.
29. Ahmad, T. and Faeth, G.M., Turbulent wall fires, *Proc. Comb. Inst.*, 1978, **17**, p. 1149–1162.
30. Turner, J. S., The 'starting plume' in neutral surroundings, *J. Fluid Mechanics*, 1962, **13**, 356.
31. Tanaka, T., Fujita, T. and Yamaguchi, J., Investigation into rise time of buoyant fire plume fronts, *Int. J. Engng Performanced-Based Fire Codes*, 2000, **2**, 14–25.
32. Fay, J.A. and Lewis, D.H., Unsteady burning of unconfined fuel vapor clouds, *Proc. Comb. Inst.*, 1976, **16**, pp. 1397–1405.

Problems

10.1 A 1.5 diameter spill of polystyrene beads burns at 38 g/m^2 s. Its flame radiative fraction is 0.54.

(a) Compute the centerline temperature at 0.5 m above the base.

(b) Compute the centerline temperature at 4.0 m above the base.

(c) Compute the flame height.

10.2 Turbulent fire plume temperatures do not generally exceed 1000 °C. This is well below the adiabatic flame temperature for fuels. Why?

10.3 A heptane pool fire of diameter 10 m burns at 66 g/m^2 s. Its radiative fraction is 0.10 and the heat of combustion is 41 kJ/g.

(a) Calculate its flame height.

(b) Calculate its centerline temperature at a height of *twice* its flame length.

10.4 Given: an axisymmetric fire burns at the following rates:

Diameter (D) in m	$\dot{m}''$ in g/m^2 s
0.5	20
1.0	45
2.0	60

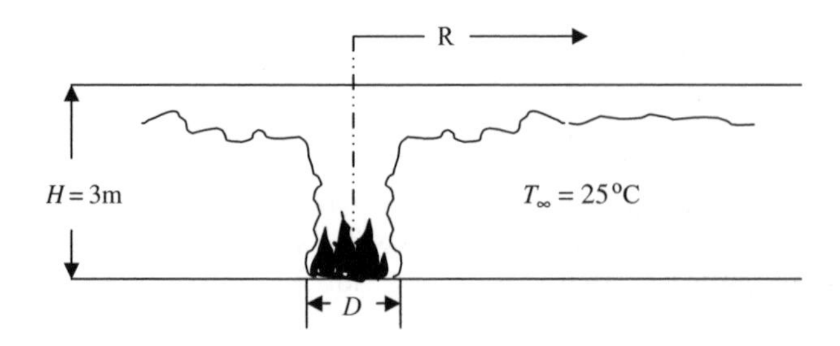

The fuel heat of combustion is 40 kJ/g and its radiative fraction (X_r) is 0.30.

For each diameter, compute the following:

(a) Gas temperature at the ceiling above the fire (maximum).

(b) Maximum velocity impinging on the ceiling above the fire.

(c) The extent of the flame, its height and any interaction with the ceiling.

(d) The maximum radiant heat flux to a target a radius r from the center of the pool fire can be estimated by $X_r\dot{Q}/(4\pi r^2)$. Compute the heat flux at 3 m.

10.5 For a constant energy release rate fire, will its flame height be higher in a corner or in an open space? Explain the reason for your answer.

10.6 Assume that the diameter of the fire as small and estimate the temperature at the ceiling when the flame height is (a) 1/3, (b) 2/3 and (c) equal to the height of the ceiling. The fire is at the floor level and the distance to the ceiling is 3 m. Also calculate the corresponding energy release rates of the fire for (a), (b) and (c). The ambient temperature, $T_\infty = 25\,°\text{C}$.

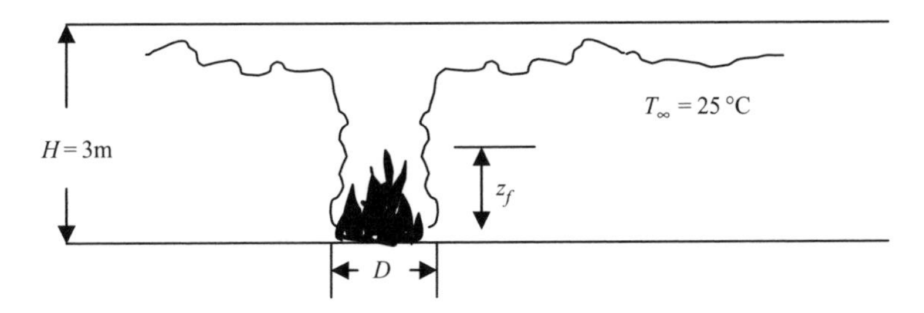

10.7 A steel sphere is placed directly above a fire source of 500 kW. The height of the sphere is 3 m. Its properties are given below:

Initial temperature, $T_o = 20\,°\text{C}$.

Specific heat, $c_p = 460\,\text{J/kg K}$

Density, $\rho = 7850\,\text{kg/m}^3$

Radius, $R = 1\,\text{cm}$

Volume of sphere $= (4/3)\pi R^3$

Suface area of sphere $= 4\pi R^2$

(a) What is the gas temperature at the sphere?

(b) What is the gas velocity at the sphere?

(c) How long will it take for the sphere to reach 100 °C?

Assume convective heating only with a constant heat transfer coefficient, $h = 25\ \mathrm{W/m^2\ K}$.

10.8 A 1 MW fire occurs on the floor of a 15 m high atrium. The diameter of the fire is 2 m. A fast response sprinkler link at the ceiling directly above the fire will activate when the gas temperature reaches 60 °C. Will it activate? Explain your answer. The ambient temperature is 20 °C.

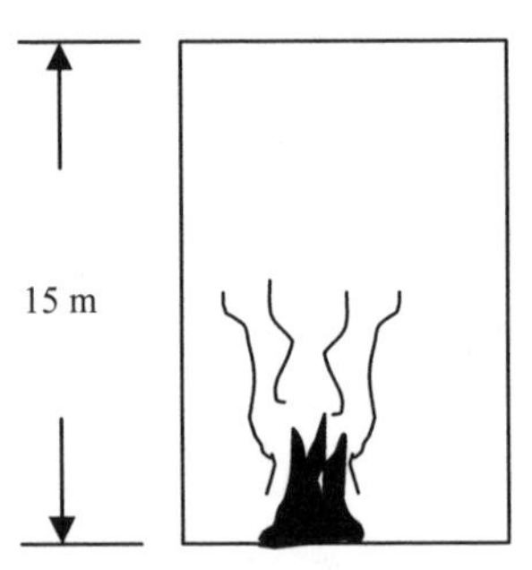

10.9 A sprinkler is 8 m directly above a fire in an atrium. It requires a minimum temperature rise of 60 °C to activate. The fire diameter is 2 m and the flame radiative fraction is 40 %.

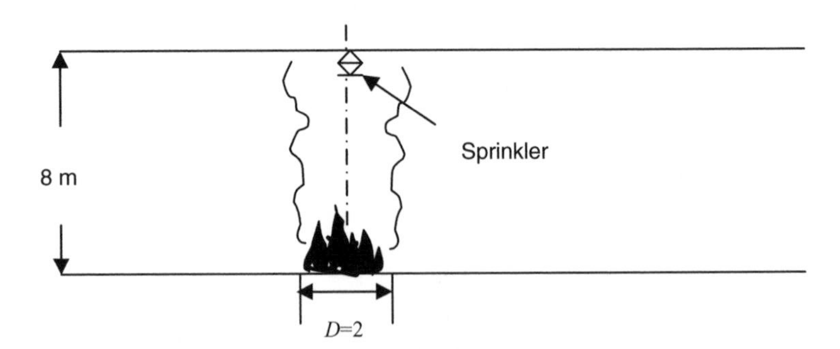

(a) Assuming no heat loss from the sprinkler, calculate the minimum energy release rate needed for sprinkler activation.

(b) What is the flame height?

(c) Estimate the maximum radiation heat flux incident to an object 6 m away.

10.10 A grease pan fire occurs on a stove top. The flames just touch the ceiling. The pan is 1 m above the floor and the ceiling is 2.4 m high. The pan is 0.3 m in diameter.

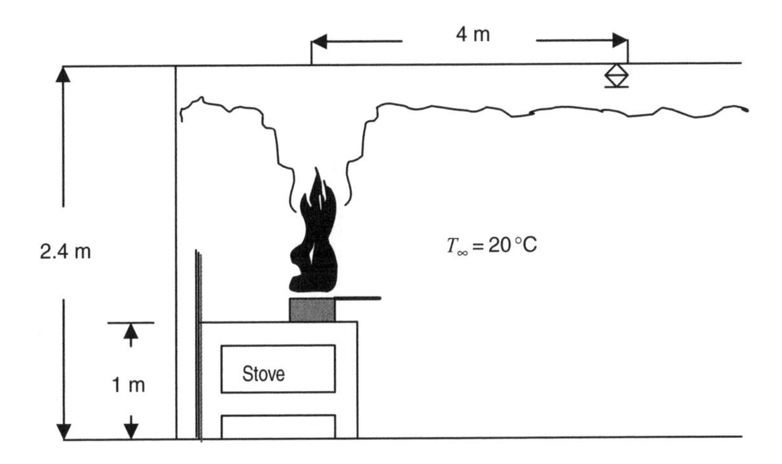

(a) Estimate the energy release rate of the fire.

(b) A sprinkler designed to activate immediately when the local temperature reaches 80 °C is located 4 m away. Will it activate? Show results to explain.

10.11 It is estimated that in the WTC disaster that the jet fuel burned in one of the towers at 10 GW. If this is assumed to burn over an equivalent circular area constituting one floor of 208×208 ft, and would produce an unimpeded fire plume, then compute the height of the flame and the temperature at two diameters above the base.

10.12 The aircraft that hit one of the WTC towers was carrying about 28 500 kg of JP4 liquid aviation fuel. Three fireballs were seen following impact. They achieved a maximum diameter of 60 m each. NIST estimates that 10–25 % of the fuel burned in the fireballs. Make a quantitative estimate of the fuel that burned in these fireballs and see if NIST is correct. Take the density of the JP4 vapor at 20 °C to be 2.5 kg/m^3. Its liquid density is 760 kg/m^3.

11
Compartment Fires

11.1 Introduction

The subject of compartment fires embraces the full essence of fire growth. The 'compartment' here can represent any confined space that controls the ultimate air supply and thermal environment of the fire. These factors control the spread and growth of the fire, its maximum burning rate and its duration. Although simple room configurations will be the limit of this chapter, extensions can be made to other applications, such as tunnel, mine and underground fires. Spread between compartments will not be addressed, but are within the scope of the state-of-the-art. A more extensive treatment of enclosure fires can be found in the book by Karlsson and Quintiere [1]. Here the emphasis will be on concepts, the interaction of phenomena through the conservation equations, and useful empirical formulas and correlations. Both thermal and oxygen-limiting feedback processes can affect fire in a compartment. In the course of fire safety design or fire investigation in buildings, all of these effects, along with fire growth characteristics of the fuel, must be understood. The ability to express the relevant physics in approximate mathematics is essential to be able to focus on the key elements in a particular situation. In fire investigation analyses, this process is useful to develop, first, a qualitative description and then an estimated time-line of events. In design analyses, the behavior of fire growth determines life safety issues, and the temperature and duration of the fire controls the structural fire protection needs. While such design analysis is not addressed in current prescriptive regulations, performance codes are evolving in this direction. The SFPE *Guide to Fire Exposures* is an example [2]. Our knowledge is still weak, and there are many features of building fire that need to be studied. This chapter will present the basic physics. The literature will be cited for illustration and information, but by no means will it constitute a complete review. However, the work of Thomas offers much, e.g. theories (a) on wood crib fires [3], (b) on flashover and instabilities [4,5], and (c) on pool fires in compartments [6].

Fundamentals of Fire Phenomena James G. Quintiere

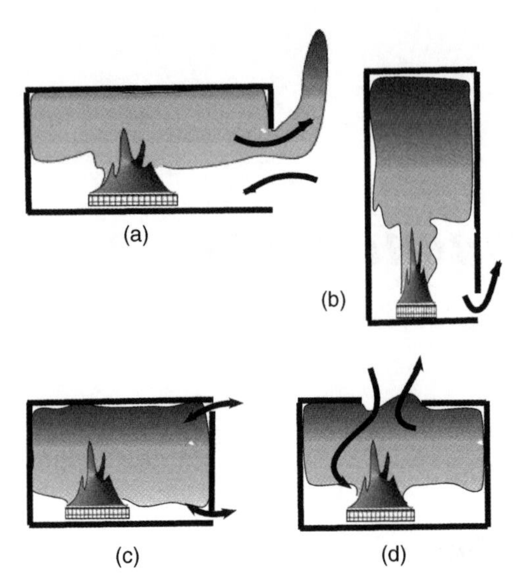

Figure 11.1 Compartment fire scenarios

11.1.1 Scope

The scope of this chapter will focus on typical building compartment fires representative of living or working spaces in which the room height is nominally about 3 m. This scenario, along with others, is depicted in Figure 11.1, and is labeled (a). The other configurations shown there can also be important, but will not necessarily be addressed here. They include:

(b) tall, large spaces where the contained oxygen reservoir is important;

(c) leaky compartments where extinction and oscillatory combustion have been observed (Tewarson [7], Kim, Ohtami and Uehara [8], Takeda and Akita [9]);

(d) compartments in which the only vent is at the ceiling, e.g. ships, basements, etc.

Scenario (a) will be examined over the full range of fire behavior beginning with spread and growth through 'flashover' to its 'fully developed' state. Here, flashover is defined as a transition, usually rapid, in which the fire distinctly grows bigger in the compartment. The fully developed state is where all of the fuel available is involved to its maximum extent according to oxygen or fuel limitations.

11.1.2 Phases of fires in enclosures

Fire in enclosures may be characterized in three phases. The first phase is fire development as a fire grows in size from a small incipient fire. If no action is taken to suppress the fire, it will eventually grow to a maximum size that is controlled by the

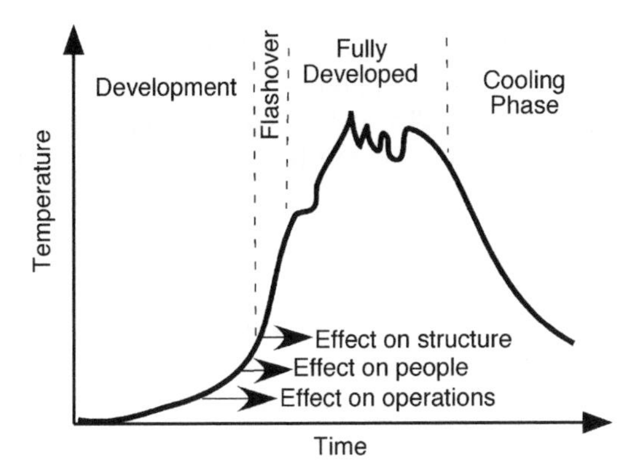

Figure 11.2 Phases of fire development

amount of fuel present (fuel controlled) or the amount of air available through ventilation openings (ventilation limited). As all of the fuel is consumed, the fire will decrease in size (decay). These stages of fire development can be seen in Figure 11.2.

The fully developed fire is affected by (a) the size and shape of the enclosure, (b) the amount, distribution and type of fuel in the enclosure, (c) the amount, distribution and form of ventilation of the enclosure and (d) the form and type of construction materials comprising the roof (or ceiling), walls and floor of the enclosure. The significance of each phase of an enclosure fire depends on the fire safety system component under consideration. For components such as detectors or sprinklers, the fire development phase will have a great influence on the time at which they activate. The fully developed fire and its decay phase are significant for the integrity of the structural elements.

Flashover is a term demanding more attention. It is a phenomenon that is usually obvious to the observer of fire growth. However, it has a beginning and an end; the former is the connotation for flashover onset time given herein. In general, flashover is the transition between the developing fire that is still relatively benign and the fully developed fire. It usually also marks the difference between the fuel-controlled or well-ventilated fire and the ventilation-limited fire. The equivalence ratio is less than 1 for the former and greater than 1 for the latter, as it is fuel-rich. Flashover can be initiated by several mechanisms, while this fire eruption to the casual observer would appear to be the same. The observer would see that the fire would 'suddenly' change in its growth and progress to involving all of the fuel in the compartment. If the compartment does not get sufficient stoichiometric air, the fire can produce large flames outside the compartment. A ventilation-limited fire can have burning mostly at the vents, and significant toxicity issues arise due to the incomplete combustion process. Mechanisms of flashover can include the following:

1. *Remote ignition.* This is the sudden ignition by autoignition or piloted ignition, due to flaming brands, as a result of radiant heating. The radiant heating is principally from the compartment ceiling and hot upper gases due to their large extent. The threshold for the piloted ignition of many common materials is about 20 kW/m^2. This value of flux, measured at the floor, is commonly taken as an operational criterion for flashover. It also corresponds to gas temperatures of 500–600 °C.

2. *Rapid flame spread.* As we know, radiant preheating of a material can cause its surface temperature to approach its piloted ignition temperature. This causes a singularity in simple flame spread theory that physically means that a premixed mixture at its lower flammability limit occurs ahead of the surface flame. Hence, a rapid spread results in the order of 1 m/s.

3. *Burning instability.* Even without spread away from a burning item, a sudden fire eruption can be recognized under the right conditions. Here, the thermal feedback between the burning object and the heated compartment can cause a 'jump' from an initial stable state of burning at the unheated, ambient conditions, to a new stable state after the heating of the compartment and the burning of the fuel come to equilibrium.

4. *Oxygen supply.* This mechanism might promote back-draft. It involves a fire burning in a very ventilation-limited state. The sudden breakage of a window or the opening of a door will allow fresh oxygen to enter along the floor. As this mixes with the fuel-rich hot gases, a sudden increase in combustion occurs. This can occur so rapidly that a significant pressure increase will occur that can cause walls and other windows to fail. It is not unlike a premixed gas 'explosion'.

5. *Boilover.* This is a phenomenon that occurs when water is sprayed on to a less dense burning liquid with a boiling temperature higher than that of water. Water droplets plunging through the surface can become 'instant'vapor, with a terrific expansion that causes spraying out of the liquid fuel. The added increase of area of the liquid fuel spray can create a tremendous increase in the fire. Even if this does not happen at the surface, the collection of the heavier water at the bottom of a fuel tank can suddenly boil as the liquid fuel recedes to the bottom. Hence, the term 'boilover'.

11.2 Fluid Dynamics

The study of fire in a compartment primarily involves three elements: (a) fluid dynamics, (b) heat transfer and (c) combustion. All can theoretically be resolved in finite difference solutions of the fundamental conservation equations, but issues of turbulence, reaction chemistry and sufficient grid elements preclude perfect solutions. However, flow features of compartment fires allow for approximate portrayals of these three elements through global approaches for prediction. The ability to visualize the dynamics of compartment fires in global terms of discrete, but coupled, phenomena follow from the flow features.

11.2.1 General flow pattern

The stratified nature of the flow field due to thermally induced buoyancy is responsible for most of the compartment flow field. Figure 11.3 is a sketch of a typical flow pattern in a compartment.

Strong buoyancy dominates the flow of the fire. Turbulence and pressure cause the ambient to mix (entrainment) into the fire plume. Momentum and thermal buoyancy

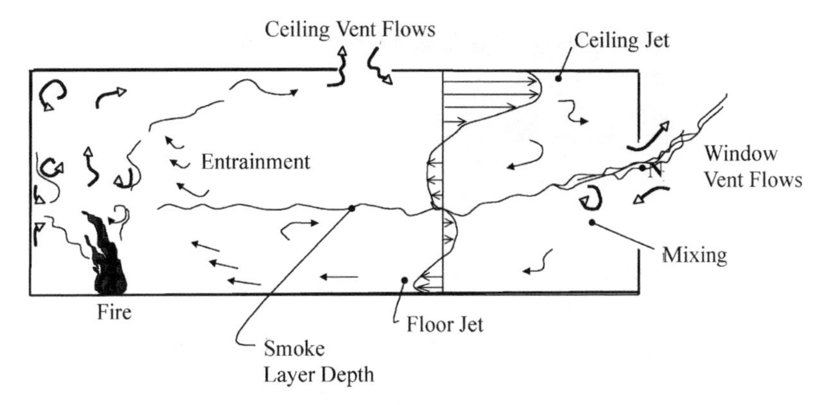

Figure 11.3 Flow pattern in a compartment fire

causes a relatively thin ceiling jet to scrub the ceiling. Its thickness is roughly one-tenth of the room height. A counterpart cold floor jet also occurs. In between these jets, recirculating flows form in the bulk of the volume, giving rise to a four-layer flow pattern. At the mid-point of this four-layer system is a fairly sharp boundary (layer interface) due to thermal stratification between the relatively hot upper layer and cooler lower layer. Many of these flow attributes are displayed in photographs taken by Rinkinen [10] for 'corridor' flows, made visible by white smoke streaks Figure 11.4. Figure 11.5 shows the sharp thermal stratification of the two layers in terms of the gas or wall temperatures [11].

11.2.2 Vent flows

The flow of fluids through openings in partitions due to density differences can be described in many cases as an orifice-like flow. Orifice flows can be modeled as inviscid Bernoulli flow with an experimental correction in terms of a flow coefficient, C, which generally depends on the contraction ratio and the Reynolds number. Emmons [12] describes the general theory of such flows. For flows through horizontal partitions, at near zero pressure changes, the flow is unstable and oscillating bidirectional flow will result. In this configuration, there is a flooding pressure above which unidirectional Bernoulli flow will occur. Epstein [13] has developed correlations for flows through horizontal partitions. Steckler, Quintiere and Rinkinen [14] were the first to measure successfully the velocity and flow field through vertical partitions for fire conditions. An example of their measured flow field is shown in Figure 11.6 for a typical module room fire.

The pressure difference promoting vertical vent flows are solely due to temperature stratification, and conform to a hydrostatic approximation. In other words, the momentum equation in the vertical direction is essentially, due to the low flows:

$$\frac{\partial p}{\partial z} = -\rho g \tag{11.1}$$

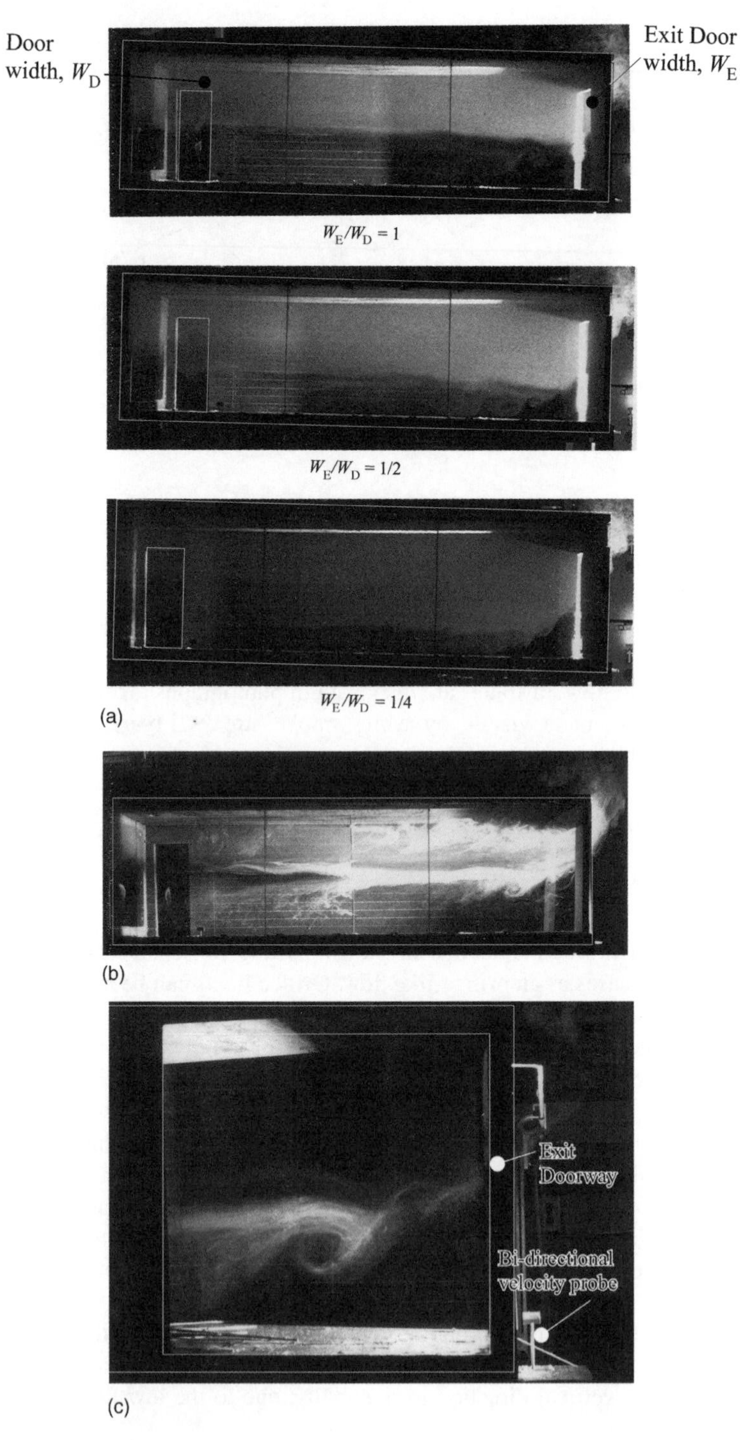

Figure 11.4 (a) Smoke layers in a model corridor as a function of the exit door width (b) Smoke streaklines showing four directional flows (c) Smoke streaklines showing mixing of the inlet flow from the right at the vent with the corridor upper layer [10]

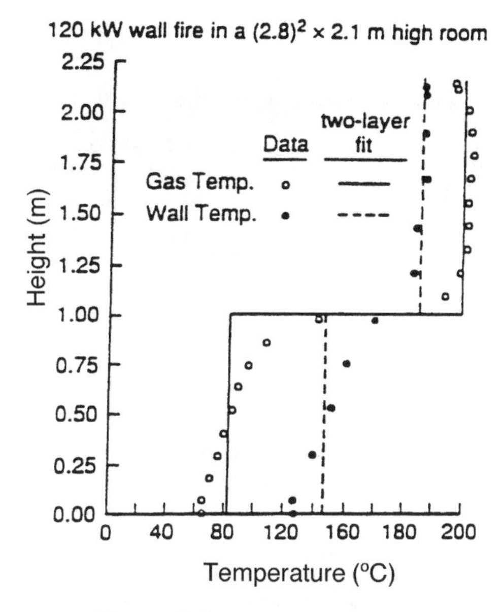

Figure 11.5 Thermal layers [11]

From Figure 11.3, under free convection, there will be a height in the vent at which the flow is zero, N; this is called the neutral plane. The pressure difference across the vent from inside (i) to outside (o) can be expressed above the neutral plane as

$$p_{\mathrm{i}} - p_{\mathrm{o}} = \int_0^z (\rho_{\mathrm{o}} - \rho_{\mathrm{i}}) g \, \mathrm{d}z \tag{11.2}$$

Figure 11.6 Measured flow field in the doorway of a room fire [14]

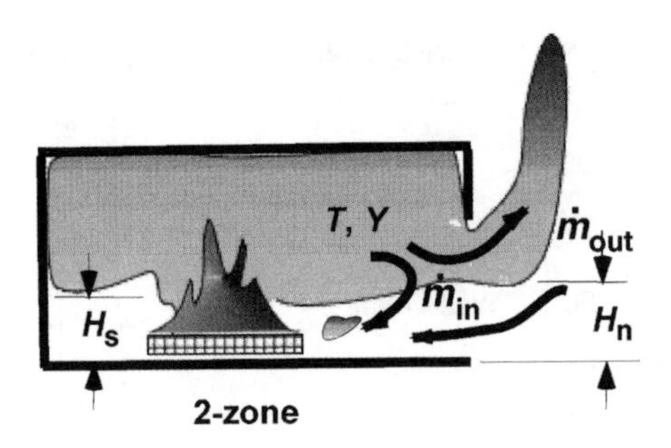

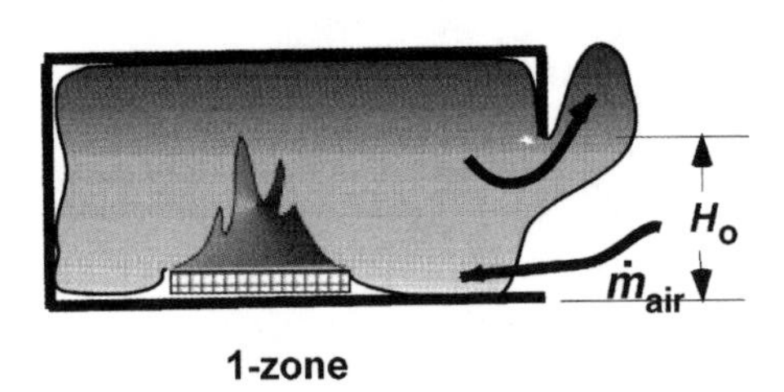

Figure 11.7 Zone modeling concepts

Such pressure differences are of the order of 10 Pa for $z \sim 1$ m and therefore these flows are essentially at constant pressure (1 atm $\approx 10^5$ Pa). The approach to computing fire flows through vertical partitions is to: (a) compute velocity from Bernoulli's equation using Equation (11.2), (b) substitute temperatures for density using the perfect gas law and, finally, (c) integrate to compute the mass flow rate. This ideal flow rate is then adjusted by a flow coefficient generally found to be about 0.7 [12,14]. Several example flow solutions for special cases will be listed here. Their derivations can be found in the paper by Rockett [15]. The two special cases consider a two-zone model in which there is a hot upper and lower cool homogeneous layer as shown in Figure 11.7(a); this characterizes a developing fire. Figure 11.7(b) characterizes a large fully developed fire as a single homogeneous zone.

The two-zone model gives the result in terms of the neutral plane height (H_n) and the layer height (H_s) for a doorway vent of area A_o and height H_o. The ambient to room temperature ratio is designated as $\theta \equiv T_o/T$:

$$\dot{m}_{\text{out}} = \tfrac{2}{3}\sqrt{2g}C\rho_o A_o\sqrt{H_o}[(\theta)(1-(\theta))]^{1/2}\left[\frac{1-H_n}{H_o}\right]^{3/2} \tag{11.3}$$

$$\dot{m}_{\text{in}} = \tfrac{2}{3}\sqrt{2g}C\rho_o A_o\sqrt{H_o}[1-(\theta)]^{1/2}\left(\frac{H_n}{H_o}-\frac{H_s}{H_o}\right)^{1/2}\left(\frac{H_n}{H_o}+\frac{H_s}{2H_o}\right) \tag{11.4}$$

The inflow is generally the ambient air and the outflow is composed of combustion products and excess air or fuel.

For a one-zone model under steady fire conditions, the rate mass of air inflow can be further computed in terms of the ratio of fuel to airflow rate supplied (μ) as

$$\dot{m}_{\text{air}} = \frac{\frac{2}{3}\rho_{\text{o}} C A_{\text{o}} \sqrt{H_{\text{o}}}\sqrt{g}\sqrt{2(1-\theta)}}{\left[1 + [(1+\mu)^2/\theta]^{1/3}\right]^{3/2}} \tag{11.5}$$

It is interesting to contrast Equation (11.5) with the air flow rate at a horizontal vent under bidirectional, unstable flow from Epstein [13]:

$$\dot{m}_{\text{air}} = \frac{0.068\rho_{\infty} A_{\text{o}}^{5/4}\sqrt{g}\sqrt{2(1-\theta)}}{(1+\theta)^{1/2}} \tag{11.6}$$

This horizontal vent flow is nearly one-tenth of that for the same size of vertical vent. Thus, a horizontal vent with such unstable flow, as in a basement or ship fire, is relatively inefficient in supplying air to the fire.

A useful limiting approximation for the air flow rate into a vertical vent gives the maximum flow as

$$\dot{m}_{\text{air}} = k_{\text{o}} A_{\text{o}} \sqrt{H_{\text{o}}}, \quad \text{where } k_{\text{o}} = 0.5 \text{ kg/s m}^{5/2} \tag{11.7}$$

This holds for $\theta < \frac{2}{3}$, and H_{s} small. In general, it is expected that $H_{\text{n}}/H_{\text{o}}$ ranges from 0.3 to 0.7 and $H_{\text{s}}/H_{\text{o}}$ from 0 to 0.6 for doorways [15].

11.3 Heat Transfer

The heat transfer into the boundary surface of a compartment occurs by convection and radiation from the enclosure, and then conduction through the walls. For illustration, a solid boundary element will be represented as a uniform material having thickness, δ, thermal conductivity, k, specific heat, c, and density, ρ. Its back surface will be considered at a fixed temperature, T_{o}.

The heat transfer path through the surface area, A, can be represented as an equivalent electric circuit as shown in Figure 11.8. The thermal resistances, or their inverses, the conductances, can be computed using standard heat transfer methods. Some will be illustrated here.

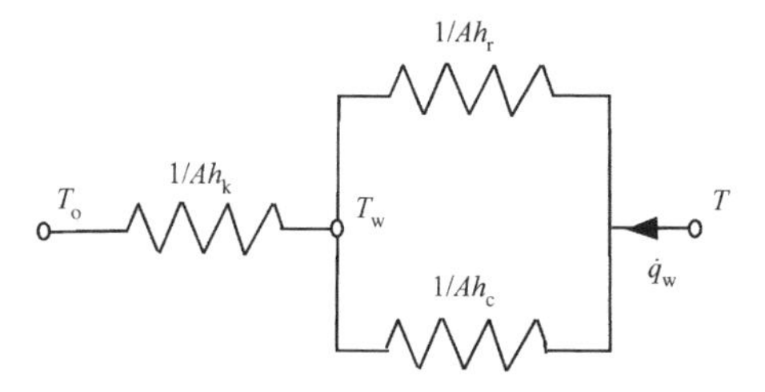

Figure 11.8 Wall heat transfer

11.3.1 Convection

Convections to objects in a fire environment usually occur under natural convection conditions. Turbulent natural convection is independent of scale and might be estimated from

$$Nu = \frac{h_c H}{K} = 0.13\left[\left(\frac{g(T - T_o)H^3}{T\nu^2}\right)Pr\right]^{1/3} \tag{11.8a}$$

where ν is the kinematic viscosity and Pr is the Prandt number. It gives h_c of about 10 W/m^2 K. Under higher flow conditions, it is possible that h_c might be as high as 40 W/m^2 K. This has been shown in measurements by Tanaka and Yamada [16] as the floor-to-ceiling coefficients vary from about 5 to 40 W/m^2 K. They found that an empirical correlation for the average compartment convection coefficient is

$$\frac{\bar{h}_c}{\rho_\infty c_p \sqrt{gH}} = \begin{cases} 2.0 \times 10^{-3}, & Q_H^* \leq 4 \times 10^{-3} \\ 0.08 Q_H^{*2/3}, & Q_H^* > 4 \times 10^{-3} \end{cases} \tag{11.8b}$$

Q_H^* is based on height, H (see Equation (10.41)).

11.3.2 Conduction

Only a finite difference numerical solution can give exact results for conduction. However, often the following approximation can serve as a suitable estimation. For the unsteady case, assuming a semi-infinite solid under a constant heat flux, the exact solution for the rate of heat conduction is

$$\dot{q}_w = A\sqrt{\frac{\pi}{4}\frac{k\rho c}{t}}(T_w - T_o) \tag{11.9}$$

or, following Figure 11.8,

$$h_k = \sqrt{\frac{\pi}{4}\frac{k\rho c}{t}} \tag{11.10}$$

For steady conduction, the exact result is

$$h_k = \frac{k}{\delta} \tag{11.11}$$

The steady state result would be considered to hold for time,

$$t > \frac{\delta^2}{4[k/\rho c]} \tag{11.12}$$

approximately, from Equation (7.35).

Table 11.1 Approximate thermal properties for typical building material

	Concrete/brick	Gypsum	Mineral wool
k (W/m K)	1	0.5	0.05
$k\rho c$ (W^2 s/m^4 K^2)	10^6	10^5	10^3
$k/\rho c$ (m^2/s)	5×10^{-7}	4×10^{-7}	5×10^{-7}

Table 11.1 gives some typical properties. An illustration for a wall six inches thick, or $\delta \approx 0.15$ m, gives, for most common building materials,

$$t = \frac{\delta^2}{4[k/\rho c]} = 3.1 \text{ hours}$$

from Equation (11.12) for the thermal penetration time. Hence, most boundaries might be approximated as thermally thick since compartment fires would typically have a duration of less than 3 hours.

Since the thermally thick case will predominate under most fire and construction conditions, the conductance can be estimated from Equation (11.10). Values for the materials characteristic of Table 11.1 are given in Table 11.2. As time progresses, the conduction heat loss decreases.

11.3.3 Radiation

Radiation heat transfer is very complex and depends on the temperature and soot distribution for fundamental computations. Usually, these phenomena are beyond the state-of-the-art for accurate fire computations. However, rough approximations can be made for homogeneous gray gas approximations for the flame and smoke regions. Following the methods in common texts (e.g. Karlsson and Quintiere [1]) formulas can be derived. For example, consider a small object receiving radiation in an enclosure with a homogeneous gray gas with gray uniform walls, as portrayed in Figure 11.9. It can be shown that the net radiation heat transfer flux received is given as

$$\dot{q}''_r = \epsilon\left[\sigma(T_g^4 - T^4) - \epsilon_{wg}\sigma(T_g^4 - T_w^4)\right] \qquad (11.13)$$

and

$$\epsilon_{wg} = \frac{(1-\epsilon_g)\epsilon_w}{\epsilon_w + (1-\epsilon_w)\epsilon_g}$$

Table 11.2 Typical wall conductance values

t(min)	h_k(W/m^2K)
10	0.8–26
30	0.3–10
120	0.2–5

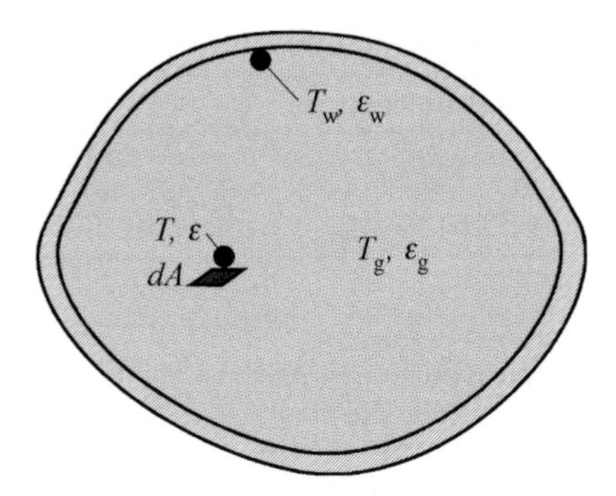

Figure 11.9 Radiation of a small object in an enclosure

where

ϵ = emissivity of the target

ϵ_w = emissivity of the wall

ϵ_g = emissivity of the gas

T = target temperature

T_w = wall temperature

T_g = gas temperature

and

σ = Stefan–Boltzmann constant $= 5.67 \times 10^{-11}$ kW/m^{2}K^4

If the object is the wall itself, then Equation (11.13) simplifies to the rate of radiation received by the wall as

$$\dot{q}_r = \frac{A\sigma(T^4 - T_w^4)}{1/\epsilon_g + 1/\epsilon_w - 1} \tag{11.14}$$

Since the boundary surface will become soot-covered as the fire moves to a fully developed fire, it might be appropriate to set $\epsilon_w = 1$.

The gas emissivity can be approximated as

$$\epsilon_g = 1 - e^{-\kappa H} \tag{11.15}$$

where H represents the mean beam length for the enclosure that could be approximated as its height. The absorption coefficient of the smoke or flames, κ, could range from about 0.1 to 1 m^{-1}. For the smoke conditions in fully developed fires, $\kappa = 1$ m^{-1} is a reasonable estimate, and hence ϵ_g could range from about 0.5 for a small-scale laboratory enclosure to nearly 1 for building fires.

Under fully developed conditions, the radiative conductance can be expressed as

$$h_r = \epsilon_g \sigma (T^2 + T_w^2)(T + T_w) \tag{11.16}$$

It can be estimated, for $\epsilon_g = 1$ and $T = T_w$, that $h_r = 104 - 725$ W/m^2 K for $T = 500 -$ 1200 °C.

11.3.4 *Overall wall heat transfer*

From the circuit in Figure 11.7, the equivalent conductance, h, allows the total heat flow rate to be represented as

$$\dot{q}_w = hA(T - T_o) \tag{11.17a}$$

where

$$\frac{1}{h} = \frac{1}{h_c + h_r} + \frac{1}{h_k} \tag{11.17b}$$

For a fully developed fire, conduction commonly overshadows convection and radiation; therefore, a limiting approximation is that $h \approx h_k$, which implies $T_w \approx T$. This result applies to structural and boundary elements that are insulated, or even to concrete structural elements. This boundary condition is 'conservative' in that it gives the maximum possible compartment temperature.

11.3.5 *Radiation loss from the vent*

From Karlsson and Quintiere [1], it can be shown that for an enclosure with blackbody surfaces ($\epsilon_w = 1$), the radiation heat transfer rate out of the vent of area A_o is

$$\dot{q}_r = A_o \epsilon_g \sigma (T_g^4 - T_o^4) + A_o (1 - \epsilon_g) \sigma (T_w^4 - T_o^4) \tag{11.18}$$

Since, for large fully developed fires, ϵ_g is near 1 or $T_w \approx T_g$, then it follows that

$$\dot{q}_r = A_o \sigma (T^4 - T_o^4) \tag{11.19}$$

and shows that the opening acts like a blackbody. This blackbody behavior for the vents has been verified, and is shown in Figure 11.10 for a large data set of crib fires in enclosures of small scale with H up to 1.5 m [17].

The total heat losses in a fully developed fire can then be approximated as

$$\dot{q} = \dot{q}_w + \dot{q}_r = h_k A (T - T_o) + A_o \sigma (T^2 + T_o^2)(T + T_o)(T - T_o) \tag{11.20}$$

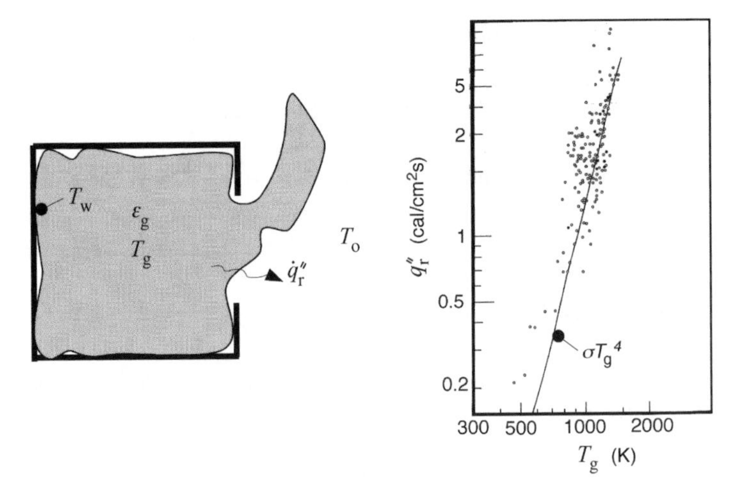

Figure 11.10 Radiation from the compartment vent [17]

11.4 Fuel Behavior

A material burning in an enclosure will depart from its burning rate in normal air due to thermal effects of the enclosure and the oxygen concentration that controls flame heating. Chapter 9 illustrated these effects in which Equation (9.73) describes 'steady' burning in the form:

$$\dot{m}'' = \frac{\dot{q}''}{L} \tag{11.21}$$

If the fuel responds fast to the compartment changes, such a 'quasi-steady' burning rate model will suffice to explain the expenditure of fuel mass in the compartment. The fuel heat flux is composed of flame and external (compartment) heating. The flame temperature depends on the oxygen mass fraction (Y_{O_2}), and external radiant heating depends on compartment temperatures.

11.4.1 Thermal effects

The compartment net heat flux received by the fuel within the hot upper layer for the blackbody wall and fuel surfaces can be expressed from Equation (11.13) as

$$\dot{q}_r'' = \epsilon_g \sigma_g (T_g^4 - T_v^4) + (1 - \epsilon_g)\sigma(T_w^4 - T_v^4) \tag{11.22}$$

where T_v is the vaporization temperature of the fuel surface. For fuel in the lower layer, an appropriate view factor should reduce the flux given in Equation (11.22).

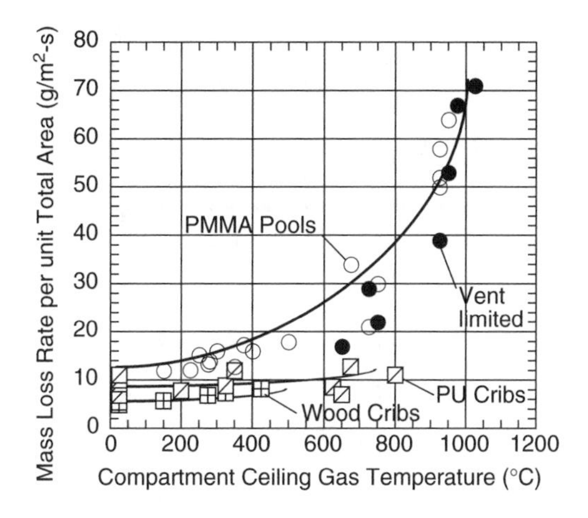

Figure 11.11 Compartment thermal effect on burning rate for wood cribs and pool fire

The quasi-steady maximum burning rate is illustrated in Figure 11.11 for wood cribs and PMMA pool fires under various fuel loading and ventilation conditions [18,19]. The cribs represent fuels controlled by internal combustion effects and radiant heat transfer among the sticks. The pool fires shown in Figure 11.11 are of small scale and have low absorptivity flames. Consequently, the small pools represent a class of fuels that are very responsive to the thermal feedback of the compartment. In general, the small-scale 'pool' fires serve to represent other fuel configurations that are equally as responsive, such as surfaces with boundary layer or wind-blown flames. The crib and pool fire configurations represent aspects of realistic building fuels. They offer two extremes of fuel sensitivity to compartment temperature.

11.4.2 *Ventilation effects*

As the vent is reduced, mixing will increase between the two layers and the oxygen feeding the fire will be reduced, and the burning rate will correspondingly be reduced. This is shown for some of the data in Figure 11.11 and represents ventilation-controlled burning. Figure 11.12 shows more dramatic effects of ventilation for experimental heptane pool fires in a small-scale enclosure [20]. These experiments included the case of a single ceiling vent or two equal-area upper and lower wall vents. For vent areas below the indicated flame extinction boundary, the fire self-extinguishes. For a fuel diameter of 95 mm, the mass loss rate associated with the ceiling vent was much lower than the wall vent case. The air supply differences indicated by Equations (11.5) and (11.6) suggest that the fuel mass loss rate for the ceiling vent case was limited by air flow.

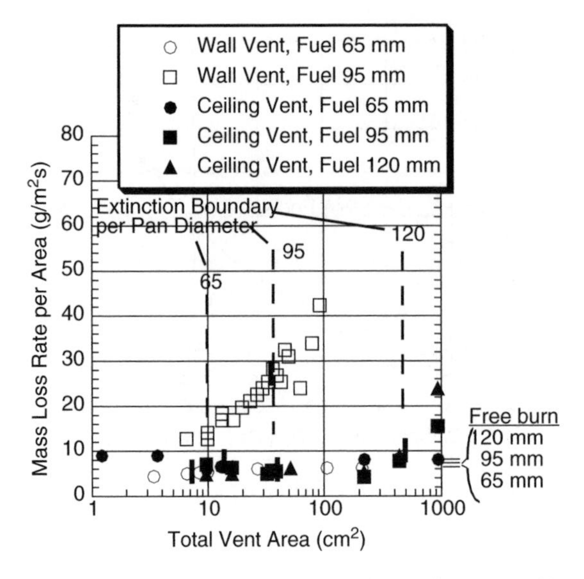

Figure 11.12 Ventilation-controlled burning for ceiling and wall vents [20]

11.4.3 Energy release rate (firepower)

The energy release rate of the fire in a compartment may occur within the compartment, and outside due to normal flame extension or insufficient air supply to the compartment. The energy release rate has been commonly labeled as the heat release rate (HRR); herein the term 'firepower' is fostered, as that is the common term adopted in energy applications. Moreover, the energy associated with chemical reaction is not heat in strict thermodynamic terminology. However, it is recognized that the fire community usage is the 'heat release rate'. Nevertheless, the important factor in compartment fire analysis is to understand the conditions that allow for burning within and outside a compartment. The heat of the flames and smoke causes the fuel to vaporize at a mass flow rate, $\dot{m}_F$. While all of the fuel may eventually burn, it may not necessarily burn completely in the compartment. This depends on the air supply rate. Either all of the fuel is burned or all of the oxygen in the incoming air is burned. That which burns inside gives the firepower within the enclosure. Thus, the firepower within the enclosure is given as

$$\dot{Q} = \begin{cases} \dot{m}_F \Delta h_c, & \phi < 1 \\ \dot{m}_{air} \Delta h_{air}, & \phi \geq 1 \end{cases} \tag{11.23}$$

where $\dot{m}_{air}$ is the mass flow rate of air supplied to the compartment and Δh_{air} is the heat of combustion per unit mass of air, an effective constant for most fuels at roughly 3 kJ/g air. The firepower is based on burning all of the gaseous fuel supplied or all of the air. The equivalence ratio, ϕ, determines the boundary between the fuel-lean and fuel-rich combustion regimes:

$$\phi = \frac{s\dot{m}_{\mathrm{F}}}{\dot{m}_{\mathrm{air}}} \tag{11.24}$$

where s is the stoichiometric air to fuel ratio. Stoichiometry is generally not obvious for realistic fuels as they do not burn completely. However, s can be computed from

$$s = \frac{\Delta h_{\mathrm{c}}}{\Delta h_{\mathrm{air}}} \tag{11.25}$$

where Δh_{c} is the heat of combustion (the chemical heat of combustion, as given in Table 2.3).

It is important to distinguish between the mass loss or supply rate of the fuel and its burning rate within the compartment. The mass loss rate in contrast to Equation (11.23) is given as

$$\dot{m}_{\mathrm{F}} = \begin{cases} \dot{m}''_{\mathrm{F,b}} A_{\mathrm{F}}, & \phi < 1 \\ \dfrac{\dot{m}_{\mathrm{air}}}{s} + \dfrac{F\dot{q}''_{r}}{L}, & \phi \geq 1 \end{cases} \tag{11.26}$$

where $\dot{m}''_{\mathrm{F,b}}$ is the fuel burn flux, A_{F} is the available fuel area, and $F\dot{q}_{\mathrm{r}}$ is the view-factor-modified net radiant flux received (Equation (11.22)). The actual area burning in a ventilation controlled fire ($\phi > 1$) would generally be less than the available area, as suggested by

$$\frac{\dot{m}_{\mathrm{air}}}{s} = \dot{m}''_{\mathrm{F,b}} A_{\mathrm{F,b}} \tag{11.27}$$

A typical transition to behavior of a ventilation-controlled fire begins with excess air as the fire feeds on the initial compartment air, but then is limited by air flow at the vent. As a consequence the fire can move to the vent and withdraw as the fuel is consumed. This might lead to two areas of 'deep' burning if the fire is extinguished before complete burnout.

11.5 Zone Modeling and Conservation Equations

The above discussion lays out the physics and chemical aspect of the processes in a compartment fire. They are coupled phenomena and do not necessarily lend themselves to exact solutions. They must be linked through an application of the conservation equations as developed in Chapter 3. The ultimate system of equations is commonly referred to as 'zone modeling' for fire applications. There are many computer codes available that represent this type of modeling. They can be effective for predictions if the

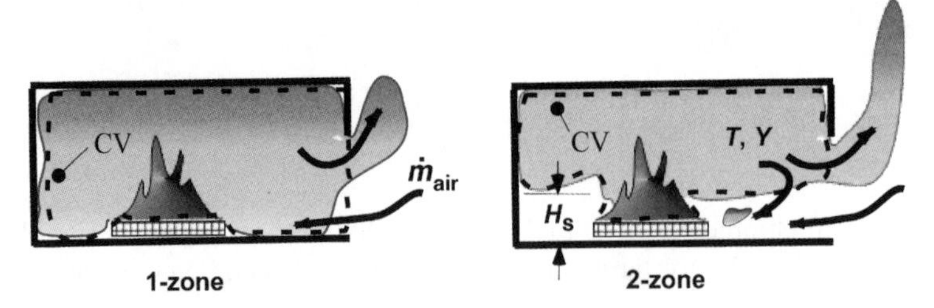

Figure 11.13 Control volume for conservation equations

fire processes are completely and accurately represented. Complete finite difference modeling of the fundamental equations is a step more refined than zone modeling, and it also has limitations, but will not be discussed here. Zone modeling is no more than the development of the conservation laws for physically justifiable control volumes. The control volume is principally the hot gas layer of the compartment fire. The fire is embedded within this volume. Figure 11.13 can be used to define a control volume for either the two-zone or one-zone approach. In the former, the lower layer would constitute another control volume. For simplicity, the one-zone model will only be considered here for illustration. This limited development is simply to present the methodology and demonstrate the basis for empirical correlations to follow.

11.5.1 Conservation relationships

Principally, conservation of energy for the compartment provides the important relationship to establish the extent of thermal feedback to the fuel. Conservation of mass and oxygen provide additional support equations. The 'process' relationships, given previously, establish the important transport rates of mass and energy. These 'constitutive' relationships may not always be complete enough to describe all fire scenarios.

The conservation equations can be developed from Chapter 3. They are listed below:

1. *Mass.* For the one-zone model with a perfect gas using Equation (3.15), the conservation of mass can be written as

$$-\frac{pV}{RT^2}\frac{\mathrm{d}T}{\mathrm{d}t}+\dot{m}=\dot{m}_{\mathrm{air}}+\dot{m}_{\mathrm{F}} \tag{11.28}$$

   The conservation of oxygen for the mass fraction designation by Y follows from Equation (3.22).

2. *Oxygen*

$$\frac{\mathrm{d}(mY)}{\mathrm{d}t}+\dot{m}Y-\dot{m}_{\mathrm{air}}(0.233)=-\frac{0.233s\dot{Q}}{\Delta h_{\mathrm{c}}} \tag{11.29}$$

Under steady conditions, or where the transient term is negligible, this is simply

$$Y = \frac{0.233(1-\phi)}{1+\phi/s} \quad \text{for } \phi \leq 1 \tag{11.30}$$

and

$$Y = 0 \quad \text{for the (global) equivalence ratio, } \phi > 1$$

The energy equation follows from Equation (3.45) where the 'loss' terms are grouped to include both heat and enthalpy transport rates as $\dot{Q}_l$.

3. *Energy*

$$c_v \frac{\mathrm{d}(mT)}{\mathrm{d}t} = \frac{c_v V}{R} \frac{\mathrm{d}p}{\mathrm{d}t} = \dot{Q} - \dot{Q}_l \tag{11.31}$$

The transient term can be neglected in most cases, but any sudden change in the energy will result in a pressure change. This can be responsible for 'puffing' effects seen in compartment fires, especially when the burning is oscillatory.

11.5.2 Dimensionless factors in a solution

By normalizing the steady equation with the air flow rate and representing all of the losses as linear in T (Equation (11.31)), the temperature rise can be solved as

$$T - T_o = \frac{\dot{Q}/(c_p \dot{m}_{air})}{x_o + x_w + x_r} \tag{11.32}$$

The dimensionless x-loss factors are parameters that effect temperature, and occur in modified forms in many correlations in the literature for compartment fire temperature.

1. *Flow-loss factor.* The flow factor, x_o, is the ratio of out-flow to airflow:

$$x_o = 1 + \frac{\dot{m}_F}{\dot{m}_{air}} = 1 + \frac{\phi}{s} \tag{11.33}$$

and is approximately 1 for $\phi < 1$ since s ranges from 3 to 13 for charring to liquid fuels respectively.

2. *Wall-loss factor.* The wall-loss factor, x_w, is given in terms of the compartment heat transfer surface area, A, and the overall heat transfer coefficient, h, as given by Equation (11.17):

$$x_w = \frac{hA}{\dot{m}_{air} c_p} \geq \frac{hA}{k_o c_p A_o \sqrt{H_o}} \tag{11.34}$$

Here, the maximum air flow rate could be inserted per Equation (11.7). In general, h depends on convection (h_c), radiation (h_r) and conduction into the walls (h_k). Their typical values range as $h_c \sim$10–30, $h_r \sim$5–100 and $h_k \sim (k\rho c/t)^{1/2} \sim$5–60 W/m^2 K; consequently, $h \sim$ 1 to 100 W/m^2 K is a typical range expected. It is important to note that the wall conductance will decrease with time for thick walls, leading to an increase in temperature.

3. *Vent-radiation factor.* As shown by Figure 11.10, the radiation flux from the vent is reasonably represented as a blackbody at the gas temperature, T. Thus, the vent radiation loss can be estimated as

$$x_\mathrm{r} = \frac{\dot{q}''_{\mathrm{r,o}}}{\dot{m}_{\mathrm{air}} c_p (T - T_\mathrm{o})}$$
$$\geq \frac{\sigma(T^4 - T_\mathrm{o}^4)A_\mathrm{o}}{c_p k_\mathrm{o} A_\mathrm{o}\sqrt{H_\mathrm{o}}(T - T_\mathrm{o})} \approx \frac{\sigma T^3}{c_p k_\mathrm{o}\sqrt{H_\mathrm{o}}} \quad \text{for } T_\mathrm{o}/T \text{ small} \tag{11.35}$$

This factor might range from roughly 10^{-2} to 10^{-1} for T ranging from 25 to 1200 °C respectively. It is the least significant of the three factors.

11.6 Correlations

While computer models exist to solve for many aspects of compartment fires, there is nothing more valuable than the foundation of experimental results. Empirical correlations from such experimental data and based on the appropriate dimensionless groups have been developed by many investigators. These can be very useful to the designer and investigator. As with computer models, these correlations can also be incomplete; they may not have tested over a sufficient range of variables, they may leave out important factors and they are still usually specific to the fuels and their configuration. Nevertheless, they are very powerful, in their algebraic or graphical simplicity, and in their accuracy. Some will be listed here. The SFPE guide addressing fully developed enclosure fires is an excellent source of most correlations [2].

11.6.1 Developing fires

The developing fire applies up to flashover, and therefore might apply to an average temperature in the upper layer of 600 °C, at most. McCaffrey, Quintiere and Harkleroad [21] developed a well-known correlation for this domain known as the MQH correlation. The MQH correlation was developed from a large database for compartment fires ranging in scale from 0.3 to 3 m high, and with fuels centered on the floor [21]. Other investigators have followed the form of this correlation for floor fires near a wall, in a corner and for burning wall and ceiling linings [22–25]. This correlation takes the form of Equation (11.32), but as a power law. The dimensionless groups that pertain are dimensionless mass and energy flow rates of the form:

$$m^* = \frac{\dot{m}}{\rho_\mathrm{o}\sqrt{g}A_\mathrm{o}\sqrt{H_\mathrm{o}}} \quad \text{and} \quad Q^* = \frac{\dot{Q}}{\rho_\mathrm{o} c_p T_\mathrm{o}\sqrt{g}A_\mathrm{o}\sqrt{H_\mathrm{o}}} \tag{11.36}$$

The dimensionless fuel supply and air inflow rate can be described, accordingly, as m_F^* and m_A^* respectively. The entrainment rate depends on the height over which entrainment

Table 11.3 Values of C_T for different fire configurations

Fire Configuration	C_T	Source
Discrete, centered	480	McCaffrey *et al.*[21]
	686	Azhakesan *et al.*[23]
Discrete, against wall	615	Mowrer *et al.*[22]
Discrete, in corner	804	Azhakesan *et al.*[23]
	856	Mowrer *et al.*[22]
Linings, wall only	1200	Azhakesan *et al.* [25]
	1060–1210	Karlsson [24]
Linings, wall and ceiling	1000	Azhakesan *et al.* [25]
	940	Karlsson [24]

occurs H_s, and on the fire configuration. This factor is simply represented here by G. Thus, the entrainment rate and air flow rate might functionally be represented as

$$m_A^* = \text{function}\left(G, \frac{H_s}{H_o}\right) \tag{11.37}$$

Eliminating the H_s dependence by Equation (11.4), the temperature of the hot layer can generally be functionally expressed

$$\frac{T}{T_o} = \text{function}\left(Q_o^*, Q_w^*, m_F^*, G\right) \tag{11.38}$$

The MQH correlation for the layer temperature rise has found the empirical fit to data:

$$\Delta T = C_T(G, m_F^*) Q_o^{*^{2/3}} Q_w^{*^{-1/3}} \tag{11.39}$$

where Q_o^* from Equation (11.36) has $\dot{Q}$ as the firepower, and Q_w^* has $\dot{Q}$ taken as $h_k A T_o$ from Equations (11.10) and (11.11). These Q^* values correspond to x_o and x_w in Equation (11.32); i.e.

$$Q_o^* = \frac{\dot{Q}}{\rho_\infty c_p T_\infty \sqrt{g} A_o \sqrt{H_o}} \quad \text{and} \quad Q_w^* = \frac{h_k A}{\rho_\infty c_p \sqrt{g} A_o \sqrt{H_o}}$$

C_T is an empirical constant that appears to increase as the entrainment rate decreases, as shown from the evidence in Table 11.3. Also from Equation (11.32), the temperature will decrease as the dimensionless fuel supply rate increases. Equation (11.39) holds for the overventilation regime, and therefore it should not be used at a point where $\dot{m}_F/\dot{m}_{air} = 1/s$ or $\phi = 1$ without at least modifying Q_o^*.

Figure 11.14 shows the original MQH data fit and Figures 11.15 and 11.16 show results for compartment lining material fires. Various lining materials were used and the firepower was measured by oxygen calorimetry. Departure from the linear slope behavior marks the onset of the ventilation-limited fires where the correlation based on the total firepower, within and outside the compartment, does not hold. The results for the lining fires also show the importance of the m_F^* factor omitted from the simple MQH correlation.

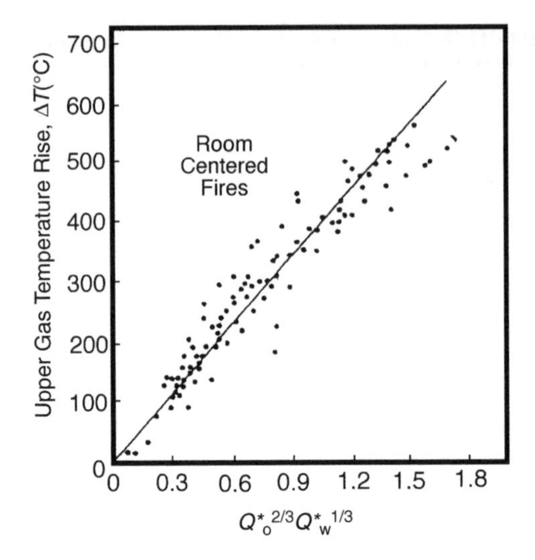

Figure 11.14 MQH correlation for discrete floor-centered fires [21]

11.6.2 *Fully developed fires*

The fully developed fire is defined as the state where all of the fuel that can get involved is involved. It does not necessarily mean that all of the fuel is burning, since the lack of air would prevent this. In most building occupancies, the fuel load is high enough to lead to a significant fire that can threaten the structure. This means a significantly high temperature is produced for a long time. Hence, many fully developed fire studies have been motivated to establish the thermal characteristics needed to assess the structural fire resistance. Consequently, these studies have aimed to establish correlations for gas temperature and the fuel burning rate (or more exactly, its mass loss rate). The latter can

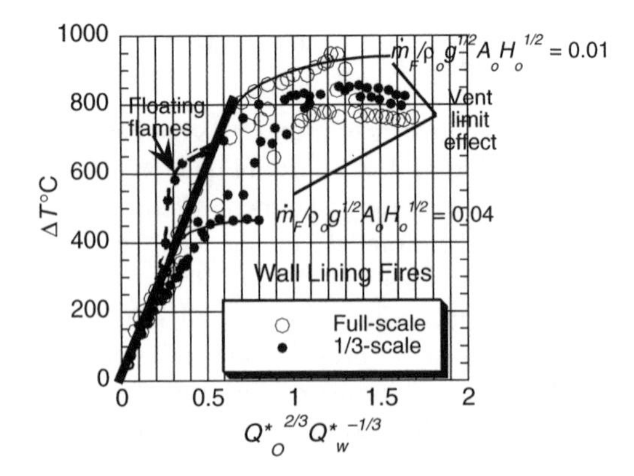

Figure 11.15 Temperature for wall lining fires [25]

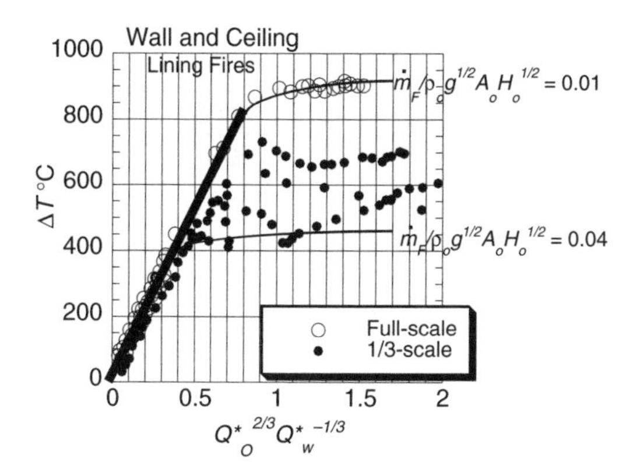

Figure 11.16 Temperatures for wall and ceiling lining fires [25]

then give the fire duration for a given fuel mass loading in the occupancy. Occupancy studies have been performed from time to time to give typical fuel loadings (usually in wood equivalent), which may range from 10 to 50 kg/m^2 (floor area).

One aspect of fully developed fires that will not be addressed here is their production of combustion products. When compartment fires become ventilation-limited, they burn incompletely, and can spread incomplete products such as CO, soot and other hydrocarbons throughout the building. It is well established that the yield of these incomplete products goes up as the equivalence ratio approaches and exceeds 1. More information on this issue can be found in the literature [1].

Fully developed fire studies have been performed over a range of fuel loadings and ventilation conditions, but primarily at scales smaller than for normal rooms. Also the fuels have been idealized as wood cribs or liquid or plastic pool fires. The results have not been fully generalized. The strength of the dimensionless theoretical implication of Equation (11.38) suggests that, for a given fuel, the fully developed, ventilation-limited fire should have dependences as

$$\frac{T}{T_\text{o}} = \text{function}\left(Q^*_\text{w}, m^*_\text{F}, G\right) \tag{11.40}$$

where Q^*_o has been eliminated since $\dot{Q}$ depends only on the air flow rate from Equation (11.23), and that depends on the temperature from the flow equation (e.g. Equation (11.5)). Thus, for a given burning configuration (represented in terms of G),

$$T \sim \frac{h_\text{k} A}{\rho_\infty c_p \sqrt{g} A_\text{o} \sqrt{H_\text{o}}}, \ \frac{\dot{m}_\text{F}}{\rho_\infty \sqrt{g} A_\text{o} \sqrt{H_\text{o}}} \tag{11.41}$$

From Equation (11.21) and recognizing that the heat flux to the fuel depends on the flame and compartment temperatures, it follows that

$$\dot{m}_\text{F} \sim \frac{A_\text{F} \dot{q}''(T)}{L} \tag{11.42a}$$

or

$$m_F^* \approx \frac{\dot{m}_F}{\rho_\infty \sqrt{g} A_o \sqrt{H_o}} \sim \left(\frac{A_F \sigma T_\infty^4}{\rho_\infty \sqrt{g} A_o \sqrt{H_o} L} \right) \left(\frac{\dot{q}''(T)}{\sigma T_\infty^4} \right) \tag{11.42b}$$

By rearrangement of these functional dependencies, it follows for a given fuel (L) that

$$T \text{ and } \frac{\dot{m}_F}{\rho_\infty \sqrt{g} A_o \sqrt{H_o}} \sim \frac{A}{A_o \sqrt{H_o}}, \frac{A_F}{A} \tag{11.43}$$

Investigators have developed correlations for experimental data in this form.

Law developed a correlation for the extensive Conseil International du Bâtiment (CIB) test series [17] involving wood cribs that covered most of the compartment floor [26]. The test series also addressed the effect of compartment shape by varying its width (W) and depth (D). The correlation for the maximum temperature was given as

$$T_{\max} = 6000 \left(\frac{1 - e^{-0.1\Omega}}{\sqrt{\Omega}} \right) (^\circ\text{C}) \qquad \Omega = \frac{A}{A_o \sqrt{H_o}} \quad \text{in m}^{-1/2} \tag{11.44}$$

Most of the CIB tests involved crib arrangements having A_F/A of mostly 0.75 [2]; therefore, leaving this parameter out of the correlation may be justified. However, for low fuel loadings, Law recommended that the maximum temperature be reduced accordingly:

$$T = T_{\max} \left(1 - e^{-0.05\Psi} \right) (^\circ\text{C}), \qquad \Psi = \frac{m_F}{\sqrt{A A_o}} \quad \text{in kg/m}^2 \tag{11.45}$$

where m_F is the fuel load in kg.

The mass loss rate is correlated as

$$\dot{m}_F = 0.18 A_o \sqrt{H_o} \left(\frac{W}{D} \right) \left(1 - e^{-0.036\Omega} \right) \quad \text{in kg/s}$$

$$\text{for } \frac{\dot{m}_F}{A_o \sqrt{(H_o)}} \left(\frac{D}{W} \right)^{1/2} < 60 \text{ kg/s m}^{5/2} \tag{11.46}$$

The data corresponding to these correlations are shown in Figures 11.17 and 11.19.

The CIB tests consisted of compartments with heights ranging from 0.5, 1.0 and 1.5 m with dimension ratios of $D/H, W/H, H/H$ coded as data sets: 211, 121, 221 and 441. A total of 321 experiments were conducted in still air conditions. The wood crib fuel loading ranged from 10 to 40 kg/m^2 with stick spacing to stick thickness ratios of $\frac{1}{3}$, 1 and 3. The data are plotted in Figures 11.17 and 11.19. The compartment surface area, A, excludes the floor area in the plot variables.

An alternative plotting format was used by Bullen and Thomas [6] for the mass loss rate, which shows the effect of fuel type in Figure 11.18. These data include an extensive compilation by Harmathy [27] for wood crib fuels (including the CIB data). Here the fuel area is included, but the compartment area is omitted. This shows the lack of

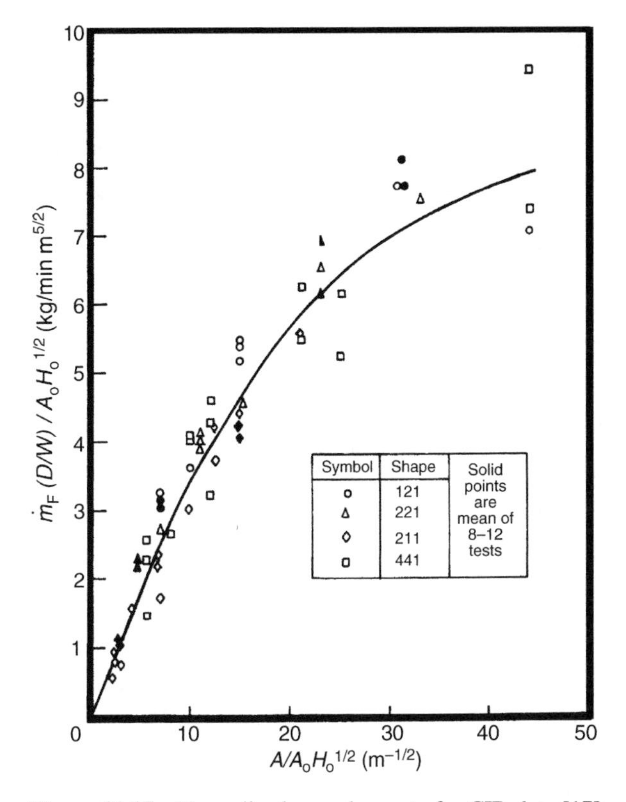

Figure 11.17 Normalized mass loss rate for CIB data [17]

completeness of the correlations. However, they are still invaluable for making accurate estimates for fully developed fires. Harmathy gives a fit to the wood crib data in Figure 11.18 as

$$\dot{m}_F = \begin{cases} (0.0062\ \text{kg/m}^2\ \text{s})A_F, & \dfrac{\rho_\infty\sqrt{g}A_o\sqrt{H_o}}{A_F} \geq 0.263\ \text{kg/s} \\ 0.0263\rho_\infty\sqrt{g}A_o\sqrt{H_o}, & \dfrac{\rho_\infty\sqrt{g}A_o\sqrt{H_o}}{A_F} < 0.263\ \text{kg/s} \end{cases} \tag{11.47}$$

The latter equation corresponds to the CIB data and Law's correlation for large $A/(A_o\sqrt{H_o})$, i.e. ventilation-limited fires. At the asymptote for large $A/(A_o\sqrt{H_o})$ from Figure 11.18, the CIB results are for $D = W$, roughly

$$\dot{m}_F \approx 0.13 A_o\sqrt{H_o} \quad \text{kg/s}$$

and for Harmathy with $\rho_\infty = 1.16\ \text{kg/m}^3$,

$$\dot{m}_F = 0.086\ A_o\sqrt{H_o}\ \text{kg/s}$$

with A_o in m^2 and H_o in m. These differences show the level of uncertainty in the results.

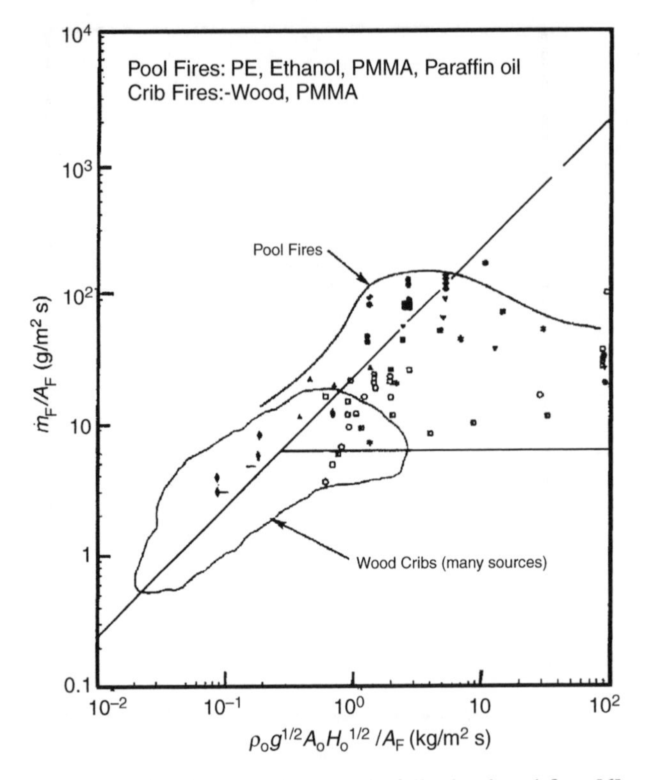

Figure 11.18 Fuel mass loss rate in fully developed fires [6]

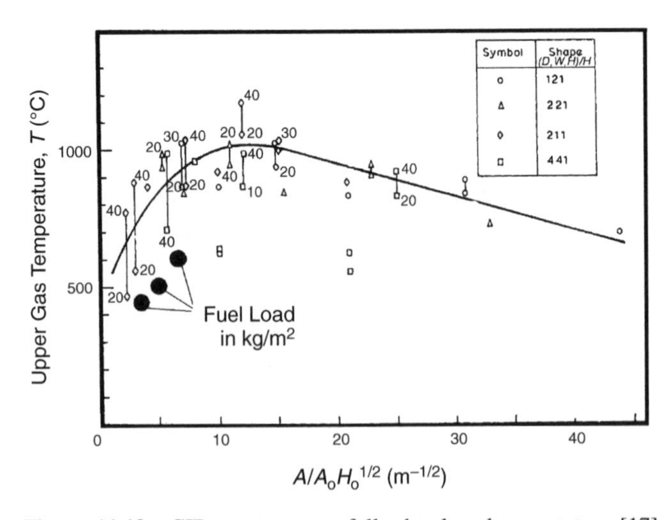

Figure 11.19 CIB compartment fully developed temperature [17]

11.7 Semenov Diagrams, Flashover and Instabilities

The onset of flashover as considered here is induced by thermal effects associated with the fuel, its configuration, the ignition source and the thermal feedback of the compartment. Several fire growth scenarios will be considered in terms of Semenov diagrams (Chapters 4 and 9) portraying the firepower $\dot{Q}$ and the loss rates $\dot{Q}_l$ as a function of compartment temperature. According to Equation (11.31), a 'thermal runaway' will occur at the critical temperature where $\dot{Q} = \dot{Q}_l$ and the two curves are tangent. The developing fire occurs for $\phi < 1$ and the result of a critical event will lead to a fully developed fire with $\phi > 1$ likely. Four scenarios will be examined:

1. Fixed fire area, representative of a fully involved fuel package, e.g. chair, crib, pool fire.
2. Ignition of a second item, representing an initial fire (e.g. chair, liquid spill) and the prospect of ignition of an adjoining item.
3. Opposed flow flame spread, representing spread on a horizontal surface, e.g. floor, large chair, mattress, etc.
4. Concurrent flow flame spread, representing vertical or ceiling spread, e.g. combustible linings.

In all of these scenarios, up to the critical condition (flashover),

$$\phi < 1 \qquad \text{and} \qquad \dot{Q} = \dot{m}''_F A_F \Delta h_c \tag{11.48}$$

Here A_F will be taken to represent the projected external fuel surface area that would experience the direct heating of the compartment. From Equation (11.21), accounting approximately for the effects of oxygen and temperature, and distribution effects in the compartment, the fuel mass flux might be represented for qualitative considerations as

$$\dot{m}''_F = \dot{m}''_{F,o}(1 - \gamma_o \phi) + \frac{\gamma_T \sigma (T^4 - T_o^4)}{L} \qquad (\phi < 1) \tag{11.49}$$

where $\dot{m}''_{F,o}$ refers to the burning flux in normal air. The distribution factors, γ_o and γ_T, are estimated to range from about 0.5 to 1 as the smoke layer descends and mixing causes homogeneous properties in the compartment. The oxygen distribution factor is very approximate and is introduced to account for the oxygen in the flow stream in the lower gas layer. The thermal parameter, γ_T, represents a radiation view factor. For all of these scenarios, the effect of oxygen on the burning rate is expected to be small, as the oxygen concentration does not significantly decrease up to flashover. The loss rate is given as, from Equations (11.31) and (11.32),

$$\dot{Q}_l = \dot{m}_{air} c_p [x_o + x_w(t) + x_r(T)](T - T_o) \tag{11.50}$$

The loss rate is nearly linear in T for low temperatures, and can depend on time, t, as well. Since T is a function of time, the normal independent variable t should be

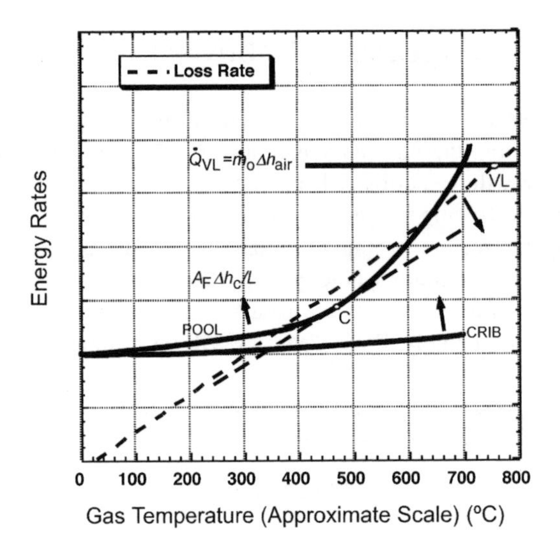

Figure 11.20 Fixed area fire

considered as dependent since the inverse function $t = t(T)$ should be available, in principle. This functional dependence will not be made explicit in the following.

11.7.1 Fixed area fire

Figure 11.20 shows the behavior of the energy release and losses as a function of compartment temperature (Semenov diagram) for fuels characteristic of cribs and 'pool' fires (representative of low flame absorbing surface fuels, e.g. walls). As the loss curve decreases due to increasing time or increasing ventilation factor $(A/A_o\sqrt{H_o})$, a critical condition (C) can occur. The upward curvature and magnitude of the energy release rate increases with $A_F\Delta h_c/L$. Thomas *et al.* [4] has estimated the critical temperatures to range from about 300 to 600 °C depending on the values of L. Thus, liquid fuels favor the lower temperature, and charring solids favor the higher. An empirical value of 500 °C (corresponding to 20 kW/m^2 blackbody flux) is commonly used as a criterion for flashover. The type of flashover depicted here is solely due to a fully involved single burning item. A steady state condition that can be attained after flashover is the ventilation-limited (VL) branch state shown in Figure 11.20 in which $\dot{Q} = \dot{m}_o\Delta h_{air}$ for $\phi = 1$. It can be shown that the state of oxygen at the critical condition is always relatively high, with $\phi < 1$. This clearly shows that oxygen depletion has no strong effect. Also, Thomas *et al.* [4] show that the burning rate enhancement due to thermal feedback at the critical condition is only about 1.5, at most.

11.7.2 Second item ignition

Figure 11.21 depicts the behavior for a neighboring item when becoming involved due to an initiating first fire. The first fire could be a fixed gas burner as in a test or a fully

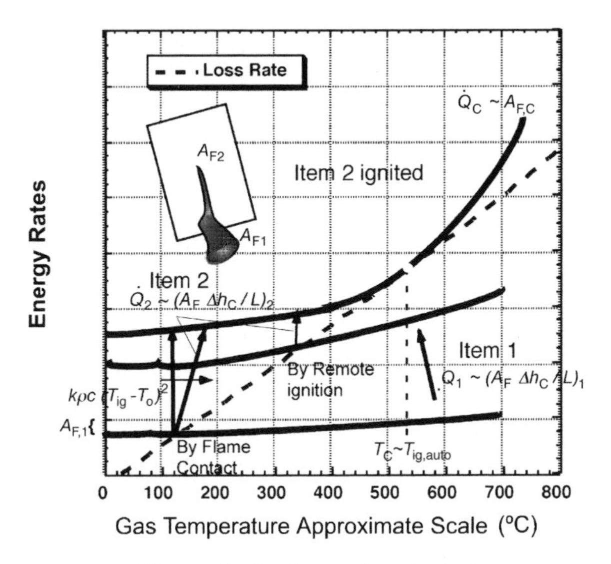

Figure 11.21 Second item ignited

involved fuel, $A_{F,1}$. For illustrative purposes, assume the second item has similar properties to $A_{F,1}$, and A_F represents the combined area after ignition. Ignition of the second item is possible if the material receives a heat flux higher than its critical flux for ignition. It can possibly receive heat by flame heating ($\dot{q}_f''$) and heating from the compartment. The critical condition for ignition occurs at a balance between the compartment heating and the objects heat loss estimated by

$$\dot{q}_f'' + \gamma_T \sigma (T^4 - T_o^4) = \sigma (T_{ig}^4 - T_o^4) + h_c (T_{ig} - T_o) \tag{11.51}$$

For remote ignition under autoignition conditions, direct flame heating is small or zero, so the critical compartment temperature needs to be greater than or equal to the autoignition temperature, e.g. about 400–600 °C for common solid fuels.

For ignition by flame contact (to a wall or ceiling), the flame heat flux will increase with D_F, the flame thickness of the igniter, estimated by

$$\dot{q}_f'' \sim 1 - \exp(-\kappa D_F) \geq 30 \text{ kW/m}^2, \text{ usually} \tag{11.52}$$

The exposed area of the second item exposed $(A_{F,2})$ depends on the flame length (z_f) and its diameter.

For the second item ignition to lead to flashover, the area involved must equal or exceed the total critical area needed for the second item. The time for ignition depends inversely on the exposure heat flux (Equation (11.51)). Figure 11.21 shows the behavior for ignition of the second item, where $A_{F,1}$ is the fixed area of the first item and $A_{F,C}$ is the critical area needed. The energy release rate of both fuels controls the size of the 'jump' at criticality and depends directly on $A_F \Delta h_c / L$. No flashover will occur if the 'jump' in energy for the second item is not sufficient to reach the critical area of fuel, $A_{F,C}$. The time to achieve the jump or to attain flashover is directly related to the fuel property,

$k\rho c(T_{ig} - T_o)^2$. For flame contact, the piloted ignition temperature applies and the ignition time is expected to be small ($\sim$ 10–100 s).

11.7.3 Spreading fires

The flame spread rate can be represented in terms of the pyrolysis front (x_p) from Equation (8.7a):

$$v = \frac{dx_p}{dt} \approx \frac{\delta_f}{t_{ig}} \tag{11.53}$$

where δ_f is a flame heating length. In opposed flow spread into an ambient flow of speed u_o, this length is due to diffusion and can be estimated by Equation (8.31) as $\delta_{f,opposed} \sim (k/\rho c)_o/u_o$. This length is small ($\sim$ 1 mm) under most conditions, and is invariant in a fixed flow field. However, in concurrent spread, this heating length is much larger and will change with time. It can be very dependent on the nature of the ignition source since here $\delta_f = z_f - x_p$. In both cases, the area of the spreading fire (neglecting burnout) grows in terms of the velocity and time as $(vt)^n$, with n ranging from 1 to 2.

The surface flame spread cases are depicted in Figure 11.22. In opposed flow spread, the speed depends on the ignition time, which decreases as the compartment heats the fuel surface. After a long time, $T_{surface} = T_{ig}$ and the growth curves approach a vertical asymptote. Consider the initial area ignited as $A_{F,i}$ and $A_{F,C}$ as the critical area needed for a fixed area fire. The area for a spreading fire depends on the time and therefore T as well. For low-density, fast-responding materials, the critical area can be achieved at a compartment temperature as low as the ignition temperature of the material.

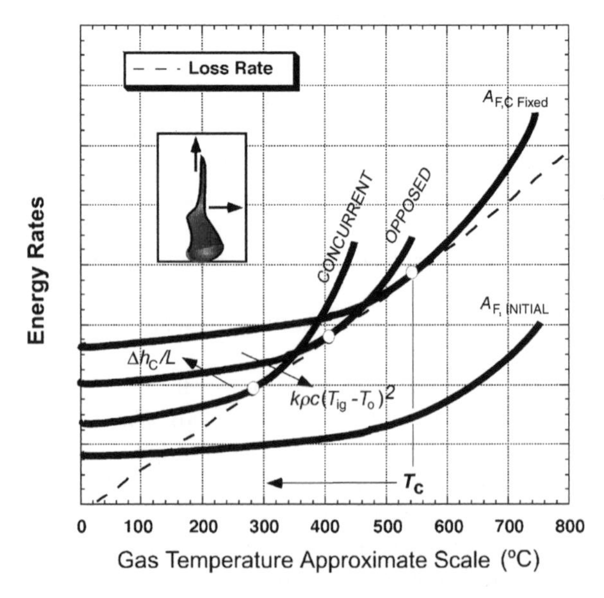

Figure 11.22 Surface spread fire growth

For concurrent spread, the growth rate can be much faster, and therefore the critical condition can be reached at lower compartment temperatures. The dependence of the concurrent flame spread area on both $\dot{Q}$ and the surface temperature of the material make this spread mode very feedback sensitive.

The Semenov criticality diagrams for fire growth are useful to understand the complex interactions of the fire growth mechanisms with the enclosure effects. These diagrams can be used qualitatively, but might also be the bases of simple quantitative graphical solutions.

References

1. Karlsson, B. and Quintiere, J. G., *Enclosure Fire Dynamics*, CRC Press, Boca Raton, Florida, 2000.
2. *Engineering Guide to Fire Exposures to Structural Elements*, Society of Fire Protection Engineers, Bethesda, Maryland, May 2004.
3. Thomas, P. H., Behavior of fires in enclosures – some recent progress, in 14th International Symposium on Combustion, The Combustion Institute, Pittsburgh, Pennsylvania, 1973, pp. 1007–20.
4. Thomas, P. H., Bullen, M. L., Quintiere, J. G. and McCaffrey, B. J., Flashover and instabilities in fire behavior, *Combustion and Flame*, 1980, **38**, 159–71.
5. Thomas, P. H., Fires and flashover in rooms – a simplified theory, *Fire Safety J.*, 1980, **3**, 67–76.
6. Bullen, M. L. and Thomas, P. H., Compartment fires with non-cellulosic fuels, in 17th International Symposium on *Combustion*, The Combustion Institute, Pittsburgh, Pennsylvania, 1979, pp. 1139–48.
7. Tewarson, A., Some observations in experimental fires and enclosures, Part II: ethyl alcohol and paraffin oil, *Combustion and Flame*, 1972, **19**, 363–71.
8. Kim, K. I., Ohtani, H. and Uehara, Y., Experimental study on oscillating behavior in a small-scale compartment fire, *Fire Safety J.*, 1993, **20**, 377–84.
9. Takeda, H. and Akita, K., Critical phenomenon in compartment fires with liquid fuels, in 18th International Symposium on *Combustion*, The Combustion Institute, Pittsburgh, Pennsylvania, 1981, 519–27.
10. Quintiere, J. G., McCaffrey, B. J. and Rinkinen, W., Visualization of room fire induced smoke movement in a corridor, *Fire and Materials*, 1978, **12**(1), 18–24.
11. Quintiere, J. G., Steckler, K. D. and Corley, D., An assessment of fire induced flows in compartments, *Fire Sci. Technol.*, 1984, **4**(1), 1–14.
12. Emmons, H. W., Vent flows, in *The SFPE Handbook of Fire Protection Engineering*, 2nd edn (eds P.J. DiNenno *et al.*), Section 2, Chapter 5, National Fire Protection Association, Quincy, Massachusetts, 1995, pp. **2**-40 to **2**-49.
13. Epstein, M., Buoyancy-driven exchange flow through small openings in horizontal partitions, *J. Heat Transfer*, 1988, **110**, 885–93.
14. Steckler, K. D., Quintiere, J. G. and Rinkinen, W. J., Flow induced by fire in a room, in 19th International Symposium on *Combustion*, The Combustion Institute, Pittsburgh, Pennsylvania, 1983.
15. Rockett, J. A., Fire induced flow in an enclosure, *Combustion Sci. Technol.*, 1976 **12**, 165.
16. Tanaka, T. and Yamada, S., Reduced scale experiments for convective heat transfer in the early stage of fires, *Int. J. Eng. Performanced-Based Fire Codes*, 1999, **1**(3), 196–203.
17. Thomas, P.H. and Heselden, A.J.M., Fully-developed fires in single compartment, a co-operative research programme of the Conseil International du Bâtiment, Fire Research Note 923, Joint Fire Research Organization, Borehamwood, UK, 1972.

18. Quintiere, J. G. and McCaffrey, B. J., *The Burning Rate of Wood and Plastic Cribs in an Enclosure*, Vol. 1, NBSIR 80-2054, National Bureau of Standards, Gaithersburg, Maryland, November 1980.
19. Quintiere, J. G., McCaffrey, B. J. and DenBraven, K., Experimental and theoretical analysis of quasi-steady small-scale enclosure fires, in 17th International Symposium on *Combustion*, The Combustion Institute, Pittsburgh, Pennsylvania, 1979, pp. 1125–37.
20. Naruse, T., Rangwala, A. S., Ringwelski, B. A., Utiskul, Y., Wakatsuki, K. and Quintiere, J. G., Compartment fire behavior under limited ventilation, in *Fire and Explosion Hazards, Proceedings of the 4th International Seminar*, Fire SERT, University of Ulster, Northern Ireland, 2004, pp. 109–120.
21. McCaffrey, B. J., Quintiere, J. G. and Harkleroad, M. F., Estimating room temperatures and the likelihood of flashover using fire test data correlations, *Fire Technology*, May 1981, **17**(2), 98–119.
22. Mowrer, F. W. and Williamson, R B., Estimating temperature from fires along walls and in corners, *Fire Technology*, 1987, **23**, 244–65.
23. Azhakesan, M. A., Shields, T. J., Silcock, G. W. H. and Quintiere, J. G., An interrogation of the MQH correlation to describe centre and near corner pool fires, *Fire Safety Sciences, Proceedings of the 7th International Symposium*, International Association for Fire Safety Science, 2003, pp. 371–82.
24. Karlsson, B., Modeling fire growth on combustible lining materials in enclosures, Report TBVV–1009, Department of Fire Safety Engineering, Lund University, Lund, 1992.
25. Azhakesan, M. A. and Quintiere, J. G., The behavior of lining fires in rooms, in *Interflam 2004*, Edinburgh, 5–7 July 2004.
26. Law, M., A basis for the design of fire protection of building structures, *The Structural Engineer*, January 1983, **61A**(5).
27. Harmathy, T. Z., A new look at compartment fires, Parts I and II, *Fire Technology*, 1972, **8**, 196–219, 326–51.

Problems

11.1 A trash fire occurs in an elevator shaft causing smoke to uniformly fill the shaft at an average temperature of 200 °C.

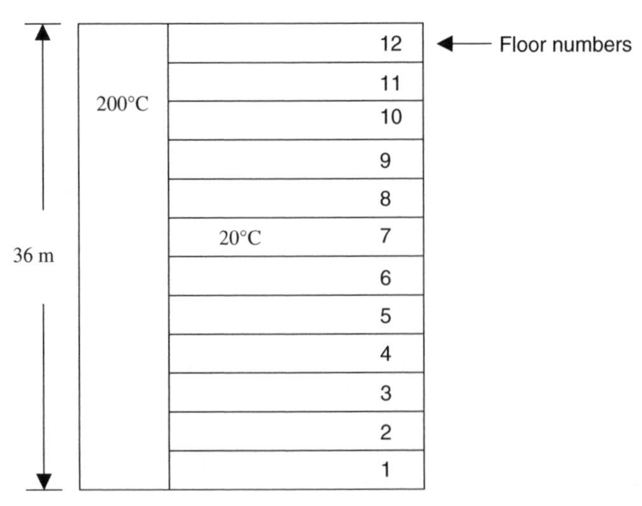

The only leakage from the shaft can occur to the building that is well ventilated and maintained at a uniform temperature of 20 °C. The leak can be approximated as a 2 cm wide vertical slit which runs along the entire 36 m tall shaft.

(a) Calculate the mass flow rate of the smoke to the floors.

(b) What floors receive smoke? Compute neutral plume height.

(c) What is the maximum positive pressure difference between the shaft and the building and where does it occur?

11.2 A 500 kW fire in a compartment causes the smoke layer to reach an average temperature of 400 °C. The ambient temperature is 20 °C. Air and hot gas flow through a doorway at which the neutral plane is observed to be 1 m above the floor. A narrow slit 1 cm high and 3 m wide is nominally 2 m above the floor.

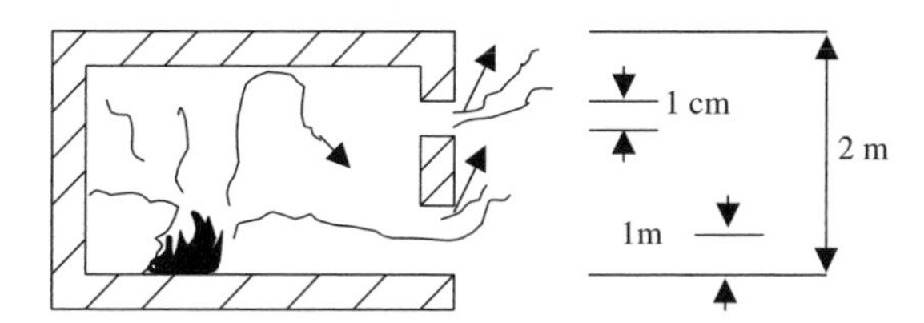

Assume the pressure does not vary over the height of the slit.

(a) Calculate the pressure difference across the slit.

(b) Calculate the mass flow rate through the slit. The slit flow coefficient is 0.7.

11.3 A fire occurs in a 3 m cubical compartment made of 2 cm thick. Assume steady state heat loss through the concrete whose thermal conductivity is 0.2 W/m^2 K. By experiments, it is found that the mass loss rate, $\dot{m}$, of the fuel depends on the gas temperature rise, ΔT, of the compartment upper smoke layer:

$$\dot{m} = \dot{m}_o + \beta(\Delta T)^{3/2}$$

where $\dot{m}_o = 10$ g/s, $\beta = 10^{-3}$ $g/sK^{3/2}$ and$\Delta T = T - T_\infty$. The compartment has a window opening 1 m × 1 m. The heat of combustion for the fuel is 20 kJ/g and ambient air is at 20 °C. Compute the gas temperature rise, ΔT.

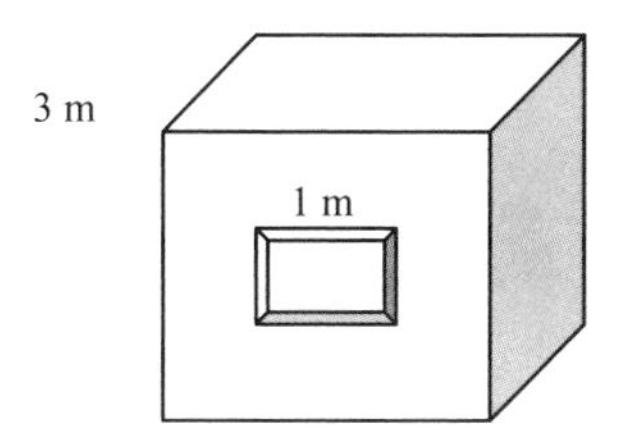

11.4 A room 3 m × 4 m × 3 m high with an opening in a wall 2 m × 2 m contains hexane fuel, which can burn in pools on the floor. The ambient temperature is 25 °C. The heat of combustion for hexane is 43.8 kJ/g. The construction material of the room is an insulator which responds quickly, and thus has a constant effective heat loss coefficient, $h_k = 0.015$ kW/m^{2}K. Use this to compute the overall heat loss to the room.

(a) A pool 0.8 m in diameter is ignited and the flame just touches the ceiling. What is the energy release rate of this fire?

(b) If this fire entrains 10 times stoichiometric air, what is the equivalence ratio for this room fire?

(c) What is the average gas temperature of the hot layer for the fire described in (a)?

(d) Flashover is said to commence if the temperature rise of the gas layer is 600 °C. At this point, more of the hexane can become involved. What is the energy release rate for this condition?

(e) What is the flame extent for the fire in (d)? Assume the diameter of the pool is now 1 m.

(f) Based on a room average gas temperature of 625 °C, what is the rate of air flow through the opening in g/s? Assume the windowsill opening is above the height of the hot gas layer interface of the room.

(g) What is the equivalence ratio for the onset of flashover where the gas temperature is 625 °C? (*Hint*: the heat of combustion per unit mass of air utilized is 3.0 kJ/g.)

(h) What is the energy release rate when the fire is just ventilation-limited, i.e. the equivalence ratio is one? Assume the air flow rate is the same as that at the onset of flashover as in (f).

(i) What is the fuel production rate when the fire is just ventilation-limited?

11.5 A fire in a ship compartment burns steadily for a period of time. The average smoke layer achieves a temperature of 420 °C with the ambient temperature being 20 °C. The compartment is constructed of 1 cm thick steel having a thermal conductivity of 10 W/m^2 K. Its open doorway hatch is 2.2 m high and 1.5 m wide. The compartment has an interior surface area of 60 m^2. The fuel stoichiometric air to fuel mass ratio is 8 and its heat of combustion is 30 kJ/g.

(a) Compute the mass burning rate

(b) If the temperature does not change when the hatch is being closed, find the width of the hatch opening (the height stays fixed) enough to just cause the fire to be ventilation-limited.

11.6 A fire burns in a compartment under steady conditions. The fuel supply rate ($\dot{m}_F$) is fixed.

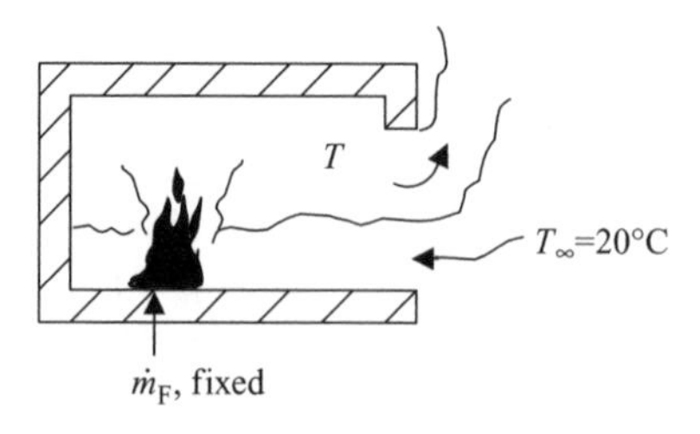

Its properties are:

Heat of combustion = 45 kJ/g

Stoichiometric air to fuel mass ratio = 15 g air/g fuel

The ambient temperature is 20 °C and the specific heat of the gases is assumed constant at 1 kJ/kg K. The fuel enters at a temperature of 20 °C. For all burning conditions 40 % of the energy released within the compartment is lost to the walls and exterior by heat transfer.

(a) The compartment has a door 2 m high, which is ajar with a slit 0.01 m wide. The fire is just at the ventilation limit for this vent opening.

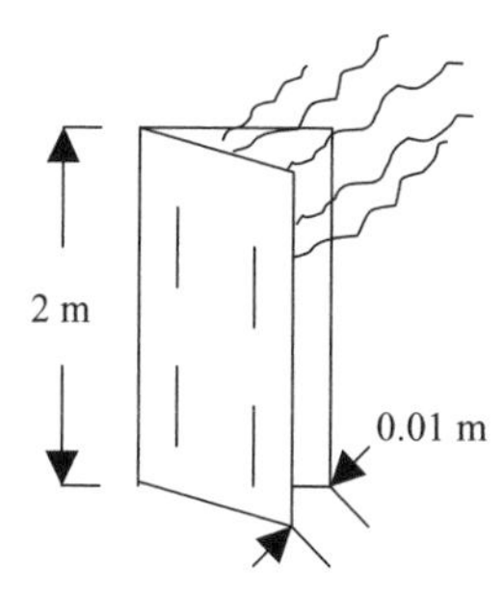

(i) Calculate the exit gas temperature.

(ii) Calculate the airflow rate.

(iii) Calculate the fuel supply rate.

(b) The door is opened to a width of 1 m. Estimate the resulting exit gas temperature. State any assumptions.

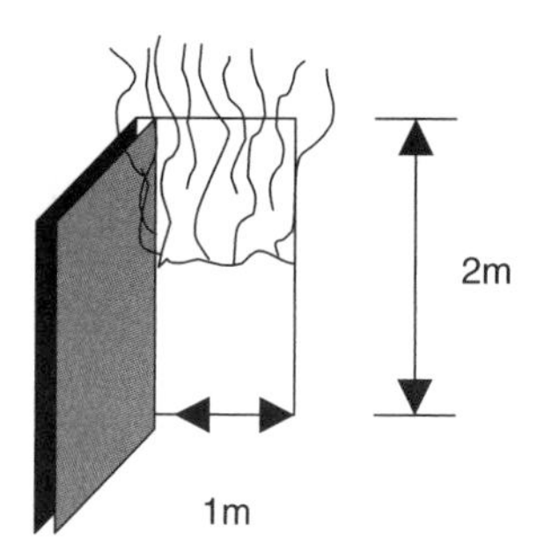

11.7 A room in a power plant has a spill of diesel fuel over a 3 m diameter diked area. The compartment is made of 20 cm thick concrete, and the properties are given below. The only opening is a 3 m by 2.5 m high doorway. The dimensions of the compartment are 10 m × 30 m × 5 m high. Only natural convection conditions prevail. The ambient air temperature is 20 °C. Other properties are given below:

Concrete:

$k = 1.0$ W/m K

$\rho = 2000$ kg/m^3

$c = 0.88$ kJ/kg K

Diesel:

Liquid density = 918 kg/m^3

Liquid specific heat = 2.1 J/g K

Heat of vaporization = 250 J/g

Vapor specific heat = 1.66 J/g K

Boiling temperature = 250 °C

Stoichiometric air to fuel mass ratio = 15

Heat of combustion = 44.4 kJ/g

The diesel fuel is ignited and spreads rapidly over the surface, reaching steady burning almost instantly. At this initial condition, compute the following:

(a) The energy release rate.

(b) The flame extent.

(c) The temperature directly above the fuel spill.

(d) The maximum ceiling jet temperature 4 m from the center of the spill.

At 100 s, the burning rate has not significantly changed, and the compartment has reached a quasi-steady condition with countercurrent flow at the doorway. At this new time, compute the following:

(e) The average smoke layer temperature.

(f) The air flow rate through the doorway.

(g) The equivalence ratio, ϕ.

By 400 s, the fuel spill 'feels' the effect of the heated compartment. At this time the fuel surface receives all the heat flux by radiation from the smoke layer, σT^4, over half of its area. For this condition, compute the following:

(h) The energy release rate as a function of the smoke layer temperature, T.

(i) Compute the total rate of losses (heat and enthalpy) as a function of the compartment smoke layer temperature. Assume the heat loss rate per unit area can be estimated by conduction only into the concrete from the gas temperature; i.e. $\sqrt{\frac{k\rho c}{t}}$ is the concrete conductance, where t is taken as 400 seconds.

(j) Plot (h) and (i) to compute the compartment gas temperature. Is the state ventilation-limited?

11.8 Compute the average steady state temperature on a floor due to the following fires in the World Trade Center towers. The building is 63.5 m on a side and a floor is 2.4 m high. The center core is not burning and remains at the ambient temperature of 20 °C. The core dimensions are 24 m × 42 m. For each fuel fire, compute the duration of the fire and the mass flow rate excess fuel or air released from the compartment space. The heat loss from the compartment fire should be based on an overall heat transfer coefficient of 20 W/m^2 K with a sink temperature of 20 °C. The vents are the broken windows and damage openings and these are estimated to be one-quarter of the building perimeter with a height of 2.4 m.

The airflow rate through these vents is given as $0.5A_oH_o^{1/2}$ kg/s with the geometric dimensions in m.

(a) 10 000 gallons of aviation fuel is spilled uniformly over a floor and ignited. Assume the fuel has properties of heptane. Heptane burns at its maximum rate of 70 g/m^2 s.

(b) After the aviation fuel is depleted, the furnishings burn. They can be treated as wood. The wood burns according to the solid curve in Figure 11.17. Assume the shape effect $W_1/W_2 = 1$, A is the floor area and $A_oH_o^{1/2}$ refers to the ventilation factor involving the flow area of the vents where air enters and the height of the vents. The fuel loading is 50 kg/m^2 of floor area.

11.9 Qualitatively sketch the loss rate and gain rate curves for these three fire states: (a) flame just touching the ceiling, (b) the onset of flashover and (c) the stoichiometric state where the fire is just ventilation-limited. Assume the L curve stays fixed in time and is nearly linear.

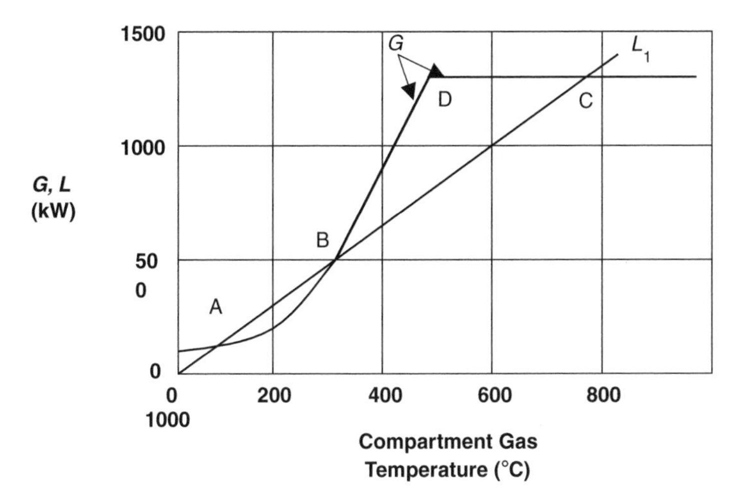

Use the graph above in which the generation rate of energy for the compartment is given by the G curve as a function of the compartment temperature and the loss rate of energy; heat from the compartment is described below.

(a) L_1 represents the loss curve shortly after the fire begins in the compartment. What is the compartment temperature and energy release rate? Identify this point on the above figure.

(b) As the compartment walls heat, the losses decrease and the L curve remains linear. What is the gas temperature just before and after the fire becomes ventilation-limited? Draw the curve and label the points VL− and VL+ respectively.

(c) Firefighters arrive after flashover and add water to the fire, increasing the losses. The loss curve remains linear. What is the gas temperature just before and after the suppression dramatically reduces the fire? Draw this L curve on the above figure. Assume the loss rate to suppression is proportional to the difference between the compartment gas temperature and ambient temperature, $T - T_\infty$.

11.10 Identify the seven elements in the graph with the best statement below:

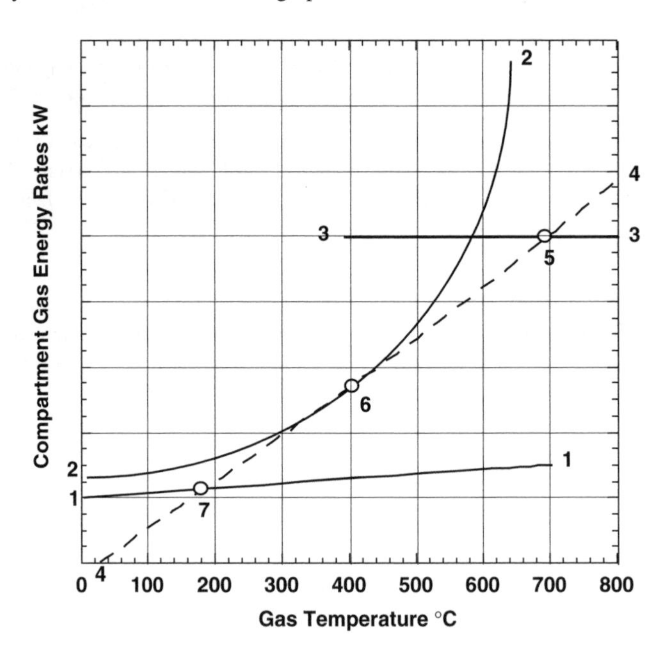

- Heat loss rate ____________
- Wood crib firepower ____________
- Unstable state ____________
- Fully developed fire ____________
- Firepower under ventilation-limited conditions ____________
- Flashover ____________
- Critical state ____________
- Stable state ____________
- Heat and enthalpy flow rate ____________
- Firepower of a burning wall ____________

12

Scaling and Dimensionless Groups

12.1 Introduction

The Wright brothers could not have succeeded with their first flight without the use of a wind tunnel. Even today, aircraft cannot be designed and developed without wind tunnel testing. While the basic equations of fluid dynamics have been established for nearly 200 years, they still cannot be solved completely on the largest of computers. Wind tunnel testing affords the investigation of flow over complex shapes at smaller geometric scales while maintaining the Reynolds number (Re) constant. The Re number represents the ratio of momentum to viscous forces, i.e.

$$\frac{\rho u^2 l^2}{\mu (u/l) l^2} = \frac{\rho u l}{\mu}$$

where Newton's viscosity law has been used (stress $= \mu \times$ velocity gradient) and μ is viscosity, ρ is density, u is velocity and l is a length scale. Although the Re is sufficient to insure similarity between the prototype and the model, many phenomena require the preservation of too many groups to ensure complete similarity. Fire falls into that category. However, that does not preclude the use of scale models or dimensionless correlations for fire phenomena to extend the generality to other scales and conditions. This incomplete process is the art of scaling.

Partial scaling requires knowledge of the physics for the inclusion and identification of the dominant variables. Many fields use this technique, in particular ship design. There the Froude number, $Fr = u^2/gl$, is preserved at the expense of letting the Re be what it will, as long as it is large enough to make the flow turbulent. As we shall see, Froude modeling in this fashion is very effective in modeling smoke flows and unconfined fire phenomena.

Dimensionless variables consisting of the dependent variables and the time and space coordinates will be designated by (^). Properties and other constants that form

Fundamentals of Fire Phenomena James G. Quintiere

dimensionless arrangements are commonly termed groups and are frequently designated by Π's. The establishment of the dimensionless variables and groups can lead to the extension of experimental results through approximate formulas. The use of such formulas has sometimes been termed fire modeling as they allow the prediction of phenomena. These formulae are usually given by power laws where the exponents might be established by theory or direct curve-fitting. The former is preferred. These formulas, or empirical correlations, are very valuable, and their accuracy can be superior to the best CFD models.

While correlations might be developed from small-scale experiments, they have not necessarily been designed to replicate a larger scale system. Such a process of replication is termed scale modeling. It can prove very powerful for design and accident investigation in fire. Again, the similarity basis will be incomplete, since all the relevant Π's cannot be preserved. Successful scale modeling in fire has been discussed by Heskestad [1], Croce [2], Quintiere, McCaffrey and Kashiwagi [3] and Emori and Saito [4]. Thomas [5,6] and Quintiere [7] have done reviews of the subject. Novel approaches have been introduced using pressure modeling as presented by Alpert [8] and saltwater models as performed by Steckler, Baum and Quintiere [9]. Even fire growth has been modeled successfully by Parker [10]. Some illustrations will be given here, but the interested reader should consult these and related publications in the literature.

12.2 Approaches for Establishing Dimensionless Groups

Three methods might be used to obtain nondimensional functionality. First is the Buckingham pi theorem, which starts with a chosen set of relevant variables and parameters pertaining to a specific modeling application. The numbers of physical independent variables are then determined. The number of Π groups is equal to the number of variables minus the dimensions. Second, the pertaining fundamental partial differential equation is identified and the variables are made dimensionless with suitable normalizing parameters. Third, the governing physics is identified in the simplest, but complete, form. Then dimensionless and dimensional relationships are derived by identities. We will use the last method.

An illustration of the three methods is given for the problem of incompressible flow over a flat plate. The flow is steady and two-dimensional, as shown in Figure 12.1, with the approaching constant flow speed designated as u_∞.

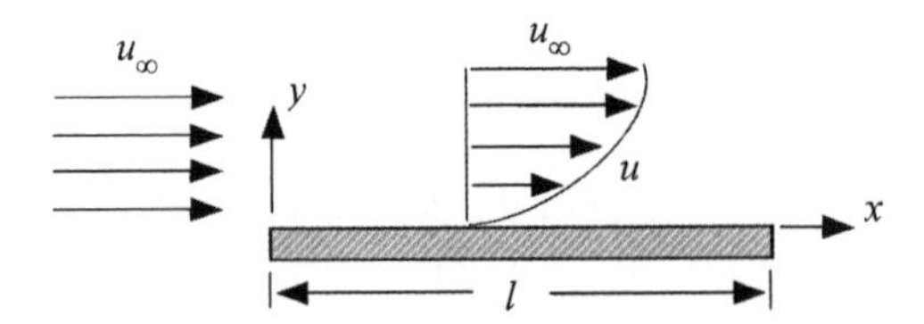

Figure 12.1 Flow over a flat plate

12.2.1 Buckingham pi method

In general, the velocity in the flow (u) is a function of x, y, l, u_∞, ρ and μ. Seven variables and parameters are identified in the problem: ρ, u, u_∞, x, y, l and μ (viscosity). Since three dimensions appear, M (mass), L (length) and T (time), three variables or parameters can be eliminated by forming four dimensionless groups: Π_1, Π_2, Π_3, Π_4.

Three repeating variables are selected, as u_∞, l and μ, to form the Π's from the other variables. Let $\Pi_1 = u_\infty^a,\ l^b,\ \mu^c,\ \rho$. Therefore by dimensional analysis Π_1 must have no dimensions. Equating powers for each dimension gives

$$\begin{array}{lcc} M: & 0 = c + 1 & \therefore c = -1 \\ L: & 0 = a + b - c - 3 & \therefore a + b = 2 \\ T: & 0 = -a - c & \therefore a = 1, b = 1 \end{array}$$

It is found that $\Pi_1 = u_\infty l\rho/\mu \equiv Re$, the Reynolds number. (Note that $\mu \sim ML^{-1}T^{-1}$.) The other Π's can be determined in a like manner, or simply by inspection:

$$\Pi_2 = u/u_\infty, \qquad \Pi_3 = x/l \qquad \text{and} \qquad \Pi_4 = y/l$$

The dimensionless functional relationship follows as

$$\frac{u}{u_\infty} = \text{function}\left(\frac{x}{l}, \frac{y}{l}, Re\right)$$

12.2.2 Partial differential equation (PDE) method

This method starts with knowledge of the governing equation. The governing equation for steady two-dimensional flow with no pressure gradient is

$$\rho\left(u\frac{\partial u}{\partial x} + v\frac{\partial u}{\partial y}\right) = \mu\frac{\partial^2 u}{\partial y^2}$$

with boundary conditions:

$$\begin{array}{ll} y = 0, & u = v = 0 \\ y \to \infty, & u = u_\infty \\ x = 0, & u = u_\infty \end{array}$$

The conservation of mass supplies another needed equation, but it essentially gives similar information for the y component of velocity. The dimensionless variables are selected as:

$$\begin{array}{ll} \hat{u} = u/u_\infty, & \hat{v} = v/u_\infty \\ \hat{x} = x/l, & \hat{y} = y/l \end{array}$$

Substituting gives

$$\hat{u}\frac{\partial \hat{u}}{\partial \hat{x}} + \hat{v}\frac{\partial \hat{u}}{\partial \hat{y}} = \left(\frac{\mu}{\rho u_\infty l}\right)\frac{\partial^2 \hat{u}}{\partial \hat{y}^2}$$

The same dimensionless functionality is apparent as from the Buckingham pi method.

12.2.3 *Dimensional analysis*

Let us use a control volume approach for the fluid in the boundary layer, and recognize Newton's law of viscosity. Where gradients or derivative relationships might apply, only the dimensional form is employed to form a relationship. Moreover, the precise formulation of the control volume momentum equation is not sought, but only its approximate functional form. From Equation (3.34), we write (with the symbol $\sim$ implying a dimensional equality) for a unit depth in the z direction

$$\rho u^2(x \times 1) \sim \mu \frac{u}{y}(x \times 1)$$

In dimensionless form:

$$\rho\left(\frac{u}{u_\infty}\right)^2\left(\frac{x}{l}\right) \sim \left(\frac{\mu}{u_\infty l}\right)\left(\frac{u/u_\infty}{y/l}\right)\left(\frac{x}{l}\right)$$

Hence,

$$\frac{u}{u_\infty} \equiv \hat{u} = \text{function}(\hat{x}, \hat{y}, Re)$$

12.3 Dimensionless Groups from the Conservation Equations

The dimensionless groups that apply to fire phenomena will be derived from the third approach using dimensional forms of the conservation equations. The density will be taken as constant, ρ_∞ without any loss in generality, except for the buoyancy term. Also, only the vertical momentum equation will be explicitly examined. Normalizing parameters for the variables will be designated as $()^*$. In some cases, the normalizing factors will not be the physical counterpart for the variable, e.g. x/l, where l is a geometric length. The pi groups will be derived and the dimensionless variables will be preserved, e.g. $\hat{u} = u/u^*$ and $\hat{x} = x/l^*$. After deriving the pi groups $\{\Pi_i\}$, we will then examine their use in correlations, and how to employ them in scale modeling applications. Common use of symbols will be made and therefore they will not always be defined here.

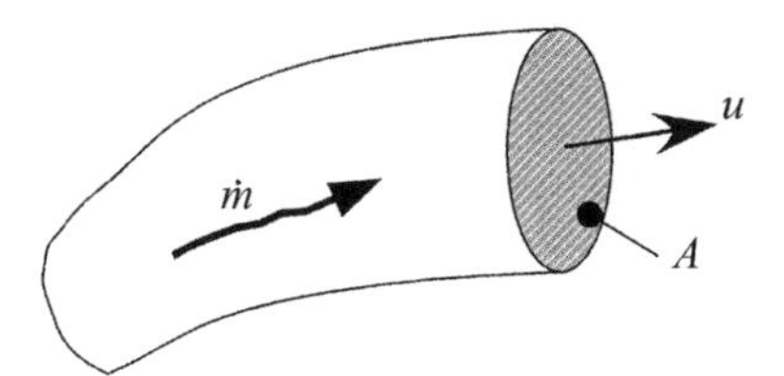

Figure 12.2 Mass flow rate

12.3.1 *Conservation of mass*

From Figure 3.5 the mass flow rate can be represented as

$$\dot{m} = \rho u A \tag{12.1}$$

as seen from Figure 12.2. The area will be represented in terms of the scale length of $A \sim l^2$ and similarly l^3 for volume.

12.3.2 *Conservation of momentum*

Consider a form of the vertical momentum equation (Equation (3.24)) with a buoyancy term and the pressure as its departure from hydrostatics. In functional form,

$$\rho V \frac{\mathrm{d}u}{\mathrm{d}t} + \dot{m}u \sim (\rho_\infty - \rho) g V + pA + \tau S \tag{12.2}$$

as demonstrated in Figure 12.3.

The relationship between the terms in Equation (12.2) can be used to establish normalizing parameters. For example, under nonforced flow conditions there is no obvious velocity scale factor (e.g. u_∞). However, one can be determined that is appropriate in natural convection conditions. Let us examine how this is done.

Relate the momentum flux term to the buoyancy force:

$$\rho u^2 l^2 \sim (\rho_\infty - \rho) g l^3 \tag{12.3}$$

By the perfect gas law (under constant pressure)

$$\frac{\rho_\infty - \rho}{\rho} = \frac{T - T_\infty}{T_\infty} \tag{12.4}$$

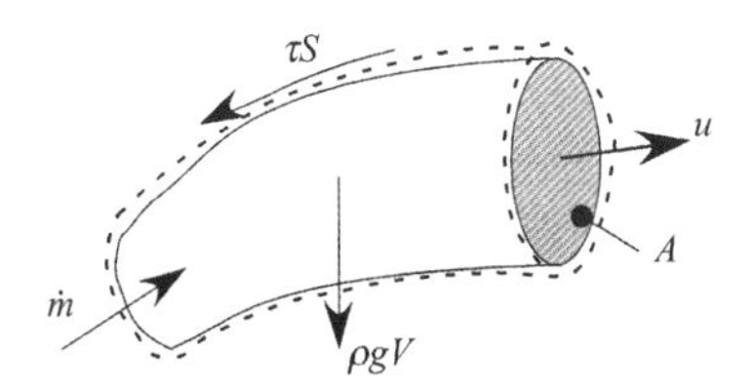

Figure 12.3 Momentum balance

Hence, it follows that

$$u \sim \sqrt{\left(\frac{T - T_\infty}{T_\infty}\right) gl} \sim \sqrt{gl} \tag{12.5}$$

and therefore an appropriate normalizing factor for velocity is

$$u^* = \sqrt{gl} \tag{12.6}$$

This need not be the only choice, but it is very proper for natural convection and represents an ideal maximum velocity due to buoyancy.

Similarly, by equating the unsteady momentum term with buoyancy gives a timescale

$$t^* \sim \frac{u^*}{g} \sim \sqrt{\frac{l}{g}} \tag{12.7}$$

The pressure and stress can be normalized as

$$\tau^* \sim p^* \sim \rho_\infty gl \tag{12.8}$$

Using Newton's viscosity law,

$$\tau^* \sim \mu \frac{\partial u}{\partial x} \sim \mu \frac{u^*}{l} \tag{12.9}$$

and equating momentum flux and stress terms gives

$$\rho_\infty u^{*2} l^2 \sim \mu \frac{u^*}{l} l^2 \tag{12.10}$$

with the ratio as the Reynolds number,

$$Re = \Pi_1 = \frac{\rho_\infty u^* l}{\mu} \sim \frac{\text{momentum}}{\text{shear force}} \tag{12.11}$$

Alternatively, writing Equation (12.10) in terms of the dimensionless velocity, and using the first part of Equation (12.5) for u^* gives (an approximate solution)

$$\hat{u} \sim \left[\frac{\mu}{\rho_\infty [(T - T_\infty)/T_\infty]^{1/2} \sqrt{gl} l}\right]^{1/2} \equiv \left(\frac{1}{Gr}\right)^{1/4} \tag{12.12}$$

where Gr denotes the Grashof number. The Gr is an alternative to Re for purely natural convection flows.

12.3.3 Energy equation

It is the energy equation that produces many more Π groups through the processes of combustion, heat transfer, evaporation, etc. We will examine these processes. Figure 12.4

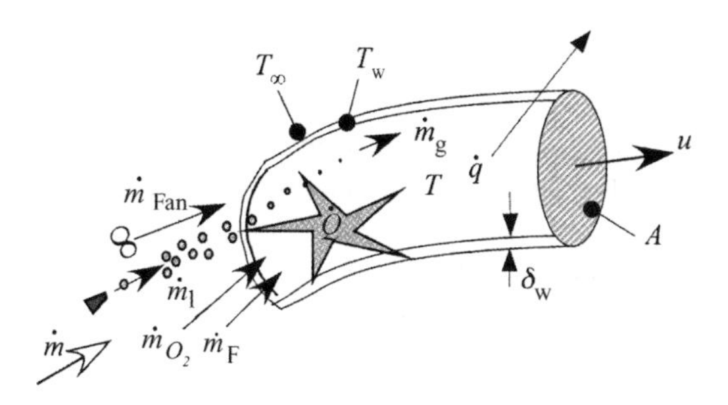

Figure 12.4 Energy transfer

is the basis of the physical and chemical processes being considered, again in functional form.

The mass flows include fuel (F), oxygen (O_2), liquid water (l), evaporated water vapor (g) and forced flows (fan). The chemical energy or firepower is designated as $\dot{Q}$ and all of the heat loss rates by $\dot{q}$. While Figure 12.4 does not necessarily represent a fire in a room, the heat loss formulations of Chapter 11 apply. From Equation (3.48), the functional form of the energy equation is

$$\rho c_p V \frac{\mathrm{d}T}{\mathrm{d}t} + \dot{m} c_p (T - T_\infty) \sim \dot{Q} - \dot{q} - \dot{m}_\mathrm{w} h_\mathrm{fg} \tag{12.13}$$

with the last term being the energy rate required for evaporation of water droplets. Recall that the firepower within the control volume is either

$$\dot{Q} = m_\mathrm{F} \Delta h_\mathrm{c}, \quad \text{fuel controlled} \tag{12.14a}$$

or

$$\dot{Q} = \frac{\dot{m}_{\mathrm{O}_2} \Delta h_\mathrm{c}}{r}, \quad \text{ventilation-limited} \tag{12.14b}$$

Relating the unsteady energy term (or advected enthalpy rate) with the firepower gives the second Π group commonly referred to as Q^* (or the Zukoski number for Professor Edward Zukoski, who popularized its use in fire):

$$Q_l^* \equiv \Pi_2 = \frac{\dot{Q}}{\rho_\infty c_p T_\infty \sqrt{g}\, l^{5/2}} \sim \frac{\text{firepower}}{\text{enthalpy flow rate}} \tag{12.15}$$

where here Q_l^* is based on the length scale l. It should be noted that this group gives rise to an inherent length scale for fire in natural convection:

$$l^* = \left(\frac{\dot{Q}}{\rho_\infty c_p T_\infty \sqrt{g}} \right)^{2/5} \tag{12.16}$$

This length scale can be associated with flame height and the size of large-scale turbulent eddies in fire plumes. It is a natural length scale for fires.

12.3.4 *Heat losses*

In Chapter 11 the heat fluxes to solid boundaries of an enclosure were described. They are summarized here for the purpose of producing the relevant Π groups. Consider first the radiant loss rate through a flow surface, A, of a control volume surrounding the enclosure gas phase in Figure 12.4:

$$\dot{q}_{\mathrm{o}} = A\sigma[\varepsilon(T^4 - T_\infty^4) + (1-\varepsilon)(T_{\mathrm{w}}^4 - T_\infty^4)] \tag{12.17}$$

where ε is the gas emissivity and the walls have been assumed as a blackbody. The gas emissivity may be represented as

$$\varepsilon \sim 1 - \mathrm{e}^{-\kappa l} \tag{12.18}$$

with κ being the absorption coefficient, and hence, another Π is

$$\Pi_3 = \kappa l \sim \frac{\text{radiation emitted}}{\text{blackbody radiation}} \tag{12.19}$$

This group is difficult to preserve for smoke and fire in scaling, and can be troublesome.

An alternate empirical radiation loss to the ambient used for unconfined fires produces the dimensionless group,

$$X_{\mathrm{r}} \equiv \Pi_4 = \frac{\dot{q}_{\mathrm{o}}}{\dot{Q}} \tag{12.20}$$

The heat transfer to the walls or other solid surfaces takes a parallel path from the gas phase as radiation and convection to conduction through the wall thickness, δ_{w}. This wall heat flow rate can be expressed as

$$\dot{q}_{\mathrm{w}} = \dot{q}_{\mathrm{k}} = \dot{q}_{\mathrm{r}} + \dot{q}_{\mathrm{c}} \tag{12.21}$$

The radiation exchange for blackbody walls can be represented as

$$\dot{q}_{\mathrm{r}} \sim \varepsilon\sigma(T^4 - T_{\mathrm{w}}^4)S \tag{12.22}$$

and convection as

$$\dot{q}_{\mathrm{c}} \sim h_{\mathrm{c}}S(T - T_{\mathrm{w}}) \tag{12.23}$$

where S is the wall surface area. The heat transfer coefficient, h_{c}, can be explicitly represented in terms of a specific heat transfer correlation, i.e.

$$h_{\mathrm{c}} \sim \frac{k}{l} Re^n Pr^m \tag{12.24}$$

The conduction loss is

$$\dot{q}_{\mathrm{k}} \sim \frac{k_{\mathrm{w}}S(T_{\mathrm{w}} - T_\infty)}{\delta} \tag{12.25}$$

where $\delta \sim \sqrt{(k/\rho c)_{\mathrm{w}} t}$ for a thermally thick wall, and $\delta = \delta_{\mathrm{w}}$ for one that is thermally thin.

Pi heat transfer groups can now be produced from these relationships. Let us consider these in terms of Q^*'s, i.e. Zukoski numbers for heat transfer. Normalizing with the advection of enthalpy, the conduction group is

$$Q_{\mathrm{k}}^* \equiv \Pi_5 = \frac{k_w T_\infty l^2 / \sqrt{(k/\rho c)_w t^*}}{\rho_\infty c_p T_\infty \sqrt{g}\, l^{5/2}} \sim \frac{\text{wall conduction}}{\text{enthalpy flow}} \tag{12.26a}$$

Substituting for t^* from Equation (12.7) gives

$$Q_{\mathrm{k}}^* = \frac{(k\rho c)_{\mathrm{w}}^{1/2}}{\rho_\infty c_p g^{1/4} l^{3/4}} \tag{12.26b}$$

Similarly, for convection,

$$Q_{\mathrm{c}}^* \equiv \Pi_6 = \frac{h_{\mathrm{c}}}{\rho_\infty c_p \sqrt{gl}} \sim \frac{\text{convection}}{\text{enthalpy flow}} \tag{12.27}$$

For radiation, we get

$$Q_{\mathrm{r}}^* \equiv \Pi_7 = \frac{\sigma T_\infty^3}{\rho_\infty c_p \sqrt{gl}} \sim \frac{\text{radiation}}{\text{enthalpy flow}} \tag{12.28}$$

In addition, for a wall of finite thickness, δ_{w}, we must include

$$\Pi_8 = \frac{\delta_{\mathrm{w}}}{\delta} = \left(\frac{\rho c}{k}\right)_{\mathrm{w}}^{1/2} \left(\frac{g}{l}\right)^{1/4} \delta_{\mathrm{w}} \sim \frac{\text{thickness}}{\text{thermal length}} \tag{12.29}$$

It is interesting to estimate the magnitude of these Zukoski numbers for typical wall materials that could be exposed to fire:

$$Q_{\mathrm{c}}^* \quad \text{and} \quad Q_{\mathrm{k}}^* \sim 10^{-3} \quad \text{to} \quad 10^{-2}$$
$$Q_{\mathrm{r}}^* \sim 10^{-2} \quad \text{to} \quad 10^{-1}$$

Hence, under some circumstances the heat losses could be neglected compared to $Q^* \sim 10^{-1}$ to 1, the combustion term.

12.3.5 Mass flows

There is a counterpart to the energy Zukoski number for mass flows. These Zukoski mass numbers can easily be shown for the phenomena represented in Figure 12.4 as

$$m_{\mathrm{Fan}}^* \equiv \Pi_9 = \frac{\dot{m}_{\mathrm{Fan}}}{\rho_\infty \sqrt{g}\, l^{5/2}} \sim \frac{\text{fan flow}}{\text{buoyant flow}} \tag{12.30}$$

Another mass flow term could be due to evaporation, e.g. water cooling from a sprinkler or that of gaseous fuel degradation from the condensed phase. From Chapter 9, we can easily deduce that

$$\dot{m}_{\mathrm{F}}^{*} \equiv \Pi_{10} = \frac{\dot{m}_{\mathrm{F}}}{\rho_{\infty}\sqrt{g}l^{5/2}} \tag{12.31}$$

in which m_{F}^{*} depends on Gr, Re, Pr, B, $\tau_{\mathrm{o}} = c_p(T_{\mathrm{v}} - T_{\infty})/L \equiv \Pi_{11}, r_{\mathrm{o}} = Y_{\mathrm{O_2,\infty}}/(rY_{\mathrm{F,o}}) = \Pi_{12}$ and other factors involving flame radiation (Π_3) and suppression.

The mass flow rate of vapor due to evaporation can be similarly represented. From Equation (6.35) it is seen that the dimensionless evaporation mass flux is

$$m_{\mathrm{g}}^{*} \equiv \hat{\dot{m}}_{\mathrm{g}}'' = \frac{\dot{m}_{\mathrm{g}}''}{\rho_{\infty}\sqrt{gl}} = Q_{\mathrm{c}}^{*}\left(\frac{M_{\mathrm{g}}}{M}\right)\frac{\mathrm{e}^{\Pi_{13,\mathrm{b}}}}{\mathrm{e}^{\Pi_{13,\mathrm{v}}}} \tag{12.32}$$

and

$$\Pi_{13,i} = \frac{M_{\mathrm{g}}h_{\mathrm{fg}}}{RT_{\mathrm{i}}} \sim \frac{\text{evaporation}}{\text{thermal energy}} \tag{12.33}$$

where b refers to boiling and v the surface vaporizing. The molecular weight ratio is a constant and might be incidental in most cases. Hence, we see that Π_{13} is the material property group that is needed for determining m_{g}^{*}.

12.3.6 *Liquid droplets*

The mass flux arising from the evaporation of liquid droplets is significant to fire scaling applications. Such scaling has been demonstrated by Heskestad [11], and specific results will be discussed later. As a first approximation, independent monodispersed droplets of spherical diameter, D_{l}, and particle volume density, n''', can be considered.

A conservation of the number of droplets, n, could be represented as a conservation equation in functional form as

$$\frac{\mathrm{d}n}{\mathrm{d}t} \sim -\dot{n}_{\mathrm{col}} \tag{12.34}$$

where $\dot{n}_{\mathrm{col}}$ is the loss rate due to droplet collisions with each other and to surfaces. A constitutive equation is needed for this loss, but is too complex to address here. Simply note that there is another Π group due to this effect:

$$\Pi_{14} = \frac{\dot{n}_{\mathrm{col}}t^{*}}{n_{\mathrm{ref}}} \tag{12.35a}$$

where the reference droplet number is based on the initial flow rate of liquid water $\dot{m}_{\mathrm{l,o}}$ through the nozzle diameter D_{o} at initial droplet size $D_{\mathrm{l,o}}$. Therefore,

$$\dot{m}_{\mathrm{l,o}} = \rho_{\mathrm{l}}u_{\mathrm{o}}\frac{\pi}{4}D_{\mathrm{o}}^{2}$$

and

$$n_{\text{ref}} \equiv \frac{\dot{m}_{\text{l,o}} t^*}{\rho_{\text{l}} D_{\text{l,o}}^3} \tag{12.35b}$$

Then

$$\Pi_{14} = \frac{\dot{n}_{\text{col}} D_{\text{l,o}}^3}{\dot{V}_{\text{l,o}}} \tag{12.35c}$$

where $\dot{V}_{\text{l,o}}$ is the initial discharge volumetric flow rate.

The initial momentum of the jet must also be considered, as it is significant. An effective nozzle diameter can be determined to account for this momentum since sprinkler discharges lose momentum when they strike the droplet dispersal plate. This effective diameter can be found by measuring the initial thrust (F_{o}) of the spray. Consequently, another group follows as

$$\Pi_{15} = \frac{F_{\text{o}}}{\rho_{\text{l}} \left(\dot{V}_{\text{l,o}} / D_{\text{o}} \right)^2} \sim \frac{\text{thrust}}{\text{spray momentum}} \tag{12.36}$$

This is an important scaling consideration as the spray momentum is a controlling factor in the droplet trajectories.

The final consideration for the droplet dynamics comes from the momentum equation applied to a single droplet, as illustrated in Figure 12.5:

$$m_{\text{l}} \frac{\mathrm{d}u_{\text{l}}}{\mathrm{d}t} \sim m_{\text{l}} g + \rho \frac{(u_{\text{l}} - u)^2}{2} C_{\text{D}} \frac{\pi}{4} D_{\text{l}}^2 \tag{12.37a}$$

with the drag force coefficient taken at the small Reynolds number regime, as the droplets are small [11]:

$$C_{\text{D}} \approx 20 \left(\frac{\rho |u_{\text{l}} - u| D_{\text{l}}}{\mu} \right)^{-1/2} \tag{12.37b}$$

The dependent variables are normalized as $\hat{u}_{\text{l}} = u_{\text{l}} / \sqrt{gl}$, $\hat{D}_{\text{l}} = D_{\text{l}} / l$ and $\hat{\rho}_{\text{l}} = \rho_{\text{l}} / \rho_{\infty}$.

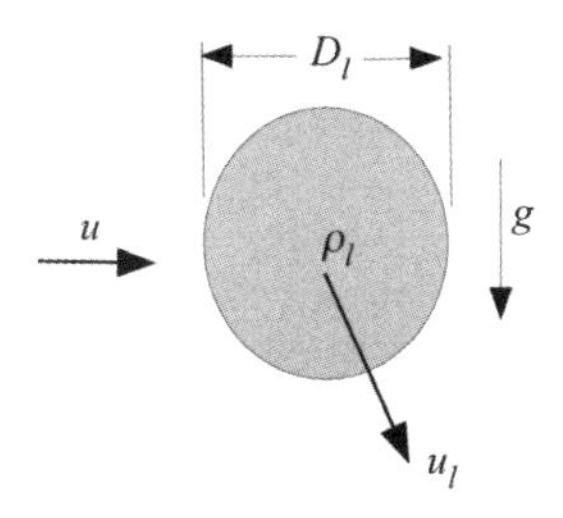

Figure 12.5 Droplet dynamics

Let us examine the droplet motion (Equation (12.37)) along with its evaporation transport (Equation (12.32)) and the conservation of mass for a single drop:

$$\frac{\rho_l}{2}\frac{dD_l}{dt} = -\dot{m}''_g \tag{12.38}$$

In functional form, and taking the ratio of both sides, this gives

$$\Pi_{16} = \frac{\dot{m}''_g l^{1/2}}{\rho_l D_l \sqrt{g}} = \frac{\hat{\dot{m}}''_g}{\hat{\rho}_l \hat{D}_l} \sim \frac{\text{evaporation rate}}{\text{droplet mass loss rate}} \tag{12.39}$$

From Equations (12.37) and (12.38), the ratio of the last two terms gives

$$\Pi_{17} = \hat{D}_\mu^{3/2} \sim \frac{\text{weight of droplet}}{\text{drag force}} \tag{12.40}$$

where $\hat{D}_\mu \equiv \hat{D}_l\, Re_l^{1/3}$, a viscous diameter.

This viscous diameter contains the important *Re* effects that must be addressed (while *Re* might be ignored in the bulk flow). Hence, let us modify Π_{16} into an alternative group:

$$\Pi_{16}\Pi_{17}^{2/3} = \Pi_{16'} \equiv \frac{m_g^*\, \mathrm{Re}_l^{1/3}}{\hat{\rho}_l} \tag{12.41}$$

Thus, we can equivalently use either Π_{16} or $\Pi_{16'}$. Finally, Equation (12.32) gives a specific group for the single droplet in $Q^*_{c,l}$ as

$$\Pi_{18} = \frac{1}{Q^*_{c,l}} = \frac{\rho_\infty C_p \sqrt{gl}}{h_c} \sim \frac{\text{advection}}{\text{mass transfer}} \tag{12.42}$$

where h_c is given from Ranz and Marshall [12] as

$$\frac{h_c D_l}{k} = 0.6\, Re_{D_l}^{1/2}\, \mathrm{Pr}^{1/3} \tag{12.43}$$

for $10 \le Re_{D_l} \le 10^3$. Consequently, Π_{18} can alternatively be expressed as

$$\Pi_{18}' = Pr^{2/3} \hat{D}_l^{1/2}\, Re_l^{1/2} \tag{12.44}$$

12.3.7 Chemical species

A conservation of species with the possibility of chemical reaction can be functionally represented from Equation (3.22) for species i with a chemical yield, y_i (mass of species i per mass of reacted fuel):

$$\rho l^3 \frac{dY_i}{dt} + \dot{m} Y_i = \frac{y_i \dot{Q}}{\Delta h_c} \tag{12.45}$$

The Π group that emerges is

$$\Pi_{19} = \frac{y_i c_p T_\infty}{\Delta h_c} \sim \frac{i\text{th enthalpy}}{\text{chemical energy}} \tag{12.46}$$

An alternative approach might be to address solely the mixture fraction f (mass of original fuel atoms per mass of mixture) since it has been established that there is a firm relationship between y_i and f for a given fuel. Note that f moves from 1 to 0 for the start and end of the fire space and f is governed by Equation (12.45) for $y_i \equiv 0$. This then conserves the fuel atoms. Under this approach it is recognized that

$$y_i = \text{function}\,(f) \tag{12.47}$$

for a given fuel under fire conditions. For the same fuel, the mass fraction varies as $Y_i \sim l^0$.

12.3.8 *Heat flux and inconsistencies*

In the implementation of scaling or even in the development of empirical correlations, it is almost universally found that radiation heat transfer is not completely included. In addition, the wall boundary thermal properties are rarely included in fire correlations. Let us consider the example of geometric scale modeling in which l is a physical dimension of the system. Then, for geometrically similar systems, we seek to maintain the same temperature at homologous points $(\hat{x}, \hat{y}, \hat{z})$ and time $\hat{t} = t\sqrt{g/l}$. Figure 12.6 shows the model and prototype systems. Let us examine the consequences of the fire heat flux for various choices in scaling. Note by preserving $Q^*(\Pi_2)$, it is required that the model firepower be selected according to

$$\dot{Q}_m = \dot{Q}_p \left(\frac{l_m}{l_p}\right)^{5/2} \tag{12.48}$$

or $\dot{Q} \sim l^{5/2}$. From Equation (12.13), if all the heat losses also go as $l^{5/2}$, then $T \sim l^0$, or temperature is invariant with scale size. Let us examine the behavior of the wall or surface heat losses under various scaling schemes to see if this can be accomplished. It will be seen that all forms of heat transfer cannot be simultaneously preserved in scaling.

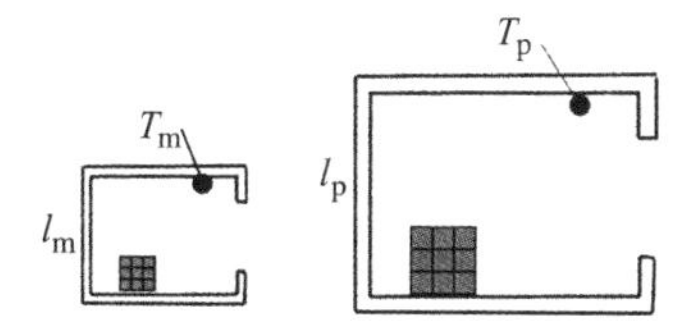

Figure 12.6 Model and prototype geometric similar systems

For complete scaling in model heat transfer, the following must be preserved:

$$\Pi_3 = \kappa l \qquad \therefore \kappa \sim l^{-1}$$

$$\Pi_5 = Q_k^* = \frac{(k\rho c)_w^{1/2}}{\rho_\infty c_p g l^{3/4}} \qquad \therefore (k\rho c)_w \sim l^{3/2}$$

$$\Pi_6 = Q_c^* = \frac{h_c}{\rho_\infty c_p \sqrt{gl}} \qquad \therefore h_c \sim l^{1/2}$$

$$\Pi_7 = Q_r^* = \frac{\sigma T_\infty^3}{\rho_\infty c_p \sqrt{gl}} \qquad \therefore T_\infty \sim l^{1/6}$$

$$\Pi_8 = \frac{(k\rho c)_w^{1/2}}{k_w}\left(\frac{g}{l}\right)^{1/4}\delta_w \qquad \therefore \delta_w \sim \left(\frac{k_w l^{1/4}}{(k\rho c)_w^{1/2}}\right)$$

In addition, for turbulent convection it is commonly found that the Nusselt number is important, $Nu \equiv h_c l/k$. Since

$$h_c \sim \frac{k}{l}\left(\frac{\rho_\infty \sqrt{g} l^{3/2}}{\mu}\right)^{4/5} \sim l^{1/5}$$

(for laminar flow we would have, instead, $h_c \sim l^{-1/4}$), we see that inconsistencies emerge: $h_c \sim l^{1/2}$ or $l^{1/5}$ and $T_\infty \sim l^{1/6}$, while it should be l^0 as it is difficult to control the ambient temperature.

Several strategies can be employed to maintain partial scaling. This is the meaning of the art of scaling. It requires some insight into the importance of competing effects. Let us first consider maintaining Π_5 and Π_8 constant. This preserves wall conduction effects. Consequently, recognizing that $k_w \sim \rho_w$, usually for materials, and then assuming $c_w \sim l^0$ (since specific heats for solids do not vary much among materials),

$$\text{from } \Pi_5: \quad k_w \sim \rho_w \sim l^{3/4}$$

and

$$\text{from } \Pi_8: \quad \delta_w \sim l^{1/4}$$

Conduction heat flux follows:

$$\dot{q}'' \sim \frac{k_w}{\delta_w}(T_w - T_\infty) \sim \frac{l^{3/4}}{l^{1/4}} \sim l^{1/2} \tag{12.49}$$

and the wall conduction heat loss rate varies as required, $l^{5/2}$.

Let us try to preserve convection. If we ignore Π_6 but maintain the more correct turbulent boundary layer convection (Equation (12.24)), then

$$\dot{q}'' \sim h_c(T - T_w) \sim l^{1/5} l^0 \sim l^{1/5} \tag{12.50}$$

Finally, consider radiation. Under optically thin conditions ($\Pi_3 = \kappa l \sim$ small),

$$\dot{q}'' \sim \kappa l \sigma T^4 \sim \kappa l \tag{12.51}$$

and for the optically thick case ($\kappa l \sim$ large),

$$\dot{q}'' \sim \epsilon \sigma T^4 \sim 1 - \mathrm{e}^{-\kappa l} \tag{12.52}$$

with Equation (12.18). If we could preserve Π_3, then

$$\dot{q}'' \sim l^0, \text{optically thin} \tag{12.53a}$$

and

$$\dot{q}'' \sim l^0, \text{optically thick} \tag{12.53b}$$

This preservation requires that $\kappa \sim l^{-1}$. It could be accomplished by changing the fuel in the model to one that would yield more soot. This may not be done with great accuracy, but is feasible.

The convection heat flux is usually much smaller than radiation flux in significant fires. Preserving Π_5, Π_6 and Π_8 yields $\dot{q}'' \sim l^{1/2}$, which is inconsistent with the radiation flux (Equation (12.53)). However, if the same fuel is used in the model as in the prototype and no change occurs in κ as $\kappa \sim Y_i$, then the radiation flux is

$$\dot{q}'' \sim l, \text{for } \kappa \text{ small}$$

and

$$\dot{q}'' \sim l^0, \text{for } \kappa \text{ large}$$

If the fuel is modified such that $\kappa \sim l^{-1/2}$, then the radiant flux for $\kappa \sim$ small becomes consistent with conduction as $\dot{q}'' \sim l^{1/2}$, and the radiation heat loss goes as required, $l^{5/2}$. Alternatively, if radiation is believed to dominate, it might be useful to consider κ large and ignore Π_5 but require that

$$\dot{q}'' \sim \left(\frac{k_\mathrm{w}}{\delta_\mathrm{w}}\right) T \sim l^0 \tag{12.54}$$

As a consequence of this choice, while still maintaining Π_8 and therefore $\delta_\mathrm{w} \sim l^{1/4}$, then $k_\mathrm{w} \sim \rho_\mathrm{w} \sim l^{1/4}$.

Hopefully, this discussion has shown some of the issues related to achieving complete scaling. However, thoughtful partial scaling is still a valid approach for obtaining fairly accurate results for complex geometries. Remember, in scaling, the turbulence and combustion phenomena are inherent in the system, and therefore no empirical models are needed as in CFD approaches. In addition to the above inconsistencies, it is also not possible to preserve the *Re* number throughout (Equation (12.11)). However, if the pressure of the system is altered in order to change ρ_∞ between the model and the prototype, Π_1 can be preserved [8], i.e.

$$Re = \frac{\rho_\infty \sqrt{g} l^{3/2}}{\mu} \sim p l^{3/2}$$

In order to preserve the *Re* number,

$$p_{\mathrm{m}} = p_p \left(\frac{l_{\mathrm{m}}}{l_{\mathrm{p}}} \right)^{-3/2} \tag{12.55}$$

Another alternative is to conduct a scale model experiment in a centrifuge in which g is now increased [13]. Modifying both p and g in the model can allow the preservation of more groups. Thus, scaling in fire is not complete, but it is still a powerful tool, and there are many ways to explore it. Illustrations will be given later of successful examples in scale modeling and correlations to specific fire phenomena.

12.3.9 Summary

Table 12.1 gives a summary of the dimensionless variables. Two additional groups have been added, the Weber number, *We*, to account for droplet formation and the Nusselt number, $Nu = h_{\mathrm{c}} l/k$, to account for gas phase convection. A corresponding Nusselt

Table 12.1 Dimensionless variables and scalling in fire

Variable/group	Dimensionless	Scaling/comment
	Dependent:	
Velocity, u	$\hat{u} = \frac{u}{\sqrt{gl}}$	$u \sim l^{1/2}$
Temperature, T	$\hat{T} = \frac{T}{T_\infty}$	$T \sim l^0$
Pressure, p	$\hat{p} = \frac{p}{\rho_\infty g l}$	$p \sim l$
Concentration, Y_i	$\frac{Y_i}{Y_{i,\infty}}$	$Y_i \sim l^0$
Droplet number, n	$\frac{n}{n_{\mathrm{ref}}}$	$n \sim l^{3/2}$
Droplet diameter, D_l	$\frac{D_l}{l}$	$\Pi_{12} \rightarrow D_l \sim l^{1/2}$
Burning rate per area, $\dot{m}''_{\mathrm{F}}$	$\frac{\dot{m}''_{\mathrm{F}} l}{\mu}$	$\dot{m}''_{\mathrm{F}} \sim \frac{h_{\mathrm{c}} l}{\mu c_p} = \frac{Nu}{Pr}$
	Independent:	
Coordinates x, y, z	$\frac{x_i}{l}$	$x_i \sim l$
Time, t	$\frac{t}{\sqrt{l/g}}$	$t \sim l^{1/2}$
	Pi groups:	
$\Pi_1 \left(\frac{\text{inertia}}{\text{viscous}} \right)$, Re	$Re = \frac{\rho_\infty \sqrt{g} l^{3/2}}{\mu}$	Usually ignored

Table 12.1 (*Continued*)

Variable/group	Dimensionless	Scaling/comment
$\Pi_2\left(\frac{\text{firepower}}{\text{enthalpy rate}}\right), Q^*$	$\frac{\dot{Q}}{\rho_\infty c_p T\sqrt{g}l^{5/2}}$	Significant in combustion
$\Pi_3\left(\frac{\text{radiant emission}}{\text{ideal emission}}\right)$	κl	$\kappa \sim l^{-1}$, gas radiation important
$\Pi_4\left(\frac{\text{radiant loss}}{\text{firepower}}\right), X_r$	$X_r = \frac{\dot{q}_r}{\dot{Q}}$	$X_r \sim l^0$, important for free burning
$\Pi_5\left(\frac{\text{conduction}}{\text{enthalpy}}\right), Q_k^*$	$\frac{(k\rho c)_w^{1/2}}{\rho_\infty c_p g^{1/4} l^{3/4}}$	$k_w \sim \rho_w \sim l^{3/4}$, conduction important
$\Pi_6\left(\frac{\text{convection}}{\text{enthalpy}}\right), Q_c^*$	$\frac{h_c}{\rho_\infty c_p\sqrt{gl}}$	$h_c \sim l^{1/2}$, convection important
$\Pi_7\left(\frac{\text{radiation}}{\text{enthalpy}}\right), Q_r^*$	$\frac{\sigma T_\infty^3}{\rho_\infty c_p\sqrt{gl}}$	$T_\infty \sim l^{1/6}$, inconsistent with others
$\Pi_8\left(\frac{\text{thickness}}{\text{thermal length}}\right)$	$\left(\frac{\rho c}{k}\right)_w^{1/2}\left(\frac{g}{l}\right)^{1/4}\delta_w$	$\delta_w \sim l^{1/4}$, thickness of boundaries
$\Pi_9\left(\frac{\text{fan flow}}{\text{advection}}\right), m_{\text{Fan}}^*$	$\frac{\dot{m}_{\text{Fan}}}{\rho_\infty\sqrt{g}l^{5/2}}$	$\dot{m}_{\text{Fan}} \sim l^{5/2}$, forced flows
$\Pi_{10}\left(\frac{\text{fuel flow}}{\text{advection}}\right), m_F^*$	$\frac{\dot{m}_F}{\rho_\infty\sqrt{g}\,l^{5/2}}$	Fuel mass flux depends on *B, Gr, Re,* etc.
$\Pi_{11}\left(\frac{\text{sensible}}{\text{latent}}\right), \tau_o$	$c_p(T_v - T_\infty)/L$	Burning rate term
$\Pi_{12}\left(\frac{\text{available } O_2}{\text{stoichiometric } O_2}\right), r_o$	$\frac{Y_{O_2,\infty}}{rY_{F,o}}$	Burning rate term
$\Pi_{13}\left(\frac{\text{evaporation energy}}{\text{sensible energy}}\right)$	$\frac{M_g h_{fg}}{RT_i}$	'Activation' of vaporization
$\Pi_{14}\left(\frac{\text{collision loss}}{\text{initial particles}}\right)$	$\hat{n}_{col} = \frac{\dot{n}_{col}}{\dot{V}_{l,o}/D_{l,o}^3}$	$\dot{n}_{col} \sim l$, collision number rate
$\Pi_{15}\left(\frac{\text{spray thrust}}{\text{jet momentum}}\right)$	$\frac{F_o}{\rho_l(\dot{V}_{l,o}/D_o)^2}$	$F_o \sim l^3, D_o$ is an effective nozzle diameter, $D_o \sim l$
$\Pi_{16}\left(\frac{\text{evaporation rate}}{\text{droplet mass loss}}\right)$	$\frac{\dot{m}_g''\sqrt{l}}{\rho_l D_l\sqrt{g}}$	$\dot{m}_g'' \sim l^0$ (see Π_{17})
$\Pi_{17}\left(\frac{\text{weight of droplet}}{\text{drag force}}\right), \hat{D}_\mu$	$\hat{D}_\mu = \hat{D}_l Re_l^{1/3}$	$D_l \sim l^{1/2}$
$\Pi_{18}\left(\frac{\text{advection}}{\text{mass transfer}}\right)$	$Pr^{2/3}\hat{D}_l^{1/2}Re_l^{1/2}$	$D_l \sim l^{-1/4}$, inconsistent with Π_{17}
$\Pi_{19}\left(\frac{\text{ith enthalpy}}{\text{chemical energy}}\right)$	$\frac{y_i c_p T_\infty}{\Delta h_c}$	$y_i \sim l^0$
$\Pi_{20}\left(\frac{\text{droplet momentum}}{\text{surface tension}}\right)$	$We = \frac{\rho_l u_l^2 D_l}{\sigma}$	$D_l \sim l^{-1}$, inconsistent with Π_{17}
$\Pi_{21}\left(\frac{\text{enthalpy}}{\text{combustion energy}}\right)$	$\frac{c_p T_\infty}{\Delta h_c/r}$	Nearly always constant
$\Pi_{22}\left(\frac{\text{convection}}{\text{conduction}}\right)$	$Nu = h_c l/k$	$h_c \sim l^{-1}$

number (or, more precisely, a Sherwood number, $Sh = hl/(\rho_\infty D_i)$, where D_i is a mass diffusion coefficient for species i in the medium) could also be added. These last two additions remind us of the explicit presence of fluid property groups, the Prandtl number, $Pr = k/(c_p\mu)$, and the Schmidt number, $Sc = \rho_\infty D_i/\mu$. For media consisting mostly of air and combustion products these property groups are nearly constant and would not necessarily appear in fire correlations. However, for analog media such as smoke to water, they could be important.

This set of dimensionless groups and variables represents a fairly complete set of dependent variables and the independent coordinates, time, property and geometric parameters for fire problems. The presentation of dimensionless terms reduces the number of variables to its minimum. In some cases, restrictions (e.g. steady state, two-dimensional conditions, etc.) will lead to a further simplification of the set. However, in general, we should consider fire phenomena, with water droplet interactions, that have the functional dependence as follows:

$$\left\{\hat{u}, \hat{u}_l, \hat{T}, \hat{Y}_i, \hat{p}, \hat{D}_l, \hat{\dot{m}}''_g, \hat{\dot{m}}''_F, \hat{n}\right\} = \text{function}\{\hat{x}_i, \hat{t}, \Pi_i, i = 1, 22\} \tag{12.56}$$

together with boundary and initial conditions included. As seen from Table 12.1, it is impossible to maintain consistency for all groups in scaling. However, some may not be so important, and others may be neglected in favor of a more dominant term. This is the art of scaling. In correlations derived from experimental (or computational) data, only the dominant groups will appear.

12.4 Examples of Specific Correlations

The literature is filled with empirical formulas for fire phenomena. In many cases, the dimensionless groups upon which the correlation was based have been hidden in favor of engineering utility, and specific dimensions are likely to apply. For example, a popular formula for flame height from an axisymmetric source of diameter D is

$$z_f = 0.23\dot{Q}^{2/5} - 1.02D$$

with z_f and D in meters and $\dot{Q}$ in kW. However, the more complete formulation from Heskestad [14] is Equation (10.50a):

$$\frac{z_f}{D} = 15.6 Q_D^{*\,2/5}\left(\frac{sc_pT_\infty}{\Delta h_c}\right)^{3/5} - 1.02$$

which involves Π_2 and Π_{21}. In contrast, an alternative, Equation (10.49), is an implicit equation for the flame height, given as

$$\frac{z_f}{D} = \text{function}\left(Q_D^*, \frac{rc_pT_\infty}{Y_{ox,\infty}\Delta h_c}, X_r\right)$$

which includes the effect of free stream oxygen concentration and the flame radiation fraction. These additional effects were included based on theory, and more data need to be established to show their complete importance. The examples presented in Chapter 10 for plumes demonstrate a plethora of equations grounded in dimensionless parameters. Some other examples for plumes will be given here.

12.4.1 Plume interactions with a ceiling

For the geometry in Figure 12.7, the maximum temperature in the ceiling jet has been correlated by Alpert [15]:

$$\frac{(T - T_\infty)/T_\infty}{Q_{\mathrm{H}}^{*2/3}} = f\left(\frac{r}{H}\right) \tag{12.57}$$

as seen in the plot found in Figure 12.7. It might not be obvious why the Zukoski number based on the height H has a 2/3 power. This follows from Equations (12.2), (12.5) and

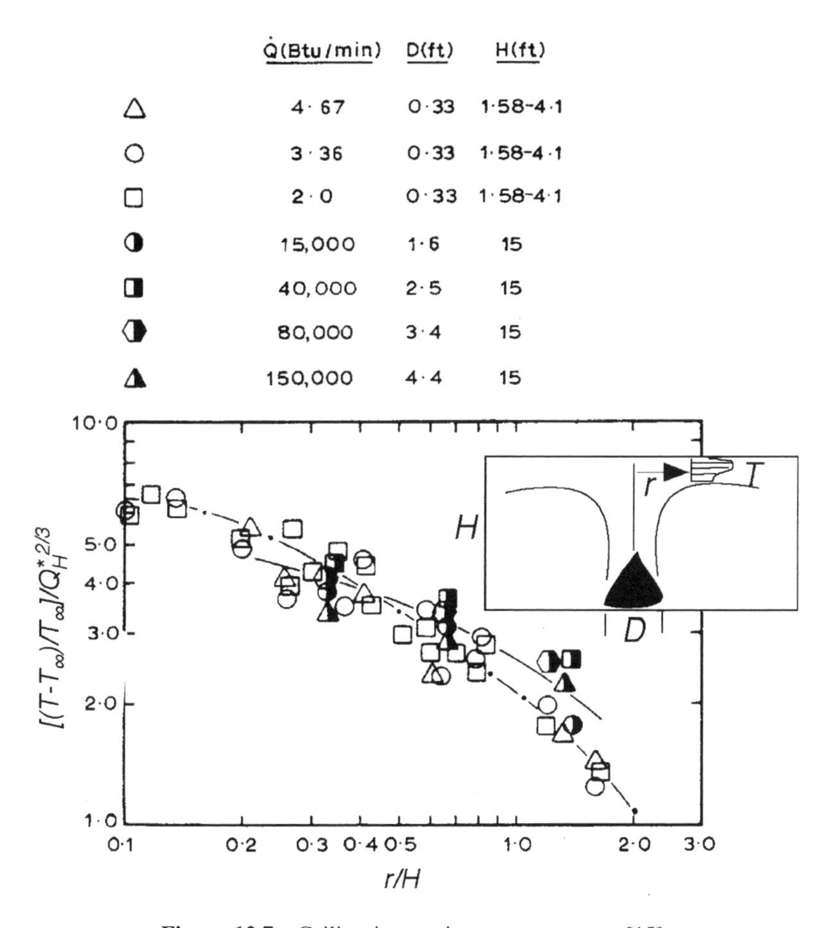

Figure 12.7 Ceiling jet maximum temperature [15]

(12.13), in which $l^* \equiv H$ and $u^* \equiv [[(T - T_\infty)/T_\infty]^* gH]^{1/2}$ based on a weakly buoyant plume. Consequently, $[(T - T_\infty)/T_\infty]^*$ is deduced as $Q_H^{*2/3}$. Alpert [15] shows similar results for velocity as

$$\frac{u/\sqrt{gH}}{Q_H^{*1/3}} = f\left(\frac{r}{H}\right) \tag{12.58}$$

12.4.2 *Smoke filling in a leaky compartment*

Zukoski [16] has shown theoretical results that have stood the test of many experimental checks for the temperature and descent of a uniform smoke layer in a closed, but leaky ($\Delta p \approx 0$), compartment. Similar to the scaling used for the ceiling jet, he found that

$$\frac{z_{\text{layer}}}{H} = \text{function}\left(\frac{t}{t^*}\right) \tag{12.59}$$

where

$$t^* \equiv \frac{\sqrt{H}}{\sqrt{g}Q_H^{*1/3}} \frac{S}{H^2}$$

where S is the floor area.

Similarly, the layer temperature follows as

$$\frac{[(T - T_\infty)/T_\infty]_{\text{layer}}}{Q_H^{*2/3}} = \text{function}\left(\frac{t}{t^*}\right) \approx \frac{(t/t^*)}{1 - z_{\text{layer}}/H} \text{ for } Q_H^* \text{ small.} \tag{12.60}$$

These relationships are shown in Figure 12.8.

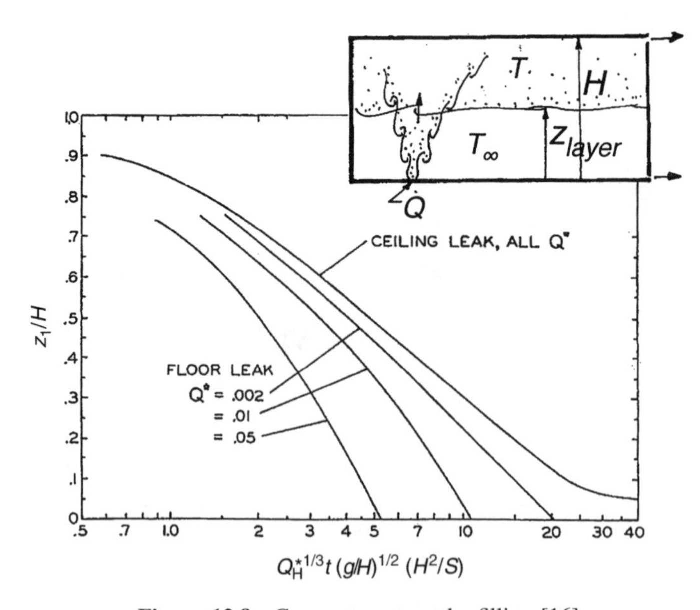

Figure 12.8 Compartment smoke filling [16]

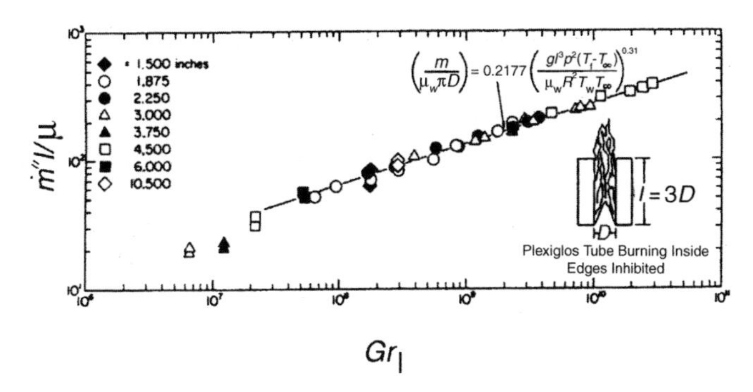

Figure 12.9 Burning rate within vertical tubes [17]

12.4.3 Burning rate

The burning rate for simple geometric surfaces burning in a forced flow (u_∞) or purely buoyant has been presented in Chapter 9. In general, it might be found that steady burning can be expressed as

$$\frac{\dot{m}''_{\mathrm{F}}x}{\mu} = \text{function}\left(\frac{x}{l}, Gr_{\mathrm{l}}, Re_{\mathrm{l}}, B_{\mathrm{m}}\right) \tag{12.61}$$

where B_{m} is the modified Spalding-B number based on L_{m} (Equation (9.72)), and primary radiation terms in the form of X_{r}, κl and Q^*_{r}, plus other effects such as water m^*_{g} (Equation (9.78)), can be included. Some additional examples of burning rate correlations are shown in Figure 12.9 for burning within vertical tubes [17] and Figure 12.10 for burning vertical cylinders and walls [18]. The latter wall data are subsumed in the work shown in Figure 9.11(b). Here the Grashof number was based on a flame temperature and was varied by pressure:

$$Gr_{\mathrm{l}} \equiv \frac{gl^3[p_\infty/(RT_\infty)]^2}{\mu^2}\left(\frac{T_{\mathrm{f}} - T_\infty}{T_\infty}\right) \tag{12.62}$$

This is an example of pressure modeling done at Factory Mutual Research (FM Global) by deRis, Kanury and Yuen [17] and Alpert [8]. In pressure modeling, Gr is preserved with $p_\infty \sim l^{-3/2}$ along. Therefore $\dot{m}''_{\mathrm{F}}l/\mu$, is also preserved and

$$Q^* \sim \frac{\dot{m}''_{\mathrm{F}}l^2}{(\rho_\infty T_\infty)l^{5/2}} \sim \frac{l^{-1}l^{-1/2}}{p_\infty} \sim \frac{l^{-3/2}}{l^{-3/2}} \sim l^0$$

but radiation is not preserved as $\kappa l \sim \rho_\infty l \sim p_\infty l \sim l^{-1/2}$ and $Q^*_{\mathrm{r}} \sim \frac{T^3_\infty}{\sqrt{l}} \sim \frac{l^0}{l^{1/2}} \sim l^{-1/2}$.

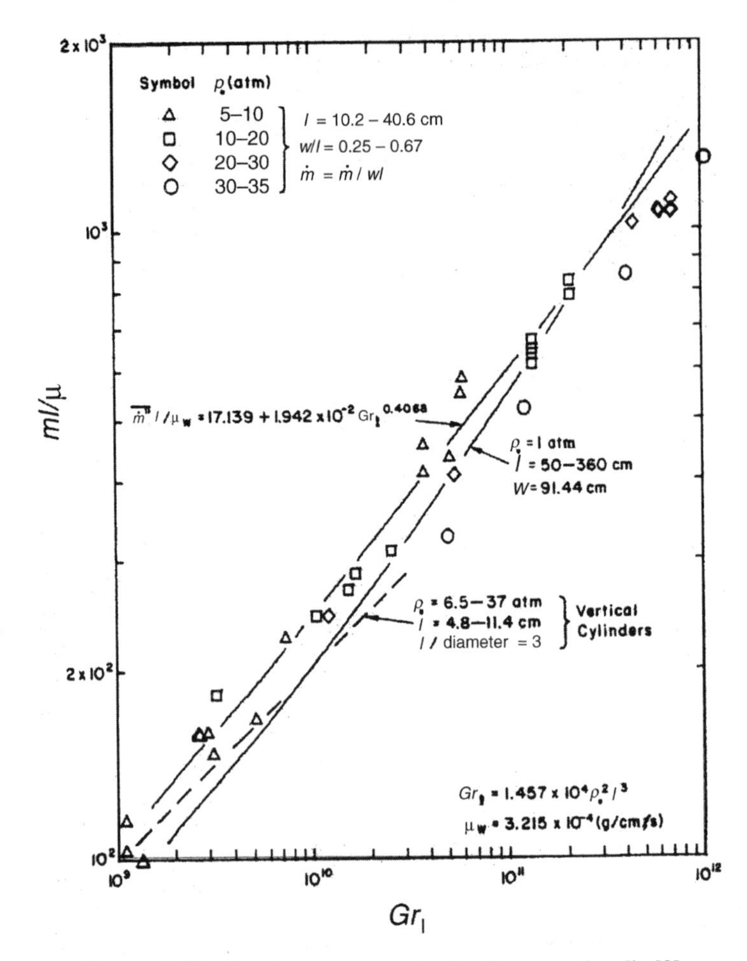

Figure 12.10 Burning rate on vertical cylinders and walls [8]

12.4.4 Compartment fire temperature

In Chapter 11 it was shown that control volume theory for the bulk compartment smoke properties could be expressed in dimensionless solutions. The characteristic length scale involves the geometric components of wall vents as $l^* = [A_o(H_o)^{1/2}]^{2/5}$. Hence, the MQH correlation [18] leads to

$$\frac{T - T_\infty}{T_\infty} = 1.6 Q_{l^*}^{*2/3} Q_w^{*-1/3} \tag{12.63}$$

for primarily, centered, well-ventilated, floor fires. Variations of this result have been shown for other fuel geometries (i.e. Equation (11.39)).

For the fully developed fire, various correlations have sought to portray the temperature in these fires in order to predict the impact on structures. Chapter 11 highlights the CIB work on wood cribs and the corresponding correlation by Law [19]. It is instructive to add the correlation by Babrauskas [20] that originated from a numerical solution to describe compartment fires. From his set of solutions, for the steady maximum temperature, he fitted parameters as

$$T - T_\infty = (1432\,^\circ\mathrm{C})\theta_1\,\theta_2\,\theta_3\,\theta_4\,\theta_5 \tag{12.64a}$$

From the dimensionless results established in Table 12.1, this result has the dimensionless counterpart as

$$\theta_1 = \begin{cases} 1 + 0.51\,\ln\phi, & \phi < 1 \text{ for wood cribs.} \\ 1 - 0.05\,(\ln\phi)^{5/3}, & \phi > 1 \end{cases} \tag{12.64b}$$

and

$$\phi = \frac{k_\mathrm{o}\rho_\infty\sqrt{g}A_\mathrm{o}\sqrt{H_\mathrm{o}}}{s\dot{m}''_{\mathrm{F},\infty}A_\mathrm{F}} \sim Q_\mathrm{o}^{*-1}$$

where $k_\mathrm{o} = 0.51$ and $s =$ stoichiometric air/fuel, or

$$\theta_1 = \frac{A_\mathrm{o}\sqrt{H_\mathrm{o}}}{A}\,\frac{0.5\Delta h_\mathrm{c}}{s\sigma\left(T^4 - T_\mathrm{boil}^4\right)} \sim Q_\mathrm{o}^{*-1}\left(\frac{A_\mathrm{F}}{A}\right)\frac{\Pi_{21}}{\Pi_7} \tag{12.64c}$$

for liquid pool fires. In addition,

$$\theta_2 = 1.0 - 0.94\exp\left[-54\left(\frac{A_\mathrm{o}\sqrt{H_\mathrm{o}}}{A}\right)^{2/3}\left(\frac{\delta}{k_\mathrm{w}}\right)^{1/3}\right] \sim Q_\mathrm{o}^{*-2/3}\left(\frac{\Pi_8}{\Pi_5}\right)^{1/3} \tag{12.64d}$$

$$\theta_3 = 1.0 - 0.92\exp\left[-150\left(\frac{A_\mathrm{o}\sqrt{H_\mathrm{o}}}{A}\right)^{3/5}\left(\frac{t}{(k\rho c)_\mathrm{w}}\right)^{2/5}\right] \sim Q_\mathrm{o}^{*-3/5}\left(\Pi_5^{-5/4}\right)^{2/5} \tag{12.64e}$$

$$\theta_4 = 1.0 - 0.205H^{-0.3} \sim \left(\Pi_7^{3/2}\right)^{1/3} \tag{12.64f}$$

and

$$\theta_5 = \text{combustion efficiency}\,\frac{\Delta h_\mathrm{c}}{\Delta h_{\mathrm{c,ideal}}} \tag{12.64g}$$

This fairly complete correlation shows that effects of fuel properties and perhaps vent radiation with the appearence of Π_7, as the physics are not evident, while these are missing in the correlation of Law [19] (Equation (11.45)). Yet Law's correlation includes the effect of a compartment shape. Also in Equation (12.64), the dimensionless groups are indicated compared to the dimensional presentation of Babrauskas, in units of kJ, kg, m, K, s, etc.

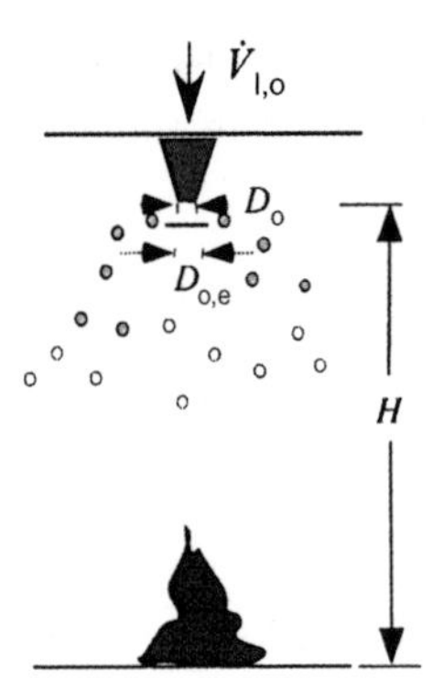

Figure 12.11 Effect of water spray on a pool fire

12.4.5 *Effect of water sprays on fire*

Heskestad [21] has shown that droplet sprays can be partially scaled in fire. He developed a correlation of the water flow rate ($\dot{V}_{l,o}$) needed for extinction of a pool fire, as shown in Figure 12.11. Heskesrtad determines that the water flow needed for extinction is

$$\dot{V}_{l,o}\left(\frac{\text{ml}}{\text{s}}\right) = 5.2\left[D_{o,e}(\text{mm})\right]^{1.08}[H(\text{m})]^{0.4}\left[\dot{Q}(\text{kW})\right]^{0.41} \tag{12.65a}$$

This followed from the data plot in Figure 12.12. The dimensionless rationale for this correlation is that, at extinction for nozzle overhead acting on an open fire, the important groups for geometric similarity are Π_2, Π_9 and Π_{15}, such that we obtain the following corresponding relationship:

$$\left(\frac{\rho_l \dot{V}_{l,o}}{\rho_\infty \sqrt{g} l^{5/2}}\right)_{\text{Extinction}} = \text{function}\left[\frac{\dot{Q}}{\rho_\infty c_p T_\infty \sqrt{g} l^{5/2}}, \frac{D_o}{l}\right] \tag{12.65b}$$

Here, Heskestad selects the nozzle diameter based on satisfying Π_{15}, i.e. the effective nozzle diameter (not its actual, due to the complex discharge configurations) is found from

$$D_{o,e} = \left(\frac{4}{\pi}\frac{\rho_l \dot{V}_{l,o}^2}{F_o}\right)^{1/2} \tag{12.66}$$

Here l is taken as H, the height of the nozzle above the fire. These data are shown to be representative for fires of about 1 to 10^3 kW, $H \sim 0.1$ to 1 m, $\dot{V}_{l,o} \sim 1$ to 50 ml/s and fire diameter, $D \sim 0.1$ to 1.0 m.

In this spray scaling, the droplet size was also a consideration, with Heskestad attempting to maintain Π_{17} constant with $D_{l,o} \sim l^{1/2}$ and l varied at 1 and 10 (or $D \sim 0.1$ and 1 m). Hence, the nozzles used were similar by preserving Π_{17}. As a

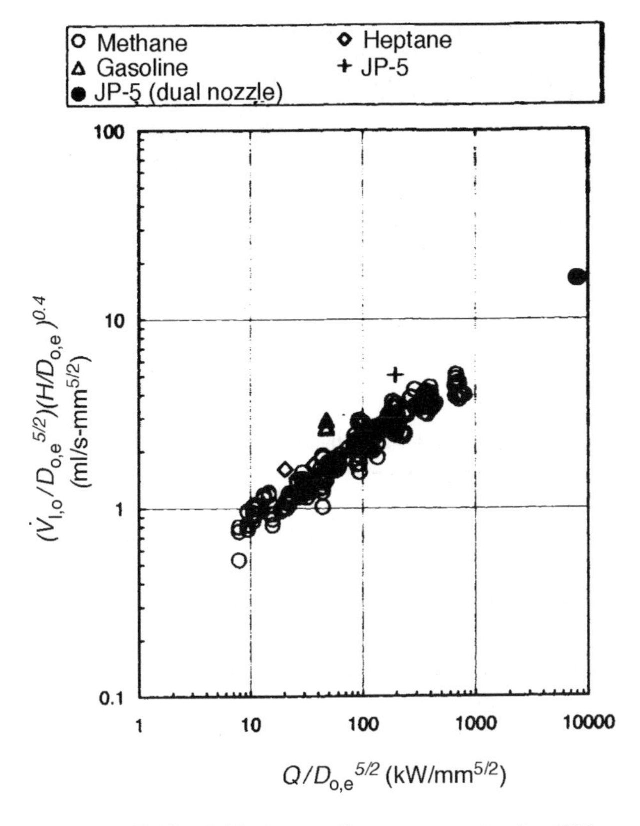

Figure 12.12 Critical water flow rate at extinction [21]

consequence, Π_{16} gives $\dot{m}''_{\rm g} \sim l^0$ for the droplet evaporation flux, and the droplet lifetime then scales as

$$t_{\rm Evap} = \frac{m_{\rm l}}{\dot{m}_{\rm g}} \sim \frac{D_{\rm l}^3}{l^0 D^2} \sim D_{\rm l} \sim l^{1/2} \tag{12.67}$$

This complies with the general timescale. While this scaling is not perfect, as $Re_{\rm l}$ and Π_{18} are not maintained, the correlation for similar nozzles is very powerful. Moreover, the basis for scale modeling fires with suppression is rational and feasible, and small-scale design testing can be done with good confidence.

12.5 Scale Modeling

The use of scale modeling is a very powerful design and analysis tool. It has been used by various fields as demonstrated in the first *International Symposium on Scale Modeling* [22] and those following. This nearly abandoned art brought together researchers examining structures, acoustics, tidal flows, aircraft, ship, vehicle design, fire and more. While computational methods have taken the forefront today, scale models can

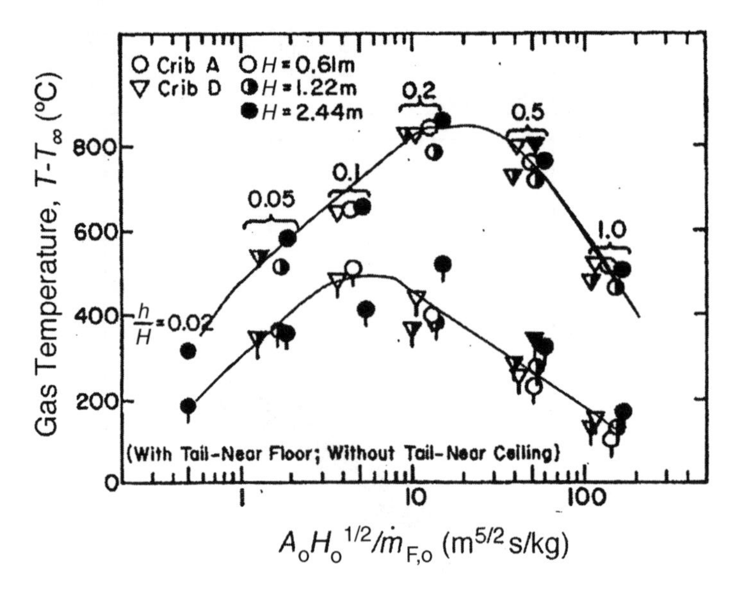

Figure 12.13 Gas temperatures in an enclosure [2]

still offer the possibility of more accuracy as well as a platform to validate these computer solutions. Some specific examples for fire will be considered for illustration. All of these are based on partial scaling with *Re* ignored (but maintained large enough for turbulent flow to prevail), and gas phase radiation is not generally preserved.

12.5.1 Froude modeling

The term 'Froude modeling' has been coined (probably from Factory Mutual researchers) as this is scaling fire by principally maintaining Q^* constant. It is called Froude modeling as that group pertains to advection and buoyancy and requires that the velocities must scale as $u \sim (gl)^{1/2}$ from Equation (12.6). Heskestad [1] and Croce [2] have used this effectively. They show that wood cribs burning in enclosures can produce scaling that obeys

$$\left\{\hat{T}, \hat{Y}_i, \frac{\dot{m}_{\mathrm{F}}}{\dot{m}_{\mathrm{F,o}}}\right\} = \text{function}\left\{\hat{x}_i, \hat{t}, m^*_{\mathrm{F,o}}, P, \frac{\Pi_8\Pi_6}{\Pi_5} = \frac{h_{\mathrm{c}}\delta_{\mathrm{w}}}{k_{\mathrm{w}}}, \frac{\Pi_8^2\sqrt{l/g}}{t_{\mathrm{F}}} = \left(\frac{\rho c}{k}\right)_{\mathrm{w}}\frac{\delta^2_{\mathrm{w}}}{t_{\mathrm{F}}}\right\} \quad (12.68)$$

where P is the crib porosity from Equation (9.83), $\dot{m}_{\mathrm{F,o}}$ is the free-burn rate of the crib and t_{F} is the burn time from $m_{\mathrm{F}}/\dot{m}_{\mathrm{F,o}}$. The results for maximum burning conditions over an ensemble of configurations and scales are shown in Figures 12.13 and 12.14. In these experiments, similar cribs and geometrically similar enclosures were used at the three geometric scales, such that $\dot{Q} \sim l^{5/2}$, and the wall material properties were changed with the scale as required. These results show the effect of $A_{\mathrm{o}}\sqrt{H_{\mathrm{o}}}/\dot{m}_{\mathrm{F,o}}$ (or the air flow rate over the fuel supply rate) and the vent height ratio, H_{o}/H.

Froude modeling was also used to prove that a code-compliant design for atrium smoke control was faulty, and consequently the code was blamed for smoke damage to a

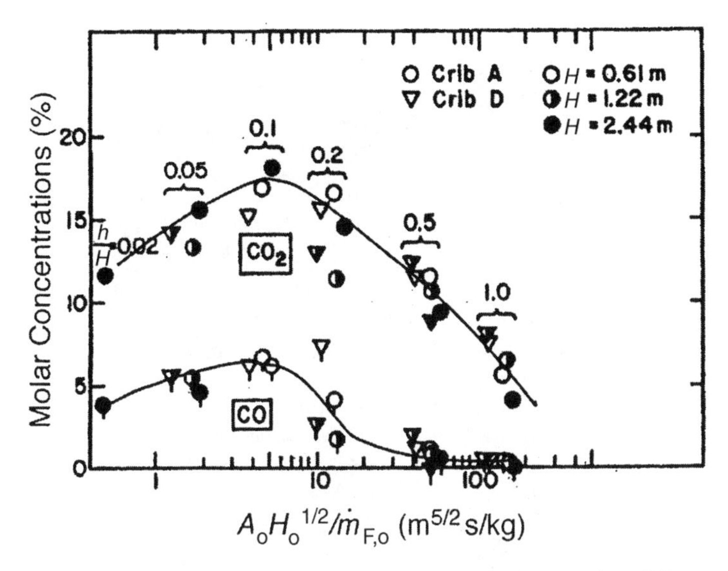

Figure 12.14 CO and CO_2 near ceiling gas concentrations [2]

large compartment store when a scale model of the conditions was introduced into a civil court case [23]. Figure 12.15 shows aspects of this scale model testing. The fire occurred in the Santa and sled that caused it to fall to the basement of the atrium of the store. By considering the fan and fire effects in the scaling, the smoke-control system was simulated. The recorded results demonstrated that the high-velocity input of the makeup air destroyed the stratification of the smoke layer, and smoke was pushed throughout the department store from the atrium.

Reviewing Table 12.1, it is seen that several groups are not preserved in Froude modeling. The *Re* (Π_1) is ignored, but the scale must be large enough to insure turbulent flow ($Re > 10^5$). This is justified away from solid boundaries where *Re* is large and inviscid flow is approached. In the Heskestad–Croce application only two conduction heat transfer groups are used: Π_5 and Π_8. The groups in Equation (12.68) were maintained in the scaling by varying the wall boundary material properties and its thickness.

Alternatively, one might satisfy convection near a boundary by invoking Π_6 and Π_8 where the heat transfer coefficient is taken from an appropriate correlation involving *Re* (e.g. Equation (12.38)). Radiation can still be a problem because re-radiation, Π_7, and flame (or smoke) radiation, Π_3, are not preserved. Thus, we have the 'art' of scaling. Terms can be neglected when their effect is small. The proof is in the scaled resultant verification. An advantage of scale modeling is that it will still follow nature, and mathematical attempts to simulate turbulence or soot radiation are unnecessary.

12.5.2 Analog scaling methods

It has been common to use two fluids of different densities to simulate smoke movement from fire. One frequently used approach is to use an inverted model in a tank of fresh water with dyed or colored salt water, for visualization, in order to simulate the hot

Figure 12.15 (a) Fire occurred on the sled. (b) Scale model of the atrium. (c) Recording smoke dynamics in the scale model

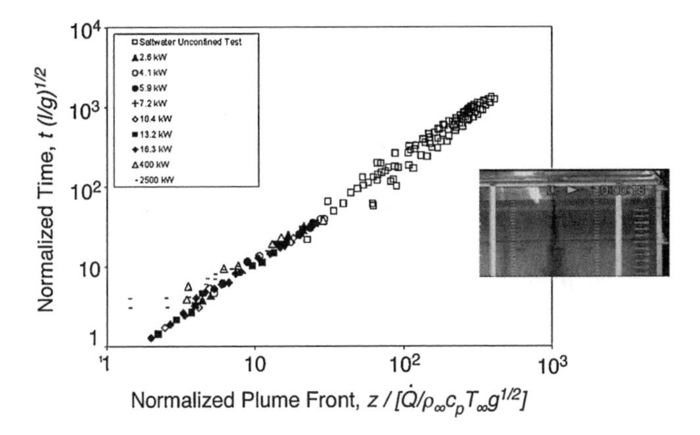

Figure 12.16 Plume front rise-time simulated by saltwater modeling [24,25]

combustion products. This follows from Equation (12.13) for combustion and Equation (12.45) for the saltwater flow rate as a source $\dot{m}_s$. We write the energy equation as

$$\rho_\infty T_\infty l^3 \frac{\mathrm{d}\left[c_p(T - T_\infty)/T_\infty\right]}{\mathrm{d}t} + \dot{m}c_p\left(\frac{T - T_\infty}{T_\infty}\right) \sim \dot{Q} \tag{12.69a}$$

for an adiabatic system and the salt species concentration equation as

$$\rho_s l^3 \frac{\mathrm{d}y_s}{\mathrm{d}t} + \dot{m}Y_s \sim \dot{m}_s \tag{12.69b}$$

The initial condition is that $(T - T_\infty)/T_\infty = Y_s = 0$ and at solid boundaries $(\partial T/\partial n = \partial Y_s/\partial n = 0$, with no heat or mass flow. Of course, radiation is ignored, but this is justified far away from the source. Note that the dimensionless source terms are Q^* and m_s^* where $t^* = (l/g)^{1/2}$ and l is chosen according to a geometric dimension, or as Equation (12.16) for an unconfined plume depending on whether the flow is confined or unconfined.

Figure 12.16 shows the rising fire plume (or falling salt plume) of Tanaka, Fujita and Yamaguchi [24] from Chapter 10 compared to the saltwater data of Strege [25]. The saltwater data correspond to the long-time regime, and follow the combustion data from Tanaka.

Figure 12.17 Saltwater smoke simulation in a two-story compartment (left) model (inverted) [26]

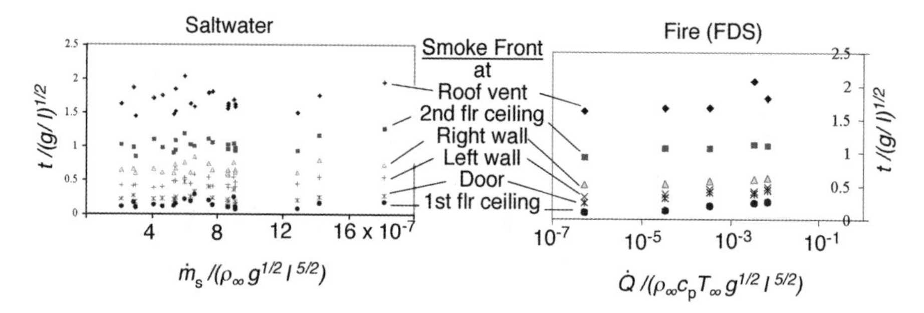

Figure 12.18 Dimensionless smoke front arrival times at various locations in the two-story compartment [26]

Another example is illustrated in Figure 12.17 for a compartment with an open first floor door and ceiling vents on both floors. This could produce a complex flow pattern for smoke from a fire. Event times were recorded, for a fire originating on the first floor as derived in dimensionless form from the saltwater model results, and compared with results from the Fire Dynamics Simulator (FDS) code from NIST. Figure 12.18 shows a comparison between event times in this two-room system for the saltwater scaled times and the corresponding time for a combustion system as computed using FDS by Kelly [26], where l was based on the height of the rooms. The FDS computations varied the fire from 10^{-1} to 10^3 kW compared to the very small salt source. It is seen that the fluid motion is simulated very well over this range of firepower, as the event times are independent of the dimensionless firepower.

Relating the fire temperatures to the saltwater concentration is problematic, but can be done by careful measurements. Yao and Marshall [27] made salt concentration measurements by using a nonintrusive laser fluorescence technique. These have been related to the temperature distribution of a ceiling jet. Figure 12.19 shows the laser signal distribution for salt concentration in the impinging region of a plume and its ceiling jet.

The saltwater, low-buoyancy system has a tendency to become laminar as the ceiling jet propagates away from the plume due to the buoyancy suppressing the turbulence.

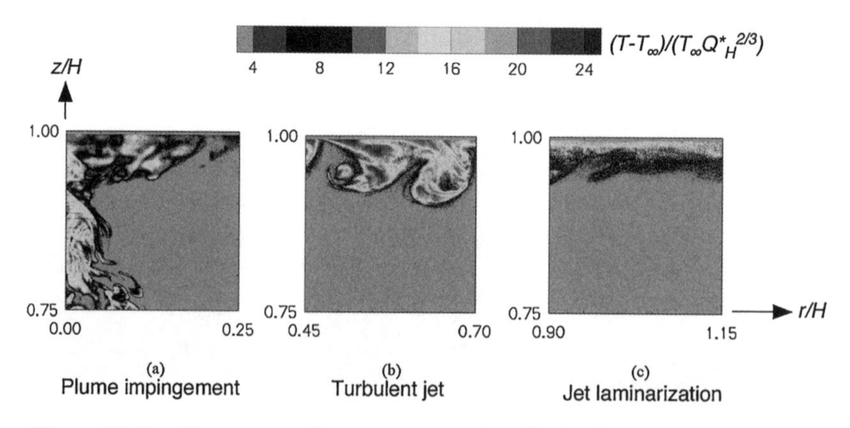

Figure 12.19 Fluorescent saltwater concentration distribution in a ceiling jet [27]

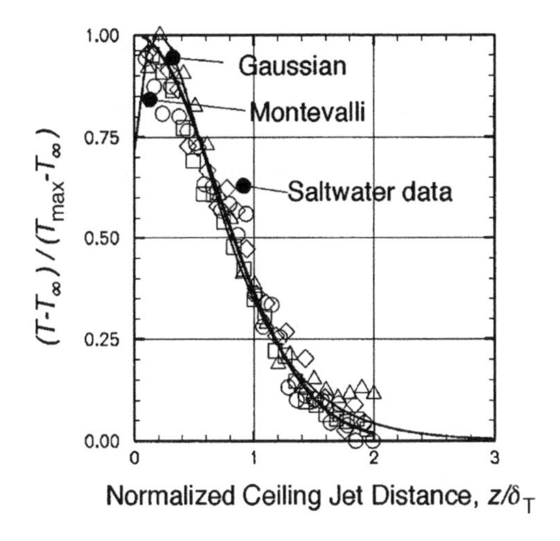

Figure 12.20 Ceiling jet temperature distribution [27]

Figure 12.20 shows the ceiling jet temperature distribution normalized by the maximum rise in the jet as a function of distance from the ceiling normalized with the jet thickness. The results apply to salt concentrations or temperature, and show that the distribution is independent of the distance from the plume. This saltwater analog technique is a powerful method for examining fire-induced flows, in particular with the technology demonstrated by Yao and Marshall [27].

References

1. Heskestad, G., Modeling of enclosure fires, *Proc. Comb. Inst.*, 1973, **14**, pp. 1021–30.
2. Croce, P. A., Modeling of vented enclosure fires, Part 1. Quasi-steady wood-crib source fires, FMRCJ.I.7A0R5.G0, Factory Mutual Research, Norwood, Massachusetts, July 1978.
3. Quintiere, J. G., McCaffrey, B. J. and Kashiwagi, T., A scaling study of a corridor subject to a room fire, *Combust. Sci. Technol.*, 1978, **18**, 1–19.
4. Emori, R. I. and Saito, K., A study of scaling laws in pool and crib fires, *Combust. Sci. Technol.*, 1983, **31**, 217–31.
5. Thomas, P. H., Modeling of compartment fires, *Fire Safety J.*, 1983, **5**, 181–90.
6. Thomas, P. H., Dimensional analysis: a magic art in fire research, *Fire Safety J.*, 2000, **34**, 111–41.
7. Quintiere, J. G., Scaling applications in fire research, *Fire Safety J.*, 1989, **15**, 3–29.
8. Alpert, R. L., Pressure modeling of fires controlled by radiation, *Proc. Comb. Inst.*, 1977, **16**, pp. 1489–500.
9. Steckler, K. D., Baum, H. R. and Quintiere, J. G., Salt water modeling of fire induce flows in multi-compartment enclosures, *Proc. Comb. Inst.*, 1986, **21**, pp. 143–9.
10. Parker, W. J. An assessment of correlations between laboratory experiments for the FAA Aircraft Fire Safety Program, Part 6: reduced-scale modeling of compartment at atmospheric pressure, NBS-GCR 83 448, National Bureau of Standards, February 1985.
11. Heskestad, G., Scaling the interaction of water sprays and flames, *Fire Safety J.*, September 2002, **37**, 535–614.

12. Ranz, W. E. and Marshall, W. R., Evaporation from drops: Part 2, *Chem. Eng. Prog.*, 1952, **48**, 173.
13. Altenkirch, R. A., Eichhorn, R. and Shang, P. C., Buoyancy effects on flames spreading down thermally thin fuels, *Combustion and Flame*, 1980, **37**, 71–83.
14. Heskestad, G., Luminous heights of turbulent diffusion flames, *Fire Safety J.*, 1983, **5**, 103–8.
15. Alpert, R. L., Turbulent ceiling-jet induced by large-scale fires, *Combust. Sci. Technol.*, 1975, **11**, 197–213.
16. Zukoski, E. E., Development of a stratified ceiling layer in the early stages of a closed-room fire, *Fire and Materials*, 1978, **2**, 54–62.
17. deRis, J., Kanury, A. M. and Yuen, M. C., Pressure modeling of polymers, *Proc. Comb. Inst.*, 1973, **14**, pp. 1033–43.
18. McCaffrey, B. J., Quintiere, J. G. and Harkleroad, M. F., Estimating room temperatures and the likelihood of flashover using fire test data correlations, *Fire Technology*, May 1981, **17**, 98–119.
19. Law, M., A basis for the design of fire protection of building structures, *The Structural Engineer*, January 1983, **61A**.
20. Babrauskas, V., A closed-form approximation for post-flashover compartment fire temperatures, *Fire Safety J.*, 1981, **4**, 63–73.
21. Heskestad, G., Extinction of gas and liquid pool fires with water sprays, *Fire Safety J.*, June 2003, **38**, 301–17.
22. Emori, R. I. (ed.), *International Symposium on Scale Modeling*, The Japan Society of Mechanical Engineers, Tokyo, Japan, 18–22 July 1988.
23. Quintiere, J. G. and Dillon, M. E., Scale model reconstruction of fire in an atrium, in Second International Symposium on *Scale Modeling* (ed. K. Saito), University of Kentucky, Lexington, Kentucky, 23–27 June 1997.
24. Tanaka, T., Fujita, T. and Yamaguchi, J., Investigation into rise time of buoyant fire plume fronts, *Int. J. Engng Performance-Based Fire Codes*, 2000, **2**, 14–25.
25. Strege, S. M., The use of saltwater and computer field modeling techniques to determine plume entrainment within an enclosure for various fire scenarios, including inclined floor fires, MS Thesis, Department of Fire Protection Engineering, University of Maryland, College Park, Maryland, 2000.
26. Kelly, A. A., Examination of smoke movement in a two-story compartment using salt water and computational fluid dynamics modeling, MS Thesis, Department of Fire Protection Engineering, University of Maryland, College Park, Maryland, 2001.
27. Yao, X. and Marshall, A. W., *Characterizing Turbulent Ceiling Jet Dynamics with Salt-Water Modeling*, Fire Safety Science-Proceedings of the Eighth International Symposium, Eds. Gottuk, G. T. and Lattimer, B. Y., pp. 927–938.

Appendix

Flammability Properties

Archibald Tewarson, FM Global, Norwood, Massachusetts, USA

These tables have been supplied by Dr Archibald Tewarson and are added as an aid in pursuing calculations for materials. Dr Tewarson has been singularly responsible for developing techniques for measuring material properties related to their fire characteristics, and for systematically compiling databases of these properties. His contributions are unique for this fledgling field of fire science. He has been a pioneer in using oxygen depletion and combustion products to measure the energy release rates in the combustion of materials under fire conditions. He has generously released these tables for use in the text.

The tables were taken from A. Tewarson, *Flammability of Polymers*, Chapter 11 in Plastics and the Environment, A. L. Andrady (editor), John Wiley and Sons, Inc. Hoboken, N.J. 2003.

Table 1 Glass transition temperature (T_{gl}), melting temperature (T_m) and melt flow index (MFI) of polymers

Polymer	T_{gl} (°C)	T_m(°C)	MFI (g/10 min)
	Ordinary polymers		
Polyethylene low density, PE-LD	−125	105–110	1.4
PE-high density, PE-HD		130–135	2.2
Polypropylene, PP (atactic)	−20	160–165	21.5
Polyvinylacetate, PVAC	28	103–106	
Polyethyleneterephthalate, PET	69	250	
Polyvinyl chloride, PVC	81		
Polyvinylalcohol, PVAL	85		
PP (isotactic)	100		
Polystyrene, PS	100		9.0
Polymethylmethacrylate, PMMA	100–120	130, 160	2.1, 6.2
	High-temperature polymers		
Polyetherketone, PEK	119–225		
Polyetheretherketone, PEEK		340	
Polyethersulfone, PES	190		
	Halogenated polymers		
Perfluoro-alkoxyalkane, PFA	75	300–310	
TFE,HFP,VDF fluoropolymer 200		115–125	20
TFE,HFP,VDF fluoropolymer 400		150–160	10
TFE,HFP,VDF fluoropolymer 500		165–180	10
Polyvinylidenefluoride, PVDF		160–170	
Ethylenechlorotrifluoroethylene, ECTFE		240	
Ethylenetetrafluoroethylene, ETFE		245–267	
Perfluoroethylene-propylene, FEP		260–270	
MFA		280–290	
Tetrafluoroethylene, TFE	130	327	

Table 2 Char yield, vaporization/decomposition temperatures (T_v/T_d), limiting oxygen index (LOI) and UL 94 ratings for polymers

Polymer	T_v/T_d (°C)	Char yield (%)	LOI[a] (%)	UL 94 ranking[b]
Ordinary polymers				
Poly(α-methylstyrene)	341	0	18	HB
Polyoxymethylene (POM)	361	0	15	HB
Polystyrene (PS)	364	0	18	HB
Poly(methylmethacrylate) (PMMA)	398	2	17	HB
Polyurethane elastomer (PU)	422	3	17	HB
Polydimethylsiloxane (PDMS)	444	0	30	HB
Poly(acrylonitrile-butadiene-styrene) (ABS)	444	0	18	HB
Polyethyleneterephthalate (PET)	474	13	21	HB
Polyphthalamide	488	3	(22)	HB
Polyamide 6 (PA6)-Nylon	497	1	21	HB
Polyethylene (PE)	505	0	18	HB
High-temperature polymers				
Cyanate ester of bisphenol-A (BCE)	470	33	24	V-1
Phenolic Triazine Cyanate Ester (PT)	480	62	30	V-0
Polyethylenenaphthalate (PEN)	495	24	32	V-2
Polysulfone (PSF)	537	30	30	V-1
Polycarbonate (PC)	546	25	26	V-2
Liquid crystal polyester	564	38	40	V-0
Polypromellitimide (PI)	567	70	37	V-0
Polyetherimide (PEI)	575	52	47	V-0
Polyphenylenesulfide (PPS)	578	45	44	V-0
Polypara(benzoyl)phenylene	602	66	41	V-0
Polyetheretherketone (PEEK)	606	50	35	V-0
Polyphenylsulfone (PPSF)	606	44	38	V-0
Polyetherketone (PEK)	614	56	40	V-0
Polyetherketoneketone (PEKK)	619	62	40	V-0
Polyamideimide (PAI)	628	55	45	V-0
Polyaramide (Kevlar®)	628	43	38	V-0
Polybenzimidazole (PBI)	630	70	42	V-0
Polyparaphenylene	652	75	55	V-0
Polybenzobisoxazole (PBO)	789	75	56	V-0
Halogenated polymers				
Poly(vinylchloride) (PVC)	270	11	50	V-0
Polyvinylidenefluoride (PVDF)	320–375	37	43–65	V-0
Poly(chlorotrifluoroethylene) (PCTFE)	380	0	95	V-0
Fluorinated Cyanate Ester	583	44	40	V-0
Poly(tetrafluoroethylene) (PTFE)	612	0	95	V-0

[a] Minimum oxygen concentration needed for burning at 20 °C.
[b] V, vertical burning; HB, horizontal burning.

Table 3 Surface re-radiation loss $\dot{q}''_{rr}$ and heat of gasification (L) of polymers

Polymer/sample	$\dot{q}''_{rr}$ (kW/m^2)	L (kJ/g)	
		DSC	ASTM E 2058
	Ordinary polymers		
Filter paper	10		3.6
Corrugated paper	10		2.2
Douglas fir wood	10		1.8
Plywood/fire retarded (FR)	10		1.0
Polypropylene, PP	15	2.0	2.0
Polyethylene, PE, low density	15	1.9	1.8
PE-high density	15	2.2	2.3
Polyoxymethylene, POM	13	2.4	2.4
Polymethylmethacrylate, PMMA	11	1.6	1.6
Nylon 6,6	15		2.4
Polyisoprene	10		2.0
Acrylonitrile-butadiene-styrene, ABS	10		3.2
Styrene-butadiene	10		2.7
Polystyrene, PS foams	10–13		1.3–1.9
PS granular	13	1.8	1.7
Polyurethane, PU, foams-flexible	16–19	1.4	1.2–2.7
PU foams-rigid	14–22		1.2-5.3
Polyisocyanurate, PIU, foams	14–37		1.2–6.4
Polyesters/glass fiber	10–15		1.4–6.4
PE foams	12		1.4–1.7
	High-temperature polymers		
Polycarbonate, PC	11		2.1
Phenolic foam	20		1.6
Phenolic foam/FR	20		3.7
Phenolic/glass fibers	20		7.3
Phenolic-aromatic polyamide	15		7.8
	Halogenated polymers		
PE/25 % chlorine (Cl)	12		2.1
PE/36 % Cl	12		3.0
PE/48 % Cl	10		3.1
Polyvinylchloride, PVC, rigid	15		2.5
PVC plasticized	10		1.7
Ethylene-tetrafluoroethylene, ETFE	27		0.9
Perfluoroethylene-propylene, FEP	38		2.4
Ethylene-tetrafluoroethylene, ETFE	48		0.8–1.8
Perfluoro-alkoxyalkane, PFA	37		1.0

Table 4 Thermophysical and ignition properties of polymers[a]

Polymer	T_v/T_d (°C)	CHF (kW/m^2)	T_{ig} (°C)	ρ (g/cm^3)	c (J/g K)	k (W/m K)	$(k\rho c)^{0.5}$ (kW s$^{1/2}$/m^2 K)	TRP$_{Exp}$	TRP$_{Cal}$ from	
									T_v/T_d[b]	T_{ig}[c]
Ordinary polymers										
PE-1	387	15	443	0.94	2.15	0.30	0.779	454	286	327
PE-2	404			0.95	2.03	0.31	0.773		297	
PE-3	356			0.95	2.12	0.37	0.863		290	
PP-1	308			0.90	2.25	0.19	0.620		179	
PP-2	293			0.90	2.48	0.21	0.685		187	
PP-3	325			1.19	1.90	0.39	0.939		286	
PP-4	298			1.21	1.76	0.39	0.911		253	
PP-5	274			1.11	1.95	0.34	0.858		218	
PP-6	310	15	443	0.93	2.20	0.20	0.640	288	186	271
PP-7	346	15	443	0.90	2.22	0.23	0.678	323	221	287
PP-8	296	10	374	1.06	2.08	0.23	0.712	277	197	252
PP-9	341	15	443	1.04	1.93	0.31	0.789	333	253	334
PE-PP-1	284			0.91	1.98	0.17	0.553		146	
PE-PP-2	305			0.88	2.15	0.21	0.630		180	
PVC-1	257	10	374	1.95	1.14	0.25	0.745	215	177	264
Nylon 66	428			1.50	1.69	0.58	1.213		495	
Nylon 12	418			1.04	1.79	0.18	0.579		230	
PMMA	330	10	374	1.19	2.09	0.27	0.819	274	254	290
POM	252	10	374	1.41	1.92	0.27	0.855	250	198	303
EPDM-1	447			1.15	1.75	0.30	0.777		332	
EPDM-2	572			1.16	1.39	0.36	0.762		421	
EPDM-3	565			1.21	1.48	0.45	0.898		489	
SMC-ester	341	20	497	1.64	1.14	0.37	0.832	483	267	397
TPO-1	334			0.97	1.87	0.33	0.774		243	
						Average	0.778			
						Standard deviation	0.140			
High temperature advanced-engineered polymers										
PSF	537	30	580	1.24	1.30	0.28	0.672	469	347	376
PEEK	606	30	580	1.32	1.80	0.25	0.771	550	452	432
PC-1	397			1.12	1.68	0.18	0.582		219	
PC-2	454	20	497	1.18	1.51	0.27	0.694	357	301	331
PC-3	411			1.19	2.06	0.20	0.700		274	

Table 4 (*Continued*)

Polymer	T_v/T_d (°C)	CHF (kW/m^2)	T_{ig} (°C)	ρ (g/cm^3)	c (J/g K)	k (W/m K)	$(k\rho c)^{0.5}$ (kW s$^{1/2}$/m^2 K)	TRP_{Exp}	TRP_{Cal} from T_v/T_d [b]	TRP_{Cal} from T_{ig} [c]
PC-4	450	20	497	1.20	2.18	0.22	0.759	434	326	362
PC-5	413			1.18	1.10	0.19	0.497		195	
PC-6	546	30	580	1.20	1.20	0.21	0.550	455	289	308
PC-7	546	30	580	1.20	1.20	0.21	0.550	455	289	308
PEI	575	25	540	1.27	1.40	0.22	0.625	435	347	325
						Average	0.640			
						Standard deviation	0.094			
Highly halogenated advanced-engineered polymers										
PTFE	401	50	700	2.18	1.00	0.25	0.738	654	281	502
FEP	401	50	700	2.15	1.20	0.25	0.803	680	306	546
ETFE	337	25	540	1.70	0.90	0.23	0.593	481	188	308
PCTFE	369	30	580	2.11	0.90	0.22	0.646	460	226	362
ECTFE	401	38	613	1.69	1.00	0.15	0.503	450	192	299
PVDF	348	40	643	1.70	1.30	0.13	0.536	506	176	334
CPVC	348	40	643	1.50	0.90	0.22	0.545	435	179	340
						Average	0.624			
						Standard deviation	0.112			
Foams and expanded elastomers										
PEU-1	248			0.11	1.77	0.04	0.088		20	
PEU-2	252			0.09	1.56	0.06	0.092		21	
PEU-3	262			0.02	1.65	0.02	0.026		6	
ABS-PVC-1	240	19	487	0.10	1.35	0.02	0.052	73	11	24
ABS-PVC-2	238			0.07	2.02	0.15	0.146		32	
PS-1	339			0.09	1.70	0.17	0.161		51	
PS-2	432	20	497	0.13	1.62	0.10	0.145	146	60	69
EPDM-1	276			0.44	2.30	0.07	0.266		68	
EPDM-2	334			0.41	1.51	0.21	0.361		113	
TPO-2	284			0.97	1.87	0.13	0.486		128	
TPO-3	396			0.93	1.96	0.20	0.604		227	

PVC-2	269	10	374	1.20	1.37	0.14	0.480	263	119	162
PVC-glass	261			1.00	1.05	0.23	0.491		118	
PVC-elastomers	267			1.60	1.24	0.10	0.445		110	
				Fabrics						
PET-1	394	10	374	0.66	1.32	0.09	0.280	174	105	99
PET-2	305			0.69	1.56	0.04	0.207		59	
Nylon 6	380	20	497	0.12	2.19	0.24	0.251	154	90	120

[a] Critical heat flux for piloted ignition (CHF), piloted ignition temperature (T_{ig}), thermal response parameter (TRP $= \sqrt{\frac{\pi}{4}(k\rho c)}(T_{ig} - T_{\infty})$);
[b] T_v, T_d values were used in the calculation of TRP values.
[c] T_{ig} values from CHF were used in the calculation of TRP values.

Table 5 Ignition-time measured in the ASTM E 1354 cone calorimeter and thermal response parameter values derived from the data

Polymers	Ignition time (seconds) Incident radiant heat flux, $\dot{q}_e''$ (kW/m^2)								TRP_{Exp}
	20	25	30	40	50	70	75	100	(kW s$^{1/2}$/m^2)
High density polyethylene, PE-HD	403		171	91	58				364
Polyethylene, PE	403			159		47			526
Polypropylene, PP	120		63	35	27				291
Polypropylene, PP	218			86		41			556
PP/glass fibers (1082)		168			47		23	13	377
Polystyrene, PS	417			97		50			556
PS foam	NI[b]		73	28	18				168
PS-fire retarded (FR)	244			90		51			667
PS foam-FR	NI		77	40	24				221
Nylon	1923			65		31			333
Nylon 6	700		193	115	74				379
Nylon/glass fibers (1077)		193			53		21	13	359
Polyoxymethylene, POM	259			74		24			357
Polymethylmethacrylate, PMMA	176			36		11			222
Polybutyleneterephthalate, PBT	609			113		59			588
Polyethyleneterephthalate, PET	718			116		42			435
Acrylonitrile-butadiene-styrene, ABS	299		130	68	43				317
ABS-FR	212			66		39			556
ABS-PVC	5198			61		39			357
Vinyl thermoplastic elastomer	NI			1271		60			294
Polyurethane, PU foam	12			1		1			76
Thermoplastic PU-FR	302			60		38			500
EPDM/Styrene acrylonitrile,SAN	486			68		36			417
Polyester/glass fibers (30 %)	NI			309	109				256
Isophthalic polyester	256		115	59	38				296
Isophthalic polyester/glass fiber (77 %)	480		172	91	77				426
Polyvinyl ester	332		120	78	38				263
Polyvinyl ester/glass fiber (69 %)	646		235	104	78				444
Polyvinyl ester/glass fiber (1031)		278			74		34	18	429

Polyvinyl ester/glass fiber (1087)		281					22	11	312
Epoxy	337		172	100	62				457
Epoxy/glass fiber (69 %)	320		120	75	57				388
Epoxy/glass fiber (1003)		198			50		73	19	555
Epoxy/glass fiber (1006)		159			49		23	14	397
Epoxy/glass fiber (1040)					18		13	9	512
Epoxy/glass fiber (1066)		140			48		14	9	288
Epoxy/glass fiber (1067)		209			63		24	18	433
Epoxy/glass fiber (1070)		229			63		30	23	517
Epoxy/glass fiber (1071)		128			34		18	10	334
Epoxy/glass fiber (1089)		535			105		60	40	665
Epoxy/glass fiber (1090)		479			120		54	34	592
Epoxy/graphite fiber (1091)		NI					53	28	484
Epoxy/graphite fiber (1092)		275			76		32	23	493
Epoxy/graphite fiber (1093)		338			94		44	28	554
Cyanate ester/glass fiber (1046)		199			58		20	10	302
Acrylic/glass fiber	553		252	148	101				180
Kydex acrylic paneling, FR	200			38		12			233
Polycarbonate, PC-1	NI			182		75			455
PC-2	6400			144		45			370
Cross linked polyethylene (XLPE)	750			105		35			385
Polyphenylene oxide, PPO-polystyrene (PS)	479			87		39			455
PPO/glass fibers	465			45		35			435
Polyphenylenesulfide, PPS/glass fibers (1069)		NI			105		57	30	588
PPS/graphite fibers (1083)		NI					69	26	330
PPS/glass fibers (1084)		NI			244		70	48	623
PPS/graphite fibers (1085)		NI			173		59	33	510
Polyarylsulfone/graphite fibers (1081)		NI			122		40	19	360
Polyethersulfone/graphite fibers (1078)		NI			172		47	21	352
Polyetheretherketone, PEEK/glass fibers (30 %)	NI			390	142				301
PEEK/graphite fibers (1086)					307		80	42	514
Polyetherketoneketone, PEKK/glass fibers (1079)		NI			223		92	53	710
Bismaleimide, BMI/graphite fibers (1095)		237					42	22	513
Bismaleimide, BMI/graphite fibers (1096)		503			141		60	36	608
Bismaleimide, BMI/graphite fibers (1097)		NI					66	37	605

Table 5 (*Continued*)

Polymers	Ignition time (seconds) Incident radiant heat flux, $\dot{q}''_e$ (kW/m^2)								TRP_{Exp}
	20	25	30	40	50	70	75	100	($kW s^{1/2}/m^2$)
Bismaleimide, BMI/graphite fibers (1098)		NI			110		32	27	515
Phenolic/glass fibers (45 %)	423			214	165				683
Phenolic/glass fibers (1099)		NI			121		33	22	409
Phenolic/glass fibers (1100)		NI			125			40	728
Phenolic/glass fibers (1101)		NI			210		55	25	382
Phenolic/glass fibers (1014)		NI			214		73	54	738
Phenolic/glass fibers (1015)		NI			238		113	59	765
Phenolic/glass fibers (1017)		NI			180		83	43	641
Phenolic/glass fibers (1018)		NI			313		140	88	998
Phenolic/graphite fibers (1102)		NI					79	45	684
Phenolic/graphite fibers (1103)		NI			104		34	20	398
Phenolic/graphite fibers (1104)		NI			187		88	65	982
Phenolic/PE fibers (1073)		714			129		28	10	267
Phenolic/aramid fibers (1074)		1110			163		33	15	278
Polyimide/glass fibers (1105)		NI			175		75	55	844
Douglas fir	254			34		12			222
Hemlock	307		73	35	19				175
Wool	24		15	11	9				232
Acrylic fiber	52		28	16	19				175
Polyvinylchloride, PVC flexible-1	117			27		11			244
PVC flexible-2	102			21		15			333
PVC flexible-3 (LOI 25 %)	119		61	41	25				285
PVC-FR flexible-1	236			47		12			222
PVC-FR flexible-2	176			36		14			263
PVC-FR (Sb_2O3) flexible-4 (LOI 30 %)	136		84	64	37				397
PVC-FR (triaryl phosphate) flexible-5 (LOI 34 %)	278		103	69	45				345
PVC-FR (alkyl aryl phosphate) flexible-6 (LOI 28 %)	114		72	49	35				401
PVC rigid-1	5159			73		45			385

PVC rigid-2	3591		85		48	417
PVC rigid-3	5171		187		43	357
PVC rigid-1 (LOI 50 %)	NI	487	276	82		388
PVC rigid-2	NI	320	153	87		390
Chlorinated PVC (CPVC)	NI		621		372	1111

NI: no ignition

Table 6 Peak heat release rate measured in the ASTM E 1354 cone calorimeter

Polymers	Ref.	$\dot{Q}''_{ch}$ (peak) (kW/m^2) Incident radiant heat flux, $\dot{q}''_e$ (kW/m^2) 20	25	30	40	50	70	75	100	$\Delta h_c/L$ (–)
		Ordinary polymers								
High-density polyethylene, HDPE	44	453		866	944	1133				21
Polyethylene, PE	42	913			1408		2735			37
Polypropylene, PP	44	377		693	1095	1304				32
Polypropylene, PP	42	1170			1509		2421			25
PP/glass fibers (1082)	43		187			361		484	432	6
Polystyrene, PS	42	723			1101		1555			17
Nylon	42	517			1313		2019			30
Nylon 6	44	593		802	863	1272				21
Nylon/glass fibers (1077)	43		67			96		116	135	1
Polyoxymethylene, POM	42	290			360		566			6
Polymethylmethacrylate, PMMA	42	409			665		988			12
Polybutyleneterephthalate, PBT	42	850			1313		1984			23
Acrylonitrile-butadiene-styrene, ABS	44	683		947	994	1147				14
ABS	42	614			944		1311			12
ABS-FR	42	224			402		419			4
ABS-PVC	42	224			291		409			4
Vinyl thermoplastic elastomer	42	19			77		120			2
Polyurethane, PU foam	42	290			710		1221			19
EPDM/Styrene acrylonitrile,SAN	42	737			956		1215			10
Polyester/glass fibers (30 %)	44	NI			167	231				6
Isophthalic polyester	44	582		861	985	985				20
Isophthalic polyester/glass fibers (77 %)	44	173		170	205	198				2
Polyvinyl ester	44	341		471	534	755				13
Polyvinyl ester/glass fibers (69 %)	44	251		230	253	222				2
Polyvinyl ester/glass fibers (1031)	43		75			119		139	166	1
Polyvinyl ester/glass fibers (1087)	43		377					499	557	2
Epoxy	44	392		453	560	706				11
Epoxy/glass fibers (69 %)	44	164		161	172	202				2

Epoxy/glass fibers(1003)	43		159			294		191	335	2
Epoxy/glass fibers(1006)	43		81			181		182	229	2
Epoxy/glass fibers (1040)	43					40		246	232	2
Epoxy/glass fibers (1066)	43		231			266		271	489	3
Epoxy/glass fibers (1067)	43		230			213		300	279	1
Epoxy/glass fibers (1070)	43		175			196		262	284	2
Epoxy/glass fibers (1071)	43		20			93		141	202	2
Epoxy/glass fibers (1089)	43		39			178		217	232	2
Epoxy/glass fibers (1090)	43		118			114		144	173	1
Epoxy/graphite fibers (1091)	43		NI					197	241	2
Epoxy/graphite fibers (1092)	43		164			189		242	242	2
Epoxy/graphite fibers (1093)	43		105			171		244	202	3
Cyanate ester/glass fibers (1046)	43		121			130		196	226	2
Kydex acrylic paneling, FR	42	117			176		242			3
		High-temperature polymers								
Polycarbonate, PC-1	42	16			429		342			21
PC-2	42	144			420		535			14
Cross linked polyethylene (XLPE)	42	88			192		268			5
Polyphenylene oxide, PPO-polystyrene (PS)	42	219			265		301			2
PPO/glass fibers	42	154			276		386			6
Polyphenylenesulfide, PPS/glass fibers (1069)	43		NI			52		71	183	3
PPS/graphite fibers (1083)	43		NI					60	80	2
PPS/glass fibers (1084)	43		NI			48		88	150	2
PPS/graphite fibers (1085)	43		NI			94		66	126	1
Polyarylsulfone/graphite fibers (1081)	43		NI			24		47	60	1
Polyethersulfone/graphite fibers (1078)	43		NI			11		41	65	0.3
Polyetheretherketone, PEEK/glass fibers (30 %)	44	NI			35	109				7
PEEK/graphite fibers (1086)	43					14		54	85	1
Polyetherketoneketone, PEKK/glass fibers (1079)	43		NI			21		45	74	1
Bismaleimide, BMI/graphite fibers (1095)	43		160					213	270	1
Bismaleimide, BMI/graphite fibers (1096)	43		128			176		245	285	2
Bismaleimide, BMI/graphite fibers (1097)	43		NI					172	168	(1)
Bismaleimide, BMI/graphite fibers (1098)	43		NI			74		91	146	1
Phenolic/glass fibers (45 %)	44			423	214	165				1
Phenolic/glass fibers (1099)	43		NI			66		102	122	1
Phenolic/glass fibers (1100)	43		NI			66		120	163	2
Phenolic/glass fibers (1101)	43		NI			47		57	96	1

Table 6 *(Continued)*

Polymers	Ref.	$\dot{Q}''_{ch}$ (peak) (kW/m^2) Incident radiant heat flux, $\dot{q}''_e$ (kW/m^2) 20	25	30	40	50	70	75	100	$\Delta h_c/L$ (−)
Phenolic/glass fibers (1014)	43		NI			81		97	133	1
Phenolic/glass fibers (1015)	43		NI			82		76	80	(1)
Phenolic/glass fibers (1017)	43		NI			190		115	141	1
Phenolic/glass fibers (1018)	43		NI			132		56	68	1
Phenolic/graphite fibers (1102)	43		NI					159	196	2
Phenolic/graphite fibers (1103)	43		NI			177		183	189	0.2
Phenolic/graphite fibers (1104)	43		NI			71		87	101	1
Phenolic/PE fibers (1073)	43		NI			98		141	234	3
Phenolic/aramid fibers (1074)	43		NI			51		93	104	1
Phenolic insulating foam	44				17	19		29		1
Polyimide/glass fibers (1105)	43		NI			40		78	85	1
Wood										
Douglas fir	42	237			221		196			(-)
Hemlock	44	233		218	236	243				(-)
Textiles										
Wool	44	212		261	307	286				5
Acrylic fibers	44	300		358	346	343				6
Halogenated polymers										
PVC flexible-3 (LOI 25 %)	44	126		148	240	250				5
PVC-FR (Sb_2O3) flexible-4 (LOI 30 %)	44	89		137	189	185				5
PVC-FR (triaryl phosphate) flexible-5 (LOI 34 %)	44	96		150	185	176				5
PVC rigid-1	42	40			175		191			3
PVC rigid-2	42	75			111		126			2
PVC rigid-3	42	102			183		190			2
PVC rigid-1 (LOI 50 %)	44	NI		90	107	155				3
PVC rigid-2	44	NI		101	137	157				3
Chlorinated PVC (CPVC)	42	25			84		93			1

Table 7 Average yields of products and heat of combustion for polymers from parts of a minivan from the data measured in the ASTM E 2058 fire propagation apparatus[a]

		y_j (g/g)				Δh_c
Part description	Polymers	CO	CO_2	HC[b]	Smoke	(kJ/g)
	Head liner					
Fabric, exposed	Nylon 6	0.086	2.09	0.001	0.045	28.8
	Instrument panel					
Cover	PVC	0.057	1.72	0.005	0.109	24.4
Shelf, main panel	PC	0.051	1.86	0.002	0.105	20.2
	Resonator					
Structure	PP	0.041	2.46	0.002	0.072	34.6
Intake tube	EPDM	0.045	2.51	0.001	0.100	33.8
	Kick panel insulation					
Backing	PVC	0.061	1.26	0.006	0.070	17.4
	Air ducts					
Large ducts	PP	0.056	2.52	0.004	0.080	35.5
	Steering column boot					
Inner boot	NR	0.061	1.87	0.003	0.130	25.6
Cotton shoddy	Cott/polyester[c]	0.039	2.17	0.002	0.087	29.4
	Brake fluid reservoir					
Reservoir	PP	0.058	2.41	0.011	0.072	33.9
	Windshield wiper tray					
Structure	SMC	0.061	1.86	0.003	0.100	25.5
	Sound reduction fender insulation					
High-density foam	PS	0.064	1.8	0.002	0.098	24.6
	Hood liner insulation					
Face	PET	0.041	1.47	0.003	0.022	20
	Wheel well cover					
Fuel tank shield	PP	0.054	2.45	0.002	0.065	34.5
	HVAC unit					
Cover	PP	0.057	2.49	0.002	0.060	35
Seal, foam	ABS-PVC	0.089	1.62	0.001	0.060	22.6
	Fuel tank					
Tank	PE	0.032	2.33	0.005	0.042	32.7
	Headlight					
Lens	PC	0.049	1.67	0.004	0.113	18.2
	Battery casing					
Cover	PP	0.045	2.59	0.004	0.071	36.2
	Bulkhead insulation engine side					
Grommet, wire harness cap	HDPF	0.064	2.67	0.012	0.058	38.2

[a] Combustion in normal air at 50 kW/m^2 of external heat flux in the ASTM E 2058 apparatus.
[b] HC-total hydrocarbons.
[c] Cotton/polyester: mixture of cotton, polyester, and other fibers.

Table 8 Average heat of combustion and yields of products for polymers from the data measured in the ASTM E 2058 fire propagation apparatus

Polymer	Composition	y_j (g/g)				Δh_c (kJ/g)
		CO	CO_2	HC[a]	Smoke	
Ordinary-polymers						
Polyethylene, PE	CH_2	0.024	2.76	0.007	0.060	38.4
Polypropylene, PP	CH_2	0.024	2.79	0.006	0.059	38.6
Polystyrene, PS	CH	0.060	2.33	0.014	0.164	27.0
Polystyrene foam	$CH_{1.1}$	0.061	2.32	0.015	0.194	25.5
Wood	$CH_{1.7}O_{0.73}$	0.004	1.30	0.001	0.015	12.6
Polyoxymethylene, POM	$CH_{2.0}O$	0.001	1.40	0.001	0.001	14.4
Polymethylmethacrylate, PMMA	$CH_{1.6}O_{0.40}$	0.010	2.12	0.001	0.022	24.2
Polyester	$CH_{1.4}O_{0.22}$	0.075	1.61	0.025	0.188	20.1
Nylon	$CH_{1.8}O_{0.17}N_{0.17}$	0.038	2.06	0.016	0.075	27.1
Flexible polyurethane foams	$CH_{1.8}O_{0.32}N_{0.06}$	0.028	1.53	0.004	0.070	17.6
Rigid polyurethane foams	$CH_{1.1}O_{0.21}N_{0.10}$	0.036	1.43	0.003	0.118	16.4
High temperature polymers						
Polyetheretherketone, PEEK	$CH_{0.63}O_{0.16}$	0.029	1.60	0.001	0.008	17.0
Polysulfone, PSO	$CH_{0.81}O_{0.15}S_{0.04}$	0.034	1.80	0.001	0.020	20.0
Polyethersulfone, PES		0.040	1.50	0.001	0.021	16.7
Polyetherimide, PEI	$CH_{0.65}O_{0.16}N_{0.05}$	0.026	2.00	0.001	0.014	20.7
Polycarbonate, PC	$CH_{0.88}O_{0.19}$	0.054	1.50	0.001	0.112	16.7
Halogenated polymers						
PE + 25 % Cl	$CH_{1.9}Cl_{0.13}$	0.042	1.71	0.016	0.115	22.6
PE + 36 % Cl	$CH_{1.8}Cl_{0.22}$	0.051	0.83	0.017	0.139	10.6
PE + 48 % Cl	$CH_{1.7}Cl_{0.36}$	0.049	0.59	0.015	0.134	5.7
Polyvinylchloride, PVC	$CH_{1.5}Cl_{0.50}$	0.063	0.46	0.023	0.172	7.7
Chlorinated PVC	$CH_{1.3}Cl_{0.70}$	0.052	0.48	0.001	0.043	6.0
Polyvinylidenefluoride, PVDF	CHF	0.055	0.53	0.001	0.037	5.4
Polyethylenetetrifluoroethylene, ETFE	CHF	0.035	0.78	0.001	0.028	7.3
Polyethylenechlorotrifluoroethylene, ECTFE	$CHCl_{0.25}F_{0.75}$	0.095	0.41	0.001	0.038	4.5
Polytetrafluoroethylene, TFE	CF_2	0.092	0.38	0.001	0.003	2.8
Polyfluoroalkoxy, PFA	$CF_{1.6}$	0.099	0.42	0.001	0.002	1.8
Polyfluorinated ethylene propylene, FEP	$CF_{1.8}$	0.116	0.25	0.001	0.003	1.0

[a] HC-total gaseous hydrocarbon.

Table 9 Average effective (chemical) heat of combustion and smoke yield calculated from the data measured in the ASTM E 1354 cone calorimeter[a]

Polymers	Δh_c (MJ/kg)	y_{smoke} (g/g)
Ordinary polymers		
High-density polyethylene, HDPE	40.0	0.035
Polyethylene, PE	43.4	0.027
Polypropylene, PP	44.0	0.046
Polypropylene, PP	42.6	0.043
PP/glass fibers (1082)	NR	0.105
Polystyrene, PS	35.8	0.085
PS-FR	13.8	0.144
PS foam	27.7	0.128
PS foam-FR	26.7	0.136
Nylon	27.9	0.025
Nylon 6	28.8	0.011
Nylon/glass fibers (1077)	NR	0.089
Polyoxymethylene, POM	13.4	0.002
Polymethylmethacrylate, PMMA	24.2	0.010
Polybutyleneterephthalate, PBT	20.9	0.066
Polyethyleneterephthalate, PET	14.3	0.050
Acrylonitrile-butadiene-styrene, ABS	30.0	0.105
ABS	29.4	0.066
ABS-FR	11.7	0.132
ABS-PVC	17.6	0.124
Vinyl thermoplastic elastomer	6.4	0.056
Polyurethane, PU foam	18.4	0.054
Thermoplastic PU-FR	19.6	0.068
EPDM/styrene acrylonitrile,SAN	29.0	0.116
Polyester/glass fibers (30 %)	16.0	0.049
Isophthalic polyester	23.3	0.080
Isophthalic polyester/glass fibers (77 %)	27.0	0.032
Polyvinyl ester	22.0	0.076
Polyvinyl ester/glass fibers (69 %)	26.0	0.079
Polyvinyl ester/glass fibers (1031)	NR	0.164
Polyvinyl ester/glass fibers (1087)	NR	0.128
Epoxy	25.0	0.106
Epoxy/glass fibers (69 %)	27.5	0.056
Epoxy/glass fibers (1003)	NR	0.142
Epoxy/glass fibers (1006)	NR	0.207
Epoxy/glass fibers (1040)	NR	0.058
Epoxy/glass fibers (1066)	NR	0.113
Epoxy/glass fibers (1067)	NR	0.115
Epoxy/glass fibers (1070)	NR	0.143
Epoxy/glass fibers (1071)	NR	0.149
Epoxy/glass fibers (1089)	NR	0.058
Epoxy/glass fibers (1090)	NR	0.086
Epoxy/graphite fibers (1091)	NR	0.082
Epoxy/graphite fibers (1092)	NR	0.049
Cyanate ester/glass fibers (1046)	NR	0.103
Acrylic/glass fibers	17.5	0.016
Kydex acrylic paneling, FR	10.2	0.095

Table 9 (*Continued*)

Polymers	Δh_c (MJ/kg)	y_{smoke} (g/g)
High-temperature polymers and composites		
Polycarbonate, PC-1	21.9	0.098
PC-2	22.6	0.087
Cross linked polyethylene (XLPE)	23.8	0.026
Polyphenylene oxide, PPO-polystyrene (PS)	23.1	0.162
PPO/glass fibers	25.4	0.133
Polyphenylenesulfide, PPS/glass fibers (1069)	NR	0.063
PPS/graphite fibers (1083)	NR	0.075
PPS/glass fibers (1084)	NR	0.075
PPS/graphite fibers (1085)	NR	0.058
Polyarylsulfone/graphite fibers (1081)	NR	0.019
Polyethersulfone/graphite fibers (1078)	NR	0.014
Polyetheretherketone, PEEK/glass fibers (30 %)	20.5	0.042
PEEK/graphite fibers (1086)	NR	0.025
Polyetherketoneketone, PEKK/glass fibers (1079)	NR	0.058
Bismaleimide, BMI/graphite fibers (1095)	NR	0.077
Bismaleimide, BMI/graphite fibers (1096)	NR	0.096
Bismaleimide, BMI/graphite fibers (1097)	NR	0.095
Bismaleimide, BMI/graphite fibers (1098)	NR	0.033
Phenolic/glass fibers (45 %)	22.0	0.026
Phenolic/glass fibers (1099)	NR	0.008
Phenolic/glass fibers (1100)	NR	0.037
Phenolic/glass fibers (1101)	NR	0.032
Phenolic/glass fibers (1014)	NR	0.031
Phenolic/glass fibers (1015)	NR	0.031
Phenolic/glass fibers (1017)	NR	0.015
Phenolic/glass fibers (1018)	NR	0.009
Phenolic/graphite fibers (1102)	NR	0.039
Phenolic/graphite fibers (1103)	NR	0.041
Phenolic/graphite fibers (1104)	NR	0.021
Phenolic/PE fibers (1073)	NR	0.054
Phenolic/aramid fibers (1074)	NR	0.024
Phenolic insulating foam	10.0	0.026
Polyimide/glass fibers (1105)	NR	0.014
Wood		
Douglas fir	14.7	0.010
Hemlock	13.3	0.015
Textiles		
Wool	19.5	0.017
Acrylic fiber	27.5	0.038
Halogenated polymers		
PVC flexible-3 (LOI 25 %)	11.3	0.099
PVC-FR (Sb_2O3) flexible-4 (LOI 30 %)	10.3	0.078
PVC-FR (triaryl phosphate) flexible-5 (LOI 34 %)	10.8	0.098
PVC rigid-1	8.9	0.103
PVC rigid-2	10.8	0.112
PVC rigid-3	12.7	0.103
PVC rigid-1 (LOI 50 %)	7.7	0.098
PVC rigid-2	8.3	0.076
Chlorinated PVC (CPVC)	5.8	0.003

Table 10 Composition, molecular weight, stoichiometric air-to-fuel ratio (s), net heat of complete combustion ($\Delta h_{c,T}$), and maximum possible stoichiometric yields of major products ($y_{i,max}$) for ordinary polymers

Polymers	Composition	M (g/mole)	s (g/g)	$\Delta h_{c,T}$ (kJ/g)	$y_{i,max}$			
					CO_2	CO	HC	Smoke
Polyethylene (PE)	CH_2	14.0	14.7	43.6	3.14	2.00	1.14	0.86
Polypropylene (PP)	CH_2	14.0	14.7	43.4	3.14	2.00	1.14	0.86
Polystyrene (PS)	CH	13.0	13.2	39.2	3.38	2.15	1.23	0.92
Polystyrene foam	$CH_{1.1}$	13.1	13.4	39.2	3.36	2.14	1.22	0.92
Wood	$CH_{1.7}O_{0.73}$	25.4	5.8	16.9	1.73	1.11	0.63	0.48
Polyoxymethylene (POM)	$CH_{2.0}O$	30.0	4.6	15.4	1.47	0.93	0.53	0.40
Polymethylmethacrylate (PMMA)	$CH_{1.6}O_{0.40}$	20.0	8.2	25.2	2.20	1.40	0.80	0.60
Polyester	$CH_{1.4}O_{0.22}$	16.9	10.1	32.5	2.60	1.65	0.95	0.71
Polyvinylalcohol	$CH_{2.0}O_{0.50}$	22.0	7.8	21.3	2.00	1.27	0.73	0.55
Polyethyleneterephthalate (PET)	$CH_{0.80}O_{0.40}$	19.2	7.2	23.2	2.29	1.46	0.83	0.63
Polyacrylonitrile-butadiene-styrene (ABS)	$CH_{1.1}N_{0.07}$	14.1	13.2	38.1	3.13	1.99	1.14	0.85
Nylon	$CH_{1.8}O_{0.17}N_{0.17}$	18.9	11.2	30.8	2.33	1.48	0.85	0.63
Flexible polyurethane foam	$CH_{1.8}O_{0.32}N_{0.06}$	19.7	9.4	25.3	2.24	1.42	0.81	0.61
Rigid polyurethane foam	$CH_{1.1}O_{0.21}N_{0.10}$	17.8	9.8	25.9	2.47	1.57	0.90	0.67

HC: hydrocarbons.

Table 11 Composition, molecular weight, stoichiometric air-to-fuel ratio (s), net heat of complete combustion ($\Delta h_{c,T}$), and maximum possible stoichiometric yields of major products ($y_{i,max}$) for high temperature polymers

Polymers	Composition	M (g/mole)	s (g/g)	$\Delta h_{c,T}$ (kJ/g)	$y_{i,\max}$(g/g) CO_2	CO	HC	Smoke	HCN	NO_2	SO_2
Polyetherketoneketone (PEKK)	$CH_{0.60}O_{0.15}$	15.0	9.8	30.3	2.93	1.87	1.07	0.80	0.00	0.00	0.00
Polyetherketone (PEK)	$CH_{0.62}O_{0.15}$	15.0	9.9	30.2	2.93	1.86	1.07	0.80	0.00	0.00	0.00
Polyetheretherketone (PEEK)	$CH_{0.63}O_{0.16}$	15.2	9.7	30.4	2.90	1.84	1.05	0.79	0.00	0.00	0.00
Polyamideimide (PAI)	$CH_{0.53}O_{0.20}$	15.7	9.0	24.2	2.80	1.78	1.02	0.76	0.00	0.00	0.00
Polycarbonate (PC)	$CH_{0.88}O_{0.19}$	15.9	9.7	30.1	2.76	1.76	1.01	0.75	0.00	0.00	0.00
Polyethylenenaphthalate (PEN)	$CH_{0.71}O_{0.29}$	17.4	8.2		2.54	1.61	0.92	0.69	0.00	0.00	0.00
Polyphenyleneoxide (PPO)	$CHO_{0.13}$	15.0	10.9		2.93	1.87	1.07	0.80	0.00	0.00	0.00
Polybutanedi olterephthalate (PBT)	$CHO_{0.33}$	18.3	8.2		2.41	1.53	0.88	0.66	0.00	0.00	0.00
Polybenzoylphenylene	$CH_{0.62}O_{0.08}$	13.9	11.1		3.18	2.02	1.16	0.87	0.00	0.00	0.00
Phenol-formaldehyzde	$CHO_{0.14}$	15.2	10.6		2.89	1.84	1.05	0.79	0.00	0.00	0.00
Polybenzimidazole (PBI)	$CH_{0.60}N_{0.20}$	15.4	12.0	30.8	2.86	1.82	1.04	0.78	0.35	0.60	0.00
Polyphenylenesulfide (PPS)	$CH_{0.67}S_{0.17}$	18.0	10.2	28.2	2.44	1.55	0.89	0.67	0.00	0.00	0.59
Polyphenylenebenzobisoxazole (PBO)	$CH_{0.43}O_{0.14}N_{0.14}$	16.6	9.7		2.65	1.68	0.96	0.72	0.23	0.39	0.00
Polyimide (PI)	$CH_{0.45}O_{0.23}N_{0.09}$	17.4	8.6	25.5	2.53	1.61	0.92	0.69	0.14	0.24	0.00
Polyetherimide (PEI)	$CH_{0.65}O_{0.16}N_{0.05}$	16.0	9.8	28.4	2.76	1.75	1.00	0.75	0.09	0.16	0.00
Polyaramide	$CH_{0.71}O_{0.14}N_{0.14}$	16.9	10.1		2.60	1.66	0.95	0.71	0.22	0.38	0.00
Polyphenylenebenzobisoxazole	$CH_{0.71}O_{0.14}N_{0.14}$	16.9	10.1		2.60	1.66	0.95	0.71	0.22	0.38	0.00
Aramid-arylester copolymer	$CH_{0.71}O_{0.14}N_{0.14}$	16.9	10.1	24.4	2.60	1.66	0.95	0.71	0.22	0.38	0.00
Polysulfone (PSF)	$CH_{0.81}O_{0.15}S_{0.04}$	16.4	9.8	29.4	2.68	1.71	0.98	0.73	0.00	0.00	0.14
Polyphenyleneethersulfone PES	$CH_{0.67}O_{0.25}S_{0.08}$	19.3	8.0	24.7	2.28	1.45	0.83	0.62	0.00	0.00	0.27

Table 12 Composition, molecular weight, stoichiometric air-to-fuel ratio (s), net heat of complete combustion ($\Delta h_{c,T}$), and maximum possible stoichiometric yields of major products ($y_{i,max}$) for halogenated polymers

				$\Delta h_{c,T}$	$y_{i,max}$ (g/g)					
Polymers	Composition	M (g/mole)	s (g/g)	(kJ/g)	CO_2	CO	HC	Smoke	HCl	HF
	Carbon-hydrogen-chlorine atoms									
PE + 25 % Cl	$CH_{1.9}Cl_{0.13}$	18.5	10.7	31.6	2.38	1.52	0.87	0.65	0.25	0.00
PE + 36 % Cl	$CH_{1.8}Cl_{0.22}$	21.5	8.9	26.3	2.05	1.30	0.74	0.56	0.37	0.00
Polychloropropene	$CH_{1.3}Cl_{0.30}$	23.8	7.2	25.3	1.85	1.18	0.56	0.50	0.45	0.00
PE + 48 % Cl	$CH_{1.7}Cl_{0.36}$	26.3	7.0	20.6	1.67	1.06	0.61	0.46	0.49	0.00
Polyvinylchloride, PVC	$CH_{1.5}Cl_{0.50}$	31.0	5.5	16.4	1.42	0.90	0.52	0.39	0.58	0.00
Chlorinated PVC	$CH_{1.3}Cl_{0.70}$	37.8	4.2	13.3	1.16	0.74	0.42	0.32	0.67	0.00
Polyvinylidenechloride, $PVCl_2$	CHCl	48.0	2.9	9.6	0.92	0.58	0.33	0.25	0.75	0.00
	Carbon-hydrogen-fluorine atoms									
Polyvinylfluoride	$CH_{1.5}F_{0.50}$	23.0	7.5	13.5	1.91	1.22	0.70	0.52	0.00	0.43
Polyvinylidenefluoride (PVDF)	CHF	32.0	4.3	13.3	1.38	0.88	0.50	0.38	0.00	0.63
Polyethylenetrifluoroethylene, ETFE	CHF	32.0	4.3	12.6	1.38	0.88	0.50	0.38	0.00	0.63
	Carbon-fluorine atoms									
Polytetrafluoroethylene, TFE	CF_2	50.0	2.7	6.2	0.88	0.56	0.32	0.24	0.00	0.00
Polyperfluoroalkoxy, PFA	$CF_{1.6}$	42.6	3.2	5.0	1.03	0.66	0.38	0.28	0.00	0.00
Polyfluorinatedethylenepropylene,FEP	$CF_{1.8}$	46.2	3.0	4.8	0.95	0.61	0.35	0.26	0.00	0.00
	Carbon-hydrogen-chlorine-fluorine atoms									
Polyethylenechlorotrifluoroethylene, ECTFE	$CHCl_{0.25}F_{0.75}$	36.0	3.8	12.0	1.22	0.78	0.44	0.33	0.25	0.42
Polychlorotrifluoroethylene, CTFE	$CCl_{0.50}F_{1.5}$	58.0	2.4	6.5	0.76	0.48	0.28	0.21	0	0

Table 13 Composition, molecular weight and combustion properties of hydrocarbons and alcohols

Fluid	Composition	M (g/mole)	$T_{ig, auto}$ (°C)	T_{boil} (°C)	Δh_c (kJ/g)	L (kJ/g)	$\Delta h_c/L$ (kJ/kJ)	y_{co} (g/g)	y_{sm} (g/g)
Gasoline	na[b]	na	371	33	41.0	0.482	85	0.010	0.038
Hexane	C_6H_{14}	86	225	69	41.5	0.500	83	0.009	0.035
Heptane	C_7H_{16}	100	204	98	41.2	0.549	75	0.010	0.037
Octane	C_8H_{18}	114	206	125	41.0	0.603	68	0.010	0.038
Nonane	C_9H_{20}	128	205	151	40.8	0.638	64	0.011	0.039
Decane	$C_{10}H_{22}$	142	201	174	40.7	0.690	59	0.011	0.040
Undecane	$C_{11}H_{24}$	156	na	196	40.5	0.736	55	0.011	0.040
Dodecane	$C_{12}H_{26}$	170	203	216	40.4	0.777	52	0.012	0.041
Tridecane	$C_{13}H_{28}$	184	na		40.3	0.806	50	0.012	0.041
Kerosine	$C_{14}H_{30}$	198	260	232	40.3	0.857	47	0.012	0.042
Hexadecane	$C_{16}H_{34}$	226	202	287	40.1	0.911	44	0.012	0.042
Mineral oil	na	466	na	360	na		72	na	na
Motor oil	na	na	na	na	29.3	0.473	62	na	na
Corn Oil	na	na	393	na	22.3	0.413	54	na	na
Benzene	C_6H_6	78	498	80	27.6	0.368	75	0.067	0.181
Toluene	C_7H_8	92	480	110	27.7	0.338	82	0.066	0.178
Xylene	C_8H_{10}	106	528	139	27.8	0.415	67	0.065	0.177
Methanol	CH_4O	32	385	64	19.1	1.005	19	0.001	0.001
Ethanol	C_2H_6O	46	363	78	25.6	0.776	33	0.001	0.008
Propanol	C_3H_8O	60	432	97	29.0	0.630	46	0.003	0.015
Butanol	$C_4H_{10}O$	74	343	117	31.2	0.538	58	0.004	0.019

na: not available.

Table 14 Asymptotic flame heat flux values for the combustion of liquids and polymers

Fuels		$\dot{q}''_f$ (kW/m^2)	
Physical state	Name	$O_2 > 30\,\%$ (small scale)	Normal air (large scale)
	Aliphatic carbon-hydrogen atom containing fuels		
Liquids	Heptane	32	41
	Hexane		40
	Octane		38
	Dodecane		28
	Kerosine		29
	Gasoline		35
	JP-4		34
	JP-5		38
	Transformer fluids	23–25	22–25
Polymers	Polyethylene, PE	61	
	Polypropylene, PP	67	
	Aromatic carbon-hydrogen atom containing fuels		
Liquids	Benzene		44
	Toluene		34
	Xylene		37
Polymers	Polystyrene, PS	75	71
	Aliphatic carbon-hydrogen -oxygen atom containing fuels		
Liquids	Methanol	22	27
	Ethanol		30
	Acetone		24
Polymers	Polyoxymethylene, POM	50	
	Polymethylmethacrylate, PMMA	57	60
	Aliphatic carbon-hydrogen -nitrogen atom containing fuels		
Liquids	Adiponitrile		34
	Acetonitrile		35
	Aliphatic carbon-hydrogen-oxygen-nitrogen atom containing fuels		
Liquid	Toluene diisocyanate		28
Polymers	Polyurethane foams (flexible)	64–76	
	Polyurethane foams (rigid)	49–53	
	Aliphatic carbon-hydrogen-halogen atom containing fuels		
Polymers	Polyvinylchloride, PVC	50	
	Ethylenetetrafluoroethylene, ETFE	50	
	Perfluoroethylene-propylene, FEP	52	

Table 15 Limited oxygen index (LOI)[a] values at 20 °C for polymers

Polymers	LOI	Polymers	LOI
Ordinary polymers		Polyetherketoneketone (PEKK)	40
Polyoxymethylene	15	Polypara(benzoyl)phenylene	41
Cotton	16	Polybenzimidazole (PBI)	42
Cellulose acetate	17	Polyphenylenesulfide (PPS)	44
Natural rubber foam	17	Polyamideimide (PAI)	45
Polypropylene	17	Polyetherimide (PEI)	47
Polymethylmethacrylate	17	Polyparaphenylene	55
Polyurethane foam	17	Polybenzobisoxazole (PBO)	56
Polyethylene	18	*Composites*	
Polystyrene	18	Polyethylene/Al_2O_3(50 %)	20
Polyacrylonitrile	18	ABS/glass fibers (20 %)	22
ABS	18	Epoxy/glass fibers (65 %)	38
Poly(α-methylstyrene)	18	Epoxy/glass fibers (65 %)-300 °C	16
Filter paper	18	Epoxy/graphite fibers (1092)	33
Rayon	19	Polyester/glass fibers (70 %)	20
Polyisoprene	19	Polyester/glass fibers(70 %)-300 °C	28
Epoxy	20	Phenolic/glass fibers (80 %)	53
Polyethyleneterephthalate (PET)	21	Phenolic/glass fibers(80 %)-100 °C	98
Nylon 6	21	Phenolic/Kevlarr (80 %)	28
Polyester fabric	21	Phenolic/Kevlarr (80 %)-300 °C	26
Plywood	23	PPS/glass fibers (1069)	64
Silicone rubber (RTV, etc.)	23	PEEK/glass fibers (1086)	58
Wool	24	PAS/graphite (1081)	66
Nylon 6,6	24-29	BMI/graphite fibers (1097)	55
Neoprene rubber	26	BMI/graphite fibers (1098)	60
Silicone grease	26	BMI/glass fibers (1097)	65
Polyethylenephthalate (PEN)	32	*Halogenated polymers*	
High-temperature polymers		Fluorinated cyanate ester	40
Polycarbonate	26	Neoprene	40
Nomex®	29	Fluorosilicone grease	31–68
Polydimethylsiloxane (PDMS)	30	Fluorocarbon rubber	41–61
Polysulfone	31	Polyvinylidenefluoride	43–65
Polyvinyl ester/glass fibers (1031)	34	PVC (rigid)	50
Polyetheretherketone (PEEK)	35	PVC (chlorinated)	45–60
Polyimide (Kapton®)	37	Polyvinylidenechloride (Saranr)	60
Polypromellitimide (PI)	37	Chlorotrifluoroethylene lubricants	67–75
Polyaramide (kevlar®)	38	Fluorocarbon (FEP/PFA) tubing	77–100
Polyphenylsulfone (PPSF)	38	Polytetrafluoroethylene	95
Polyetherketone (PEK)	40	Polytrichlorofluorethylene	95

[a] Oxygen concentration needed to support burning.

Table 16 Smoke yield for various gases, liquids and solid fuels

State	Generic name	Generic nature	Smoke yield (g/g)
Gas	Methane to butane	Aliphatic-saturated	0.013–0.029
	Ethylene, propylene	Aliphatic-unsaturated	0.043–0.070
	Acetylene, butadiene	Aliphatic-highly unsaturated	0.096–0.125
Liquid	Alcohols, ketones	Aliphatic	0.008–0.018
	Hydrocarbons	Aliphatic	0.037–0.078
		Aromatic	0.177–0.181
Solid	Cellulosics	Aliphatic (mostly)	0.008–0.015
	Synthetic Polymers	Aliphatic-oxygenated	0.001–0.022
		Aliphatic-highly fluorinated	0.002–0.042
		Aliphatic-unsaturated	0.060–0.075
		Aliphatic unsaturated-chlorinated	0.078–0.099
		Aromatic	0.131–0.191

Index